Reference Manual for Telecommunications Engineering

1995 Update

Reference Manual for Telecommunications Engineering

SECOND EDITION

1995 Update

Roger L. Freeman

A WILEY-INTERSCIENCE PUBLICATION
John Wiley & Sons, Inc.
New York · Chichester · Brisbane · Toronto · Singapore

This text is printed on acid-free paper.

Library of Congress Cataloging in Publication Data:

Freeman, Roger L.
Reference manual for telecommunications engineering/Roger L. Freeman.—2nd ed.
p. cm.
"A Wiley-Interscience publication."
ISBN 0-471-57960-2 (alk. paper)
ISBN 0-471-047562 (update)
1. Telecommunication systems—Design and construction—Handbooks, manuals, etc. I. Title.
TK5102.5.F68 1993
621.382—dc20 92-35910
CIP

ISBN 0-471-047562
Printed in the United States of America

10 9 8 7 6 5 4 3 2

Preface

This first update to the *Reference Manual for Telecommunications Engineering, Second Edition* consists of abstracts of current engineering information that were not available when the *Second Edition* was prepared. Literally hundreds of publications: standards, research papers, books, and other documents have appeared since its early 1994 publication date. I have selected a small group of these for this update based upon my 50 years of experience in the industry and included materials I deemed most useful to the largest group of users of the *Manual*.

Each new section consists of highlights of the selected material and is cross-referenced to the *Manual* chapter, section number and page number to which it relates. Some material presented here is replacing or correcting existing sections but most is new and useful updates to the comprehensive Manual. The source of each new section is clearly documented and the reader should refer to the original publication for complete information if a firm engineering project is envisioned.

During the production of the *Second Edition*, an organizational change was taking place at the largest standardization organization—the International Telecommunications Union (ITU). One-third of the information published in the *Manual* was derived from two of the subsidiary organizations of the ITU—the International Consultive Committee for Telegraph and Telephone (CCITT) and the International Consultive Committee for Radio (CCIR). Since January 1, 1993, the CCITT has been known as the ITU-T, the Telecommunications Standardization Sector of the ITU; the CCIR has been known as ITU-R, the Radio Communications Sector of the ITU. For ITU recommendations published prior to January 1, 1993, CCITT and CCIR are used as reference citations. Publications issued after this date use ITU-T and ITU-R, respectively, as reference citations.

I acknowledge with gratitude the following people who advised and helped me compile this update: Joseph Golden of Nynex, Framingham, MA, and these independent consultants: Dr. Ronald Brown of Melrose, MA; John Lawlor of Sharon, MA; and Bill Otaski of Beverly Farms, MA.

I appreciate the support of my old friend, Frank Zawislan of the USAF Rome Laboratory, NY, and my son, Bob Freeman of Chipcom, Southborough, MA. Without the devotion and patience of my wife, Paquita, this work would have never been completed.

Roger L. Freeman

Scottsdale, Arizona
April 1995

Contents to the 1995 Update

Note to the reader: Materials that appear only in this *Update* and not in the main volume are indicated by (New) after the title.

6. Transmission Factors in Telephony 1

6-10.4. Examples of Values of *R* and *T* Pads Adopted by Some Administrations **(New)** 1
6-14.1. Delay Estimation for Circuits **(New)** 3
6-14.1.1. National Extensions **(New)** 3
6-17. Transmission Impairments due to Digital Processes **(New)** 5
6-17.1. Quantization Distortion **(New)** 5
6-17.2. Sources of Quantization Distortion **(New)** 6
6-17.3. Attenuation Distortion and Group-Delay Distortion **(New)** 6
6-17.4. Provisional Planning Rule **(New)** 8
6-17.5. Limitations of the Provisional Planning Rule **(New)** 8
6-18. Information for Planning Purposes Concerning Attenuation Distortion and Group-Delay Distortion Introduced by Circuits and Exchanges in the Switched Telephone Network **(New)** 10

7. Multiplexing Techniques 11

7-2.9.6. SDH Pointers **(New)** 11
7-2.9.7. Mapping of Tributaries into VCs **(New)** 17
7-2.15. Transmission Performance Characteristics of Pulse Code Modulation **(New)** 32
7-2.15.1. Introduction **(New)** 32
7-2.15.2. Relative Levels at Voice-Frequency Ports **(New)** 36
7-2.15.3. Adjustment of Actual Relative Levels **(New)** 36
7-2.15.4. Short- and Long-Term Variation of Level with Time **(New)** 37
7-2.15.5. Impedance Unbalance about Earth **(New)** 37
7-2.15.6. Group Delay **(New)** 41
7-2.15.7. Idle Channel Noise **(New)** 43
7-2.15.8. Discrimination Against Out-of-Band Signals **(New)** 43
7-2.15.9. Spurious Signals at the Channel Output Port **(New)** 45
7-2.15.10. Total Distortion, Including Quantizing Distortion **(New)** 46
7-2.15.11. Variation of Gain with Input Level **(New)** 48
7-2.15.12. Crosstalk **(New)** 48

7-2.15.13. Interference from Signaling **(New)** 52
7-2.15.14. Echo and Stability at 2-Wire Ports, E2 **(New)** 53

8. Outside Plant—Metallic Pair Systems 55

8-7.20. Local Area Network Twisted Pair Data Communications Cable **(New)** 55
8-7.20.1. Preferred Ratings and Characteristics **(New)** 55
8-7.21. Cable Specifications for Commercial Building Telecommunication Applications **(New)** 57
8-7.21.1. Introduction **(New)** 57
8-7.21.2. Backbone Cabling Distances **(New)** 57
8-7.21.3. 100-Ω Unshielded Twisted Pair (UTP) Cabling Systems **(New)** 58
8-7.21.4. 150-ohm Shielded Twisted Pair (STP) Cable **(New)** 61
8-7.21.5. Backbone UTP Cable **(New)** 64
8-7.21.6. Horizontal 62.5/125 μm Optical Fiber Cable **(New)** 67

9. Fiber Optics Transmission 69

9-5.10. Single-Mode Optical Fiber Cable **(New)** 69
9-5.10.1. Scope **(New)** 69
9-5.10.2. Fiber Characteristics **(New)** 69
9-5.10.3. Factory Length Specifications **(New)** 72
9-5.10.4. Elementary Cable Sections **(New)** 73
9-5.11. Dispersion-Shifted Single-Mode Fiber Optic Cable **(New)** 73
9-5.11.1. Introduction and Scope **(New)** 73
9-5.11.2. Fiber Characteristics **(New)** 74
9-5.11.3. Factory Length Specifications **(New)** 76
9-5.11.4. Elementary Cable Sections **(New)** 77

11. Electromagnetic Wave Propagation 79

11-7.2.4. Excess Attenuation Due to Rainfall, CCIR Method for Satellite Paths **(New)** 79
11-7.3. Attenuation in Vegetation **(New)** 85
11-7.4. Signal Reception Inside Buildings **(New)** 85
11-7.5. Conversion of Annual Statistics to Worst-Month Statistics **(New)** 86
11-7.6. Ionospheric Effects Influencing Radio Systems Involving Spacecraft **(New)** 88
11-7.6.1. Ionospheric Effects **(New)** 88
11-7.6.2. Scintillation **(New)** 89
11-7.6.3. Frequency Dependence of Scintillation **(New)** 89
11-7.6.4. Instantaneous Statistics and Spectrum Behavior **(New)** 90
11-7.6.5. Geometric Considerations **(New)** 93
11-7.6.6. Morphology **(New)** 94

11-7.6.7. Cumulative Statistics **(New)** 95

11-7.6.8. Simultaneous Occurrence of Ionospheric Scintillation and Rain Fading **(New)** 96

13. Radio Systems 97

New text added to the end of Section 13-2.1.1 97

Replacement text for Section 13-2.2.2, page 901, of the *Manual*, 104

13-2.11. Typical Link Parameters: Selected Examples from the Fixed Service **(New)** 108

13-4.7.8. Contributions to Noise Temperature of an Earth-Station Receiving Antenna **(New)** 108

13-4.7.9. Method of Determining Earth-Station Antenna Characteristics at Frequencies Above 10 GHz **(New)** 115

13-4.12.8. Maximum Permissible Level of Spurious Emissions from Very-Small-Aperture Terminals (VSATs) **(New)** 119

13-4.12.9. Maximum Permissible Levels of Interference in a Geostationary-Satellite Network **(New)** 120

13-6.2. Communication by Meteor-Burst Propagation: ITU-R Rec. 843 **(New)** 122

13-6.2.1. Temporal Variations in Meteor Flux **(New)** 122

13-6.2.2. Spatial Variation in Meteor Flux **(New)** 123

13-6.2.3. Underdense and Overdense Trails **(New)** 123

13-6.2.4. Effective Length and Radius of Meteor Trails **(New)** 123

13-6.2.5. Received Power and Basic Transmission Loss **(New)** 124

13-6.2.6. The Underdense Echo Ceiling and Average Trail Height **(New)** 127

13-6.2.7. Positions of Regions of Optimum Scatter **(New)** 128

13-6.2.8. Estimating the Useful Burst Rate **(New)** 128

13-6.2.9. Antenna Considerations **(New)** 129

13-6.2.10. Considerations in the Choice of Frequency **(New)** 130

14. Antennas, Passive Repeaters, and Towers 133

14-8.3. Design Wind Load on Typical Microwave Antennas/Reflectors 133

14-8.3.1. Determination of Allowable Beam Twist and Sway for Cross-Polarization Limited Systems 141

17. Data/Telegraph Transmission 149

A Data Modem Operating at Data Signaling Rates Up to 14,400 bps (new section that follows Section 17-10.1.7 of the *Manual*) 149

17-12.2.6. Nine-Position Nonsynchronous Interface Between Data Terminal Equipment and Data

Circuit-Terminating Equipment Employing Serial Binary Data Interchange **(New)** 157
17-12.2.7. Simple 8-Position Nonsynchronous Interface Between Data Terminal Equipment and Data Circuit-Terminating Equipment Employing Serial Binary Data Interchange **(New)** 159
17-12.2.8. High-Speed 25-Position Interface for Data Terminal Equipment and Data Circuit-Terminating Equipment Including Alternative 26-Position Connector **(New)** 161
17-12.2.9. Electrical Characteristics for an Unbalanced Digital Interface **(New)** 169
17-14. Adapting Standard Data Rates to the 64-kbps Digital Channel **(New)** 176
17-14.1. Digital Data Systems (DDS) 176
17-14.2. Adaptation Based on CCITT Rec. V.110 **(New)** 177
17-14.3. Transmission of Start-Stop Characters Over Synchronous Bearer Channels **(New)** 197

18. Data Networks 201

18-5.2. Transmission Control Protocol/Internet Protocol TCP/IP **(New)** 201
18-5.2.1. Background and Application **(New)** 201
18-5.2.2. TCP/IP and Data-Link Layers **(New)** 201
18-5.2.3. The IP Routing Function **(New)** 204
18-5.2.4. Detailed IP Operation **(New)** 205
18-5.2.5. The Transmission Control Protocol (TCP) **(New)** 211
18-7. The ISO Transport Protocol **(New)** 219
18-7.1. Services Provided and Assumed **(New)** 220
18-7.2. Functions of the Transport Layer **(New)** 220
18-7.3. Classes and Options **(New)** 222
18-7.4. Model of the Transport Layer **(New)** 223
18-7.5. Structure and Coding of TPDUs **(New)** 224
18-8. IBM System Network Architecture (SNA) **(New)** 228
18-8.1. Background **(New)** 228
18-8.2. The SNA Layered Architecture **(New)** 230
18-8.3. Advanced Peer-to-Peer Networking (APPN) **(New)** 232
18-8.4. Architectural Components of an SNA Network **(New)** 232
18-8.5. Nodes with Both Hierarchical and Peer-Oriented Function **(New)** 236
18-8.6. Links and Transmission Groups **(New)** 237
18-8.7. SNA Network Configurations **(New)** 238
18-8.8. The Transport Network and Network Accessible Units **(New)** 240
18-8.9. SNA Logical Units **(New)** 242
18-8.10. Some SNA Data Formats **(New)** 244
18-9. International Numbering Plan for Public Data Networks **(New)** 248
References (Updated) 261

20. Broadband ISDN (B-ISDN) 263

20-3.1. Introduction to ATM **(New)** 263
20-3.2. User-Network Interface (UNI) Configuration and Architecture **(New)** 264
20-3.3. The ATM Cell: Key to Operation **(New)** 266
20-3.4. Cell Delineation and Scrambling **(New)** 270
20-3.5. ATM Layering and B-ISDN **(New)** 273
20-3.6. Services: Connection-Oriented and Connectionless **(New)** 283
20-3.7. Some Aspects of a B-ISDN/ATM Network **(New)** 287
20-3.8. Signaling Requirements **(New)** 289
20-3.9. Quality of Service (QoS) **(New)** 291
20-3.10. Traffic Control and Congestion Control **(New)** 294
20-3.11. Transporting ATM Cells **(New)** 301
References to Section 20-3 (Updated) 307

22. Television Transmission 309

22-9.3. Conference Television: Video CODEC at $p \times 64$ kbps **(New)** 309

25. Standard Time and Frequency 327

25-6.2. Standard Frequencies and Time Signals 327
25-10. A Note on Standard Time and Date Conventions **(New)** 327

Index 341

Reference Manual for Telecommunications Engineering

1995 Update

6

Transmission Factors in Telephony

This new section follows Section 6-14, which ends on page 395 of the Manual.

6-10.4. Examples of Values of *R* and *T* Pads Adopted by Some Administrations

This section gives values of R and T pads that have been adopted by some administrations for their digital networks. The values given are those appropriate for digital connections between subscribers with existing 2-wire subscriber lines on digital local exchanges. It should be recognized that different values may be appropriate for connections in the evolving mixed analog/digital network.

These values are given by the ITU-T organization as guidance to developing countries that are considering the planning of new networks. If similar values are adopted for new networks then, in association with adequate echo and stability balance return losses, there are unlikely to be difficulties in meeting the requirements of ITU-T Rec. G.122.

Some administrations consider losses in terms of the input and output relative levels. These values can be derived from Table 6-U.1 using the relationship given in Figure 6-U.1.

In this circuit, it is assumed that the relative levels of the encoder input and the decoder output are 0 dBr, that the T pad represents all the loss between the 2-wire point, t, and the encoder input, and that the R pad represents all the loss between the decoder output and t. Accordingly, the relation between relative levels and losses is

$$L_i = T, \qquad L_0 = -R$$

Note: The modern trend is to use a complex nominal impedance at the 2-wire port.

In exceptional cases, some of the R and T losses may be achieved by digital pads. See ITU-T Rec. G.101, 6.2/G.101 and 2.8/G.101 for a discussion.

In general, the range of input levels has been derived assuming that speech powers in the network are close to the conventional load assumed in the design of FDM systems. However, actual measurements reveal that this load is not being attained [see Supplement No. 5 to Fascicle III.2 of the *Red Book* (1985)]. For this reason, it may be that there is some advantage in adopting different input (and output) levels for future designs of exchange. However, any possible changes need to take into account:

- **(i)** The range of speech powers encountered on an individual channel at the exchange input and the subjective effects of any peak clipping, noting that any impairment is confined to that channel
- **(ii)** Levels of nonspeech analogue signals (e.g., from data modems or multifrequency signaling devices), particularly from customers on short exchange lines

TABLE 6-U.1
Values of *R* and *T* for Various Countries

Country	Connection Type					
	Own Exchange		Local via Digital Junctions (Digital Trunks)		Trunk Via Digital Trunk Exchange	
	R dB	*T* dB	*R* dB	*T* dB	*R* dB	*T* dB
Germany (F.R.) (For subscribers on short lines: R = 10 dB, T = 3 dB)	7	0	7	0	7	0
Australia	6	0	6	0	6	0
Austria	7	0	7	0	7	0
Belgium	7	0	7	0	7	0
Canada	0	0	3	0	6	0
Denmark	6	0	6	0	6	0
Spain	7	0	7	0	7	0
United States	0	0	3	0	6	0
Finland	7	0	7	0	7	0
France	7	0	(Not used)	(Not used)	7	0
India	6	0	6	0	6	0
Italy	7	0	7	0	7	0
Japan	4	0	8	0	8	0
The Netherlands	4.5	1.5	4.5	1.5	4.5 (National) 10.5 (International)	1.5
Norway	5	2	5	2	5	2
United Kingdom (values shown are for median lines; additional loss is introduced on short local lines in both directions of transmission)	6	1	6	1	6	1
Sweden	5	0	5	0	5 (National 7 (International)	0 (National) 0 (International)
USSR	7	0	7	0	7	0
Yugoslavia	7	0	7	0	7	0
New Zealand	7	0.5	7	0.5	7	0.5

Source: Table C.1/G.121, ITU-T Rec. G.121, Annex C, 3/93.

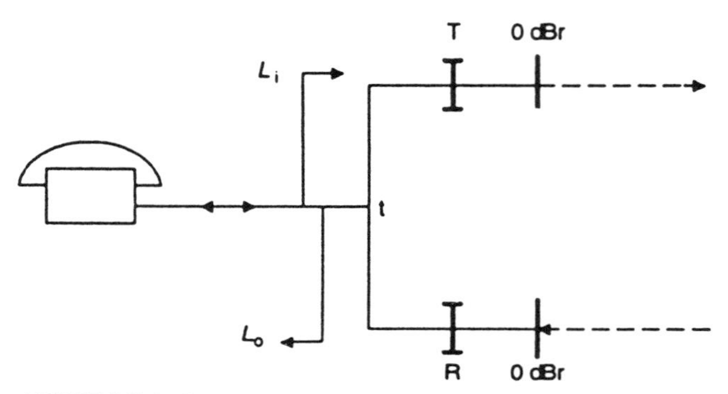

FIGURE 6-U.1. Relation between relative levels and *R* and *T* pads.

(iii) The need to meet the echo and stability requirements of ITU-T Rec. G.122, particularly when the sum of R and T is less than 6 dB

(iv) The need to consider the difference in loss between the two directions of transmission, as required by ITU-T Rec. G.121, paragraph 2.2

At this stage, administrations should note that there may be some advantage in considering a range of level adjustment for future designs of digital local exchange.

From ITU-T Rec. G.121, Annex C, 3/93.

This new section follows Section 6-14, which ends on page 395 of the Manual.

6-14.1. Delay Estimation for Circuits

The conventional planning values for propagation time are shown in Table 6-16. For national extension circuits, the following guidelines may be used.

6-14.1.1. National Extensions

The main arteries of the national network should consist of high-velocity propagation lines. In these conditions, the transmission time between the international center and the subscriber farthest away from it in the national network can be estimated as follows:

(a) In purely analog networks, the transmission time will probably not exceed:

$$12 + (0.004 \times \text{distance in kilometers})\ \text{msec}$$

Here the factor 0.004 is based on the assumption that national trunk circuits will be routed over high-velocity plant (250 km/msec). The 12-msec constant term makes allowance for terminal equipment and for the probable presence in the national network of a certain quantity of loaded cables (e.g., three pairs of channel translating equipments plus about 160 km of H 88/36 loaded cables). For an average-sized country the one-way propagation time will be less than 18 msec.

(b) In mixed analog/digital networks, the transmission time can generally be estimated by the equation given for purely analog networks. However, under certain unfavorable conditions increased delay may occur compared with the purely analog case. This occurs in particular when digital exchanges are connected with analog transmission systems through PCM/FDM equipments in tandem, or transmultiplexers. With the growing degree of digitization the transmission time will gradually approach the condition of purely digital networks.

(c) In purely digital networks between local exchanges (e.g., an IDN), the transmission time as defined above will probably not exceed

$$3 + (0.004 \times \text{distance in kilometers})\ \text{msec}$$

The 3-msec constant term makes allowance for one PCM coder or decoder and five digitally switched exchanges.

TABLE 6-16
Planning Values for Propagation Time

Transmission Medium	Contribution to One-Way Propagation Time	Remarks
Terrestrial coaxial cable or radio relay system; FDM and digital transmission	4 μsec/km	Allows for delay in repeaters and regenerators
Optical fiber cable system; digital transmission	5 μsec/km	Allows for delay in repeaters and regenerators
Submarine coaxial cable system	6 μsec/km	
Satellite system 14,000-km altitude 36,000-km altitude	 110 msec 260 msec	Between earth stations only
FDM channel modulator or demodulator	0.75 msec[a]	Half the sum of propagation times in both directions of transmission
FDM compandored channel modulator or demodulator	0.5 msec[b]	
PCM coder or decoder	0.3 msec[a]	
PCM/ADPCM/PCM transcoding	0.5 msec	
Transmultiplexer	1.5 msec[c]	
Digital transit exchange, digital-digital	0.45 msec[d]	
Digital local exchange, analog–analog	1.5 msec[d]	
Digital local exchange, analog subscriber line-digital junction	0.975 msec[d]	
Digital local exchange, digital subscriber line-digital junction	0.825 msec[d]	
Echo cancelers	1 msec[e]	

[a]These values allow for group delay distortion around frequencies of peak speech energy and for delay of intermediate higher order multiplex and through-connecting equipment.

[b]This value refers to FDM equipments designed to be used with a compandor and special filters.

[c]For satellite digital communications where the transmultiplexer is located at the earth station, this value may be increased to 3.3 msec.

[d]These are mean values: Depending on traffic loading, higher values can be encountered—for example, 0.75 msec (1.950 msec, 1.350 msec, or 1.250 msec) with 0.95 probability of not exceeding. (For details, see Rec. Q.551.)

[e]Echo cancelers, when placed in service, will add a one-way propagation time of up to 1 msec in the send path of each echo canceler. This delay excludes the delay through any codec in the echo canceler. No significant delay should be incurred in the receive path of the echo canceler.

Source: Table 1/G.114, CCITT Rec. G.114, Fascicle III.1, page 86, CCITT Blue Books, 1988 [Ref. 3].

Note: The value 0.004 is a mean value for coaxial cable systems and radio-relay systems; for optical fiber systems, 0.005 is to be used.

(d) In purely digital networks between subscribers (e.g., an ISDN), the delay of (c) above has to be increased by up to 3.6 msec if burst-mode (time compression multiplexing) transmission is used on 2-W local subscriber lines.

These values do not cover the additional delays introduced by PABXs and Private Branch Networks (PBNs).

From ITU-T Rec. G.114, 3/93, Annex A.

This new section precedes Reference section on page 406 of the Manual.

6-17. TRANSMISSION IMPAIRMENTS DUE TO DIGITAL PROCESSES

6-17.1. Quantization Distortion

From the point of view of quantizing distortion, it is recommended that no more than 14 units of quantizing distortion (qdu) or equivalent total code distortion should be introduced in an international telephone connection.

For telephone connections which incorporate unintegrated digital processes, it is permissible to simply add the units of quantizing distortion that have been assigned to the individual digital processes to determine the total or overall quantizing distortion. Some sources of quantizing distortion and the units tentatively assigned to them are given in Section 6-17.2.

By definition, an average 8-bit codec pair (A/D and D/A conversions, A-law or μ-law) which complies with Rec. G.711 introduces 1 quantizing distortion unit (1 qdu). An average codec pair produces about 2 dB less quantizing distortion than the limits indicated in Rec. G.712. This would correspond to a signal-to-distortion ratio of 35 dB for the sine-wave test method and approximately 36 dB for the noise test method. (A total of fourteen 8-bit PCM processes each of which just comply with the limits for signal-to-distortion ratio in Rec. G.712 would be unacceptable.) The same principle should be applied when proposing planning values of quantizing distortion units for other digital processes.

In principle, the number of units for other digital processes are determined by comparison with an 8-bit PCM codec pair such that the distortion of the digital process being evaluated is assigned n quantizing distortion units if it is equivalent to n unintegraded 8-bit PCM process in tandem. Several methods of comparison are possible; these include objective measurements (or equivalent analysis), subjective tests, and data tests in which the effect on the bit error ratio at the output of a voice-band data modem receiver is used as a criterion.

At the present time no objective measurement capability exists which can produce results (e.g., SNR) that correlate closely with results obtained from subjective measurement of the effect of many of the digital processes now being studied on speech performance. Therefore, the number of units of quantization distortion for digital processes should, in general, be determined by subjective measurement methods, such as those found in Rec. P.83. In some instances the number of units of quantization distortion for a digital process can be determined without subjective measurement by decomposing a digital process into two or more parts and allocating to the parts suitable fractions of the total number of units assigned to the digital process. However, while this method may be considered an objective method for determining the qdu assignments for the parts, it uses as a starting point a subjectively determined value. Furthermore, except for relatively simple digital processes where the decomposition is uncomplicated, this method may not be reliable and should be used with care.

Planning rules should be applicable to all signals transmitted in the voice-frequency band. Therefore, in general, both speech quality and data performance must be considered. Speech quality should be evaluated by subjective tests and data performance should be evaluated by objective measurements which provide estimates of the expected bit error ratio and signaling performance. At present, however, because of the lack of an objective method for evaluating the effect of digital processes on voice-band data performance, the planning rule in this Recommendation is limited to voice connection planning purposes only. Clause 4 discusses some of the problems associated with developing a planning rule for connections carrying voice-band data and other nonspeech signals. Such a rule would be based on a unit reflecting the contribution digital processes make to the impairment or impairments that affect voice-band data modems and/or signaling systems. Such a unit does not exist yet.

Note: qdu is defined in terms of quantizing distortion as present in PCM and other waveform coders and assumes that the quantizing distortion adds on a 15 $\log_{10}(n)$ law for n codec pairs in tandem. There is some evidence to suggest that while the 32-kbps ADPCM codec which complies with Rec. G.726 exhibits the same distortion and additivity as PCM, the 16-kbps LD-CELP codec as tested and studied in 1991 exhibits additivity closer to 20 $\log_{10}(n)$. However, subjective tests carried out under the guidance of Study Group XII Experts Group on Speech Quality indicate that the 16-kbps codec pair closely tracks the subjective quality of the G.726 codec for up to 4 codecs in tandem. Beyond 4 in tandem, the 16-kbps codec performance decreases more rapidly than that of the G.726 codec. Thus, it is proposed that the 16-kbps codec be treated the same as the G.726 codec determining network performance with the stipulation that no more than three 16-kbps codecs be allowed in the worldwide connection and noting that the LD-CELP distortion is not additive with the qdus of other codecs.

6-17.2. Sources of Quantization Distortion

The units of quantizing distortion (qdu) tentatively assigned to a number of digital processes are given in Table 6-U.2. Background information on these assignments is given in Supplement Nos. 21 and 22, *Red Book*, Fascicles III.1 and III.2 (1985), respectively, and in the notes associated with Table 6-U.2.

Conceptually the number of qdus assigned to a particular digital process should reflect the effect of only the quantization noise produced by the process on speech. In practice, the qdu must be determined from subjective measurements of real or simulated processes, where subjects will be exposed to not only the quantization noise but other impairments produced by the digital process tested.

Therefore, the subjective test results will be biased by these other impairments if the levels of these other impairments differ to a greater or lesser extent from the levels produced by PCM (the reference). Such biases will cause the derived qdu to not be a true measure of the effect of quantization distortion. The qdu assignment will instead reflect the effect of all the impairments on speech quality. Thus, to reduce the chance for such a bias to occur when determining the qdu assignments for digital processes, it is important to design the subjective test so as to (a) minimize the contributions of impairments other than quantization distortion to the subjective test results or (b) equalize the levels of these other impairments in the test and reference conditions.

6-17.3. Attenuation Distortion and Group-Delay Distortion

The provisional recommendation made in Section 6-17.1 specifies that the total quantizing distortion introduced by unintegrated digital processes in international telephone connections should be limited to a maximum of 14 units. It is expected that if this provisional recommendation is complied with, the accumulated attenua-

TABLE 6-U.2
Planning Values for Quantizing Distortion (Valid for Speech Service Only)

Digital Process	Quantizing Distortion Units (qdu)	Notes
Processes involving A/D conversion		
8-bit PCM codec pair (according to Rec. G.711.A- or μ-law)	1	2, 3
7-bit PCM codec pair (A- or μ-law)	3	3, 4, 5
Transmultiplexer pair based on 8-bit PCM, A- or μ-law (according to Rec. G.792)	1	3
32-kbit/s ADPCM (with adaptive predictor) (combination of an 8-bit PCM codec pair and a PCM-ADPCM-PCM tandem conversion) (according to Rec. G.726 or G.727)	3, 5	6
16 kbit/s LD-CELP codec pair (according to Rec. G.728)	3, 5	13
Purely digital processes		
Digital loss pad (8-bit PCM, A- or μ-law)	0, 7	7
A/μ-law or μ/A-law converter (according to Rec. G.711)	0, 5	10
A/μ/A-law tandem conversion	0, 5	
μ/A/μ-law tandem conversion	0, 25	
PCM to ADPCM to PCM conversion (according to Rec. G.721 or G.727)	2, 5	8, 9
8–7–8 bit transcoding (A- or μ-law)	3	9

Note 1: As a general remark, the number of units of quantizing distortion entered for the different digital processes is that value which has been derived at a mean Gaussian signal level of about −20 dBm0. The cases dealt with in Supplement No. 21, *Red Book*, Fascicle III.1 are in accordance with this approach.

Note 2: By definition.

Note 3: For general planning purposes, half the value indicated may be assigned to either of the send or receive parts.

Note 4: This system is not recommended by CCITT but is in use by some administrations in their national networks.

Note 5: The impairment indicated for this process is based on subjective tests.

Note 6: Recommendations G.726 and G.727 perform equivalently at corresponding bit rates, including 24 and 40 kbps/s. However, qdu values cannot be assigned for 24- and 40-kbps operation, at this time.

Note 7: The impairment indicated is about the same for all digital pad values in the range 1–8 dB. One exception is the 6-dB A-law pad which introduces negligible impairment for signals down to about −30 dBm0 and thus attracts 0 units for quantizing distortion.

Note 8: The value of 2.5 units was derived by subtracting the value for an 8-bit PCM codec pair from the 3.5 units determined subjectively for the combination of an 8-bit PCM code pair and a PCM/ADPCM/PCM conversion. Multiple synchronous digital conversions, such as PCM/ADPCM and PCM/ADPCM/PCM, are assigned a value of 2.5 units.

Note 9: This process might be used in a digital speech interpolation system.

Note 10: The qdu contribution made by coding law converters (e.g., μ-law to A-law) are assigned to the international part.

Note 11: The qdu assignments to these digital processes reflect, to the extent possible, only the effect of quantization distortion on speech performance. Other impairments, such as circuit noise, echo, and attenuation distortion, also affect speech performance. The effect of these other impairments must therefore be taken into account in the planning process.

Note 12: The qdu impairments in this table are derived under the assumption of negligible bit error.

Note 13: The distortion produced in the 16-kbps LD-CELP codec appears to be of a different nature than that denoted in terms of qdu in that it appears to add on a $20 \log_{10}(n)$ basis. It is noted that one 16-kbps codec pair produces a speech quality subjectively equivalent to that of one 32-kbps ADPCM codec pair, while three 16-kbps codec pairs produce a speech quality approximating that produced by four 32-kbps ADPCM codec pairs. Thus, on the basis of this equivalence, one 16-kbps LD-CELP codec (according to Rec. G.728) is assigned 3.5 qdu. It should be recognized that the qdu of the 16-kbps codec is not strictly additive with qdus of the other entries in the table.

Source: Table 1/G.113, ITU-T Rec. G.113, 3/93.

tion distortion and the accumulated group-delay distortion introduced by unintegrated digital processes in such connections would also be kept within acceptable limits.

6-17.4. Provisional Planning Rule

As a consequence of the relationship indicated in Section 6-17.3 above concerning quantization distortion, attenuation distortion, and group-delay distortion, it is possible to recommend a provisional planning rule governing the incorporation of unintegrated digital processes in international telephone connections. This provisional planning rule is in terms of units of transmission impairment which numerically are the same as the units of quantizing distortion allocated to specific digital processes as indicated in Table 6-U.2. The provisional planning rule is as follows:

> *The number of units of transmission impairment in an international telephone connection should not exceed* 5 + 4 + 5 = 14 *units.*

Under the above rule, each of the two national portions of an international telephone connection is permitted to introduce up to a maximum of 5 units of transmission impairment, and the international portion is permitted to introduce up to a maximum of 4 units.

Note: It is recognized that in the mixed analog/digital period, it might for a time not be practical for some countries to limit their national contributions to a maximum of 5 units of transmission impairment. To accommodate such countries, a temporary relaxation of the provisional planning rule is being permitted. Through this relaxation, the national portion of an international telephone connection would be permitted to introduce up to 7 units of transmission impairment. Theoretically, this could result in international telephone connections with a total of 18 qdus of transmission impairment. Such connections would introduce an additional transmission penalty insofar as voice telephone service is concerned. Administrations which find it indispensable to have a national allowance of more than 5 units (but no more than 7 units) should ensure that not more than a small percentage of traffic on national extensions exceeds 5 units.

6-17.5. Limitations of the Provisional Planning Rule

In Section 6-17.4, it is assumed that for estimating the transmission impairment due to the presence of unintegrated digital processes in international telephone connections, the units of transmission impairment correspond to the units of quantizing distortion, and it is also assumed that the simple addition of such units would apply.

For international telephone circuits that include tandem digital processes in an all-digital environment, adding the individual units of quantizing distortion might not accurately reflect the accumulated quantizing distortion (and, consequently, the accumulated units of transmission impairment). This could be the case because the individual amounts of quantizing distortion power produced by the individual digital processes might not be uncorrelated, and therefore the addition of individual units of quantizing distortion might, under some circumstances, indicate totals that could be different from those actually in effect. This is explained in some detail in Supplement No. 21, *Red Book*, Fascicle III.1.

Although the 5 + 4 + 5 = 14 rule given in Section 6-17.4 might under some conditions provide only approximate results, the rule, nevertheless, is considered to be suitable for most planning purposes, particularly in cases involving unintegrated digital processes. Examples of tandem digital processes are A–μ–A code conversion, μ–A–μ code conversion, and PCM–ADPCM–PCM conversion.

From ITU-T Rec. G.113, 3/93.

TABLE 6-U.3
Two-Wire Local and Primary Exchanges

Frequency (Hz)	Attenuation Distortion		Group-Delay Distortion	
	Mean Value (dB)	Standard Deviation (dB)	Mean Value (msec)	Standard Deviation (msec)
200	1.69	1.20	0.56	0.07
300	0.63	0.81	0.28	0.05
400	0.30	0.43	0.23	0.05
600	0	0.28	0.11	0.03
800	0	0	0.05	0.02
1000	−0.05	0.11	0.03	0.01
2000	−0.04	0.35	0	0
2400	−0.29	0.45	0	0
2800	−0.45	0.50	0	0
3000	−0.24	0.65	0	0
3400	−0.29	0.63	0	0

Note: The group-delay distortion may be taken to be with respect to about 2000 Hz.
Source: Table A.1/G.113, ITU-T Rec. G.113, Annex A, 3/93.

TABLE 6-U.4
Four-Wire Exchanges

Frequency (Hz)	Attenuation Distortion		Group-Delay Distortion	
	Mean Value (dB)	Standard Deviation (dB)	Mean Value (msec)	Standard Deviation (msec)
200	0.32	0.14	0.40	0.02
300	0.16	0.28	0.14	0.02
400	0.13	0.21	0.14	0.03
600	0.02	0	0.07	0.02
800	0	0	0.03	0.01
1000	0	0	0.02	0.01
2000	0.01	0.14	0	0
2400	0.06	0.21	0	0
2800	0.02	0.02	0	0
3000	0.10	0.07	0	0
3400	0.20	0.50	0	0

Note: The group-delay distortion may be taken to be with respect to about 2000 Hz.
Source: Table A.2/G.113, ITU-T Rec. G.113, Annex A, 3/93.

TABLE 6-U.5
Trunk Junctions

Frequency (Hz)	Attenuation Distortion		Group-Delay Distortion	
	Mean Value (dB)	Standard Deviation (dB)	Mean Value (msec)	Standard Deviation (msec)
200	4.29	1.95	3.05	0.36
300	0.86	0.49	1.42	0.18
400	0.36	0.31	0.78	0.09
600	0.09	0.17	0.34	0.06
800	0	0.03	0.16	0.02
1000	−0.03	0.04	0.08	0.02
2000	0.14	0.20	0.02	0.01
2400	0.33	0.29	0.06	0.03
2800	0.58	0.35	0.18	0.06
3000	0.88	0.55	0.31	0.11
3400	2.21	1.06	0.92	0.26

Note 1: The group-delay distortion may be taken to be with respect to about 1500 Hz.
Note 2: The sample of trunk junctions included those on metallic lines, FDM and PCM systems.
Note 3: PCM circuits may exhibit a somewhat lower attenuation distortion at 2000 Hz than that indicated above.
Note 4: The values for trunk junctions are inclusive of 2-wire/4-wire terminations.
Source: Table A.3/G.113, ITU-T Rec. G.113, Annex A, 3/93.

6-18. INFORMATION FOR PLANNING PURPOSES CONCERNING ATTENUATION DISTORTION AND GROUP-DELAY DISTORTION INTRODUCED BY CIRCUITS AND EXCHANGES IN THE SWITCHED TELEPHONE NETWORK

The information given in Tables 6-U.3 through 6-U.5 is derived from measurements on modern equipment. The performance of actual connections in the switched telephone network can be expected to be worse than would be calculated from the tabulated data because of:

- Mismatch and reflection
- Unloaded subscriber lines
- Loaded trunk junctions with a low cutoff frequency
- Older equipment

7

Multiplexing Techniques

The title of Table 7-63 is corrected to: "SDH Bit Rates (Prior to 1993)."

TABLE 7-63
SDH Bit Rates (Prior to 1993)

SDH Level	Hierarchical Bit Rate (kbps)
1	155,520
4	622,080

Replace Table 7-64 (p. 509 of the *Manual*) with the following:

TABLE 7-64
SDH Bit Rates (Post 1993)

Synchronous Digital Hierarchy Level	Hierarchical Bit Rate (kbps)
1	155 520
4	622 080
16	2 488 320

Note: The specification of levels higher than 16 requires further study.

This new section follows Section 7-2.9.5, which ends on page 517 of the Manual.

7-2.9.6. SDH Pointers

AU-*n* Pointer. The AU-*n* pointer provides a method of allowing flexible and dynamic alignment of the VC-*n* within the AU-*n* frame. Dynamic alignment means that the VC-*n* is allowed to "float" within the AU-*n* frame. Thus, the pointer is able to accommodate differences not only in the phases of the VC-*n* and the SOH but also in the frame rates.

AU-n Pointer Location. The AU-4 pointer is contained in bytes H1, H2, and H3 as shown in Figure 7-U.1. The three individual AU-3 pointers are contained in three separate H1, H2, and H3 bytes as shown in Figure 7-U.2.

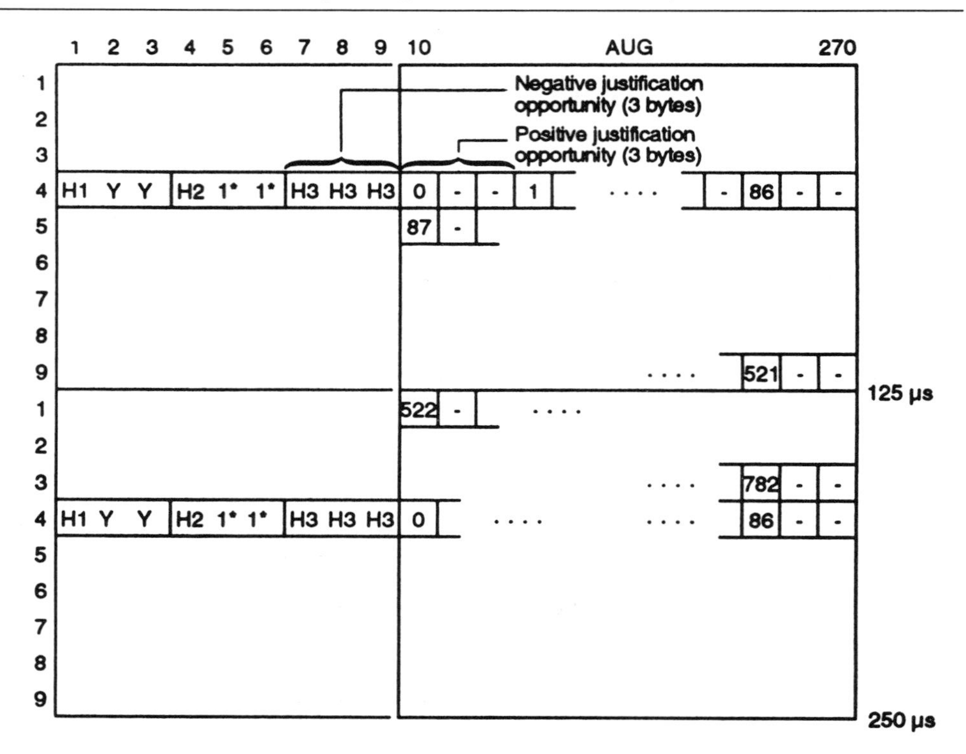

FIGURE 7-U.1. AU-4 pointer offset numbering. (From Figure 3-1 / G.709, 3 / 93. ITU-T Rec. G.709, page 11.)

AU-n Pointer Value. The pointer contained in H1 and H2 designates the location of the byte where the VC-*n* begins. The two bytes allocated to the pointer can be viewed as one word as shown in Figure 7-U.3. The last 10 can be viewed as one word as shown in Figure 7-U.3. The last ten bits (bits 7–16) of the pointer word carry the pointer value.

As shown in Figure 7-U.3, the AU-4 pointer value is a binary number in the range of 0–782, which indicates the offset, in three byte increments, between the pointer and the first byte of the VC-4 (see Figure 7-U.1). Figure 7-U.3 also indicates one additional valid pointer, the concatenation indication. The concatenation indication is given by "1001" in bits 1–4 (bits 5 and 6 are unspecified) and by 10 "1s" in bits 7–16. The AU-4 pointer is set to concatenation indication for AU-4 concatenation.

As shown in Figure 7-U.3, the AU-3 pointer value is also a binary number with a range of 0–782. Because there are three AU-3s in the AUG, each AU-3 has its own associated H1, H2, and H3 bytes. As shown in Figure 7-U.2, the H bytes are shown in sequence. The first H1, H2, H3 set refers to the first AU-3, the second set refers to the second AU-3, and so on. For the AU-3s, the point operates independently.

In all cases, the AU-*n* pointer bytes are not counted in the offset. For example, in an AU-4 the pointer value of 0 indicates that the VC-4 starts in the byte location that immediately follows the last H3 byte, whereas an offset of 87 indicates that the VC-4 starts three bytes after the K2 byte.

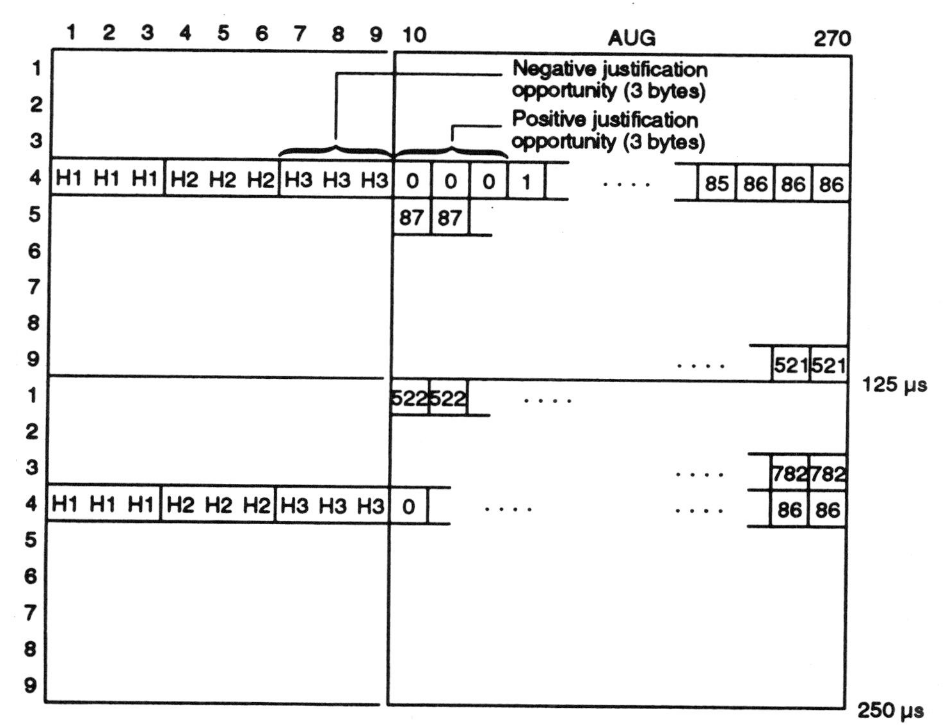

FIGURE 7-U.2. AU-3 pointer offset numbering. (From Figure 3-2/G.709. ITU-T Rec. G.709, 3/93, page 12.)

Frequency Justification. If there is a frequency offset between the frame rate of the AUG and that of the VC-*n*, then the pointer value will be incremented or decremented as needed, accompanied by a corresponding positive or negative justification byte or bytes. Consecutive pointer operations must be separated by at least three frames (i.e., every fourth frame) in which the pointer value remains constant.

If the frame rate of the VC-*n* is too slow with respect to that of the AUG, the alignment of the VC-*n* must periodically slip back in time and the pointer value must be incremented by one. This operation is indicated by inverting bits 7, 9, 11, 13, and 15 (I bits) of the pointer word to allow 5-bit majority voting at the receiver. Three positive justification bytes appear immediately after the last H3 byte in the AU-4 frame containing inverted I bits. Subsequent pointers will contain the new offset. This is shown in Figure 7-U.4.

For AU-3 frames, a positive justification byte appears immediately after the individual H3 byte of the AU-3 frame containing inverted I bits. Subsequent pointers will contain the new offset. This is illustrated in Figure 7-U.5.

If the frame rate of the VC-*n* is too fast with respect to that of the AUG, then the alignment of the VC-*n* must periodically be advanced in time and the pointer value must be decremented by one. This operation is indicated by inverting bits 8, 10, 12, 14, and 16 (D bits) of the pointer word to allow 5-bit majority voting at the receiver. Three negative justification bytes appear in the H3 bytes in the AU-4 frame containing the inverted D bits. Subsequent points will contain the new offset. This is shown in Figure 7-U.6.

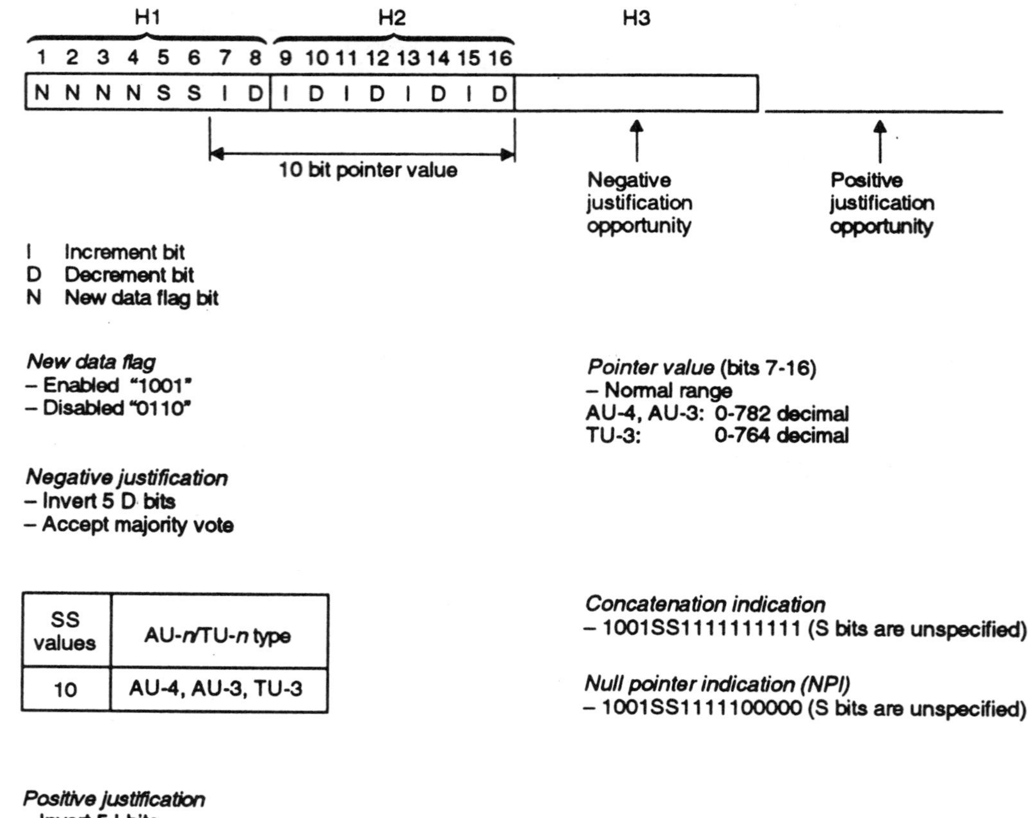

SS values	AU-n/TU-n type
10	AU-4, AU-3, TU-3

NOTES

1 NPI value applies only to TU-3 pointers.

2 The pointer is set to all "1"s when an AIS occurs.

FIGURE 7-U.3. AU-*n* / TU-3 pointer (H1, H2, H3) coding. (From Figure 3-3 / G.709, ITU-T Rec. G.709, 3 / 93, page 13.)

For the AU-3 frame, a negative justification byte appears in the individual H3 byte of the AU-3 frame containing the inverted D bits. Subsequent pointers will contain the new offset. This is shown in Figure 7-U.7.

New Data Flag (NDF). Bits 1–4 (N bits) of the pointer carry an NDF which allows an arbitrary change of the pointer value if that change is due to a change in the payload.

Four bits are allocated to the flag to allow error correction. The decoding may be performed by accepting NDF enabled if at least three bits match. Normal operation is indicated by a "0110" code in the N bits. NDF is indicated by inversion of the N bits to "1001." The new alignment is indicated by the pointer value accompanying the NDF and takes effect at the offset indicated.

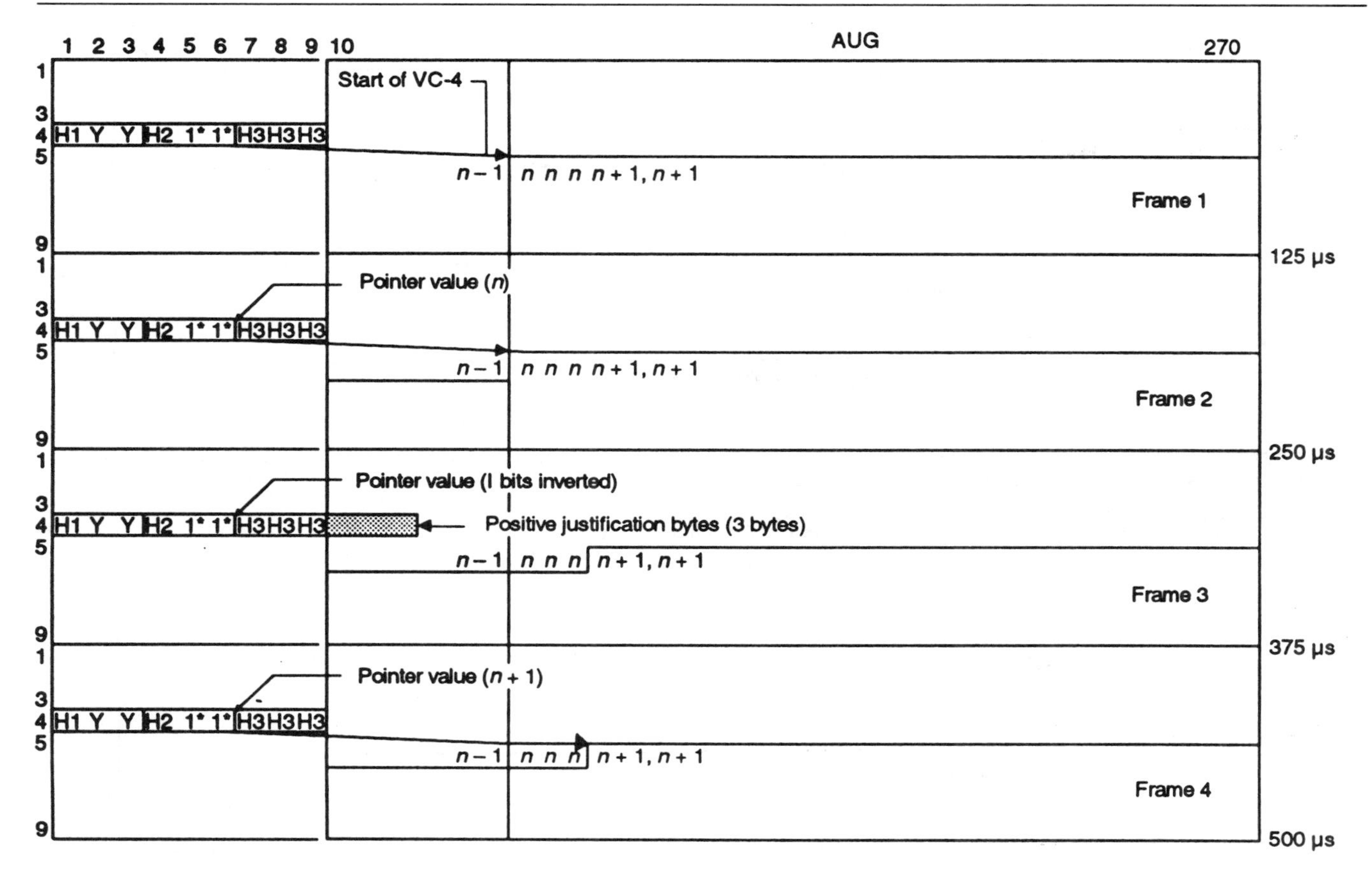

FIGURE 7-U.4. AU-4 pointer adjustment operation—positive justification. (From Figure 3-4/G.709, ITU-T Rec. G.709, 3/93, page 14.)

Pointer Generation. The following summarizes the rules for generating the AU-*n* pointers.

1. During normal operation, the pointer locates the start of the VC-*n* within the AU-*n* frame. The NDF is set to "0110."
2. The pointer value can only be changed by operation 3, 4, or 5.
3. If a positive justification is required, the current pointer value is sent with the I bits inverted and the subsequent positive justification opportunity is filled with dummy information. Subsequent pointers contain the previous pointer value incremented by one. If the previous pointer is at its maximum value, the subsequent pointer is set to zero. No subsequent increment or decrement operation is allowed for at least three frames following this operation.
4. If a negative justification is required, the current pointer value is sent with the D bits inverted and the subsequent negative justification opportunity is overwritten with actual data. Subsequent pointers contain the previous pointer value decremented by one. If the previous pointer value is zero,

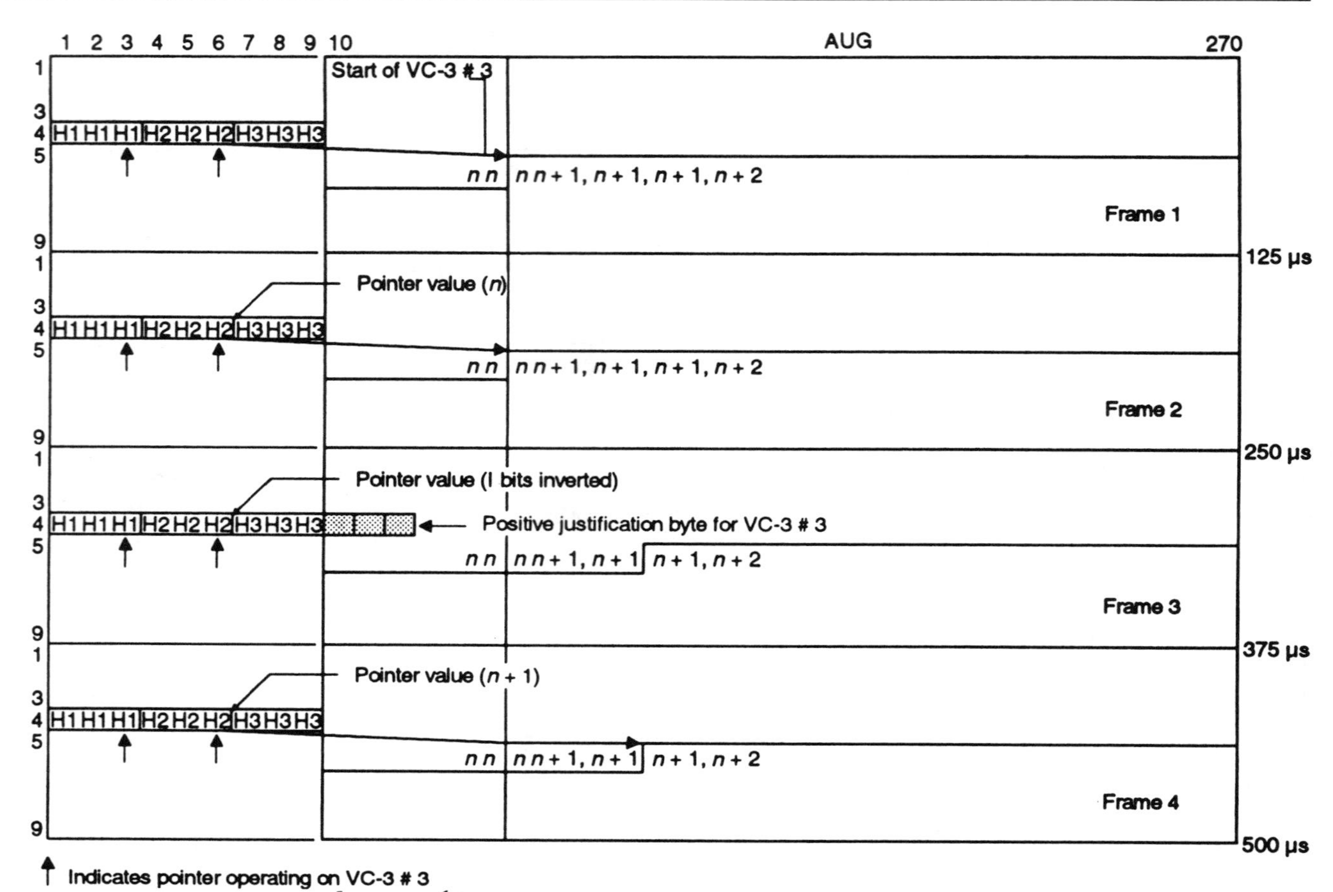

FIGURE 7-U.5. AU-3 pointer adjustment operation — positive justification. (From Figure 3-5/G.709, ITU-T Rec. G.709, 3/93, page 15.)

the subsequent pointer is set to its maximum value. No subsequent increment or decrement operation is allowed for at least three frames following this operation.

5. If the alignment of the VC-*n* changes for any reason other than rule 3 or 4, the new pointer value shall be sent accompanied by NDF set to "1001." The NDF only appears in the first frame that contains the new values. The new location of the VC-*n* begins at the first occurrence of the offset indicated by the new pointer. No subsequent increment or decrement operation is allowed for at least three frames following this operation.

Pointer Interpretation. The following summarizes the rules for interpreting the AU-*n* pointers.

1. During normal operation, the pointer locates the start of the VC-*n* within the AU-*n* frame.
2. Any variation from the current pointer value is ignored unless a consistent new value is received three times consecutively or it is preceded by one of the rules 3, 4, or 5. Any consistent new value received three times consecutively overrides (i.e., takes priority over) rule 3 or 4.

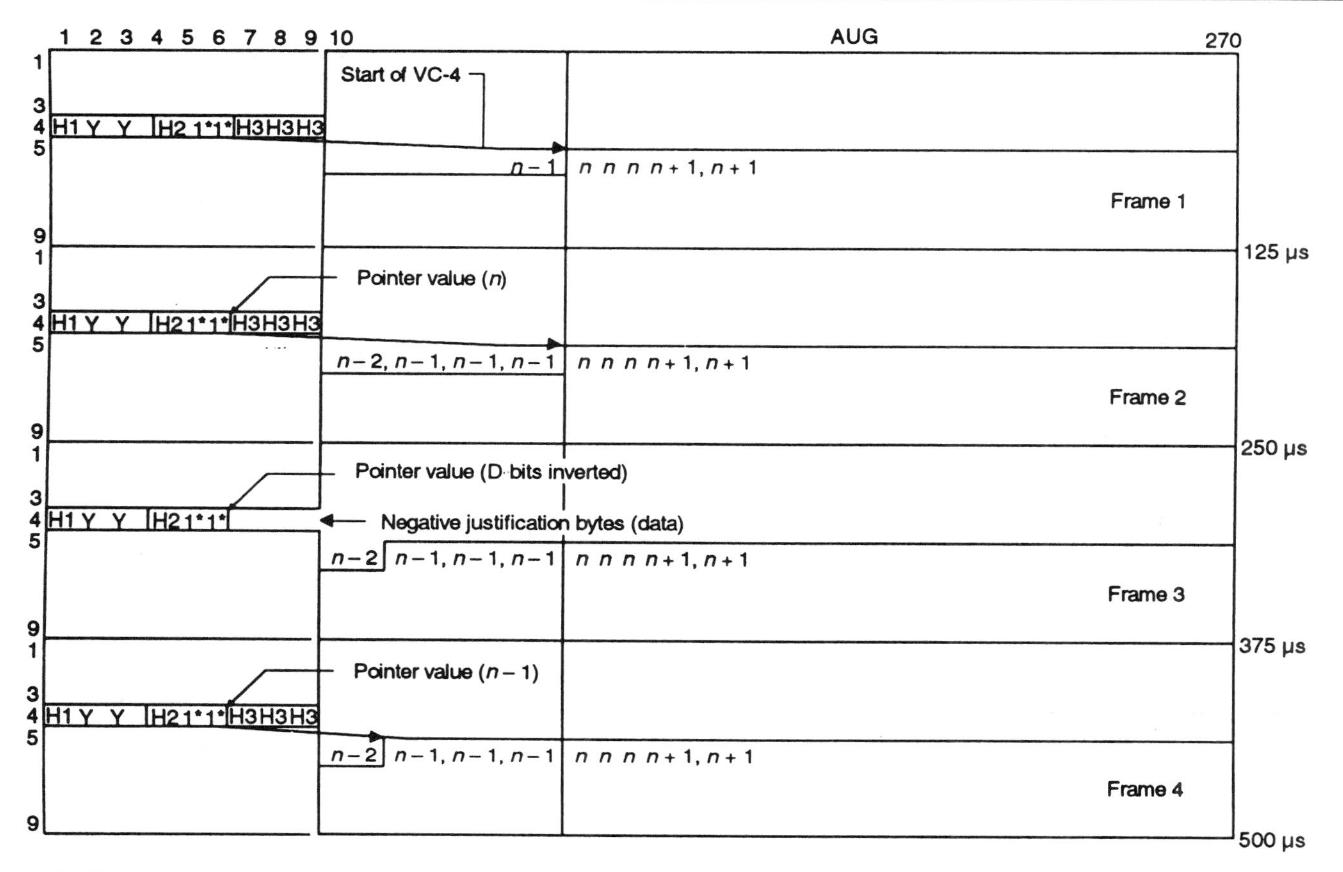

FIGURE 7-U.6. AU-4 pointer adjustment operation — negative justification. (From Figure 3-6 / G.709, ITU-T Rec. G.709, 3 / 93, page 16.)

3. If the majority of the I bits of the pointer word are inverted, a positive justification operation is indicated. Subsequent pointer values shall be incremented by one.
4. If the majority of the D bits of the pointer word are inverted, a negative justification operation is indicated. Subsequent pointer values shall be decremented by one.
5. If the NDF is set to "1001," then the coincident pointer value shall replace the current one at the offset indicated by the new pointer value unless the receiver is in a state that corresponds to a loss of pointer.

7-2.9.7. Mapping of Tributaries into VCs

Accommodation of asynchronous and synchronous tributaries presently defined in ITU-T Rec. G.702 is possible. At the TU-1/TU-2 level, asynchronous accommodation utilizes the floating mode, whereas synchronous accommodation utilizes both

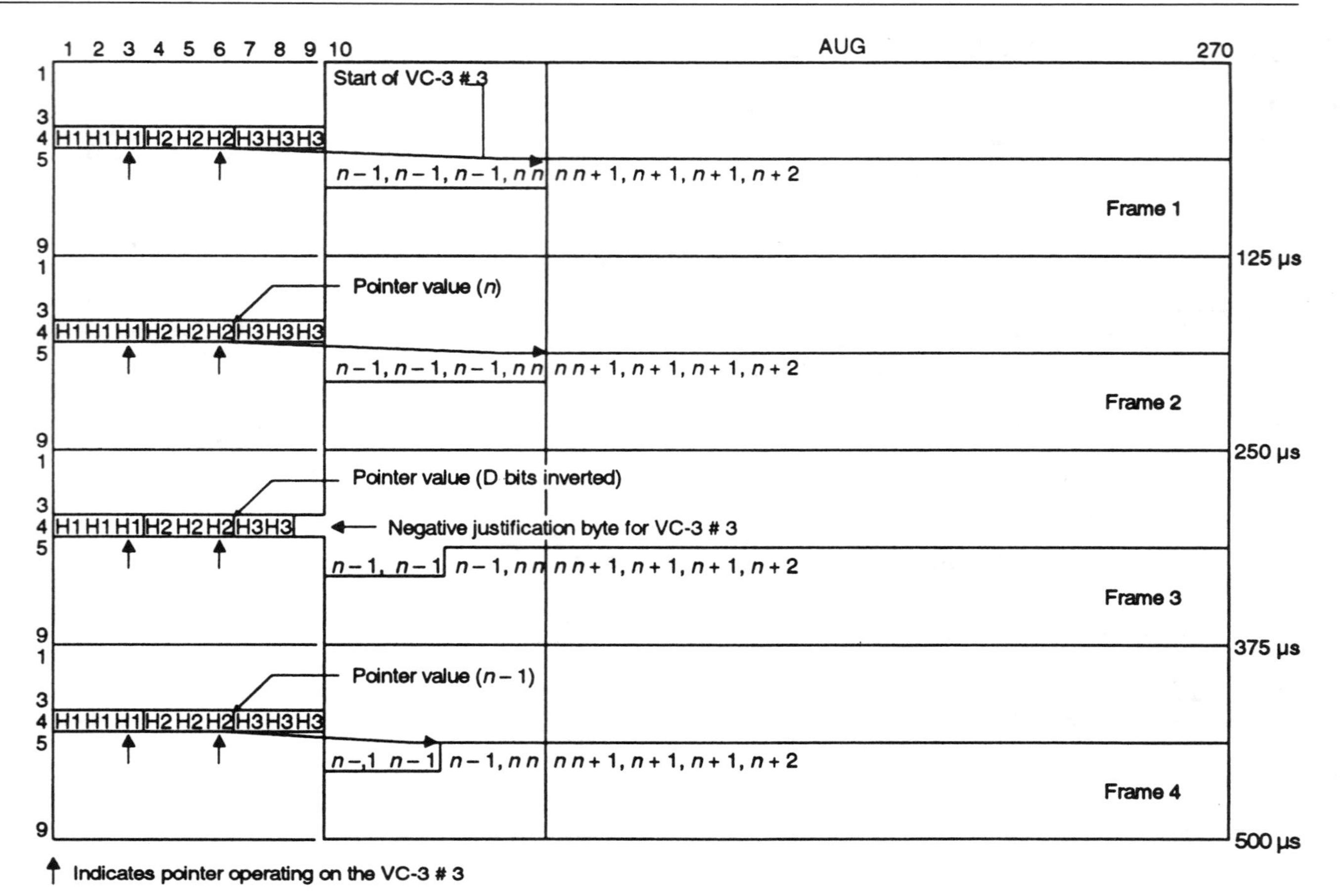

FIGURE 7-U.7. AU-3 pointer adjustment operation — negative justification. (From Figure 3-7 / G.709, ITU-T Rec. G.709, 3 / 93, page 17.)

the locked and floating mode. Figure 7-U.8 shows the TU-1 and TU-2 sizes and formats.

Mapping of Tributaries into VC-4. One 139,264-kbps signal can be mapped into a VC-4 of an STM-1 frame as shown in Figures 7-U.9 and 7-U.10. This payload can be used to carry one 139,264-kbps signal:

1. Each of the nine rows is partitioned into 20 blocks consisting of 13 bytes each (Figure 7-U.9).
2. In each row, one justification opportunity (S) and five justification control bits (C) are provided (Figure 7-U.10.).
3. The first byte consists of:
   a. either eight information bits (I) (byte W), or
   b. eight fixed stuff bits (R) (byte Y), or
   c. one justification control bit (C) plus five fixed stuff bits (R) plus two overhead bits (O) (byte X), or
   d. six information bits (I) plus one justification opportunity bit (S) plus one fixed stuff bit (R) (byte Z).
4. The last 12 bytes of one block consist of information bits (I).

The sequence of these bytes is shown in Figure 7-U.11.

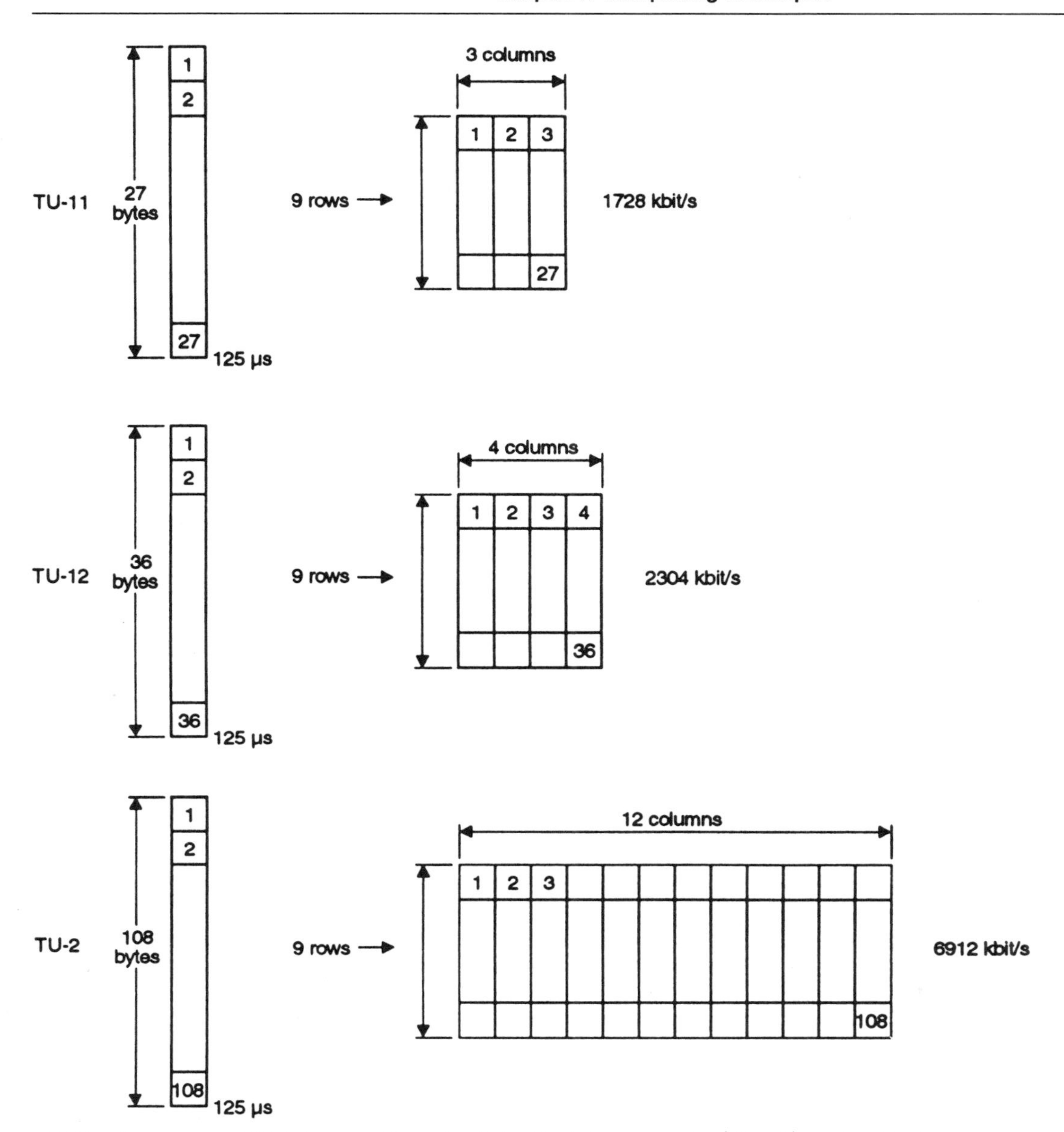

FIGURE 7-U.8. TU-1 and TU-2 sizes and formats. *Note*: The TU pointer bytes (V1 – V4) are located in byte 1 (using a 4-frame multiframe). (From Figure 5-1 / G.709, ITU-T Rec. G.709, 3 / 93, page 33.)

The set of five justification control bits (C) in every row is used to control the corresponding justification opportunity bit (S). CCCCC = 00000 indicates that the S bit is an information bit, whereas CCCCC = 11111 indicates that the S bit is a justification bit.

Majority vote should be used to make the justification decision in the desynchronizer for protection against single and double bit errors in the C bits.

The value contained in the S bit when used as justification bit is not defined. The receiver is required to ignore the value contained in this bit whenever it is used as a justification bit.

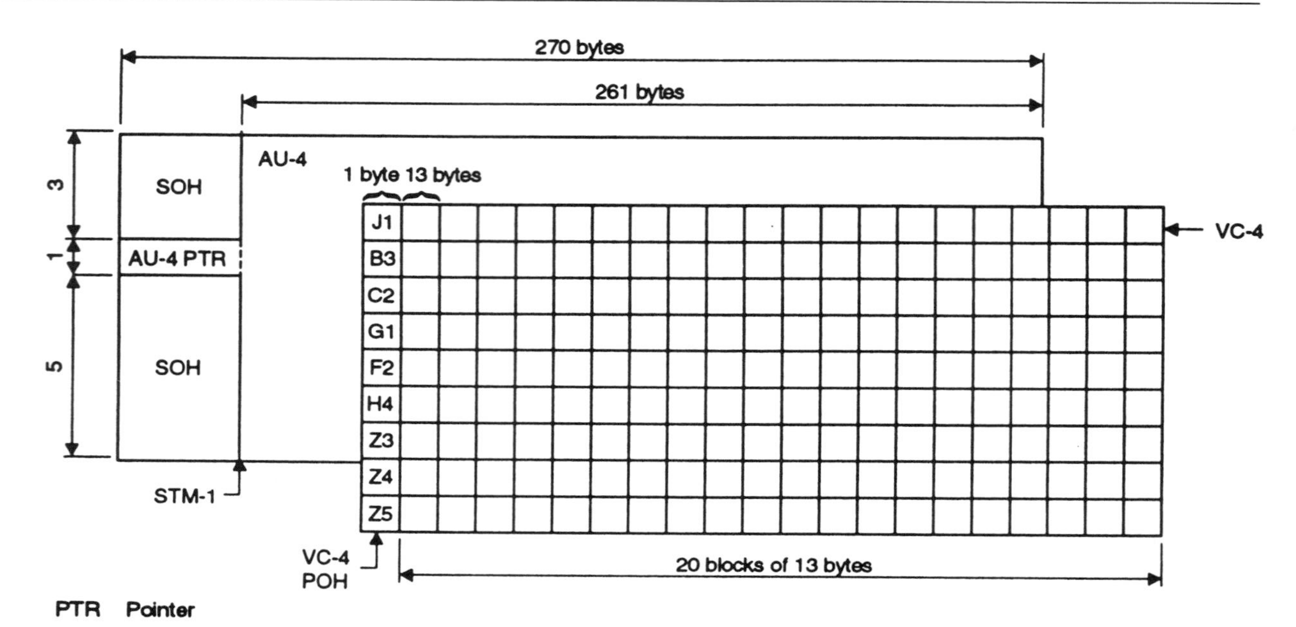

FIGURE 7-U.9. Mapping of VC-4 into STM-1 and block structure of VC-4 for synchronous mapping of 139,264 kbps. (From Figure 5-2/G.709, ITU-T Rec. G.709, 3/93, page 34.)

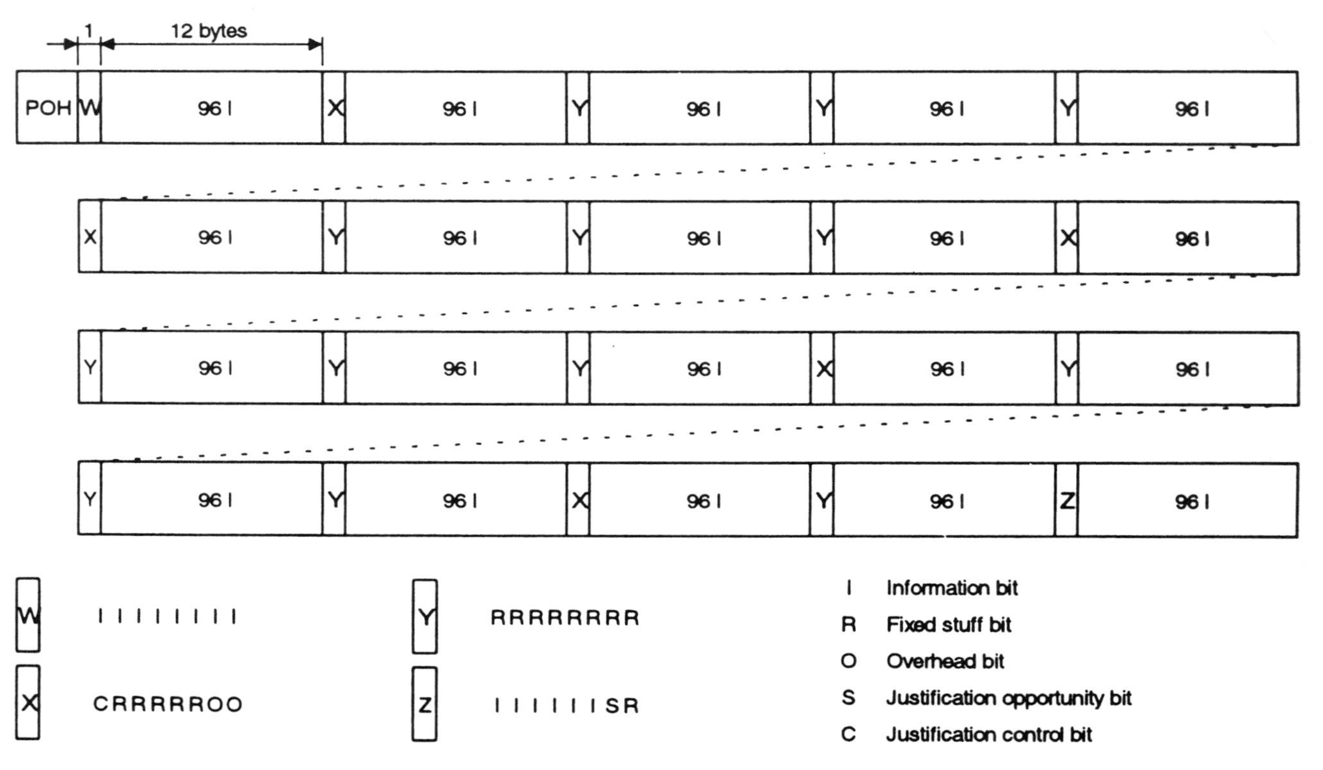

NOTE – This figure shows one row of the nine-row VC-4 container structure.

FIGURE 7-U.10. Asynchronous mapping of 139,264-kbps tributary into VC-4. (From Figure 5-3 / G.709, ITU-T Rec. G.709, 3 / 93, page 35.)

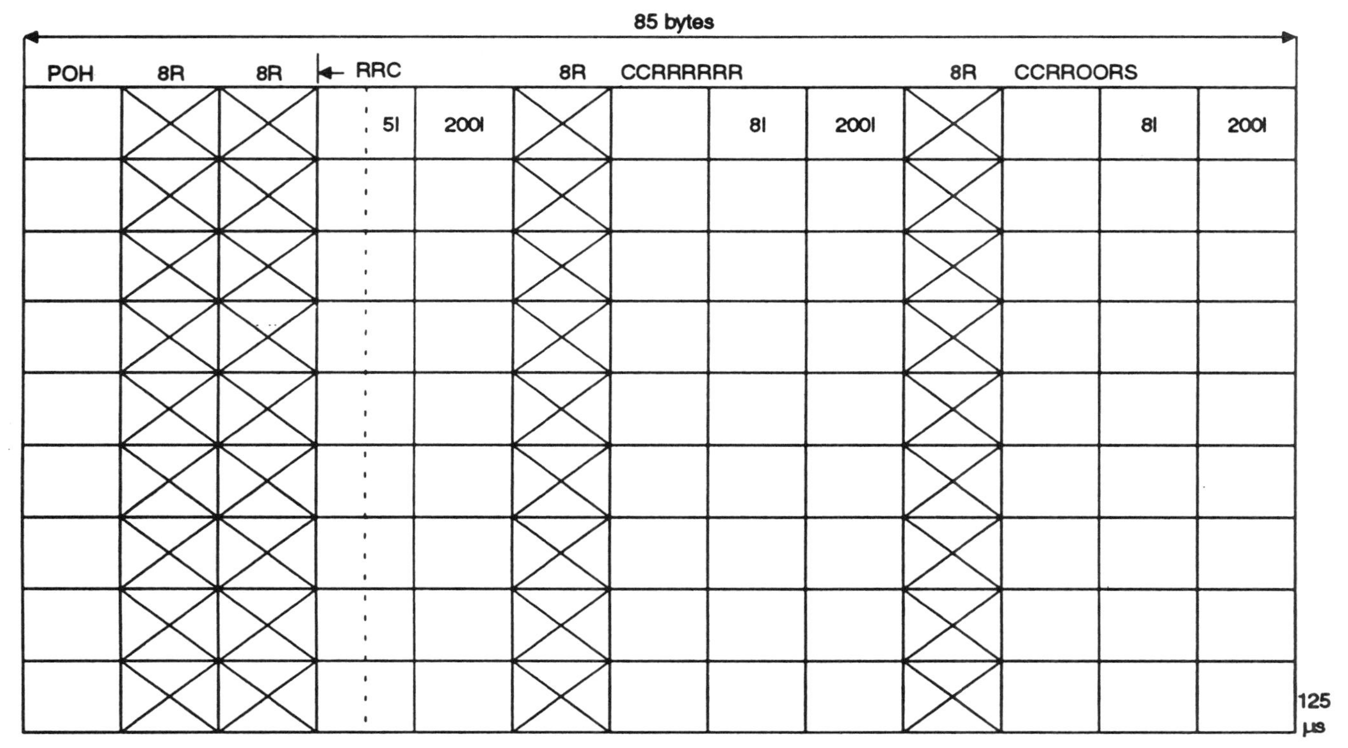

R Fixed stuff bit

C Justification control bit

S Justification opportunity bit

I Information bit

O Overhead bit

FIGURE 7-U.11. Asynchronous mapping of 44,736-kbps tributary into VC-3. (From Figure 5-4 / G.709, ITU-T Rec. G.709, 3 / 93, page 36.)

Mapping of Tributaries into VC-3

Asynchronous Mapping of 44,736 kbps. One 44,736 kbps signal can be mapped into a VC-3 as shown in Figure 7-U.11.The VC-3 consists of nine subframes every 125 μsec. Each subframe consists of one byte of VC-3 POH, 621 data bits, a set of five justification control bits, one justification opportunity bit, and two overhead communication channel bits. The remaining bits are fixed stuff (R) bits. The O bits are reserved for future overhead communication purposes.

The set of five justification control bits is used to control the justification opportunity (S) bit. CCCCC = 00000 indicates that the S bit is a data bit, whereas CCCCC = 11111 indicates that the S bit is a justification bit. Majority vote should be used to make the justification decision in the desynchronizer for protection against single and double bit errors in the C bits.

The value contained in the S bit when used as a justification bit is not defined. The receiver is required to ignore the value contained in this bit whenever it is used as a justification bit.

Asynchronous Mapping of 34,368 kbps. One 34,368-kbps signal can be mapped into a VC-3 as shown in Figure 7-U.12. In addition to the VC-3 POH, the VC-3 consists

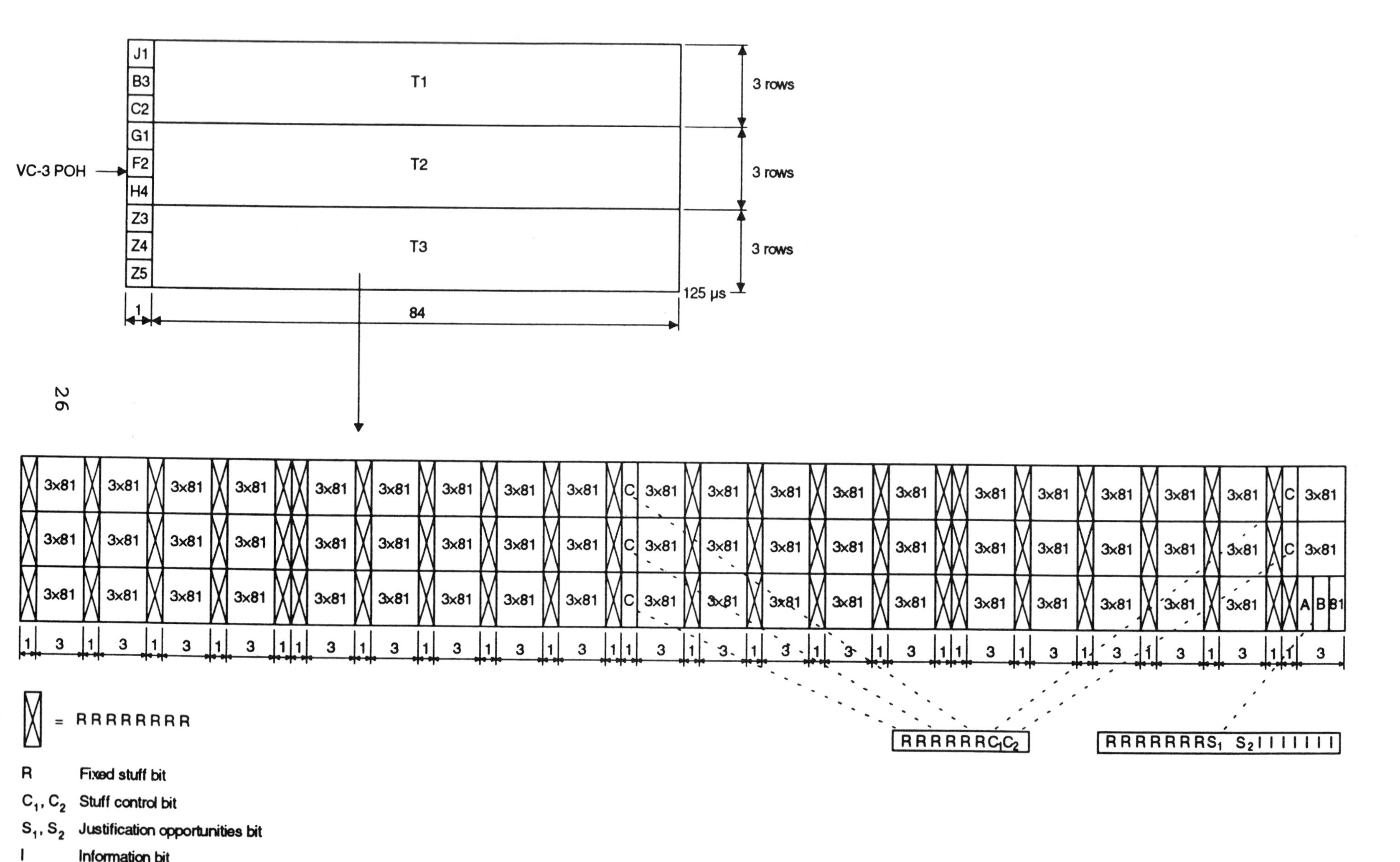

FIGURE 7-U.12. Asynchronous mapping of 34,368-kbps tributary into a VC-3.

of a payload of 9 × 84 bytes every 125 μsec. This payload is divided in three subframes, each subframe consisting of:

- 1431 information bits (I)
- Two sets of five justification control bits (C_1, C_2)
- Two justification opportunity bits (S_1, S_2)
- 573 fixed stuff bits (R)

Two sets of five justification control bits C_1 and C_2 are used to control the two justification opportunity bits S_1 and S_2, respectively.

$C_1C_1C_1C_1C_1$ = 00000 indicates that S_1 is a data bit, while $C_1C_1C_1C_1C_1$ = 11111 indicates that S_1 is a justification bit. C_2 bits control S_2 in the same way. Majority vote should be used to make the justification decision in the desynchronizer for protection against single and double bit errors in the C bits.

The value contained in S_1 and S_2 when they are justification bits is not defined. The receiver is required to ignore the value contained in these bits whenever they are used as justification bits.

Note: The same mapping could be used for bit or byte synchronous 34,368 kbps. In these cases, S_1 bit should be a fixed stuff and S_2 bit an information bit. By setting the C_1 bits to 1 and the C_2 bits to 0, a common desynchronizer could be used for both asynchronous and synchronous 34,368-kbps mappings.

Asynchronous Mapping of 6312 kbps. One 6312-kbps signal can be mapped into a VC-2. Figure 7-U.13 shows this over a period of 500 μsec. In addition to the VC-2

V5	I I I I I I I R	(24 x 8) I	R	
R	C_1C_2OOOOI R	(24 x 8) I	R	
I I I I I I I I	C_1C_2 OOOOI R	(24 x 8) I	R	
R	C_1C_2 I I I S_1S_2R	(24 x 8) I		125 μs
J2 a)	I I I I I I I R	(24 x 8) I	R	
R	C_1C_2OOOOI R	(24 x 8) I	R	
I I I I I I I I	C_1C_2OOOOI R	(24 x 8) I	R	
R	C_1C_2 I I I S_1S_2R	(24 x 8) I		250 μs
Z6 a)	I I I I I I I R	(24 x 8) I	R	
R	C_1C_2OOOOI R	(24 x 8) I	R	
I I I I I I I I	C_1C_2OOOOI R	(24 x 8) I	R	
R	C_1C_2 I I I S_1S_2R	(24 x 8) I		375 μs
Z7 a)	I I I I I I I R	(24 x 8) I	R	
R	C_1C_2OOOOI R	(24 x 8) I	R	
I I I I I I I I	C_1C_2OOOOI R	(24 x 8) I	R	
R	C_1C_2 I I I S_1S_2R	(24 x 8) I		500 μs

a) Provisional allocation.

R Fixed stuff

C Justification control bit

S Justification opportunity bit

I Information bit

O Overhead bit

FIGURE 7-U.13. Asynchronous mapping of 6312-kbps tributary. (From Figure 5-6 / G.709, ITU-T Rec. G.709, 3 / 93, page 38.)

V5	II I I I I I I R	(24 × 8) I	R	
R	10 OOOO I R	(24 × 8) I	R	
I I I I I I I I I	10 OOOO I R	(24 × 8) I	R	
R	10 I I I R I R	(24 × 8) I		
J2 [a]	II I I I I I I R	(24 × 8) I	R	125 µs
R	10 OOOO I R	(24 × 8) I	R	
I I I I I I I I I	10 OOOO I R	(24 × 8) I	R	
R	10 I I I R I R	(24 × 8) I		
Z6 [a]	II I I I I I I R	(24 × 8) I	R	250 µs
R	10 OOOO I R	(24 × 8) I	R	
I I I I I I I I I	10 OOOO I R	(24 × 8) I	R	
R	10 I I I R I R	(24 × 8) I		
Z7 [a]	II I I I I I I R	(24 × 8) I	R	375 µs
R	10 OOOO I R	(24 × 8) I	R	
I I I I I I I I I	10 OOOO I R	(24 × 8) I	R	
R	10 I I I R I R	(24 × 8) I		500 µs

[a] Provisional allocation.

R Fixed stuff

I Information bit

O Overhead bit

FIGURE 7-U.14. Bit synchronous mapping of 6312-kbps tributary. (From Figure 5-7/G.709, ITU-T Rec. G.709, 3/93, page 39.)

POH, the VC-2 consists of 3152 data bits, 24 justification control bits, eight justification opportunity bits, and 32 overhead communication channel bits. The remaining are fixed stuff bits (R). The O bits are reserved for future overhead communication purposes.

Two sets (C_1, C_2) of three justification control bits are used to control the two justification opportunities S_1 and S_2, respectively.

$C_1C_1C_1 = 000$ indicates that S_1 is a data bit, while $C_1C_1C_1 = 111$ indicates that S_1 is a justification bit. C_2 bits control S_2 in the same way. Majority vote should be used to make the justification decision in the desynchronizer for protection against single bit error in the C bits.

The value contained in S_1 and S_2 when they are justification bits is not defined. The receiver is required to ignore the value contained in these bits whenever they are used as justification bits.

Bit Synchronous Mapping of 6312 kbps. The bit synchronous mapping of 6312-kbps tributaries is shown in Figure 7-U.14.

Mapping of Tributaries into VC-12

Asynchronous Mapping of 2048 kbps. One 2048-kbps signal can be mapped into a VC-12. Figure 7-U.15 shows this over a period of 500 μsec. In addition to the VC-1 POH, the VC-12 consists of 1023 data bits, six justification control bits, two justification opportunity bits, and eight overhead communication channel bits. The remaining are fixed stuff bits (R). The O bits are reserved for future overhead communication purposes.

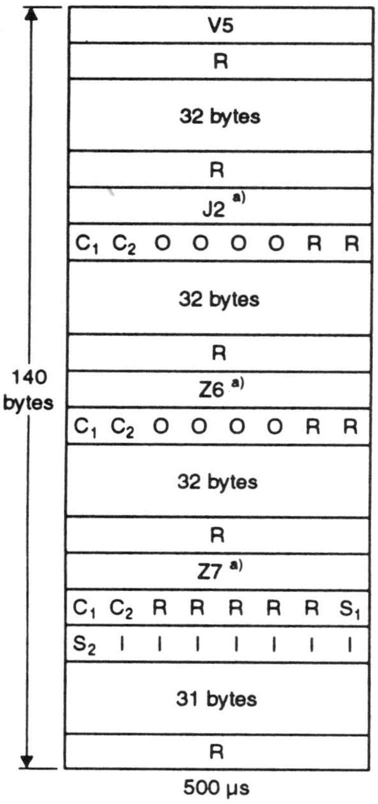

140 bytes

V5
R
32 bytes
R
J2 [a]
C_1 C_2 O O O O R R
32 bytes
R
Z6 [a]
C_1 C_2 O O O O R R
32 bytes
R
Z7 [a]
C_1 C_2 R R R R R S_1
S_2 I I I I I I I
31 bytes
R

500 μs

[a] Provisional allocation.

I Information bit
O Overhead
C Justification control
S Justification opportunity
R Fixed stuff

FIGURE 7-U.15. Asynchronous mapping of 2048-kbps tributary. (From Figure 5-8/G.709, ITU-T Rec. G.709, 3/93, page 40.)

Two sets (C_1, C_2) of three justification control bits are used to control the two justification opportunities S_1 and S_2, respectively. $C_1C_1C_1 = 000$ indicates that S_1 is a data bit, while $C_1C_1C_1 = 111$ indicates that S_1 is a justification bit. C_2 controls S_2 in the same way. Majority vote should be used to make the justification decision in the desynchronizer for protection against single bit errors in the C bits.

The value contained in S_1 and S_2 when they are justification bits is not defined. The receiver is required to ignore the value contained in these bits whenever they are used as justification bits.

Bit Synchronous Mapping of 2048 kbps. The bit synchronous mapping of 2048-kbps tributaries is shown in Figure 7-U.16. Note that a common desynchronizer can be used for both asynchronous and bit synchronous mappings.

Byte Synchronous Mapping of 2048 kbps. Figure 7-U.17 shows byte synchronous mapping for 30-channel 2048-kbps tributaries employing channel-associated signaling (CAS). Signaling is carried out in bytes 19, 54, 89, and 124 in the floating point mode and in byte 19 in the locked mode.

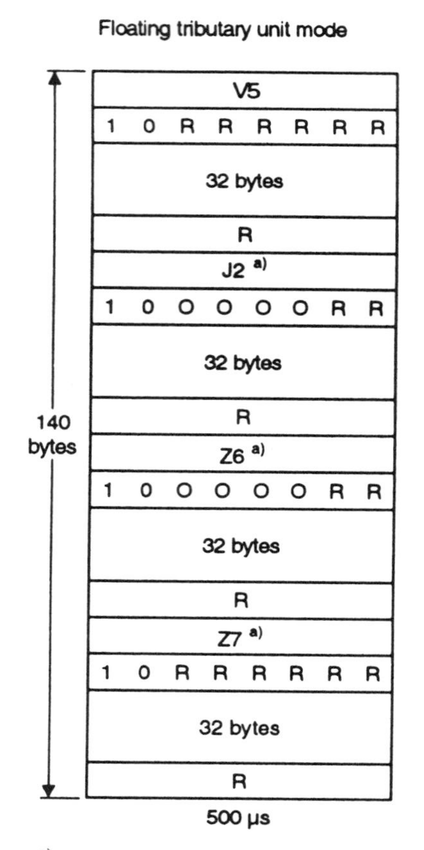

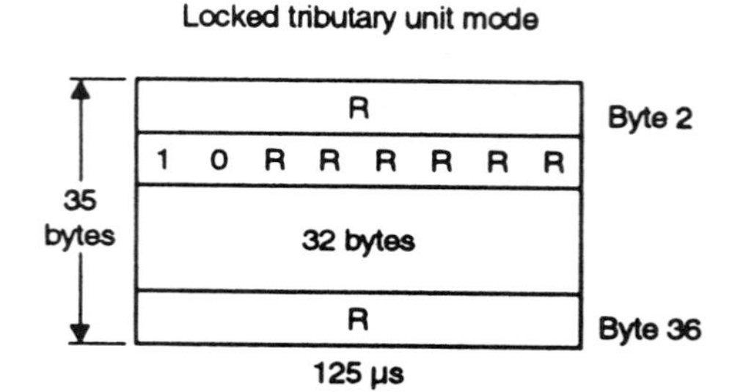

a) Provisional allocation.

O Overhead

R Fixed stuff

FIGURE 7-U.16. Bit synchronous mapping for 2048-kbps tributary. (From Figure 5-9/G.709, ITU-T Rec, G.709, 3/93, page 41.)

The S_1, S_2, S_3, and S_4 bits contain the signaling for the 30 × 64-kbps channels. The phase of the signaling bits is by position indicator byte (H4) in the located tributary unit mode. This is illustrated in Figure 7-U.18.

Byte synchronous mapping of 31-channel tributaries is shown in Figure 7-U.19. Byte 19 carries tributary channel 16.

Mapping Tributaries into VC-11

Asynchronous Mapping of 1544 kbps. One 1544-kbps signal can be mapped in a VC-11. Figure 7-U.20 shows this over 500 μsec. In addition to the VC-1 POH, the VC-11 consists of 771 data bits, six justification control bits, two justification opportunity bits, and eight overhead communication channel bits. The remaining are fixed stuff bits (R). The O bits are reserved for future communication purposes.

Two sets (C_1, C_2) of three justification control bits are used to control the two justification opportunities, S_1 and S_2, respectively.

$C_1C_1C_1 = 000$ indicates that S_1 is a data bit, while $C_1C_1C_1 = 111$ indicates that S_1 is a justification bit. C_2 controls S_2 in the same way. Majority vote should

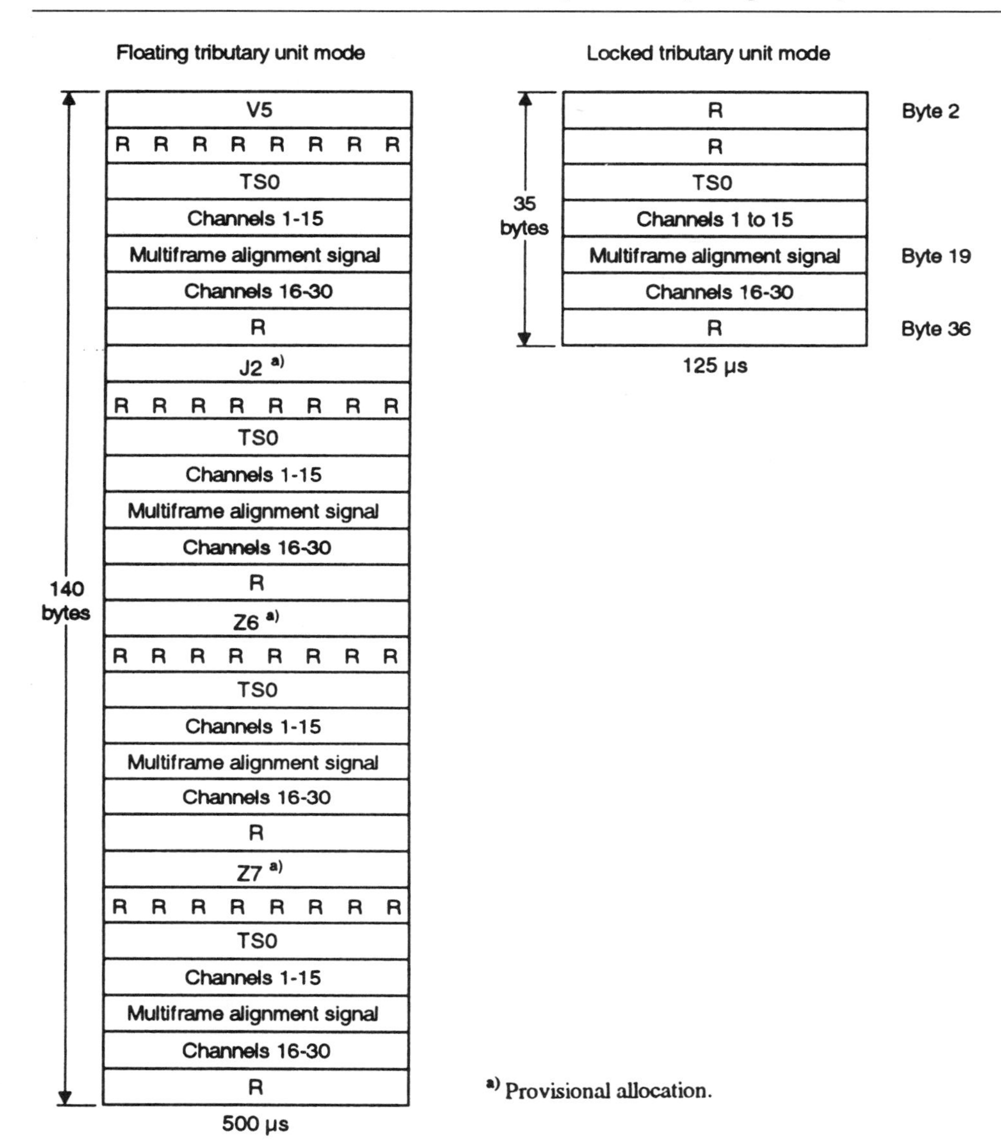

FIGURE 7-U.17. Byte synchronous mapping for 2048-kbps tributary (30 channels with channel-associated signaling). (From Figure 5-10/G.709, ITU-T Rec. G.709, 3/93, page 42.)

Locked												
H4 value				CAS format								Channel
C_3	C_2	C_1	T	S_1	S_2	S_3	S_4	S_1	S_2	S_3	S_4	
0	0	0	0	0	0	0	0	x	y	x	x	None
0	0	0	1	a	b	c	d	a	b	c	d	1/16
0	0	1	0	a	b	c	d	a	b	c	d	2/17
1	1	1	1	a	b	c	d	a	b	c	d	15/30

FIGURE 7-U.18. Out slot signaling assignments (30-channel signaling operations). (From Figure 5-11/G.709, ITU-T Rec. G.709, 3/93, page 43.)

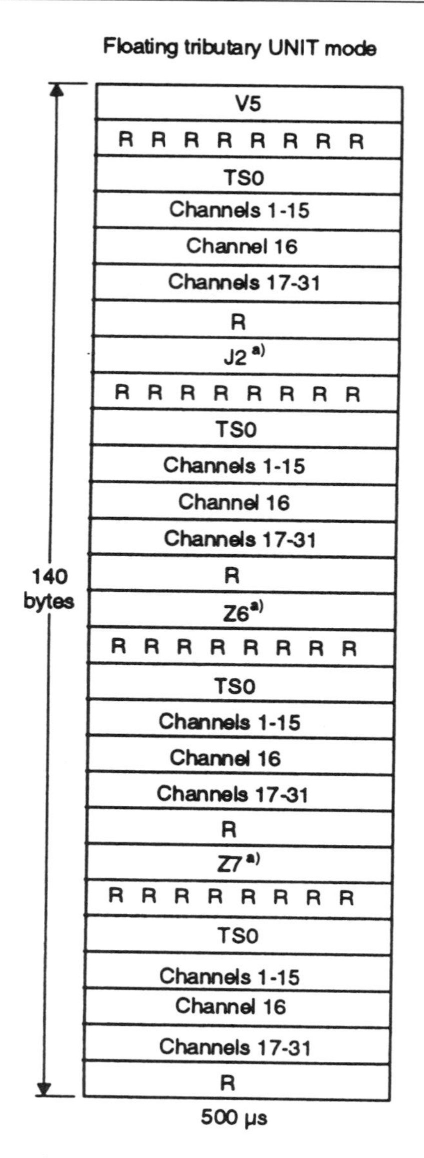

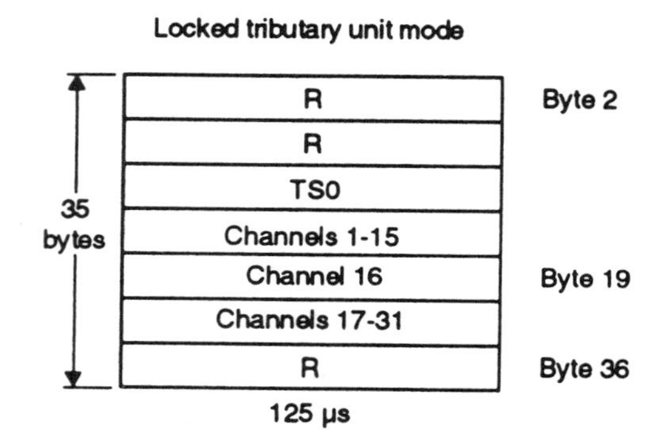

a) Provisional allocation.

FIGURE 7-U.19. Byte synchronous mapping for 2048-kbps tributary (31 channels with common channel signaling). (From Figure 5-12/G.709, ITU-T Rec. G.709, 3/93, page 44.)

be used to make the justification decision in the desynchronizer for protection against single bit errors in the C bits.

The value contained in S_1 and S_2 when they are justification bits is not defined. The receiver is required to ignore the value contained in these bits whenever they are used as justification bits.

Bit Synchronous Mapping of 1544 kbps. The bit synchronous mapping of 1544-kbps tributaries is shown in Figure 7-U.21.

Byte Synchronous Mapping of 1544 kbps. The byte synchronous mapping of 1544 kbps is shown in Figure 7-U.22.

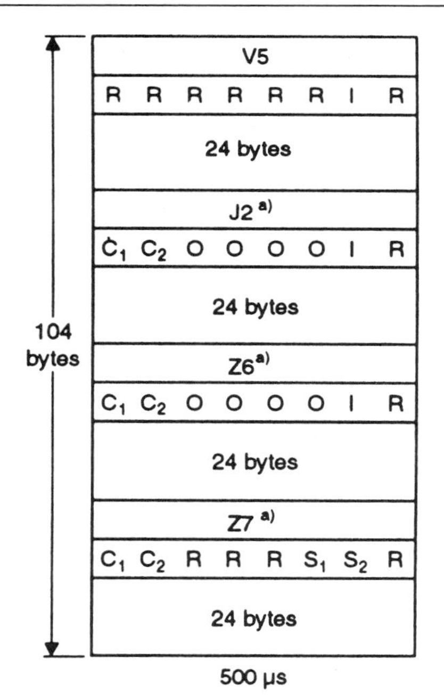

a) Provisional allocation.
I Information bit
O Overhead
C Justification control
S Justification opportunity
R Fixed stuff

FIGURE 7-U.20. Asynchronous mapping of 1544-kbps tributary. (From Figure 5-13/G.709, ITU-T Rec. G.709, 3/93, page 45.)

Floating tributary unit mode

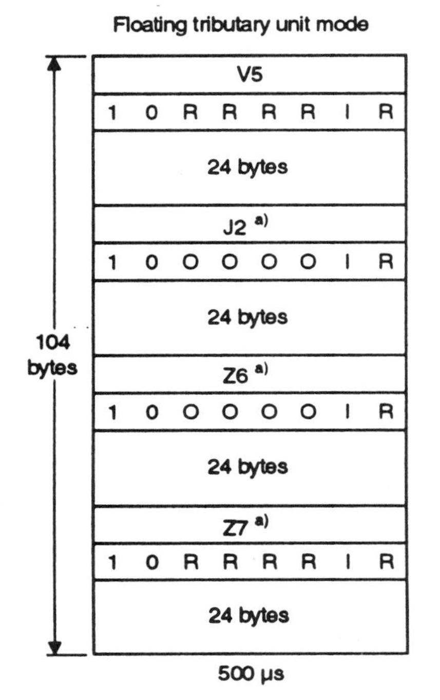

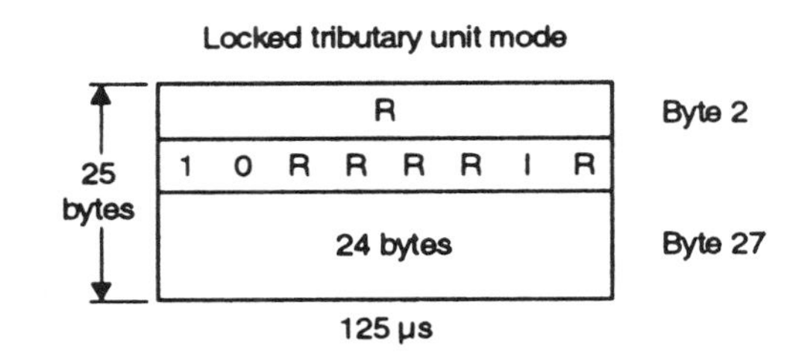

a) Provisional allocation.
I Information bit
O Overhead
R Fixed stuff

NOTE – O bits are currently not defined in the locked tributary unit mode.

FIGURE 7-U.21. Bit synchronous mapping for 1544-kbps tributary. (From Figure 5-14/G.709, ITU-T Rec. G.709, 3/93, page 46.)

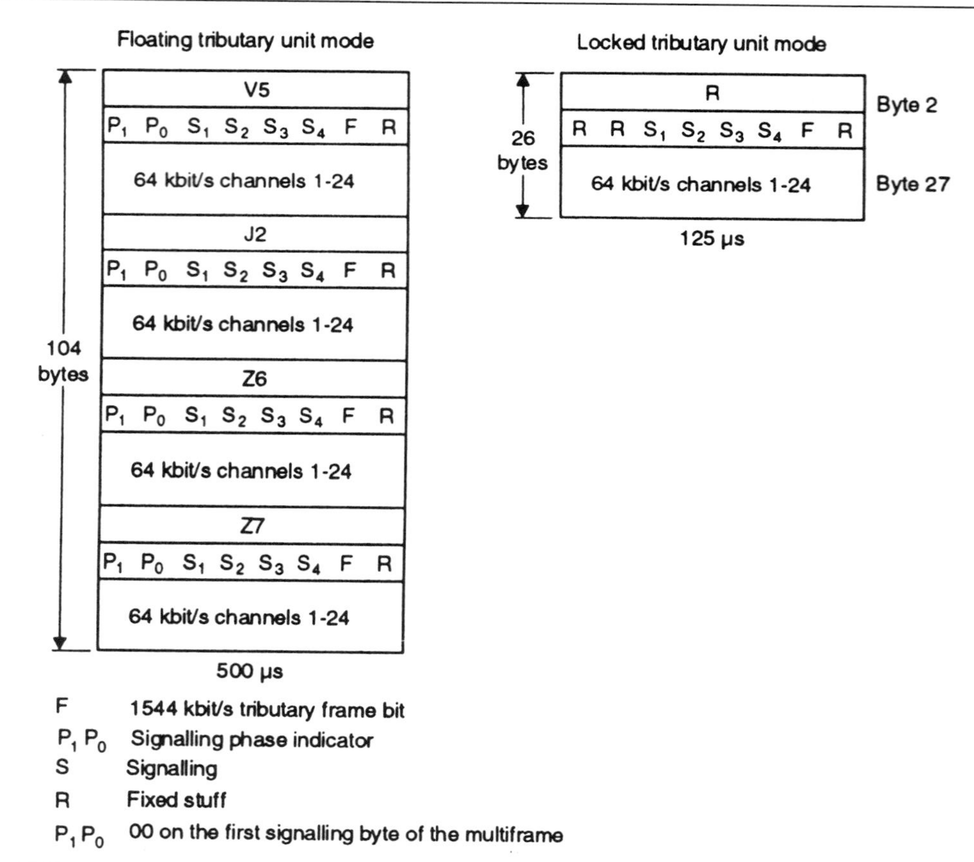

FIGURE 7-U.22. Byte synchronous mapping of 1544-kbps tributary. (From Figure 5-15/G.709, ITU-T Rec. G.709, 3/93, page 47.)

The S_1, S_2, S_3, and S_4 bits contain the signaling for the 24×64-kbps channels. The phase of the signaling bits can be indicated by the P_1 and P_0 bits in the floating tributary unit mode, as well as by the position indicator byte (H4) in the locked mode. This is illustrated in Figure 7-U.23. The usage of the P bits is optional, because the common channel signaling methods and some of the channel-associated signaling methods (refer to ITU-T Rec. G.704) do not need the P bits. The out slot signaling assignments for one of the channel-associated signaling methods is shown in Figure 7-U.24.

VC-11 to VC-12 Conversion for Transport by a TU-12. When transporting a VC-11 in a TU-12, the VC-11 is adapted by adding fixed stuff with even parity, as shown in Figure 7-U.25. Thus the resulting TU-12 payload can be monitored and cross-connected in the network as though it were a VC-12 with its BIP value unchanged while preserving end-to-end integrity of the real VC-11 path.

Floating- and Locked-Mode Conversion. There are two possible multiplexing modes of the tributary unit structures: floating and locked.

In the floating tributary unit mode four consecutive 125 μsec VC-n frames ($n = 11, 12, 2$) are organized into a 500 μsec multiframe, the phase of which is

Locked																				
					Floating															
					Signalling															
H4 value					2 state				4 state				16 state							
P_1	P_0	S_2	S_1	T	S_1	S_2	S_3	S_4	S_1	S_2	S_3	S_4	S_1	S_2	S_3	S_4	P_1	P_0		
0	0	0	0	0	A_1	A_2	A_3	A_4	A_1	A_2	A_3	A_4	A_1	A_2	A_3	A_4	0	0		
0	0	0	0	1	A_5	A_6	A_7	A_8	A_5	A_6	A_7	A_8	A_5	A_6	A_7	A_8	0	0		
0	0	0	1	0	A_9	A_{10}	A_{11}	A_{12}	A_9	A_{10}	A_{11}	A_{12}	A_9	A_{10}	A_{11}	A_{12}	0	0		
0	0	0	1	1	A_{13}	A_{14}	A_{15}	A_{16}	A_{13}	A_{14}	A_{15}	A_{16}	A_{13}	A_{14}	A_{15}	A_{16}	0	0		
0	0	1	0	0	A_{17}	A_{18}	A_{19}	A_{20}	A_{17}	A_{18}	A_{19}	A_{20}	A_{17}	A_{18}	A_{19}	A_{20}	0	0		
0	0	1	0	1	A_{21}	A_{22}	A_{23}	A_{24}	A_{21}	A_{22}	A_{23}	A_{24}	A_{21}	A_{22}	A_{23}	A_{24}	0	0		
0	1	0	0	0	A_1	A_2	A_3	A_4	B_1	B_2	B_3	B_4	B_1	B_2	B_3	B_4	0	1		
0	1	0	0	1	A_5	A_6	A_7	A_8	B_5	B_6	B_7	B_8	B_5	B_6	B_7	B_8	0	1		
0	1	0	1	0	A_9	A_{10}	A_{11}	A_{12}	B_9	B_{10}	B_{11}	B_{12}	B_9	B_{10}	B_{11}	B_{12}	0	1		
0	1	0	1	1	A_{13}	A_{14}	A_{15}	A_{16}	B_{13}	B_{14}	B_{15}	B_{16}	B_{13}	B_{14}	B_{15}	B_{16}	0	1		
0	1	1	0	0	A_{17}	A_{18}	A_{19}	A_{20}	B_{17}	B_{18}	B_{19}	B_{20}	B_{17}	B_{18}	B_{19}	B_{20}	0	1		
0	1	1	0	1	A_{21}	A_{22}	A_{23}	A_{24}	B_{21}	B_{22}	B_{23}	B_{24}	B_{21}	B_{22}	B_{23}	B_{24}	0	1		
1	0	0	0	0	A_1	A_2	A_3	A_4	A_1	A_2	A_3	A_4	C_1	C_2	C_3	C_4	1	0		
1	0	0	0	1	A_5	A_6	A_7	A_8	A_5	A_6	A_7	A_8	C_5	C_6	C_7	C_8	1	0		
1	0	0	1	0	A_9	A_{10}	A_{11}	A_{12}	A_9	A_{10}	A_{11}	A_{12}	C_9	C_{10}	C_{11}	C_{12}	1	0		
1	0	0	1	1	A_{13}	A_{14}	A_{15}	A_{16}	A_{13}	A_{14}	A_{15}	A_{16}	C_{13}	C_{14}	C_{15}	C_{16}	1	0		
1	0	1	0	0	A_{17}	A_{18}	A_{19}	A_{20}	A_{17}	A_{18}	A_{19}	A_{20}	C_{17}	C_{18}	C_{19}	C_{20}	1	0		
1	0	1	0	1	A_{21}	A_{22}	A_{23}	A_{24}	A_{21}	A_{22}	A_{23}	A_{24}	C_{21}	C_{22}	C_{23}	C_{24}	1	0		
1	1	0	0	0	A_1	A_2	A_3	A_4	B_1	B_2	B_3	B_4	D_1	D_2	D_3	D_4	1	1		
1	1	0	0	1	A_5	A_6	A_7	A_8	B_5	B_6	B_7	B_8	D_5	D_6	D_7	D_8	1	1		
1	1	0	1	0	A_9	A_{10}	A_{11}	A_{12}	B_9	B_{10}	B_{11}	B_{12}	D_9	D_{10}	D_{11}	D_{12}	1	1		
1	1	0	1	1	A_{13}	A_{14}	A_{15}	A_{16}	B_{13}	B_{14}	B_{15}	B_{16}	D_{13}	D_{14}	D_{15}	D_{16}	1	1		
1	1	1	0	0	A_{17}	A_{18}	A_{19}	A_{20}	B_{17}	B_{18}	B_{19}	B_{20}	D_{17}	D_{18}	D_{19}	D_{20}	1	1		
1	1	1	0	1	A_{21}	A_{22}	A_{23}	A_{24}	B_{21}	B_{22}	B_{23}	B_{24}	D_{21}	D_{22}	D_{23}	D_{24}	1	1		

FIGURE 7-U.23. Out slot signaling assignments (24-channel signaling operations). (From Figure 5-16 / G.709, ITU-T Rec. G.709, 3 / 93, page 48.)

indicated by the position indicator byte (H4) in the VC-*m* POH ($m = 3, 4$). This 500 μsec tributary unit multiframe is shown in Figure 7-U.26.

Locked tributary unit mode of transport is a fixed mapping of synchronous structured payloads into a VC-*m*. This provides a direct correspondence between subtending tributary information and the location of that information within the VC-*m*. Because the tributary information is fixed and immediately identifiable with respect to the AU-*m* pointer associated with the VC-*m*, no tributary unit pointers are required. All bytes of a tributary unit or TUG are available for payload usage.

Figure 7-U.26 illustrates the conversion between floating and locked TU modes for each of the three tributary unit sizes. Note that certain bytes (R) in the current set of mapping are not used in the floating mode in order that those mappings can be used in both floating and locked modes. Because the V1–V4 and V5 bytes are reserved, the 500-μsec virtual container multiframe is unnecessary. Therefore the role of the multiframe indicator byte (H4) in locked mode is to define 2- and 3-msec signaling frames for byte synchronous mappings.

This new section, beginning on the next page, follows Section 7-2.14.2, which ends on page 542 of the Manual.

Frame number	n	$n+1$	$n+2$	$n+3$	$n+4$	$n+5$	$n+6$	$n+7$
Use of S_i bit (i = 1, 2, 3, 4)	Fs	Y_1	Y_2	Y_3	Y_4	Y_5	Y_6	X
(Note 1)	(Note 2)	(Note 3)						(Note 5)

NOTES

1 Each S_i (i = 1, 2, 3, 4) constitutes an independent signalling multiframe over eight frames. S_i includes the phase indicator in itself, so that the PP-bits cannot be used for the phase indicator.

2 The Fs bit is either alternate 0, 1 or the following 48 bit digital pattern:

A101011011 0000011001 1010100111 0011110110 10000101

For the 48-bit digital pattern, the "A" bit is usually fixed to state 1 and is reserved for optional use. The pattern is generated according to the following primitive polynomial (refer to Recommendation X.50):

$$x^7 + x^4 + 1$$

3 Y_j bit (j = 1 to 6) carries channel associated signalling or maintenance information. When the 48 bit pattern is adopted as Fs frame alignment signal, each Y_j bit (j = 1 to 6) can be multiframed, as follows.

$$Y_{j1}, Y_{j2}, \ldots, Y_{j12}$$

Y_{j1} bit carries the following 16-bit frame alignment pattern generated according to the same primitive polynomial as for the 48-bit pattern.

A011101011011000

The "A" bit is usually fixed to 1 and is reserved for optional use. Each Y_{ji} (i = 2 to 12) bit carries channel associated signalling for sub-rate circuits and/or maintenance information.

4 S_i bits (Fs, Y_1, . . ., Y_6 and X), all at state 1 indicate alarm indication signal (AIS) for six 64 kbit/s channels.

5 The X-bit is usually fixed to state 1. When backward AIS for six 64 kbit/s channels is required to be sent, the X-bit is set to state 0.

FIGURE 7-U.24. Out slot signaling assignments (24-channel signaling operations). (From Figure 5-17/G.709, ITU-T Rec. G.709, 3/93, page 49.)

7-2.15. Transmission Performance Characteristics of Pulse Code Modulation

7-2.15.1. Introduction

Performance characteristics which follow should be met between voice-frequency ports or between the voice-frequency and digital ports of PCM channels coded in accordance with Rec. G.711.

Equipment which meets the analog-to-analog requirements, but not the analog-to-digital requirements, may only be used as permanently connected pairs of equipment.

The parameters and values specified in this Recommendation apply to the use of PCM equipment connected to analog trunks or to analog and digital exchanges. When PCM equipment is connected directly to analog subscriber lines, different values for some of the parameters may be required. Recommendation Q.552 contains those values. The requirements in this Recommendation may also be applied if the PCM equipment is directly connected to an analog local exchange that is virtually transparent with regard to the impedances connected to its ports and the subscriber lines are short (e.g., less than 500 meters).

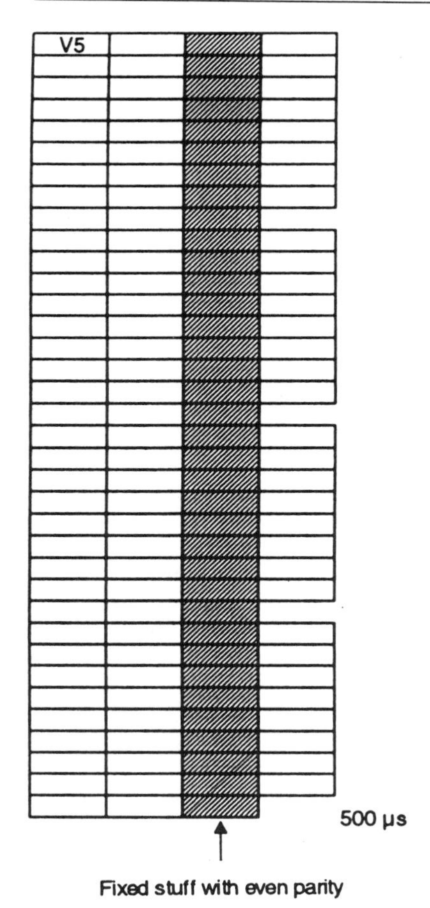

FIGURE 7-U.25. Conversion of VC-11 to VC-12 for transport by TU-12. (From Figure 5-18/G.709, ITU-T Rec. G.709, 3/93, page 50.)

In deriving the limits, an allowance has been included for the effect of possible signaling functions and/or line current feeding on the transmission performance. The limits should be met when any signaling function is in the normal speaking condition, but excluding any dynamic signaling conditions (e.g., metering).

The limits do not, in general, have any allowance for the effects of line current noise. The permissible amount of line current noise and the need for allowances are under study.

Measurements. When a nominal reference frequency of 1020 Hz is indicated (e.g., measurement of attenuation/frequency distortion and adjustment of relative levels), the actual frequency should be 1020 Hz, +2 Hz, and −7 Hz in accordance with Rec. O.6. For an interim period, administrations may, for practical reasons, need to use a reference frequency of nominally 800 Hz. To avoid level errors produced as a result of the use of test frequencies which are submultiples of the PCM sampling rate, the use of integer submultiples of 8 kHz should be avoided.

In the following sections, the concepts of a "standard digital generator" and "a standard digital analyzer" should be assumed, and these are defined as follows:

A *standard digital generator* is a hypothetical device which is absolutely ideal [i.e., a perfect analog-to-digital converter preceded by an ideal low pass filter (assumed to have no attenuation/frequency distortion and no group-delay distortion)] and which may be simulated by a digital processor.

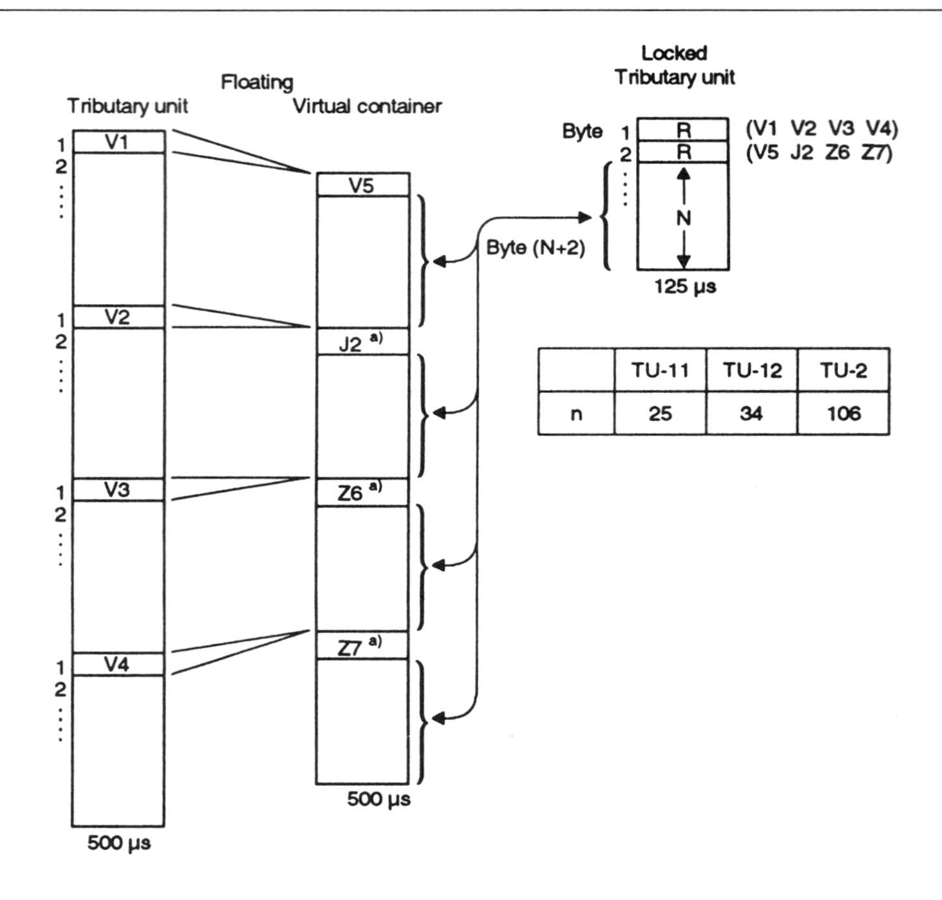

a) Provisional allocation.

V1 PTR-1
V2 PTR-2
V3 PTR-3 (action)
V4 Reserved
V5 VC-1/VC-2 POH
PTR Pointer

FIGURE 7-U.26. Conversion between floating and locked tributary unit modes. (From Figure 5-19/G.709, ITU-T Rec. G.709, 3/93, page 52.)

A *standard digital analyzer* is a hypothetical device which is absolutely ideal [i.e., a perfect digital-to-analog converter followed by an ideal low pass filter (assumed to have no attenuation/frequency distortion and no group-delay distortion)] and which may be simulated by a digital processor.

Recommendation O.133 contains information about test equipment based on these concepts. Account should be taken of the measurement accuracy provided by test equipment designed in accordance with that Recommendation.

A PCM quiet signal is defined to be an idle character signal corresponding to decoder output value number 0 for μ-law or decoder output value number 1 for A-law with the sign bit in a fixed state.

The following specifications are based on ideal measuring equipment. Therefore, they do not include any margin for measurement errors.

Definitions of Ports. The term *port* in this section is defined as a functional unit (e.g., a connector) of the PCM equipment through which signals can enter or leave

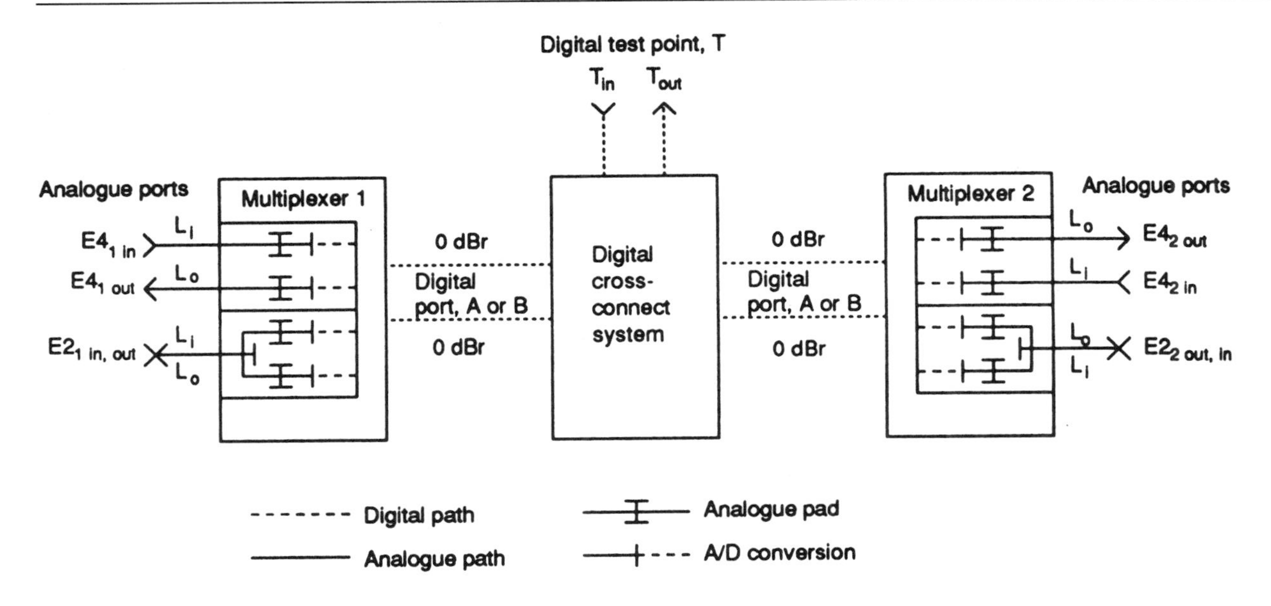

Note – **Numbers 1 or 2 in the subscript indicate multiplexers 1 or 2.**

FIGURE 7-U.27. PCM equipment and test ports. (From Figure 1/G.712, ITU-T Rec. G.712, 9/92, page 3.)

the unit under test. The measurements are made at the equipment, and the requirements do not include an allowance for wiring to a distribution frame.

Figure 7-U.27 shows two PCM equipments connected via a digital cross-connect (DXC) system. Each PCM equipment is shown to have a 4-wire analog voice-frequency (VF) port, E4, and a 2-wire analog VF port, E2. Each of the PCM equipments is connected to the DXC by digital ports A or B, which consists of transparent 64-kbps PCM channels within a higher-order digital signal conforming to applicable ITU-T recommendations. The DXC implements transparent 64-kbps cross connections between its own A or B ports, or to the digital test point T. The T point consists of a 64-kbps channel within a digital signal suitable for connection to the standard digital generator and analyzer. Because the connections provided by the DXC are 64-kbps transparent, the paths within the DXC are assumed not to affect the recommended transmission performance characteristics except for absolute delay. The test points T are defined for specification purposes. They may not physically exist in a DXC but may be accessed via the DXC network.

Unless stated otherwise, measurements between 2-wire ports ($E2_{1in}$ to $E2_{2out}$ connections) should be made with the 4-wire loop opened in such a way that the impedances presented to the 4-wire port of the 2-wire/4-wire terminating unit are representative of those that will occur in normal operation. This condition may be achieved by interrupting the digital signal in the opposite direction to the direction of measurement and injecting a PCM quiet signal into the appropriate channel. It should be noted that the opening of the 4-wire loop is considered necessary to determine the intrinsic performance of the equipment. In normal operation, where the loop is not opened, account needs to be taken of the impact on overall performance of the terminating impedances connected to the 2-wire ports.

Figure 7-U.27 identifies ports, transmission levels, and test points of a digital exchange. Table 7-U.1 shows correspondence between corresponding previous ITU-T recommendations and that specified herein.

TABLE 7-U.1
Equivalence Between CCITT Recommendations and the Connections Specified Herein

Channel	Recommendation
4-wire analog to 4-wire analog ($E4_1$ to $E4_2$ channels)	G.712
2-wire analog to 2-wire analog ($E2_1$ to $E2_2$ channels)	G.713
4-wire analog to digital (E4 to A or B for a primary multiplexer)	G.714
2-wire analog to digital (E2 to A or B for a primary multiplexer)	G.715

7-2.15.2. Relative Levels at Voice-Frequency Ports

Relative Levels at 4-Wire Ports (E4). When attenuation pads are set to zero loss, L_o at $E4_{2out}$ and L_i at $E4_{1in}$ must have one of the following two series of nominal values: Maximum $L_o = +4$ dBr and minimum $L_i = -14$ dBr; or maximum $L_o = +7$ dBr and minimum $L_i = -16$ dBr. See Rec. G.232, Section 11.

Relative Levels at 2-Wire Ports (E2). Because of differences in network transmission plans and equipment utilization, administrations have differing requirements for the range of relative levels to be provided. The following ranges would encompass the requirements of a large number of administrations:

Input level (L_i encoding side) 0 to -5 dBr in 0.5-dB steps
Output level (L_o decoding side) -2 to -7.5 dBr in 0.5-dB steps

It is recognized that it is not necessarily appropriate for a particular design of equipment to be capable of operating over the entire range.

7-2.15.3. Adjustment of Actual Relative Levels

Adjustment of Decoding Side (T_{in} to E_{out}). The gain of the decoding side should be adjusted by connecting T_{in} to a standard digital generator and applying a 1020-Hz sinusoidal test signal at a level of 0 dBm0. The adjustment should result in an output level of 0 dBm0 $\pm$ 0.3 dB for 4-wire ports (T_{in} to $E4_{out}$) or an output level of 0 dBm0 $\pm$ 4 dB for 2-wire ports (T_{in} to $E2_{out}$) and should be made under typical conditions of power supply voltage, humidity, and temperature.

Adjustment of the Encoding Side (E_{in} to T_{out}). The gain of the encoding side should be adjusted by connecting T_{out} to a standard digital analyzer and applying a 1020-Hz test signal at a level of 0 dBm0 to E_{in}. The adjustment should result in an output level of 0 dBm0 $\pm$ 0.3 dB for 4-wire ports ($E4_{in}$ to T_{out}) or an output level of 0 dBm0 $\pm$ 0.4 dB for 2-wire ports ($E2_{in}$ to T_{out}) and should be made under typical conditions of power supply voltage, humidity, and temperature.

Load Capacity (Overload Point). The load capacity of the encoding side may be checked by applying a 1020-Hz sinusoidal test signal at E_{in}. The level of this signal should initially be well below T_{max} and should then be slowly increased. The input level should be measured at which the first occurrence is observed of the character signal corresponding to the extreme quantizing interval for both positive and negative values. T_{max} is taken as being 0.3 dB greater than the measured input level.

TABLE 7-U.2
Requirements for Short- and Long-Term Variation of Level with Time

Measurement Configurations	Maximum Permitted Variation		
	Ten-Minute Interval (dB)	One-Year Interval (dB)	
4-wire to 4-wire ($E4_{1in}$ to $E4_{2out}$)	±0.2	±0.5	
2-wire to 2-wire ($E2_{1in}$ to $E2_{2out}$)	±0.2	±0.6	
4-wire to digital ($E4_{in}$ to T_{out})	±0.1	±0.3	
Digital to 4-wire (T_{in} to $E4_{out}$)	±0.1	±0.3	(See Note)
2-wire to digital ($E2_{in}$ to T_{out})	±0.1	±0.3	
Digital to 2-wire (T_{in} to $E2_{out}$)	±0.1	±0.3	(See Note)

Note: The 0 dBm0 sequence of Rec. G.711, Tables 5/G.711 and 6/G.711 may be used.
Source: Table 2/G.712, ITU-T Rec. G.712, 9/92, page 4.

This method allows T_{max} to be checked for both positive and negative amplitudes, and the values thus obtained should be within 0.4 dB of the theoretical load capacity (i.e., +3.14 dBm0 for the A-law or +3.17 dBm0 for the μ-law).

7-2.15.4. Short- and Long-Term Variation of Level with Time

When a 1020-Hz sinusoidal test signal at a level of −10 dBm0 (preferred value; a level of 0 dBm0 may be used) is applied to any VF input, the level measured at the corresponding output should not vary by more than the limits shown in Table 7-U.2 during any 10-minute interval of typical operation, nor by more than the limits shown during any one year under the permitted variations of power supply voltage and temperature.

Nominal Impedance. The nominal impedance at the 4-wire voice-frequency input and output ports, $E4_{in}$ and $E4_{out}$, should be 600 Ω, balanced.

For 2-wire voice-frequency ports, E2, no single value of impedance is recommended. The following values can be found in practice:

600 Ω resistive, balanced
900 Ω resistive, balanced
600 Ω + 2.16 μF, balanced
900 Ω + 2.16 μF, balanced

Note: Some examples of complex impedances used in connection with subscriber lines can be found in Rec. Q.552, Section 2.2.1.

Return Loss. The return loss measured against the nominal impedance should meet the return loss requirements shown in Table 7-U.3 over the frequency range 300–3400 Hz.

7-2.15.5. Impedance Unbalance about Earth

The longitudinal conversion loss parameters referred to below are defined in ITU-T Rec. O.9, which also gives information regarding requirements of test circuits (see Note 1 in Table 7-U.4). The value Z in the driving test circuit should

TABLE 7-U.3
Return Loss Requirements for E4 and E2 Ports

Analog Port	Frequency Range		Notes
	From 300 Hz to 600 Hz	From 600 Hz to 3400 Hz	
4-wire, E4	> 20-dB return loss	> 20-dB return loss	1
2-wire, E2	> 12-dB return loss	> 15-dB return loss	2

Note 1: The return loss limit should be met when the adjusting pads are set to 0 dB (see Rec. G.232, Figure 5/G.232).

Note 2: Reflections due to impedance mismatches at 2-wire/4-wire ports may cause severe sidetone and echo problems in the network. Administrations need to adopt a suitable impedance strategy, including tolerances, to ensure an adequate transmission quality. (For further information, see Rec. G.121, Section 5.)

Source: Table 3/G.712, ITU-T Rec. G.712, 9/92, page 5.

TABLE 7-U.4
Longitudinal Conversion Loss Requirements for E4 and E2 Ports

Measured Port	Z (Ω)	Longitudinal Conversion Loss Requirement			Notes
		300 Hz to 600 Hz	600 Hz to 2400 Hz	2400 Hz to 3400 Hz	
4-wire, $E4_{in}$	600	> 46 dB	> 46 dB	> 41 dB	1, 2, 5
4-wire, $E4_{out}$	600	> 46	> 46	> 41	1, 2, 5
2-wire, E2	600	> 40	> 46	> 41	1, 2, 3, 4
2-wire, E2	750	> 40	> 46	> 41	1, 2, 3, 4

Note 1: Attention is drawn to Rec. O.9, Section 3, which shows the equivalence between a number of different driving test circuits and also includes information concerning the inherent balance requirements of the test bridge.

Note 2: Attention is drawn to the fact that these values represent minimum requirements. The magnitude of potential longitudinal signal voltages depends, for example, on system use, the system environment, and the location of hybrid transformers and attenuators and may therefore vary for different administrations. Some administrations have found it necessary to specify higher values for longitudinal conversion loss and longitudinal conversion transfer loss to ensure that transverse voltages caused by possible longitudinal signal voltages are sufficiently small.

Note 3: The possible need to introduce limits for frequencies below 300 Hz, in particular at 50 Hz or 60 Hz, is under study. Overall rejection of longitudinal interference can be achieved by a combination of good longitudinal balancing and high-pass filtering (see Section 10.2).

Note 4: The measurements should be made selectively.

Source: Table 4/G.712, ITU-T Rec. G.712, 9/92, page 6.

be within ±20% of the values in Tables 7-U.4 and 7-U.5. The other port is terminated in its nominal characteristic impedance.

Longitudinal Conversion Loss. The longitudinal conversion loss should not be less than the limits set forth in Table 7-U.4.

Longitudinal Conversion Transfer Loss. The difference between the longitudinal conversion transfer loss (ITU-T Rec. O.9, paragraph 2.3) at the specified frequencies and the insertion loss at the same frequencies should not be less than the limits set forth in Table 7-U.5 as specified from input to output ports. The measurement is only applicable to the configuration where the driving test circuit is applied to one of the VF ports and a measurement is made at the other VF port.

TABLE 7-U.5
Longitudinal Conversion Transfer Loss Requirements for 4-Wire and 2-Wire Analog-to-Analog Channels

Connected Channel of Two Primary Multiplexers	Z (Ω)	Requirements on the Difference Between the Longitudinal Conversion Transfer Loss and the Insertion Loss			Notes
		300 Hz to 600 Hz	600 Hz to 2400 Hz	2400 Hz to 3400 Hz	
4-wire, $E4_{1in}$ to $E4_{2out}$	600	> 46 dB	> 46 dB	> 41 dB	1, 2
2-wire, $E2_{1in}$ to $E2_{2out}$	600	> 40	> 46	> 41	1, 2, 3, 4

Note: See Notes to Table 4/G.712.
Source: Table 5/G.712, ITU-T Rec. G.712, 9/92, page 6.

The measurement should be made with the switch S (refer to Figure 3/O.9 in ITU-T Rec. O.9) closed.

The variation with frequency of the attenuation of any channel should lie within the limits shown in the masks of Figure 7-U.28, Figure 7-U.29, Figure 7-U.30, or Figure 7-U.31. The nominal reference frequency is 1020 Hz.

The preferred input power level is −10 dBm0, in accordance with Rec. O.6. As an alternative, a level of 0 dBm0 may be used. If complex nominal impedances are used at 2-wire analog ports, the attenuation/frequency distortion is the logarithmic ratio of output voltage at the reference frequency (nominally 1020 Hz), U(1020 Hz), divided by its value at frequency f, $U(f)$:

$$\text{Attenuation/frequency distortion} = 20 \log[U(1020 \text{ Hz})/U(f)]$$

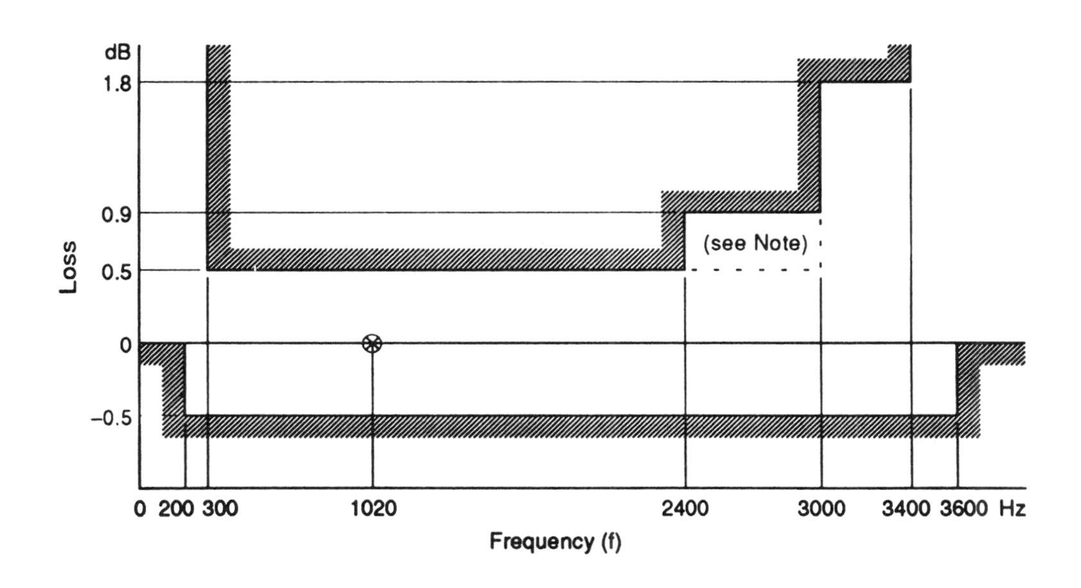

Note – In some applications in which several PCM channels may be connected in tandem, it may be necessary to extend the +0.5 dB limit from 2400 Hz to 3000 Hz.

FIGURE 7-U.28. Attenuation/frequency distortion for analog-to-analog channels between 4-wire ports ($E4_{1in}$ to $E4_{2out}$). (From Figure 2/G.712, ITU-T Rec. G.712, 9/92, page 7.)

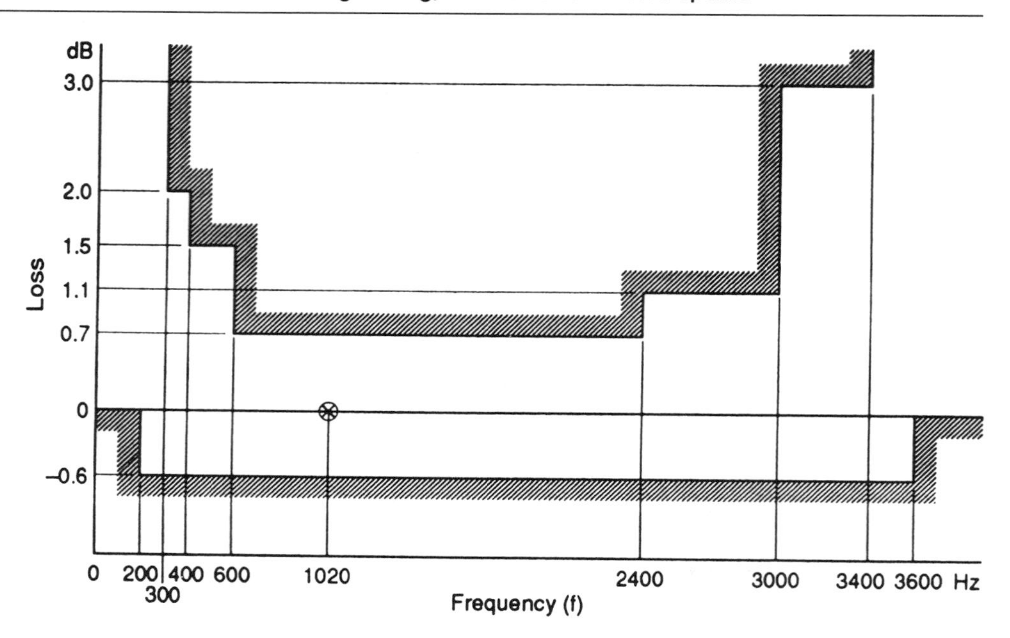

Note – Some Administration apply a limit of 1 dB maximum loss for the frequency range 300 Hz to 3000 Hz.

FIGURE 7-U.29. Attenuation/frequency distortion for analog-to-analog channels between 2-wire ports ($E2_{1in}$ to $E2_{2out}$). (From Figure 3/G.712, ITU-T Rec. G.712, 9/92, page 8.)

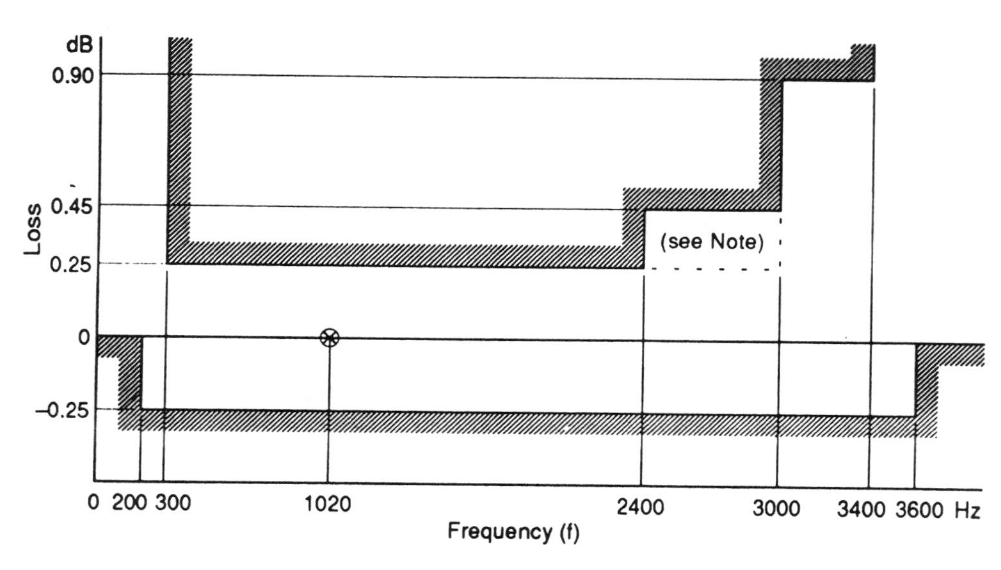

Note – In some applications in which several PCM channels may be connected in tandem, it may be necessary to extend the +0.25 dB limit from 2400 Hz to 3000 Hz.

FIGURE 7-U.30. Attenuation/frequency distortion for 4-wire analog-to-digital channels ($E4_{in}$ to T_{out} or T_{in} to $E4_{out}$). (From Figure 4/G.712, ITU-T Rec. G.712, 9/92, page 8.)

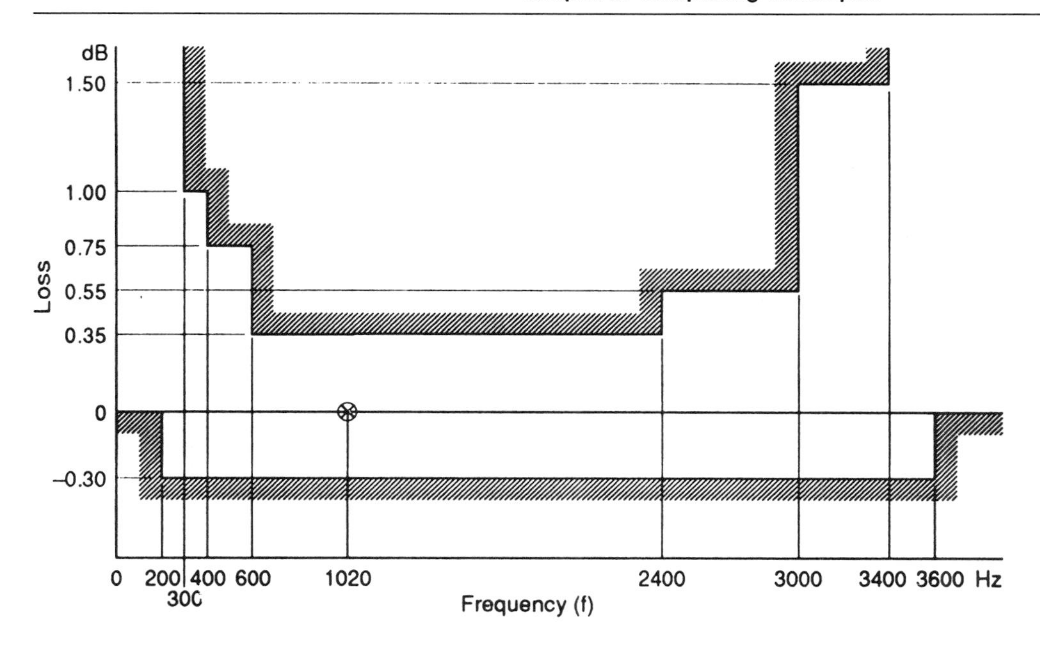

Note – Some Administrations apply a limit of 0.5 dB maximum loss for the frequency range 300 Hz to 3000 Hz.

FIGURE 7-U.31. Attenuation/frequency distortion for 2-wire analog-to-digital channels ($E2_{in}$ to T_{out} or T_{in} to $E2_{out}$). (From Figure 5/G.712, ITU-T Rec. G.712, 9/92, page 9.)

7-2.15.6. Group Delay

Specifications of absolute group delay and group-delay distortion between analog and digital ports are design objectives only.

The requirements on absolute group delay and group-delay distortion should be met at an input power level of −10 dBm0 (preferred value). As an alternative, a level of 0 dBm0 may be used.

Absolute Group Delay. The absolute group delay at the frequency of minimum group delay should not exceed the limit in Table 7-U.6. Note that absolute delay is

TABLE 7-U.6
Requirements for Absolute Group Delay

Measurement Configuration	Absolute Group Delay (μsec)
4-wire analog-to-analog channel ($E4_{1in}$ to $E4_{2out}$)	< 600
2-wire analog-to-analog ($E2_{1in}$ to $E2_{2out}$)	< 750
4-wire analog-to-digital ($E4_{in}$ to A_{out} to B_{out})	< 360
Digital to 4-wire analog (A_{in} or B_{in} to $E4_{out}$)	< 240
2-wire analog-to-digital ($E2_{in}$ to A_{out} or B_{out})	< 450
Digital to 2-wire analog (A_{in} or B_{in} to $E2_{out}$)	< 300

Source: Table 6/G.712, ITU-T Rec. G.712, 9/92, page 9.

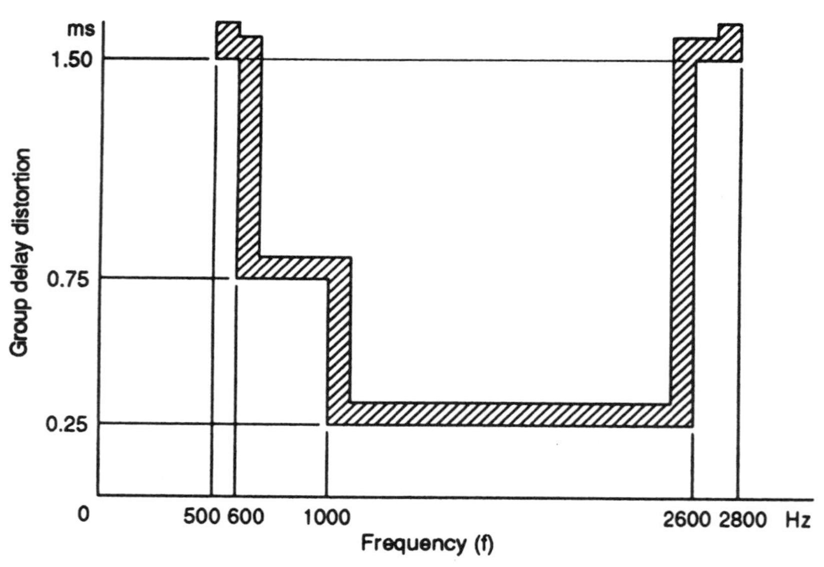

FIGURE 7-U.32. Group-delay distortion with frequency for analog-to-analog channels between 4-wire ports ($E4_{1in}$ to $E4_{2out}$). (From Figure 6/G.712, ITU-T Rec. G.712, 9/92, page 10.)

specified to the A or B port, because the digital cross-connect system will contribute additional delay. These are design objectives only.

Group-Delay Distortion with Frequency. The group-delay distortion should lie within the limits shown within the mask of Figure 7-U.32, Figure 7-U.33, Figure 7-U.34, or Figure 7-U.35.

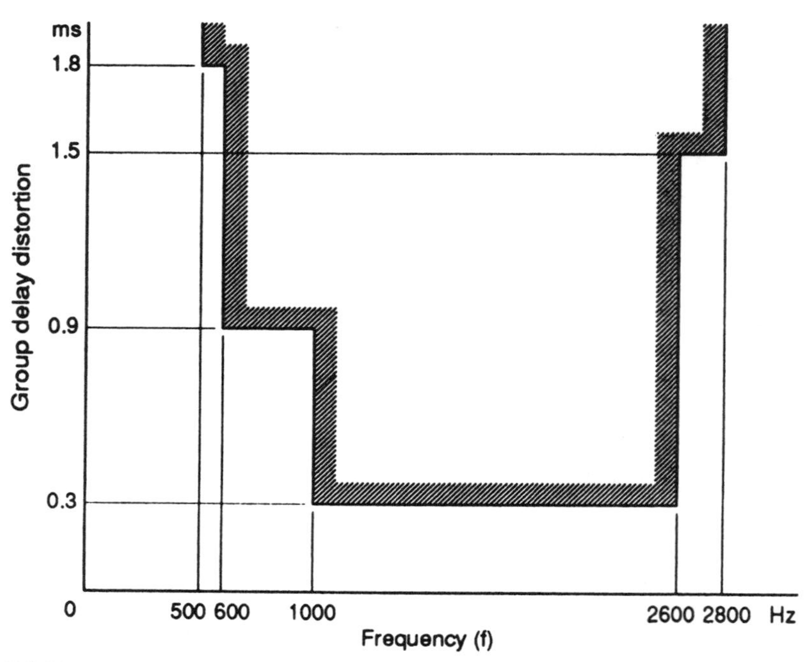

FIGURE 7-U.33. Group-delay distortion with frequency for analog-to-analog channels between 2-wire ports ($E2_{1in}$ to $E2_{2out}$). (From Figure 7/G.712, ITU-T Rec. G.712, 9/92, page 10.)

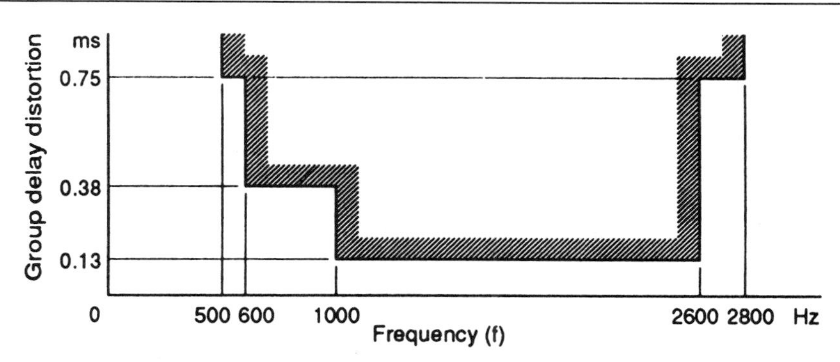

FIGURE 7-U.34. Group-delay distortion with frequency for 4-wire analog-to-digital channels ($E4_{in}$ to T_{out} or T_{in} to $E4_{out}$). (From Figure 8/G.712, ITU-T Rec. G.712, 9/92, page 11.)

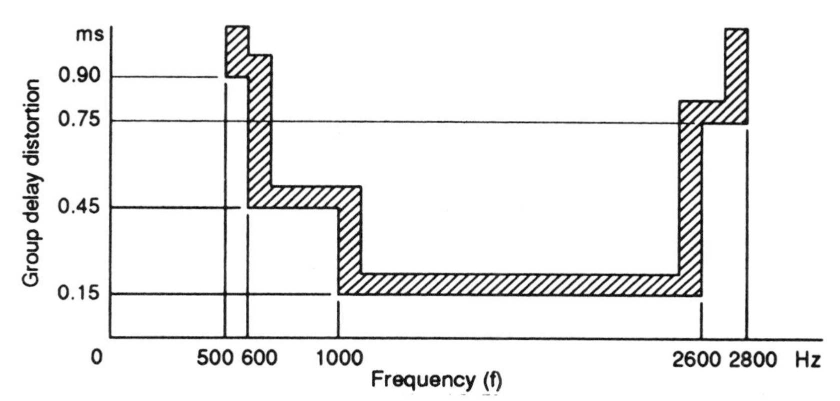

FIGURE 7-U.35. Group-delay distortion with frequency for 2-wire analog-to-digital channels ($E2_{in}$ to T_{out} or T_{in} to $E2_{out}$). (From Figure 9/G.712, ITU-T Rec. G.712, 9/92, page 11.)

The minimum value of absolute group delay is taken as the reference for the group-delay distortion.

7-2.15.7. Idle Channel Noise

Weighted Noise. With the input and output ports of the channel terminated in the nominal impedance, the idle channel noise should not exceed the limits given in Table 7-U.7.

Single-Frequency Noise. The level of any single frequency measured selectively (in particular the sampling frequency and its multiples at the 4-wire analog, $E4_{out}$, and 2-wire analog, $E2_{out}$ ports) should not exceed −50 dBm0. Between 300 Hz and 3400 Hz the level of any single frequency measured selectively and corrected by the psophometric weighting factor should not exceed −73 dBm0.

7-2.15.8. Discrimination Against Out-of-Band Signals

Input Signals Above 4600 Hz at Analog Ports E4 and E2. With any sine-wave signal in the range from 4600 Hz to 72 kHz applied to the input voice-frequency port of the channel at a suitable level, the level of any image frequency produced

TABLE 7-U.7
Requirements for Weighted Idle Channel Noise

Port Terminated	Port Measured	Weighted Noise Limit (dBm0p)	Notes
4-wire analog, $E4_{1in}$	4-wire analog, $E4_{2out}$	< -65	
2-wire analog, $E2_{1in}$	2-wire analog, $E2_{2out}$	< -65	1, 4
4-wire analog, $E4_{in}$	Digital, T_{out}	< -67	2
Digital, T_{in}	4-wire analog, $E4_{out}$	< -70	3
2-wire analog, $E2_{in}$	Digital, T_{out}	< -67	2
Digital, T_{in}	2-wire analog, $E2_{out}$	< -70	3, 5

Note 1: This limit does not include any allowance for additional noise which might be present when signaling takes place on the two wires. The derivation of limits for this case, taking account of the philosophy adopted in Rec. Q.551, is under study. Because of the effects of quantization, it is not necessarily the case that noise powers can be added.

Note 2: Weighted noise measured at the encoding side.

Note 3: Noise measured at the decoding side. The digital port is driven by a PCM signal (quiet code) corresponding to the decoder output value number 0 for the μ-law or decoder output value number 1 for the A-law.

Note 4: Below -5 dBr and -8 dBr the noise limit is -64 dBm0p.

Note 5: Below -5 dBr the noise limit is -75 dBmp.

Note 6: Psophometric measurements of composite signals at ports with complex impedances should be performed with a psophometer having an input impedance equal to the nominal complex impedance specified for that port. The psophometer has to be calibrated accordingly (see also Rec. O.41 and Annex A to Rec. G.100).

Source: Table 7/G.712, ITU-T Rec. G.712, 9/92, page 13.

at an output E or T port should, as a minimum requirement, be at least 25 dB below the level of the test signal.

Note: It has been found that a suitable test level is -25 dBm0.

Input Signals Below 300 Hz at Analog Port E2. No particular value is recommended.

Note 1: While some administrations have no particular requirement in this respect, some other administrations have found it necessary to provide at least 20–26 dB of rejection at the encoding side at frequencies across the band 15–60 Hz.

Note 2: Overall rejection of longitudinal interference can be achieved by a combination of good longitudinal balancing (see Section 6) and high-pass filtering.

Overall Requirements (4-Wire Only). Under the most adverse conditions encountered in a national network, the 4-wire PCM channel ($E4_{in}$ to $E4_{out}$ or $E4_{in}$ to T_{out}) should not contribute more than 100 pW0p of additional noise in the band 10 Hz to 4 kHz at the channel output, as a result of the presence of out-of-band signals at the 4-wire analog channel input.

Note 1: The discrimination required depends on the performance of frequency division multiplex (FDM) channel equipments and telephone instruments in national networks; and individual administrations should carefully consider the requirements they should specify, taking into account the comments above and the requirement given above. In all cases at least the minimum requirement of this section should be met.

Note 2: Attention is drawn to the importance of the attenuation characteristic in the range 3400–4600 Hz. Although other attenuation characteristics can satisfy the following requirements, the filter template of Figure 7-U.36 gives

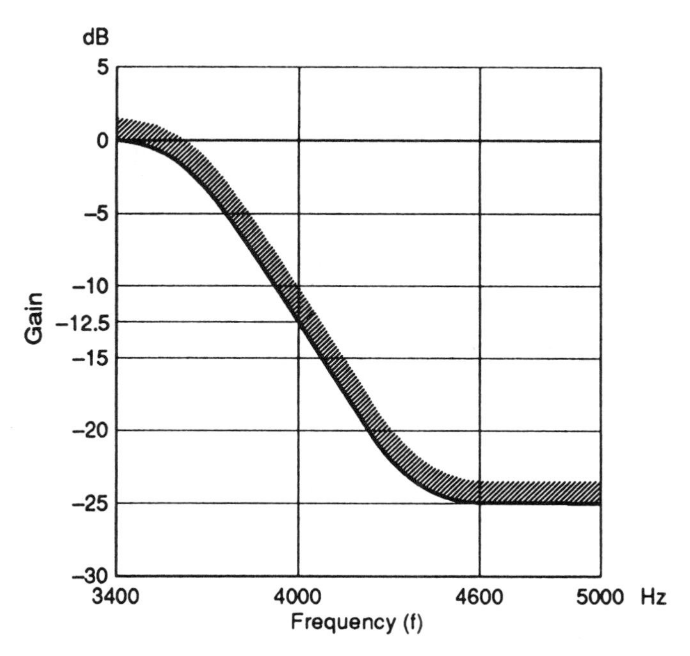

Note – The curved portion of the graph conforms to the equation:

$$G = 12.5\left[\sin\frac{\pi(4000-f)}{1200} - 1\right] \text{ dB}$$

for the range $3400 \leq f \leq 4600$.

FIGURE 7-U.36. Gain relative to gain at 1020 Hz. (From Figure 10 / G.712, ITU-T Rec. G.712, 9 / 92, page 13.)

adequate protection against the out-of-band signals at 4-wire and 2-wire analog channel inputs.

7-2.15.9. Spurious Signals at the Channel Output Port

Spurious Out-of-Band Signals at the Channel Output Port

In-Band Input Signal. With any sine-wave test signal in the frequency range 300–3400 Hz and at a level of 0 dBm0 applied to the digital or analog input port of a channel (T_{in} or a connected $E4_{in}$ or $E2_{in}$), the level of spurious out-of-band image signals measured selectively at the 4-wire or 2-wire analog output port ($E4_{out}$ or $E2_{out}$) should be lower than −25 dBm0.

Note: Attention is drawn to the importance of the attenuation characteristic in the range 3400–4600 Hz. Although other attenuation characteristics can satisfy the above requirement, the filter template of Figure 7-U.36 gives adequate protection against out-of-band signals.

Overall Requirement. The spurious out-of-band signals should not give rise to unacceptable interference in equipment connected to the PCM channel. In particular, the intelligible or unintelligible crosstalk in a connected FDM channel should not exceed a level of −65 dBm0 as a consequence of the spurious out-of-band signals at the PCM channel output.

Note: The discrimination required depends on the performance of FDM channel equipment and telephone instruments in national networks; and individual

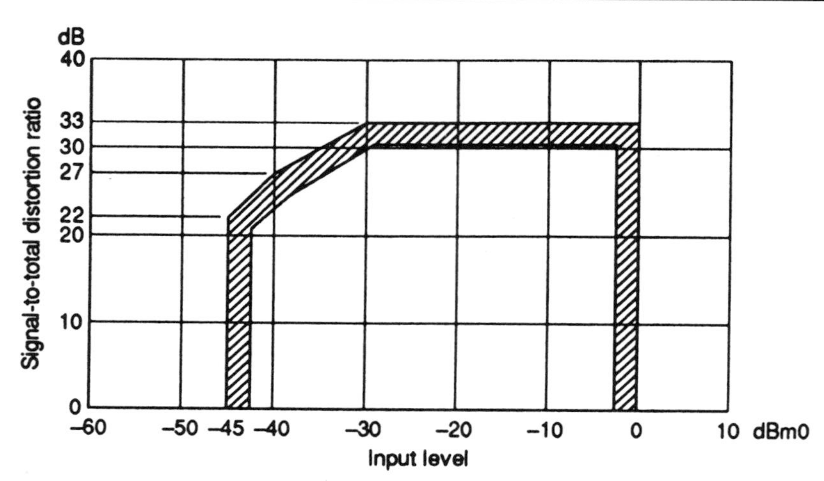

FIGURE 7-U.37. Signal-to-total distortion ratio as a function of input level for analog 4-wire to 4-wire and 2-wire to 2-wire channels ($E4_{1in}$ to $E4_{2out}$ and $E2_{1in}$ to $E2_{2out}$). (From Figure 11/G.712, ITU-T Rec. G.712, 9/92, page 14.)

administrations should carefully consider the requirements they should specify, taking into account the comments and requirements given above. In all cases at least the minimum requirement stated herein should be met.

Spurious In-Band Signals at the Channel Output Port. With any sine-wave test signal in the frequency range 700–1100 Hz and at a level of 0 dBm0 applied to the analog input port of a channel ($E4_{1in}$ or $E2_{1in}$), the output level at any frequency other than the frequency of the test signal, measured selectively in the frequency band 300–3400 Hz at the 4-wire or 2-wire analog output port ($E4_{2out}$ or $E2_{2out}$), should be less than −40 dBm0.

7-2.15.10. Total Distortion, Including Quantizing Distortion

With a sine-wave test signal at the nominal reference frequency of 1020 Hz applied to the input port of a channel, the ratio of signal-to-total distortion power

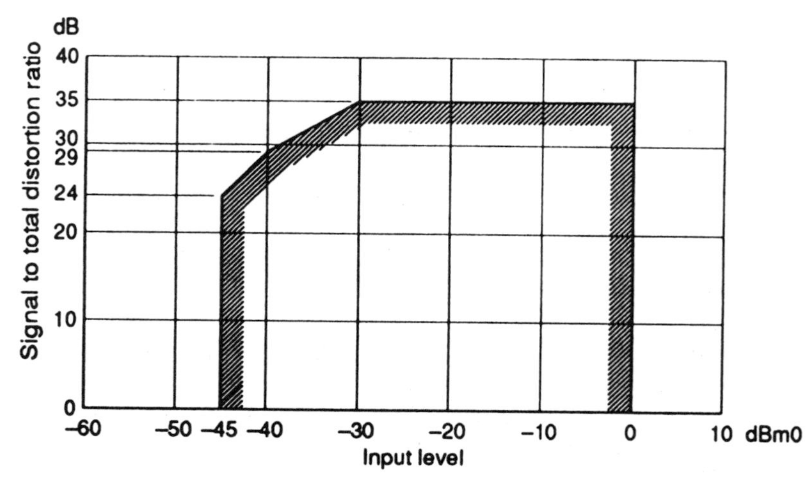

FIGURE 7-U.38. Signal-to-total distortion ratio as a function of input level for analog-to-digital channels ($E4_{in}$ to T_{out}, T_{in} to $E4_{out}$, $E2_{in}$ to T_{out}, and T_{in} to $E2_{out}$). (From Figure 12/G.712, ITU-T Rec. G.712, 9/92, page 15.)

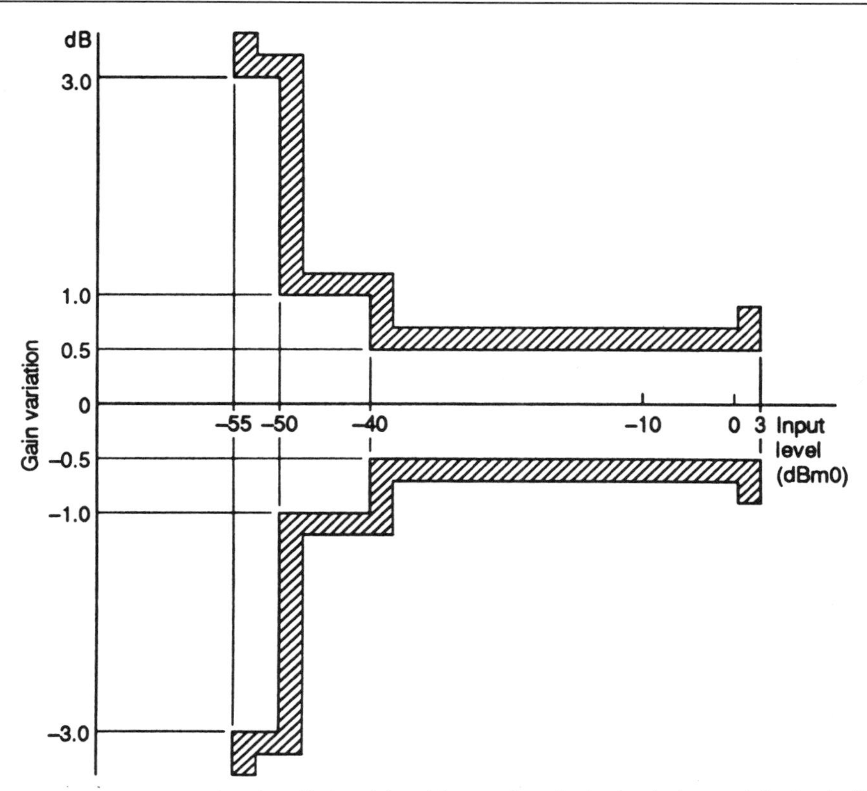

FIGURE 7-U.39. Variation of gain with input level for analog 4-wire to 4-wire and 2-wire to 2-wire channels ($E4_{1in}$ to $E4_{2out}$ and $E2_{1in}$ to $E2_{2out}$). (From Figure 13/G.712, ITU-T Rec. G.712, 9/92, page 15.)

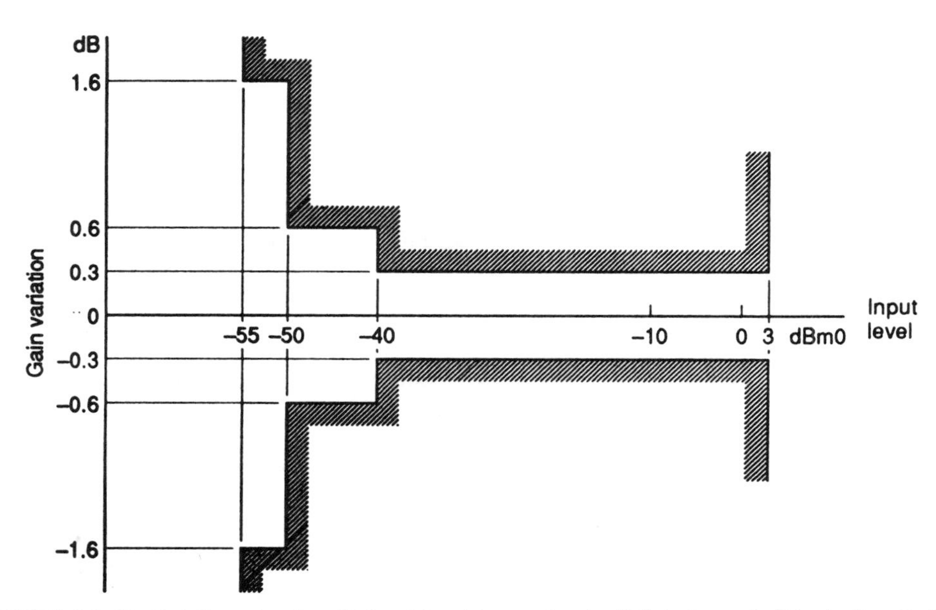

FIGURE 7-U.40. Variation of gain with input level for analog-to-digital channels ($E4_{in}$ to T_{out}, T_{in} to $E4_{out}$, $E2_{in}$ to T_{out}, and T_{in} to $E2_{out}$). (From Figure 14/G.712, ITU-T Rec. G.712, 9/92, page 16.)

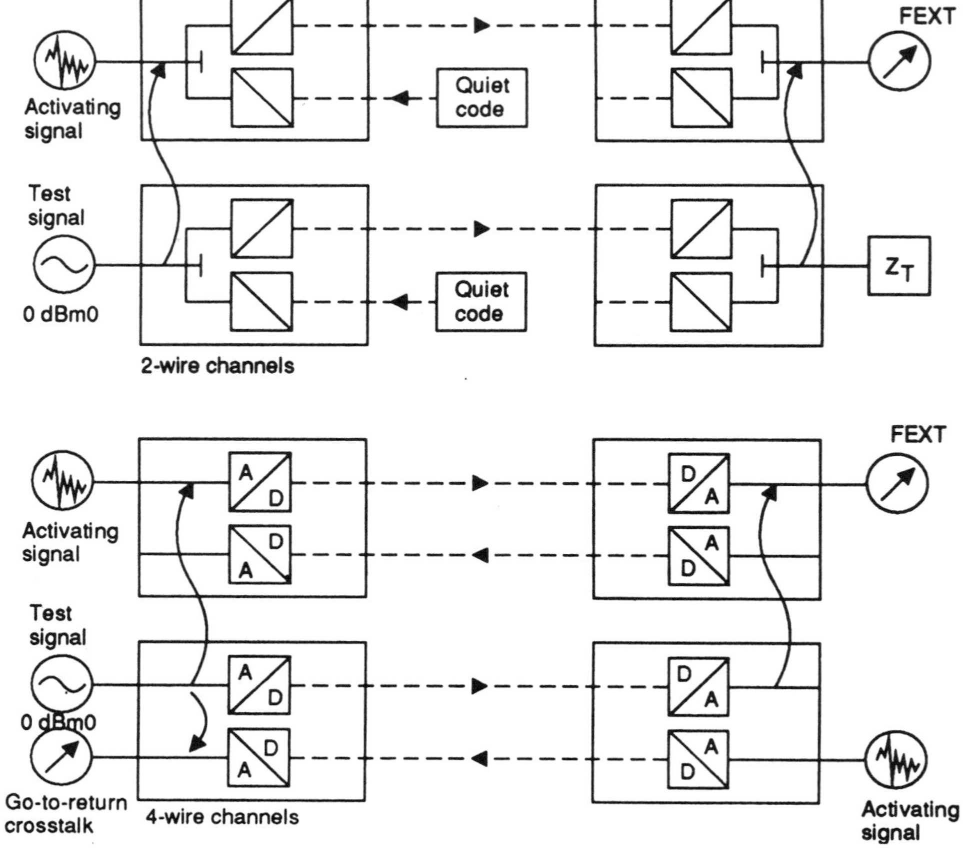

FIGURE 7-U.41. Measurement of crosstalk between two channels. (From Figure 15/G.712, ITU-T Rec. G.712, 9/92, page 17.)

measured with the appropriate noise weighting at the output port should lie above the limits given in Figures 7-U.37 and 7-U.38.

7-2.15.11. Variation of Gain with Input Level

With a sine-wave test signal at the nominal reference frequency of 1020 Hz applied to the input port of any channel at a level between −55 dBm0 and +3 dBm0, the gain variation of that channel relative to the gain at an input level of −10 dBm0 should be within the limits of Figures 7-U.39 and 7-U.40.

7-2.15.12. Crosstalk

Introduction. For crosstalk measurements, auxiliary signals are injected as shown in Figures 7-U.41 through 7-U.47. These signals are:

- The quiet code [i.e., a PCM signal corresponding to decoder output value 0 (μ-law) or output value 1 (A-law)(with the sign bit in a fixed state)].
- A low-level activating signal, a sine wave at a level in the range of −33 dBm0 to −40 dBm0. Care must be taken in the choice of frequency and filtering characteristics of the measuring equipment in order that the activating signal does not significantly affect the accuracy of the crosstalk measurement.

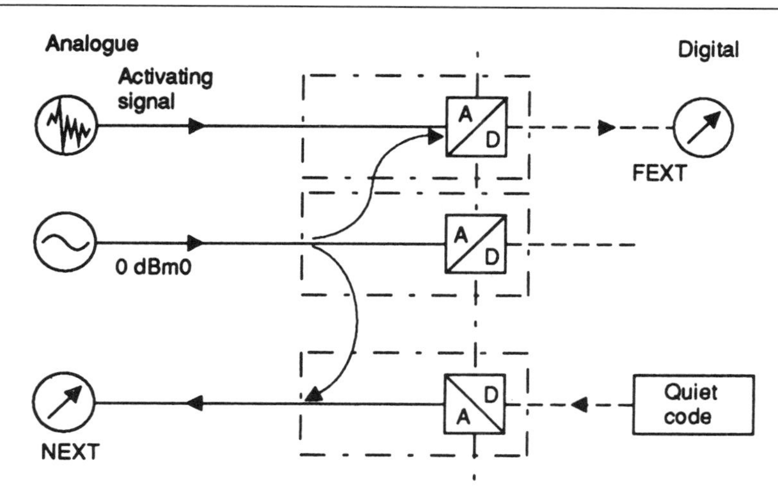

FIGURE 7-U.42. Measurements on 4-wire (E4) ports with an analog test signal between different channels. (From Figure 16/G.712, ITU-T Rec. G.712, 9/92, page 18.)

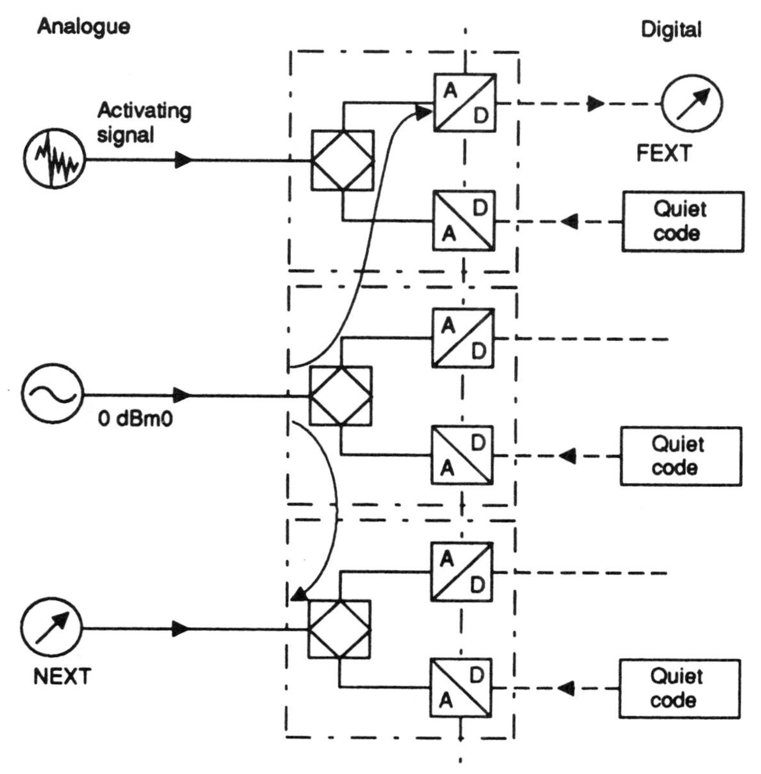

FIGURE 7-U.43. Measurements on 2-wire (E2) ports with an analog test signal between different channels. (From Figure 17/G.712, ITU-T Rec. G.712, 9/92, page 18.)

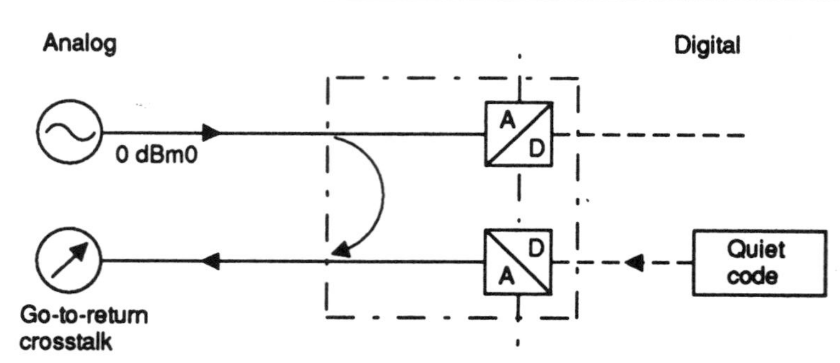

FIGURE 7-U.44. Measurements on 4-wire (E4) port with an analog test signal between go and return directions of the same channel. (From Figure 18/G.712, ITU-T Rec. G.712, 9/92, page 19.)

Interchannel Crosstalk, Analog-to-Analog Channels

Far-End Crosstalk Measured with Analog Test Signal. The crosstalk between individual transmission paths of a PCM equipment should be such that with a sine-wave signal at the nominal reference frequency of 1020 Hz and at a level of 0 dBm0 applied to a 4-wire or 2-wire analog input port ($E4_{1in}$ or $E2_{1in}$), the crosstalk level received at the 4-wire or 2-wire analog output of any other transmission path ($E4_{2out}$ or $E2_{2out}$) should not exceed -65 dBm0 far-end crosstalk (FEXT). See Figure 7-U.41 for measurements of 4-wire and 2-wire channels.

Go-to-Return Crosstalk for 4-Wire to 4-Wire Analog Channels. The crosstalk between a channel and its associated return channel should be such that with a sine-wave signal at any frequency in the range of 300–3400 Hz and at a level of 0 dBm0 applied to a 4-wire analog input port $E4_{1in}$ or $E4_{2in}$, the crosstalk level measured at the 4-wire analog output port $E4_{1out}$ or $E4_{2out}$, respectively, of the same channel of the same primary multiplexer should not exceed -60 dBm0 when the channel is connected through to another primary multiplexer. See Figure 7-U.41.

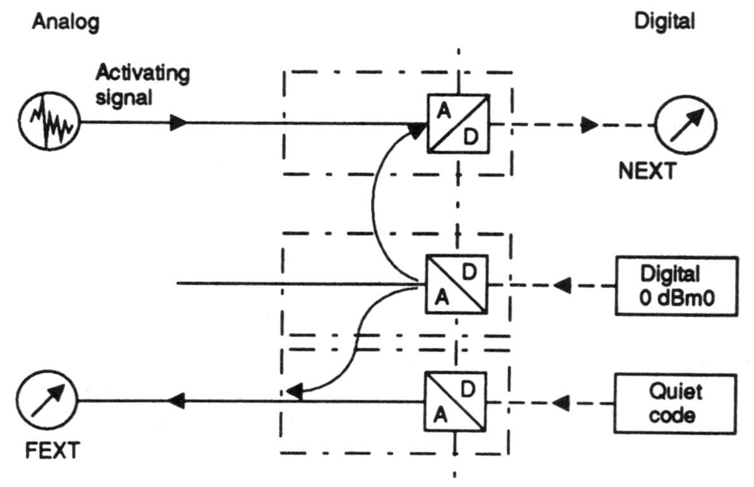

FIGURE 7-U.45. Measurements on 4-wire (E4) ports with a digital test signal between different channels. (From Figure 19 / G.712, ITU-T Rec. G.712, 9 / 92, page 19.)

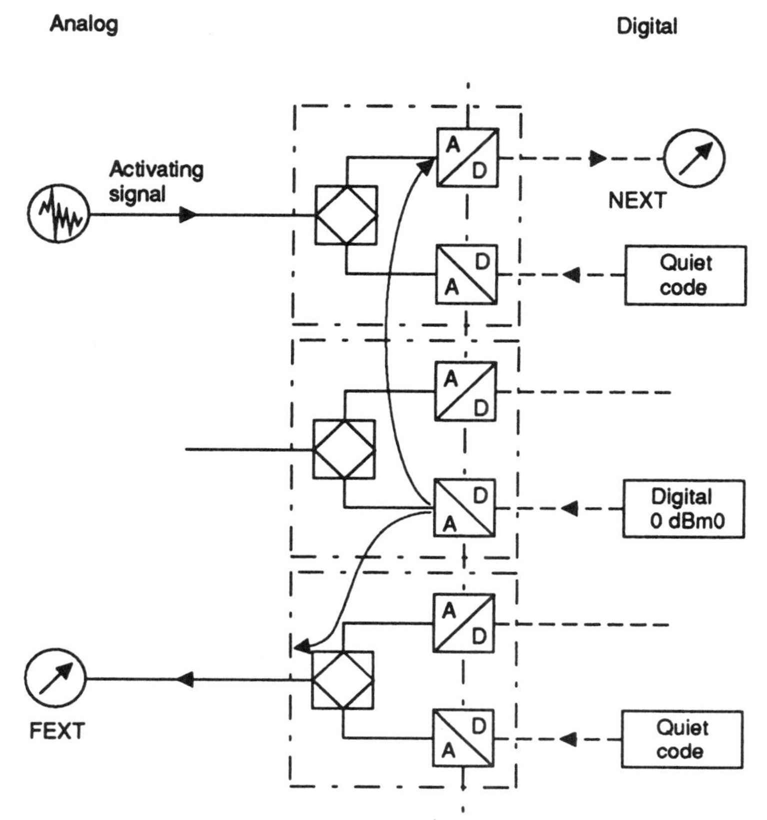

FIGURE 7-U.46. Measurements on 2-wire (E2) ports with a digital test signal between different channels. (From Figure 20 / G.712, ITU-T Rec. G.712, 9 / 92, page 20.)

Crosstalk Measurements of Individual Primary Multiplexers

Far-End and Near-End Crosstalk Measured with an Analog Test Signal. The crosstalk between individual channels of a multiplex should be such that with a sine-wave signal at the nominal reference frequency of 1020 Hz and at a level of 0 dBm0 applied to a VF input port, the crosstalk level produced in any other channel should not exceed −73 dBm0 for near-end crosstalk (NEXT) and −70

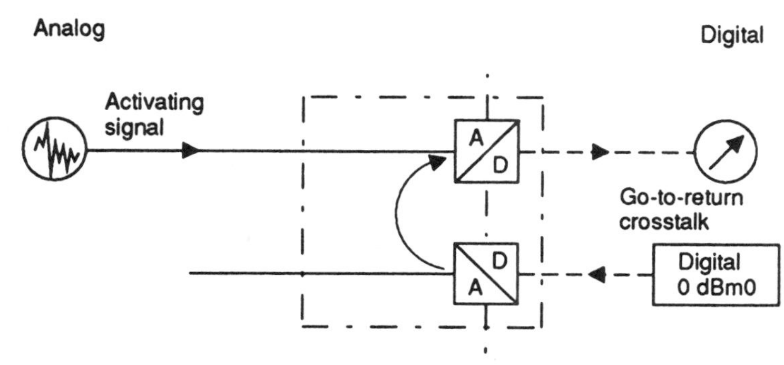

FIGURE 7-U.47. Measurements on 4-wire (E4) port with a digital test signal between go and return directions of the same channel. (From Figure 21/G.712, ITU-T Rec. G.712, 9/92, page 20.)

dBm0 for far-end crosstalk (FEXT). See Figures 7-U.42 and 7-U.43 for measurements of 4-wire and 2-wire channels, respectively.

Go-to-Return Crosstalk Measured with an Analog Test Signal. The crosstalk between a channel and its associated return channel should be such that with a sine-wave signal at any frequency in the range of 300–3400 Hz and at a level of 0 dBm0 applied to an input port, the crosstalk level measured at the output port of the corresponding return channel should not exceed −66 dBm0. See Figure 7-U.44.

7-2.15.13. Interference from Signaling

4-Wire Analog to 4-Wire Analog Channels. The maximum level of any interference into a channel should not exceed −60 dBm0p when signaling (10-Hz signal with a 50/50 ratio) is active simultaneously on all other channels.

4-Wire Analog-to-Digital Channels. The characterization of such interference by separate measurements requires four different types of measurement for crosstalk (see Figure 7-U.48). In each case the maximum level of interference in one channel should not exceed −63 dBm0p when signaling (10 Hz with 50/50 duty ratio) is active simultaneously on all other channels.

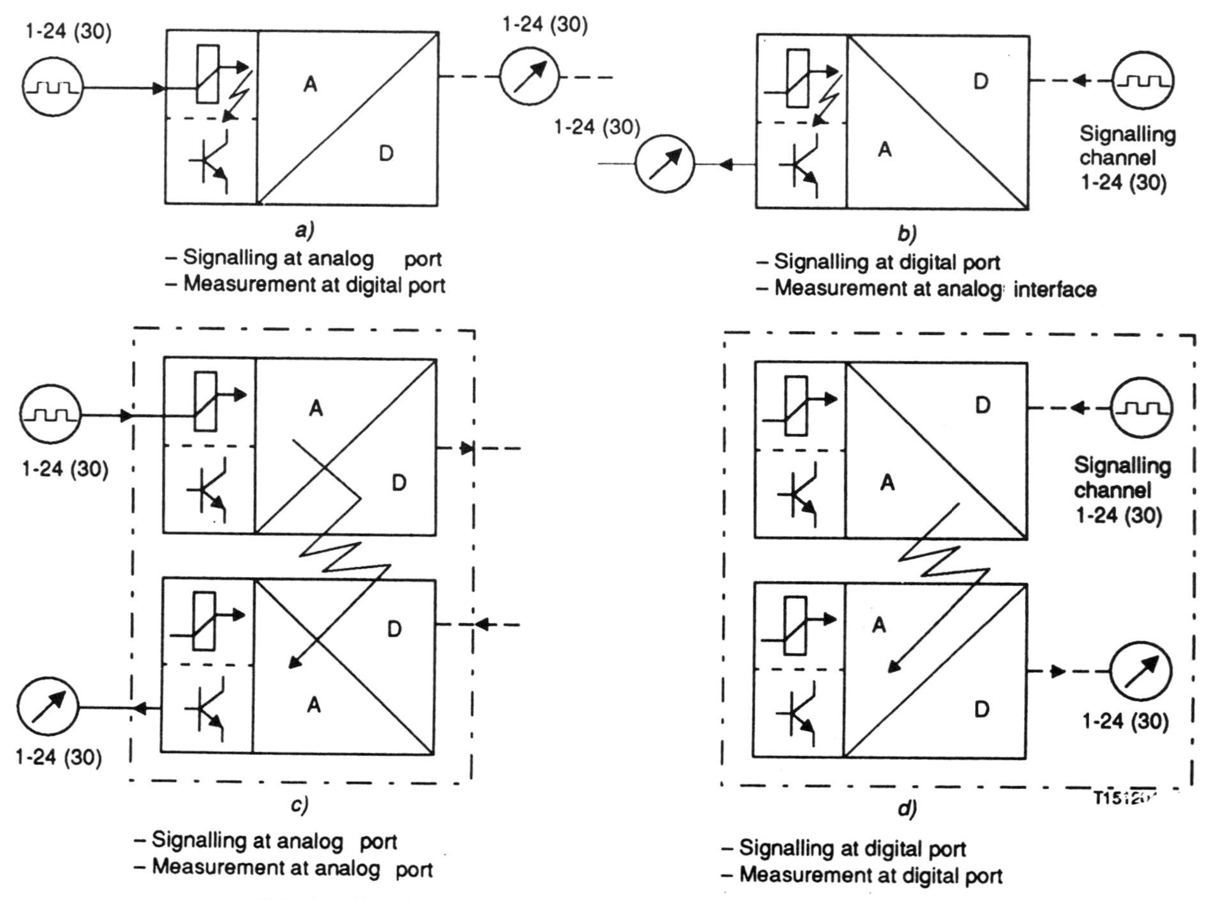

FIGURE 7-U.48. Measurement of signaling interference contributions. (From Figure 22 / G.712, ITU-T Rec. G.712, 9 / 92, page 21.)

2-Wire Analog-to-Digital Channels. The characterization of such interference by separate measurements requires two different types of measurements for crosstalk (see Figure 7-U.48, parts a and b). In each case the maximum level of interference in one channel should not exceed X dBm0p when signaling (10 Hz with 50/50 duty ratio) is active simultaneously on all other channels. *Note*: The value of X is under study by the ITU-T organization.

7-2.15.14. Echo and Stability at 2-Wire Ports, E2

Terminal Balance Return Loss (TBRL). This quantity characterizes the equipment performance required to comply with the network performance objective given in ITU-T Rec. G.122 (see Section 6-5) with respect to echo. The terminal balance return loss is defined as the balance return loss measured against a balance test network. It is related to the loss between the digital test input point, T_i, and the digital test output point, T_o (see Figure 7-U.49), as follows:

$$a_{io} = T_i \text{ to } T_o \text{ loss} = P_i + P_o + \text{TBRL} \qquad (\text{dB})$$

where P_i and P_o are the measured values of loss in the equivalent circuit of Figure 7-U.49 which represent all loss between the digital test point and the 2-wire point or, conversely, at the measurement frequency.

The TBRL should be measured in the arrangement shown in Figure 7-U.49 with a sinusoidal test signal at frequencies across the telephone band covering the bandwidth 300–2400 Hz.

Values for the nominal balance impedance and for the maximum deviation of this impedance from the nominal value differ from one administration to another. The range of impedances presented at the 2-wire port during normal operation also varies considerably. Administrations will need to establish their own requirements for TBRL, taking into account national and international transmission plans. As a minimum requirement, the TBRL limits in Figure 7-U.50 should be met when the 2-wire port is terminated with a balance test network which is representative of the impedance conditions expected in the speaking condition from a population of 2-wire trunks connected to the PCM equipment. The ITU-T organization reports that the limits are provisional.

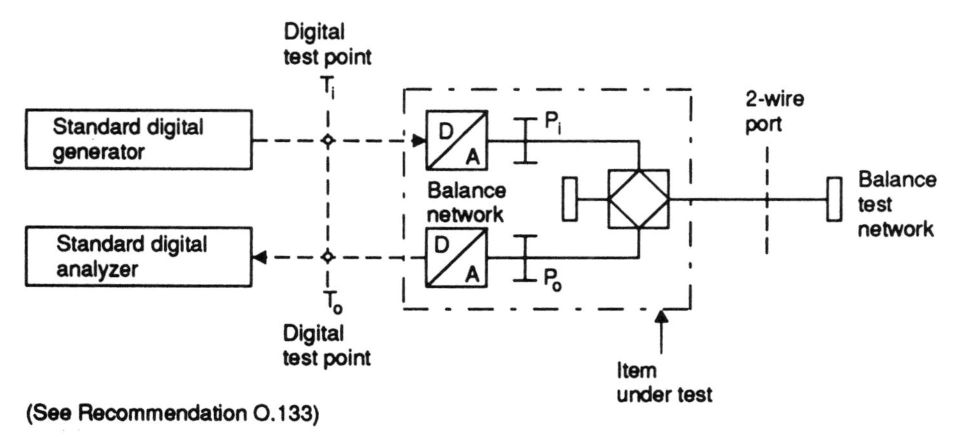

FIGURE 7-U.49. Arrangement for measuring half-loop loss. (From Figure 23/G.712, ITU-T Rec. G.712, 9/92, page 22.)

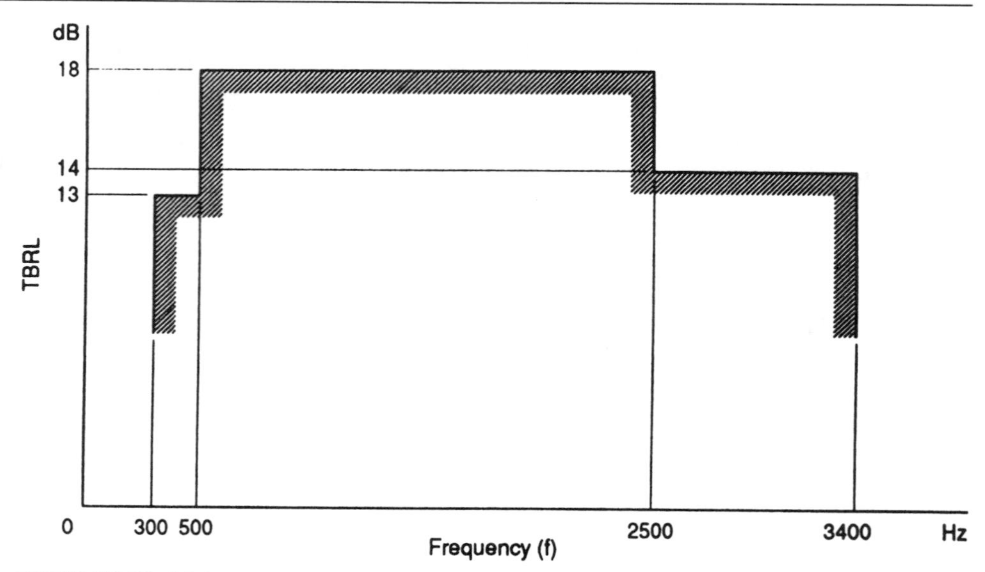

FIGURE 7-U.50. Minimum values of terminal balance return loss (provisional). (From Figure 24/G.712, ITU-T Rec. G.712, 9/92, page 23.)

Stability Loss (SL). The stability loss (SL) is defined as the minimum value of the loss a_{io} measured in the arrangement of Figure 7-U.49. The stability loss should be measured between T_i and T_o by terminating the 2-wire port with stability test networks representing the worst-case terminating condition encountered in normal operation. Some administrations may find that open-circuit and short-circuit terminations are sufficiently representative of worst-case conditions. Other administrations may need only to specify, for example, the inductive termination to represent the worst-case condition.

The stability loss at any frequency can be expressed as follows:

$$\text{SL} \geq P_i + P_o - X \qquad \text{(dB)}$$

where P_i and P_o are measured values of loss, at the measurement frequency, under normal terminating conditions at the 2-wire port. X has a value dependent on the interaction between the 2-wire input impedance, the 2-wire balance impedance and the impedance actually applied at the 2-wire port. X can be computed or measured by the method described in ITU-T Rec. Q.552.

The 2-wire input and balance impedances at a 2-wire/4-wire port usually have to be optimized by administrations with regard to echo and sidetone. The worst-case terminations depend on the actual network conditions. Thus, the value of X is fully determined by network conditions and the impedance strategy. Values between 0 and 3 dB have been observed in practice.

Administrations should choose the nominal values of P_i and P_o, taking account of the value of X for their particular operating conditions and of national and international transmission plans for overall network stability (see Rec. G.122).

Section 7.2.15 was extracted from ITU-T Rec. G.712, 9/92.

8

Outside Plant—Metallic Pair Systems

This new section follows Section 8-7.19, which ends on page 575 of the Manual.

8-7.20. Local Area Network Twisted Pair Data Communications Cable

8-7.20.1. Preferred Ratings and Characteristics

Conductors. Solid copper conductors used in the construction of these cables shall conform to the latest revision of ASTM-B3. Conductors, as measured in the completed cable, shall conform to the following dimensions:

Gauge (AWG)	Dimensions (mm)
22 solid	0.643 ± 0.02
24 solid	0.510 ± 0.03
26 solid	0.400 ± 0.03
26 stranded (7×34)	0.490 ± 0.03

Voice Grade Media (VGM)

Configuration. Cables may contain twisted pairs of VGM conductors using either a polyethylene or fluorocarbon material.

The wire type, insulation, and insulation thickness are as follows:

Wire Type	Insulation/Wire Diameter (mm)
AWG 22 solid copper	Solid fluorocarbon/1.16 ± 0.21
AWG 22 solid copper	Solid polyethylene/1.32 ± 0.20

The VGM pairs must be located between the cable outer jacket and the braided shield of the DGM media.

	Tip		Ring	
Pair	Color of Insulation	Color of Tracer Marking	Color of Insulation	Color of Tracer Marking
1	White	Blue	Blue	White
2	White	Orange	Orange	White
3	White	Green	Green	White
4	White	Brown	Brown	White

Twist of Pairs (VGM). No pair shall have the same lay or be an integral multiple of any other pair. The average length of pair twist in any pair in the finished cable, when measured on any 3.05-m (10-ft) length, shall not exceed 15.24 cm (6 in).

Shields. All indoor cables utilizing braided shields require 65% (minimum) braid over aluminum shielding. The percent coverage shall be calculated per MIL-C-915E. Where multiple shielding elements are used (such as braided shields and aluminum-backed polyester tape), all elements must be in electrical contact continuously along the entire cable length. Refer to the appropriate detail specification for a description of the shielding.

Protective Tapes. Cables containing VGM pairs shall require a protective tape over the braided shield. This tape is necessary in order to prevent shield abrasion of the VGM insulation. This material may be polyester or other suitable material. Cables may contain an optional heat barrier tape under the jacket. This tape may be used because of manufacturing considerations or to meet other requirements such as those of the National Electrical Code.

Rip Cord and Binder. A rip cord, if required by the detail specification, shall be laid longitudinally beneath the outer jacket. The cord shall be continuous throughout the length of the cable and shall have sufficient strength to open the jacket without fracture to the cord itself. A yellow or natural white color is required. Binders, if used, are required to be blue.

Jacket. Jackets shall be constructed of PVC or fluorocarbon material as appropriate for the particular application. Outdoor cable not requiring U.L. (Underwriter Laboratories) approval may use other material such as polyethylene.

Fluorocarbon Materials. Cables requiring fluorocarbon material shall not use any reground material in the extrusion of primary insulation or jackets.

Electrical Specifications. All cables shall meet their electrical specifications across a temperature range of −40°C to +80°C. Reference should be made to the applicable EIA/IS-43 test descriptions.

Crosstalk. Specified in detail specification. Crosstalk above a frequency of 1 MHz shall be specified over a frequency range corresponding to the application rather than at individual frequency points.

Impedance. Specified in detail specification. Impedance above a frequency of 1 MHz shall be specified over a frequency range corresponding to the application rather than at individual frequency points.

Attenuation. Specified in detail specification.

Capacitive Unbalance. The maximum capacitive unbalance (pF/km) of any twisted pair in a completed cable shall be less than the following:

Capacitive Unbalance	DGM	VGM
Pair to ground	1500	1500
Pair to pair	NS	181

NS, not specified.

Resistance. DC resistance of all conductors in the completed cable must be specified.

Resistance Unbalance for VGM Pairs. The difference in DC resistance between the two conductors of a pair as measured in the completed cable shall not exceed the following values:

Resistance Unbalance—Maximum for Any Reel

Individual Pair: 4.0%

where:

$$\% \text{ Resistance unbalance} = \frac{(\text{Max. Res.} - \text{Min. Res.}) \times 100}{\text{Min. Res.}}$$

Note: The resistance unbalance between tip and ring conductors shall be random with respect to the direction of unbalance. That is, the resistance of the tip conductors shall not be consistently higher with respect to the ring conductors and vice versa.

High-Voltage Withstanding VGM Pairs

Conductor to Conductor. In each length of completed cable, the insulation between conductors shall be capable of withstanding for three seconds a DC potential whose value is not less than 3600 volts.

Conductor to Shield. In each length of completed cable, the dielectric between the shield and conductors shall withstand for 3 seconds a DC potential whose value is not less than 3600 volts.

Insulation Resistance

VGM. Each insulated conductor in each length of completed cable, when measured with all other insulated conductors and the shield grounded, shall have an insulation resistance of not less than 2000 MΩ-km at 23°C ±1°C.

Conductance shall not exceed 6.0 micromhos/km when tested at a frequency of 1000 ± 100 Hz and a temperature of 23°C ±3°C.

DGM. Each insulated conductor in each length of completed cable, when measured with all other insulated conductors and the shield grounded, shall have an insulation resistance of not less than 16,000 MΩ-km at 23°C ±1°C.

From EIA/IS-43, Electronic Industries Association, Washington, DC, September 1987.

8-7.21. Cable Specifications for Commercial Building Telecommunication Applications

8-7.21.1. Introduction

This section is based on EIA/TIA SP2840-A, proposed commercial building telecommunications cabling standard. At or near the publication of this update, the underlying standard document name will change to EIA/TIA-568-A. The primary thrust of this section is to provide technical data on twisted-pair copper cables and connecting hardware with enhanced transmission characteristics and the use of optical fiber cabling for system interconnection. Emphasis is provided for 100-ohm UTP (unshielded twisted pair) cable, categories 3, 4, and 5. It also covers 150-ohm STP (shielded twisted pair) cable.

8-7.21.2. Backbone Cabling Distances

Intra and Inter Building Distance Guidelines. Category 3 multipair UTP-backbone cabling, for applications whose spectral ranges are from 5 to 16 MHz, should be limited to a total of 90 m (295 ft).

Category 4 UTP backbone cabling, for applications whose spectral bandwidth ranges from 10 MHz to 20 MHz, should be limited to a total of 90 m (295 ft).

Category 5 multipair backbone cabling, for applications whose spectral bandwidth ranges from 20 MHz to 100 MHz, should be limited to a total distance of 90 m (295 ft).

150-ohm STP-A backbone cabling, for applications whose spectral bandwidth ranges from 20 MHz to 300 MHz, should be limited to a total distance of 90 m (295 ft). The 90 m (295 ft) distance assumes that 5 m (16 ft) are needed at each end for equipment cables connecting to the backbone.

8-7.21.3. 100-Ω Unshielded Twisted Pair (UTP) Cabling Systems

Recognized Categories

(a) Category 3. 100-Ω UTP cables and associated connecting hardware whose transmission characteristics are specified up to 16 MHz.
(b) Category 4. 100-Ω UTP cables and associated connecting hardware whose transmission characteristics are specified up to 20 MHz.
(c) Category 5. 100-Ω UTP cables and associated connecting hardware whose transmission characteristics are specified up to 100 MHz.

Horizontal UTP Cable. The cable consists of 24 AWG (0.5 mm) thermoplastic insulated solid conductors formed into four individually twisted pairs and enclosed by a thermoplastic jacket. The cable shall meet all requirements of ANSI/ICEA Publication S-80-576 that are applicable to four-pair inside wiring cable for plenum or general cabling within a building.

Mechanical Design. In addition to the requirements of ANSI/ICEA Publication S-80-576, the physical design of the cable shall meet the following specifications.

Insulated Conductor. The diameter over the insulation shall be 1.22 mm (0.048 in) maximum.

Pair Assembly. The cable shall be restricted to four-pair size to support a broad range of applications. The pair twists of any pair shall not be exactly the same as any other pair. The pair twist lengths shall be selected by the manufacturer to ensure compliance with the crosstalk requirements stated herein.

Color Codes. The color codes shall be as shown in Table 8-U.1.

Cable Diameter. The diameter of the completed cable shall be less than 6.35 mm (0.25 in).

Breaking Strength. The ultimate breaking strength of the completed cable, measured in accordance with ASTM D 4565 shall be 400 N (90 lb) minimum.

Bending Radius. The cable tested in accordance with ASTM D 4565 shall withstand a bending radius of 25.4 mm (1 in) at a temperature of −20°C (±1°C) without jacket or insulation cracking.

Note: For certain applications (e.g., prewiring buildings in cold climate), a cable with a lower temperature bending performance of −30°C (±1°C) may be required.

Transmission Requirements

DC Resistance. The resistance of any conductor, measured in accordance with ASTM D 4566, shall not exceed 9.38 Ω per 100 m (328 ft) at or corrected to a temperature of 20°C.

DC Resistance Unbalance. The resistance unbalance between the two conductors of any pair shall not exceed 5% when measured at or corrected to a temperature of 20°C in accordance with ASTM D 4566.

Mutual Capacitance. The mutual capacitance at 1 kHz, measured at or corrected to a temperature of 20°C, should not exceed 6.6 nF per 100 m(328 ft) for category 3 cable, and should not exceed 5.6 nF per 100 m (328 ft) for category 4 and category 5 cables. The measurements are performed in accordance with ASTM D 4566.

Capacitance Unbalance: Pair-to-Ground. The capacitance unbalance to ground at 1 kHz of any pair, measured in accordance with ASTM D 4566 and "Measurement Precaution" section which follows below, shall not exceed 330 pF per 100 m (328 ft) at or corrected to a temperature of 20°C.

Attenuation. The attenuation is commonly derived from swept signal-level measurements at the output of cable length greater than or equal to 100 m (328 ft) of cable. The maximum attenuation of any pair, in dB per 100 m, measured at or corrected to 20°C, in accordance with ASM D 4566 and the "Measurement Precaution" section that follows, shall be less than or equal to the value determined using the formula:

$$\text{Attenuation}(f) \leq k_1\sqrt{(f)} + k_2 f + k_3/\sqrt{(f)}$$

for all frequencies (f) in MHz from 0.772 MHz to the highest references frequency. The value of the constants given in Table 8-U.2 shall be used with the formula to compute attenuation values.

Table 8-U.3 gives values of attenuation at specific frequencies in the band of interest. These values are provided for engineering information and are derived from the above formula rounded to the nearest decimal place.

TABLE 8-U.1
Color Codes

Conductor Identification	Color Code	Abbreviation
Pair 1	White-blue[a]	W-BL
	Blue[b]	(BL)
Pair 2	White-orange[a]	W-O
	Orange[b]	O
Pair 3	White-green[a]	W-G
	Green[b]	G
Pair 4	White-brown[a]	W-BR
	Brown[b]	BR

[a]The wire insulation is white, and a colored marking is added for identification. For cables with tightly twisted pairs [all less than 38.1 mm (1.5 in.) per twist] the mate conductor may serve as the marking for the white conductor.

[b]A white marking is optional.

TABLE 8-U.2
Constants for Attenuation Formula

	k_1	k_2	k_3
Category 3	2.320	0.238	0.000
Category 4	2.050	0.043	0.057
Category 5	1.967	0.023	0.050

TABLE 8-U.3
Cable Attenuation, Horizontal UTP Cable (per 100 m (328 ft) at 20 °C)

Frequency (MHz)	Category 3 (dB)	Category 4 (dB)	Category 5 (dB)
0.064	0.9	0.8	0.8
0.256	1.3	1.1	1.1
0.512	1.8	1.5	1.5
0.772	2.2	1.9	1.8
1.0	2.6	2.2	2.0
4.0	5.6	4.3	4.1
8.0	8.5	6.2	5.8
10.0	9.7	6.9	6.5
16.0	13.1	8.9	8.2
20.0	—	10.0	9.3
25.0	—	—	10.4
31.25	—	—	11.7
62.5	—	—	17.0
100.0	—	—	22.0

Note The attenuation of some category 3 UTP cables, such as those with PVC insulation, exhibits a significant temperature dependence. A temperature coefficient of attenuation of 1.5% per °C is not uncommon for such cables. In particular installations where the cable will be subjected to higher temperatures, a less-temperature dependent cable may be required.

Characteristic Impedance And Structural Return Loss (SRL). The different categories of horizontal UTP cable specified in this document shall all have a characteristic impedance of 100 Ω ± 15% in the frequency range from 1 MHz up to the highest reference frequency when measured in accordance with ASTM D 4566 Method 3. Characteristic impedance has a specific meaning for an ideal transmission line (i.e., a cable whose geometry is fixed and does not vary along the length of cable).

Note. The characteristic impedance is commonly derived from swept frequency input impedance measurements using a network analyzer or an *S*-parameter test set. As a result of structural nonuniformities, the measured input impedance for an electrically long length of cable (greater than 1/8 of a wavelength) will fluctuate as a function of frequency. These random fluctuations are superimposed on the curve for characteristic impedance which asymptotically approaches a fixed value at frequencies above 1 mHz. The characteristic impedance can be obtained from these measurements by using a smoothing function over the bandwidth of interest.

The fluctuation in input impedance is related to the SRL for a cable that is terminated in its own characteristic impedance. The values of SRL are dependent on frequency and cable construction.

The SRL, measured in accordance with ASTM D 4566 Method 3, shall be greater than or equal to the values in Table 8-U.4 for all frequencies from 1 MHz to the highest references frequency for a length of 100 m (328 ft) or longer.

Near-End Crosstalk (NEXT). The NEXT loss is commonly derived from swept frequency measurements using a network analyzer or an *S*-parameter test set. A balanced input signal is applied on a disturbing pair while the crosstalk signal is measured in accordance with ASTM D 4566 at the output of the disturbed pair at the near-end of the cable.

Table 8-U.5 gives values of worst-pair NEXT loss at specific frequencies in the band of interest. These values are provided for engineering information and are derived from the following formula truncated to the nearest dB.

NEXT loss decreases as the frequency increases. The minimum NEXT loss for any pair combination at room temperature shall be greater than the value

TABLE 8-U.4
Horizontal UTP Cable SRL (Worst Pair)

Frequency (f)	Category 3 (dB)	Category 4 (dB)	Category 5 (dB)
1–10 MHz	12	21	23
10–16 MHz	$12 - 10 \log(f/10)$	$21 - 10 \log(f/10)$	23
16–20 MHz	—	$21 - 10 \log(f/10)$	23
20–100 MHz	—	—	$23 - 10 \log(f/20)$

Where f = frequency in MHz.

TABLE 8-U.5
Horizontal UTP Cable Next Loss (Worst-Pair Combination) [$\geq$ 100 M (328 ft)]

Frequency (MHz)	Category 3 (dB)	Category 4 (dB)	Category 5 (dB)
0.150	53	68	74
0.772	43	58	64
1.0	41	56	62
4.0	32	47	53
8.0	27	42	48
10.0	26	41	47
16.0	23	38	44
20.0	—	36	42
25.0	—	—	41
31.25	—	—	39
62.5	—	—	35
100.0	—	—	32

Note 0.150 MHz is for reference purposes only.

determined using the formula:

$$\text{NEXT}(f) \geq \text{NEXT}(0.772) - 15 \log(f/0.772)$$

for all frequencies (f) in MHz in the range from 0.772 MHz to the highest referenced frequency for a length of 100 m (328 ft) or longer. The NEXT value at 0.772 MHz shall be 43 dB for category 3 cable, 58 dB for category 4 cable, and 64 dB for category 5 cable.

Measurement Precaution. The transmission measurements of mutual capacitance, capacitance unbalance, characteristic impedance and NEXT shall be performed on cable samples removed from the reel or packages. The test sample shall be stretched along a nonconducting surface or supported in aerial spans such that a minimum 25.4 mm (1.0 inch) separation exists between any convolutions. On-reel or packaged cable measurements that satisfy the performance requirements of Section 8-7.21.3 are acceptable. In case of conflict, the off-reel or unpackaged method shall provide conformance to the minimum requirement stated herein.

8-7.21.4. 150-ohm Shielded Twisted Pair (STP) Cable

A new and improved specification for 150-Ω STP cable has been developed to provide stable performance criteria for higher frequency applications. It has been denominated 150-Ω STP-A cable.

TABLE 8-U.6
Color Codes for Horizontal 150-U STP-A Cable

Conductor Identification	Color Code
Pair 1	Red
	Green
Pair 2	Orange
	Black

Note Color coding may be accomplished by use of helical striping, band marking, solid color skin, or pigment in the insulation itself. Helical striping shall complete a 360° rotation at least every 2.5 cm (1 in). Band spacing shall be less than or equal to 1.25 cm (0.492 in). Bands and stripes shall adhere to the insulation during cable preparation and termination.

Horizontal 150-Ω STP-A Cable

Application. These specifications cover cables consisting of two individually twisted pairs of 22 AWG thermoplastic insulated solid conductors enclosed by a shield and an overall thermoplastic jacket.

Mechanical Requirements

Insulated Conductor. The diameter of the insulated conductor shall not exceed 2.6 mm (0.103 in) maximum.

Pair Assembly. The cable is restricted to two-pair size. The pair twist lengths shall be selected by the manufacturer to ensure compliance with the crosstalk requirements stated herein.

Color Codes. The color codes shall be as shown in Table 8-U.6.

Core Wrap. The core may be covered with one or more layers of dielectric material of adequate thickness to ensure compliance with dielectric strength requirements.

Core Shield. An electrically continuous shield shall be applied over the core wrap. The shield shall consist of a plastic- and aluminum-laminated tape with an aluminum side facing out and a braid of tin-coated copper wires with a minimum of 65% coverage in contact with the aluminum. The tape may be applied to isolate the two pairs within the cable core.

Braided-Shielded Core Sizing. The braided-shielded core assembly shall be capable of being formed without the use of tools to fit through the gauge block shown in Figure 8.U.1. In addition, the angle of the braided shield shall be such that it can be pushed back over the gauge block. The diameter of the braided-shielded core shall not exceed 8.6 mm (0.34 in) when measured with a diameter tape.

Jacket and Jacket Slitting Cord. The core shield shall be enclosed by a uniform, continuous thermoplastic jacket. The jacket slitting cord shall be laid longitudinally beneath the outer jacket.

Cable Diameter. The overall diameter of the completed nonplenum cable shall be less than 11 mm (0.433 in). The overall diameter of the completed plenum cable shall be less than 10 mm (0.394 in) when measured with a diameter tape.

Breaking Strength. The ultimate breaking strength of the completed cable, measured in accordance with ASTM D 4565, shall be 780 N (175 lb) minimum.

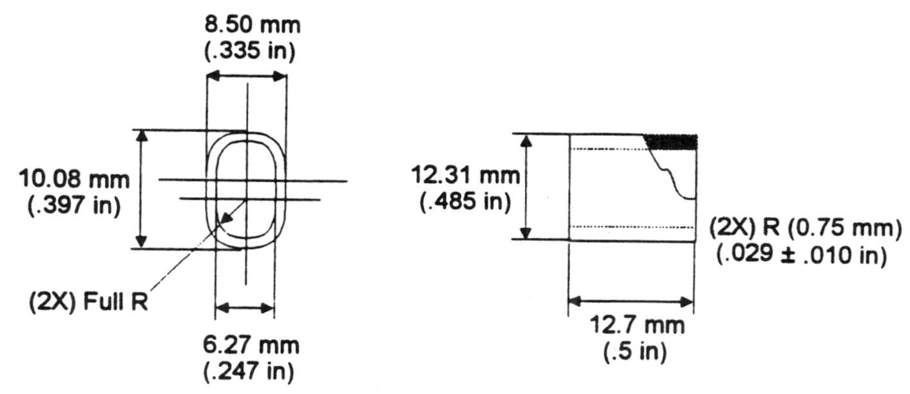

FIGURE 8-U.1. Gauge block.

Low Temperature Bending Radius. The cable shall be tested in accordance with ASTM D 4565. The cable shall withstand a bend radius of 7.5 cm (3 in) for nonplenum, 15 cm (6 in) for plenum, at a temperature of −20°C (±1°C) without jacket or insulation cracking.

Transmission Requirements

DC Resistance. The resistance of any conductor, measured in accordance with ASTM D 4566 and corrected to a temperature of 25°C, shall not exceed 5.71 Ω/100 m.

DC Resistance Unbalance. The resistance unbalance between two conductors of any pair, at or corrected to a temperature of 25°C (±3°C) and measured in accordance with ASTM D 4566, shall not exceed 4%.

Capacitance Unbalance: Pair-to-Ground. The capacitance unbalance to ground at 1 kHz of any pair, at a temperature of 25°C (±3°C) and measured in accordance with ASTM D 4566, shall not exceed 100 pF/100 m (328 ft)

Balanced Mode Attenuation. The balanced mode attenuation of any pair, at a temperature of 25°C (±3°C) and measured in accordance with ASTM 4566, shall not exceed 0.30 dB/100 m (328 ft) at 9.6 kHz, 0.50 dB/100 m at 38.4 kHz $2.2\sqrt{f/4}$ dB/100 m (328 ft), for frequencies from 4.0 MHz to 20.0 MHz, and $9.75\sqrt{f/62.5}$ dB/100 m (328 ft), for frequencies from 20 MHz to 300 MHz; where f = frequency in MHz.

The attenuation measurements from 9.6 kHz to 20 MHz shall be performed on cable lengths of 305 m (1000 ft) or greater. The attenuation measurements from 20 MHz to 300 MHz shall be performed on cable lengths of 100 m (328 ft) to 305 m (1000 ft) using a balun meeting the specification in Annex C of the referenced publication.

Table 8-U.7 provides data for reference purposes only.

Common Mode Attenuation. The common mode attenuation of any pair, at or corrected to a temperature of 25°C (±3°C) and measured in accordance with ASTM D 4566, shall not exceed $9.5\sqrt{f/50}$ dB/km, for frequencies from 50 MHz to 600 MHz, where f = frequency in MHz. The attenuation measurements from 50 MHz to 600 MHz shall be performed on cable lengths of 100 m (328 ft) to 305 m (1000 ft).

Characteristic Impedance. The characteristic impedance shall be per the requirements in Table 8-U.8 when measured in accordance with ASTM D 4566.

The structural return loss (SRL), measured and normalized to the characteristic impedance value in accordance with ASTM D 4566 and performed on cable lengths of 100 m (328 ft) to 305 m (1000 ft) using a balun, meeting that specified in Annex C of the referenced publication, for the entire frequency range from 3 MHz to 300 MHz shall exceed 24 dB up to 20 MHz and $24 - 10\,\mathrm{Log}(f/20)$ dB above 20 MHz, where f = frequency in MHz.

NEXT Loss. The NEXT loss between two pairs within a cable, measured in accordance with ASTM D 4566, shall exceed 58.0 dB for frequencies from 9.6 kHz to 5 MHz and $58.0\ \mathrm{dB} - 15\,\mathrm{Log}(f/5)$ for frequencies from 5 MHz to 300 MHz, where f = frequency in MHz.

The NEXT loss measurements from 9.6 kHz to 20 MHz shall be performed on cable lengths of 305 m (1000 ft) or greater. The NEXT loss measurements from 20 MHz to 300 MHz shall be performed on cable lengths of 100 m (328 ft) to 305 m (1000 ft) using a balun specified in Annex C of the referenced publication. The NEXT loss values in Table 8.U.9 are given for reference purposes only.

Dielectric Strength. The insulation between each conductor and core shield shall be capable of withstanding a minimum DC potential of 5 kV for 3 seconds when tested in accordance with ASTM D 4566.

8-7.21.5. Backbone UTP Cable

100-Ohm UTP Multipair Backbone Cable. The cables consist of 24 AWG thermoplastic insulated conductors formed into binder groups of 25 pairs. The groups are identified by distinctly colored binders and assembled to form a single compact core. The core is covered by a protective sheath. The sheath consists of an overall thermoplastic jacket and may contain an underlying metallic shield and one or more layers of dielectric material applied over the core.

Mechanical Design

Insulated Conductor. The diameter over the insulation shall be 1.22 mm (0.048 in) maximum.

Pair Assembly. The pair twist lengths shall be selected by the manufacturer to ensure compliance with the crosstalk requirements of this standard.

TABLE 8-U.7
Cable Balanced Mode Attenuation for Horizontal STP-A Cable

Frequency (MHz)	Maximum Attenuation [dB/100 m (328 ft)]
0.0096	0.30
0.0384	0.50
4.0	2.2
8.0	3.1
10	3.6
16	4.4
20	4.9
25	6.2
31.25	6.9
62.5	9.8
100	12.3
300	21.4

TABLE 8-U.8
Characteristic Impedance of Horizontal STP-A Cable

Frequency (MHz)	Characteristic Impedance (Ω)
0.0096	270 ± 10%
0.0384	185 ± 10%
3.0–20[a]	150 ± 10%
20–300[a]	Under study

[a] The specification shall be met over the entire frequency range specified.

TABLE 8-U.9
Horizontal STP-A Cable NEXT

Frequency (MHz)	NEXT Loss Worst Pair (dB)
0.0096	58.0
0.0384	58.0
4.0	58.0
8.0	54.9
10	53.5
16	50.4
20	49.0
25	47.5
31.25	46.1
62.5	41.5
100	38.5
300	31.3

Color Code. The conductor identification shall be indicated by coloring the insulation used on each conductor of a twisted pair. The color code shall follow the industry standard color code composed of 10 distinctive colors to identify 25 pairs. (Refer to ICEA Publication S-80-576) for appropriate colors). Marking of each conductor of a pair using its mate's color is optional.

Core Assembly. When cable sizes larger than 25 pairs are required, the core shall be assembled in units or subunits of 25 pairs. Each 25-pair unit shall be identified by color-coded binders in accordance with ICEA Publication S-80-576 or the manufacturer's specifications. Binder color-code integrity shall be maintained whenever cables are spliced.

Core Wrap. The core may be covered with one or more layers of dielectric material of adequate thickness to ensure compliance with the dielectric strength requirements.

Core Shield. When an electrically continuous shield is applied over the core wrap, it shall comply with shield resistance requirements.

Note: Detailed information regarding shield criteria can be found in UL Subject 444, ANSI/ICEA STD S-84-608 and Bellcore Reference TR-TSY-000421.

Jacket. The core shall be enclosed by a uniform continuous thermoplastic jacket.

Transmission Requirements

DC Resistance. The resistance of any conductor, measured in accordance with ASTM D 4566, shall not exceed 9.38 Ω per 100 m (328 ft) at or corrected to a temperature of 20°C.

DC Resistance Unbalance. The resistance unbalance between the two conductors of any pair shall not exceed 5% measured at or corrected to a temperature of 20°C in accordance with ASTM D 4566.

Mutual Capacitance. The mutual capacitance of any pair at 1 kHz, measured in accordance with ASTM D 4566 and "Measurement Precaution" in the previous subsection, should not exceed 6.6 nF per 100 m (328 ft) for category 4 and 5 cables at or corrected to a temperature of 20°C. Mutual capacitance values are provided for engineering design purposes only and are not a requirement for conformance testing.

TABLE 8-U.10
Backbone UTP Cable SRL (Worst Pair)

Frequency (f)	Category 3 (dB)	Category 4 (dB)	Category 5 (dB)
1 – 10 MHz	12	21	23
10 – 16 MHz	$12 - 10 \log(f/10)$	$21 - 10 \log(f/10)$	23
16 – 20 MHz	—	$21 - 10 \log(f/10)$	23
20 – 100 MHz	—	—	$23 - 10 \log(f/20)$
Where f = frequency in MHz			

Capacitance Unbalance: Pair-to-Ground. The capacitance unbalance to ground at 1 kHz of any pair, measured at or corrected to 20°C in accordance with ASTM D 4566 shall not exceed 300 nF per 100 m (328 ft).

Characteristic Impedance and Structural Return Loss. The backbone UTP cable shall have a characteristic impedance of 100 Ω ± 15% in the frequency range from 1 mHz up to the highest measured frequency when measured in accordance with ASTM D 4566. Characteristic impedance has a specific meaning for an ideal transmission line (i.e. a cable whose geometry is fixed and does not vary along the length of cable).

Note: The characteristic impedance is commonly derived from swept frequency input impedance measurements using a network analyzer with an S parameter test set. As a result of structural nonuniformities, the measured input impedance for an electrically long length of cable (greater than 1/8 of a wavelength) will fluctuate as a function of frequency. These random fluctuations are superimposed on the curve for characteristic impedance which asymptotically approaches a fixed value at frequencies above 1 MHz. The characteristic impedance can be obtained from these measurements by using a smoothing function over the bandwidth of interest.

The fluctuation in input impedance is related to the SRL for a cable that is terminated in its own characteristic impedance. The values of SRL are dependent on frequency and cable construction.

The SRL, when measured in accordance with ASTM D 4566 Method 3, shall be greater than or equal to the values given in Table 8-U.10 for all frequencies from 1 MHz to the highest frequency for a length of 100 m (328 ft) or longer.

Attenuation . The attenuation of backbone cables shall meet the requirements of horizontal UTP cables given in Table 8-U.11. The maximum attenuation given in Table 8-U.11 shall be adjusted to elevated temperatures using a factor of 0.4% increase per °C for category 4 and 5 cables. The cable attenuation shall be verified at a temperature of 40°C and 60°C and shall meet the requirements specified here after adjusting for temperature.

NEXT Loss. NEXT loss is commonly derived from swept frequency measurements using a network analyzer or an S parameter test set. A balanced input signal is applied on a disturbing pair while the crosstalk signal is measured in accordance with ASTM D 4566 at the output port on a disturbed pair at the near end of the cable. NEXT loss decreases as the frequency increases. The minimum power sum NEXT loss, tested in accordance with ASTM D 4566, of backbone UTP cable shall be greater than the value determined using the formula:

$$\text{NEXT}(f) \geq \text{NEXT}(0.772) - 15 \log(f/0.772)$$

for all frequencies (f) in MHz in the range from 0.772 MHz to the highest reference frequency for a length of 100 m (328 ft) or longer. The NEXT value of

TABLE 8-U.11
Backbone UTP Cable Attenuation

Frequency (MHz)	Maximum Attenuation (dB per .305 m) (dB per 1000 ft)
0.064	2.8
0.256	4.0
0.512	5.6
0.772	6.7
1.0	7.6
4.0	15.4
8.0	22.3
10.0	25.0
16.0	32.0

TABLE 8-U.12
Backbone UTP Cable Power Sum NEXT Loss [≥ 100 M (328 FT)]

Frequency (MHz)	Category 3 (dB)	Category 4 (dB)	Category 5 (dB)
0.150	53	68	74
0.772	43	58	64
1.0	41	56	62
4.0	32	47	53
8.0	27	42	48
10.0	26	41	47
16.0	23	38	44
20.0	—	36	42
25.0	—	—	41
31.25	—	—	39
62.5	—	—	35
100.0	—	—	32

Note = 0.150 MHz is for reference purposes only.

0.772 MHz shall be 43 dB for category 3 cable, 58 dB for category 4 cable, and 64 dB for category 5 cable.

Table 8-U.12 gives values of power sum NEXT loss at specific frequencies in the band of interest for all pairs. These values are provided for engineering information and are derived from the above formula truncated to the nearest dB. These measurements should be swept-frequency and should not be restricted to 25-pair binder groups.

Dielectric Strength. The insulation between each conductor and the core shield, when present, shall be capable of withstanding a minimum DC potential of 5 kV for 3 seconds in accordance with ASTM D 4566.

Core Shield Resistance. When a shield is present around the core, the DC resistance of the core shield shall not exceed the value given by the following equation:

$$R\ (\Omega/\text{km}) = 62.5/D\ (\text{mm})$$

or

$$R\ (\Omega/1000\ \text{ft}) = 0.75/D\ (\text{in})$$

where R = maximum core shield resistance
D = outside diameter of the shield

This requirement is applicable to outside plant cables or inside building cables having their shields bonded to the shields of outside plant cables at building entrances. The electrical and physical requirements of the shields of inside building cables contained within a building are under study.

8-7.21.6. Horizontal 62.5/125 μm Optical Fiber Cable

Introduction. The optical fiber cable shall consist of a minimum of two 62.5/125 μm optical fibers enclosed by a protective sheath. This cable has a bandwidth capacity in excess of 1 GHz for the 90 m (295 ft) distance specified for horizontal cabling. The fiber shall be multimode, graded-index optical fiber with a nominal 62.5/125 μm core/cladding diameter. The fiber complies with ANSI-EIA/TIA-492AAAA.

Cable Transmission Performance Specifications. Each cable fiber shall meet the performance specifications in Table 8-U.13. Attenuation shall be measured in

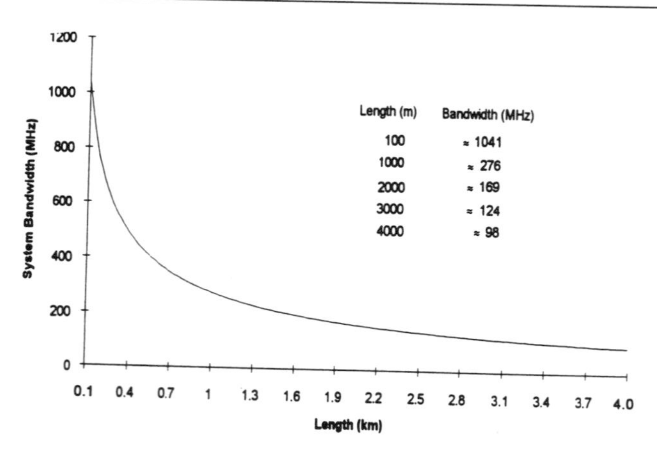

FIGURE 8-U.2. Typical system bandwidth utilizing ANSI/EIA-TIA-568A 62.5/125 μm optical fiber cable and a 1300 nm LED.

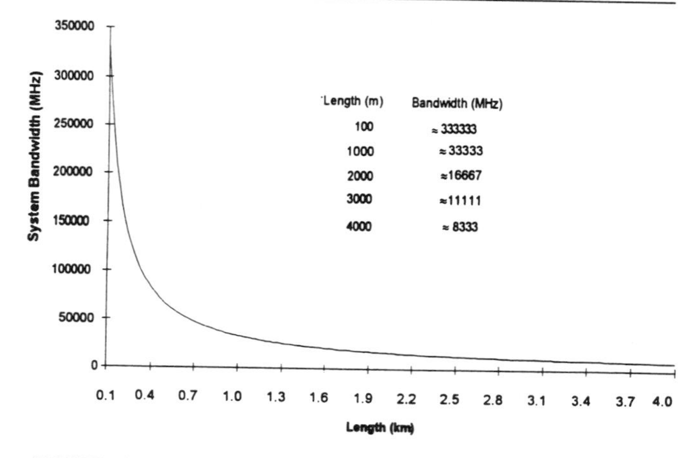

FIGURE 8-U.3. Typical system bandwidth utilizing ANSI/EIA/TIA-568A single-mode optical fiber cable and a 1310 nm laser.

accordance with ANSI/EIA/TIA-455-46, -53, or -61. Information transmission capacity shall be measured in accordance with ANSI/EIA/TIA-455-51 or -30. The cable shall be measured at 23°C (±5°C)

TABLE 8-U.13
Fiber Optic Cable Performance Specifications

Wavelength (nm)	Maximum Attenuation (dB/km)	Minimum Information Capacity MHz/km)
850	3.75	160
1300	1.5	500

Applicability. Customer premises optical fiber backbone cabling has been and continues to be primarily 62.5/125 μm multimode fiber-based because this fiber can use LED transmitters. With rapidly growing bandwidth requirements, more customer premises are installing single-mode optical fiber in addition to multimode optical fiber for present and future bandwidth requirements. Single-mode optical fiber systems inherently have higher bandwidth and longer distance capabilities than 62.5/125 μm optical fiber.

System bandwidth is not only a function of the fiber but also distance and transmitter characteristics, specifically center wavelength and spectral width, and optical rise time. Therefore, only typical system bandwidth values can be given. Figure 8-U.2 depicts typical system bandwidth relative to distance for a 62.5/125 μm system utilizing ANSI/EIA/TIA-568 specified fiber with a typical LED transmitter operating at 1300 nm wavelength. Figure 8-U.3 depicts typical system bandwidth relative to distance for a single-mode system utilizing the specified fiber with a typical laser transmitter operating at 1310 wavelength.

While it is recognized that the capabilities of single-mode optical fiber may allow for backbone link distances of up to 60 km (37 miles), this distance is generally considered to extend outside the scope of this standard.

Section 8-7.21 consists of abstracts of Standards Proposal No. 2840-A, Proposed Revision of EAI/TIA-568, "Commercial Building Telecommunications Cabling Standard," EIA/TIA, Washington, DC, February 1995.

9

Fiber Optics Transmission

This new section follows Section 9.5.9, which ends on page 635 of the Manual.

9-5.10. Single-Mode Optical Fiber Cable

9-5.10.1. Scope

This section describes single-mode (mono-mode) optical fiber which has the zero dispersion wavelength around 1310 nm and which is optimized for use in the 1310-nm wavelength region and which can also be used in the 1550-nm wavelength region, where the fiber is not optimized. Only those characteristics of the fiber providing the minimum essential design framework for fiber manufacture are provided. Of these, the cabled fiber cutoff wavelength may be significantly affected by cable manufacture and installation. Otherwise, the recommended characteristics apply equally to individual fibers, fibers incorporated into a cable wound on a drum or reel, and fibers in installed cable.

9-5.10.2. Fiber Characteristics

Mode Field Diameter. The nominal value of the mode field diameter at 1310 nm shall lie within the range of 9–10 μm. The mode field diameter deviation should not exceed the limits of $\pm 10\%$ of the nominal value.

Note 1: A value of 10 μm is commonly employed for matched cladding designs, and a value of 9 μm is commonly employed for depressed cladding designs. However, the choice of a specific value within the above range is not necessarily associated with a specific fiber design.

Note 2: It should be noted that the fiber performance required for any given application is a function of essential fiber and systems parameters (i.e., mode field diameters, cutoff wavelength, total dispersion, system operating wavelength, and bit rate/frequency of operation) and not primarily of the fiber design.

Note 3: The mean value of the mode field diameter, in fact, may differ from the above nominal values provided that all fibers fall within $\pm 10\%$ of the specified nominal value.

Cladding Diameter. The recommended nominal value of the cladding diameter is 125 μm. The cladding deviation should not exceed the limits of ± 2 μm.

Mode Field Concentricity Error. The recommended mode field concentricity error at 1310 nm should not exceed 1 μm.

Note 1: For some particular jointing techniques and joint loss requirements, tolerances up to 3 μm may be appropriate.

Note 2: The mode field concentricity error and the concentricity error of the core represented by the transmitted illumination using wavelengths different from

1310 nm (including white light) are equivalent. In general, the deviation of the center of the refractive index profile and the cladding axis also represents the mode field concentricity error; but if any inconsistency appears between the mode field concentricity error, measured according to the reference test method (RTM), and the core concentricity error, the former will constitute the reference.

Noncircularity

Mode Field Noncircularity. In practice, the mode field noncircularity of fibers having nominally circular mode fields is found to be sufficiently low that propagation and jointing are not affected. It is therefore not considered necessary to recommend a particular value for the mode field noncircularity. It is not normally necessary to measure the mode field noncircularity for acceptance purposes.

Cladding Noncircularity. The cladding noncircularity should be less than 2%. For some particular jointing techniques and joint loss requirements, other tolerances may be appropriate.

Cutoff Wavelength. Two useful types of cutoff wavelength can be distinguished:

1. The cutoff wavelength λ_c of a primary coated fiber according to the relevant fiber RTM
2. The cutoff wavelength λ_{cc} of a cabled fiber in a deployment condition according to the relevant cable RTM.

The correlation of the measured values of λ_c and λ_{cc} depends on the specific fiber and cable design and the test conditions. While in general $\lambda_{cc} < \lambda_c$, a quantitative relationship cannot easily be established. The importance of ensuring single-mode transmission in the minimum cable length between joints at the minimum system operating wavelength is paramount. This can be approached in two alternate ways:

1. Recommending λ_c to be less than 1280 nm: when a lower limit is appropriate, λ_c should be greater than 1100 nm.
2. Recommending the maximum value of λ_{cc} to be either 1260 nm or 1270 nm.

Note 1: A sufficient wavelength margin should be ensured between the lowest-permissible system operating wavelength λ_s and the highest-permissible cable cutoff wavelength λ_{cc}.

Note 2: To prevent modal noise effects and ensure single-mode transmission in fiber jumpers of any length and under any deployment condition, fibers should be selected with sufficiently low cutoff wavelength. Considering the worst-case conditions, the maximum λ_c for fibers to be used in jumpers should not be higher than 1240 nm when measured under the conditions of the relevant RTM of Rec. G.650.

These two specifications need not both be invoked. Because specification of λ_{cc} is a more direct way of ensuring single-mode cable operation, it is the preferred option. When circumstances do not readily permit the specification of λ_{cc} (e.g., in single-fiber cables such as jumper cables or cables to be deployed in a significantly different manner than in the λ_{cc} RTM), then the specification of λ_c is appropriate.

When the user chooses to specify λ_{cc} as in specification 2, it should be understood that λ_c may exceed 1280 nm.

When the user chooses to specify λ_c as in specification 1, then λ_{cc} need not be specified.

In the case where the user chooses to specify λ_{cc}, it may be permitted that λ_c be higher than the minimum system operating wavelength, relying on the effects of cable fabrication and installation to yield λ_{cc} values below the minimum system operating wavelength for the shortest length of cable between two joints.

In the case where the user chooses to specify λ_{cc}, a qualification test may be sufficient to verify that the λ_{cc} requirement is being met.

1550-Nanometer Loss Performance. In order to ensure low-loss operation of deployed 1310-nm optimized fibers in the 1550-nm wavelength region, the loss increase of 100 turns of fiber loosely wound with a 37.5-mm radius and measured at 1550 nm shall not be less than 1.0 dB.

Note 1: A qualification test may be sufficient to ensure that this requirement is being met.

Note 2: The above value of 100 turns corresponds to the approximate number of turns deployed in all splice cases of a typical repeater span. The radius of 37.5 mm is equivalent to the minimum bend-radius widely accepted for long-term deployment of fibers in practical systems installations to avoid static-fatigue failure.

Note 3: If for practical reasons fewer than 100 turns are chosen to implement this test, it is suggested that not less than 40 turns and a proportionately smaller loss increase be used.

Note 4: If bending radii smaller than 37.5 mm are planned to be used in splice cases or elsewhere in the system (e.g., $R = 30$ mm), it is suggested that the same loss value of 1.0 dB shall apply to 100 turns of fiber deployed with this smaller radius.

Note 5: The 1550-nm bend-loss recommendation relates to the deployment of fibers in practical single-mode fiber installations. The influence of the stranding-related bending radii of cabled single-mode fibers on the loss performance is included in the loss specification of the cabled fiber.

Note 6: In the event that routine tests are required, a small-diameter loop with one or several turns can be used instead of the 100-turn test, for accuracy and measurement ease of the 1550-nm bend sensitivity. In this case the loop diameter, number of turns, and the maximum permissible bend loss for the several-turn test should be chosen, so as to correlate with the 1.0-dB loss recommendation of the 37.5-mm radius 100-turn functional test.

Material Properties of Fiber

Fiber Materials. The substances of which the fiber is made should be indicated.

Note: Care may be needed when fusion splicing fibers of different substances. The ITU-T organization reports that provisional results indicate that adequate splice loss and strength can be achieved when splicing different high-silica fibers.

Protective Materials. The physical and chemical properties of the material used for the fiber primary coating, and the best way of removing it (if necessary), should be indicated. Single-jacket fiber should be handled in a similar way.

Proofstress Level

- The proofstress σ_p shall be at least 0.35 GPa (which approximately corresponds to a proofstrain $\sim 0.5\%$).
- The dwell-time t_d shall be 1 sec. A shorter alternate dwell-time t_a may be chosen; then a larger alternate proofstress σ_a must be chosen according to the following equation:

the following equation:

$$\sigma_a = \sigma_p \left[\frac{t_d}{t_a} \right]^{1/n_d}$$

- The value of the dynamic fatigue parameter n_d is determined by a dynamic fatigue test method.
- For some applications, such as local networks or submarine systems, higher values of proofstress (or proofstrain) may be desired. Values such as 0.7 GPa or 1.4 GPa (or ~ 1% and ~ 2%) are for further study.

Refractive Index Profile. The refractive index profile of the fiber does not generally need to be known. If it must be measured, a test method is provided in ITU-T Rec. G.651.

9-5.10.3. Factory Length Specifications

Attenuation Coefficient. Optical fiber cables covered by this section generally have attenuation coefficients below 1.0 dB/km in the 1310-nm wavelength region and below 0.5 dB/km in the 1500-nm wavelength region.

Note 1: The lowest values depend on the fabrication process, fiber composition and design, and cable design. Values in the range 0.3–0.4 dB/km in the 1310-nm region and 0.15–0.25 dB/km in the 1550-nm region have been achieved.

Note 2: The attenuation coefficient may be calculated across a spectrum of wavelengths, based on measurements at a few (3–5) predictor wavelengths.

Chromatic Dispersion Coefficient. The maximum chromatic dispersion coefficient shall be specified by:

- The allowed range of the zero-dispersion wavelength between $\lambda_{0\,\min}$ = 1300 nm and $\lambda_{0\,\max}$ = 1324 nm
- The maximum value $S_{0\,\max}$ – 0.093 psec/(nm^2 · km) of the zero-dispersion slope.

The chromatic dispersion coefficient limits for any wavelength λ within the range 1260–1360 nm shall be calculated as

$$D_1(\lambda) = \frac{S_{0\,\max}}{4} \left[\lambda - \frac{\lambda_{0\,\min}^4}{\lambda^3} \right]$$

$$D_2(\lambda) = \frac{S_{0\,\max}}{4} \left[\lambda - \frac{\lambda_{0\,\max}^4}{\lambda^3} \right]$$

Note 1: As an example, the values of $\lambda_{0\,\min}$, $\lambda_{0\,\max}$, and $S_{0\,\max}$ yield chromatic dispersion coefficient magnitudes $|D_1|$ and $|D_2|$ equal to or smaller than the maximum chromatic dispersion coefficients in the following table:

Wavelength (nm)	Maximum Chromatic Dispersion Coefficient [psec/(nm · km)]
1288–1339	3.5
1271–1360	5.3
1550	20 (approx.)

Note 2: Use of these equations in the 1550-nm region should be approached with caution.

Note 3: For high-capacity or long systems, a narrower range of $\lambda_{0\,\min}$, $\lambda_{0\,\max}$ may need to be specified; or, if possible, a smaller value of $S_{0\,\max}$ may need to be chosen.

Note 4: It is not necessary to measure chromatic dispersion coefficient of single-mode fiber on a routine basis.

9-5.10.4. Elementary Cable Sections

An elementary cable section usually includes a number of spliced factory lengths. The requirements for factory lengths are given in Section 9-5.10.3. The transmission parameters for elementary cable sections must take into account not only the performance of the individual cable lengths, but also, amongst other factors, such things as splice losses and connector losses (if applicable).

In addition, the transmission characteristics of the factory length fibers as well as such items as splices and connectors, and so on, will all have a certain probability distribution which often needs to be taken into account if the most economic designs are to be obtained. The following subsections should be read with this statistical nature of the various parameters in mind.

Attenuation. The attenuation A of an elementary cable section is given by

$$A = \sum_{n=1}^{m} \alpha_n \cdot L_n + \alpha_s \cdot \chi + \alpha_c \cdot y$$

where

α_n is the attenuation coefficient of nth fiber in elementary cable section
L_n is the length of nth fiber
m is the total number of concatenated fibers in elementary cable section
α_s is the mean splice loss
χ is the number of splices in elementary cable section
α_c is the mean loss of line connectors
y is the number of line connectors in elementary cable section (if provided)

A suitable allowance should be allocated for a suitable cable margin for future modifications of cable configurations (additional splices, extra cable lengths, aging effects, temperature variations, etc.). The above equation does not include the loss of equipment connectors.

The mean loss is used for the loss of splices and connectors. The attenuation budget used in designing an actual system should account for the statistical variations in these parameters.

Chromatic Dispersion. The chromatic dispersion (in picoseconds) can be calculated from the chromatic dispersion coefficients of the factory lengths, assuming a linear dependence on length and with due regard for the signs of the coefficients and system source characteristics. See subsection entitled "Chromatic Dispersion Coefficient" in Section 9-5.10.3.

The material in Section 9-5.10 has been derived from ITU-T Rec. G.952, Helsinki, 3/93.

9-5.11. Dispersion-Shifted Single-Mode Fiber Optic Cable

9-5.11.1. Introduction and Scope

A dispersion-shifted single-mode fiber has (a) a nominal zero-dispersion wavelength close to 1550 nm and (b) a dispersion coefficient which is monotonically increasing with wavelength. The fiber is optimized for use at wavelengths in the region between 1500 nm and 1600 nm, but may also be used in the region around 1310 nm subject to the constraints outlined below. Its geometrical, optical, transmission, and mechanical parameters are described in this section.

Only those characteristics of the fiber providing a minimum essential design framework for fiber manufacturers are recommended in this section. Of these, the cabled fiber cutoff wavelength may be significantly affected by cable manufacture and installation. Otherwise, the characteristics given below will apply equally to individual fibers, fibers incorporated into a cable wound on a drum or reel, and fibers in an installed cable.

9-5.11.2. Fiber Characteristics

Mode Field Diameter. The nominal value of the mode field diameter at 1550 nm shall lie within the range of 7.0–8.3 μm. The mode field diameter deviation should not exceed the limits of $\pm 10\%$ of the nominal value.

Note 1: The choice of a specific value within the above range is not necessarily associated with a specific fiber design.

Note 2: It should be noted that the fiber performance required for any given application is a function of essential fiber and systems parameters (i.e., mode field diameters, cutoff wavelength, chromatic dispersion, system operating wavelength, and bit rate/frequency of operation) and not primarily of the fiber design.

Cladding Diameter. The recommended nominal value of the cladding diameter is 125 μm. The cladding deviation should not exceed ± 2 μm. For some particular jointing techniques and joint loss requirements, other tolerances may be appropriate.

Mode Field Concentricity Error. The recommended mode field concentricity error at 1550 nm should not exceed 1 μm.

Note: For some particular jointing techniques and joint loss requirements, tolerances up to 3 μm may be appropriate.

Noncircularity

Mode Field Noncircularity. In practice, the mode field noncircularity of fibers having nominally circular mode fields is found to be sufficiently low that propagation and jointing are not affected. It is therefore not considered necessary to recommend a particular value for the mode field noncircularity. It is not normally necessary to measure the mode field noncircularity for acceptance purposes.

Cladding Noncircularity. The cladding noncircularity should be less than 2%. For some particular jointing techniques and joint loss requirements, other tolerances may be appropriate.

Cutoff Wavelength. Two useful types of cutoff wavelength can be distinguished:

1. The cutoff wavelength λ_c of a primary coated fiber according to the relevant fiber RTM
2. The cutoff wavelength λ_{cc} of a cabled fiber in a deployment condition according to the relevant cable RTM.

The correlation of the measured values of λ_c and λ_{cc} depends on the specific fiber and cable design and the test conditions. While in general $\lambda_{cc} < \lambda_c$, a quantitative relationship cannot easily be established.

Single-mode transmission in the 1550-nm region can be ensured by recommending λ_{cc} to be less than 1270 nm.

Note: The above recommendation is not sufficient to ensure 1310-nm region single-mode operation in any possible combination of system operating wavelength, cable length, and cable deployment conditions. Suitable limits on λ_c or λ_{cc} should be set in case 1310-nm region operation is foreseen, with particular attention to prevent modal noise effects in minimum cable lengths between repair joints and cable jumpers.

1550-Nanometer Bend Performance. The loss increase for 100 turns of fiber, loosely wound with 37.5-mm radius and measured at 1550 nm, shall be less than 0.5 dB.

Note 1: A qualification test may be sufficient to ensure that this requirement is being met.

Note 2: The above value of 100 turns corresponds to the approximate number of turns deployed in all splice cases of a typical repeater span. The radius of 37.5 mm is equivalent to the minimum bend-radius widely accepted for long-term deployment of fibers in practical systems installations to avoid static-fatigue failure.

Note 3: If for practical reasons fewer than 100 turns are chosen to implement this test, it is suggested that not less than 40 turns and a proportionately smaller loss increase be used.

Note 4: If bending radii smaller than 37.5 mm are planned to be used in cases or elsewhere in the system (e.g., $R = 30$ mm), it is suggested that the same loss value of 0.5 shall apply to 100 turns of fiber deployed with this smaller radius.

Note 5: The 1550-nm bend-loss recommendation relates to the deployment of fibers in practical single-mode fiber installations. The influence of the stranding-related bending radii of cabled single-mode fibers on the loss performance is included in the loss specification of the cabled fiber.

Note 6: In the event that routine tests are required, a small-diameter loop with one or several turns can be used instead of the 100-turn test, for accuracy and measurement ease of the 1550-nm bend sensitivity. In this case the loop diameter, number of turns, and the maximum permissible bend loss for the several-turn test should be chosen, so as to correlate with the 0.5-dB loss recommendation of the 37.5-mm radius 100-turn functional test.

Material Properties of the Fiber

Fiber Materials. The substances of which the fibers are made should be indicated.

Note: Care may be needed in fusion splicing fibers of different substances. The ITU-T organization reports that provisional results indicate that adequate splice loss and strength can be achieved when splicing different high-silica fibers.

Protective Materials. The physical and chemical properties of the material used for the fiber primary coating, and the best way of removing it (when necessary), should be indicated. In the case of single-jacket fiber, similar indications should be given.

Proofstress Level

- The proofstress σ_p shall be at least 0.35 GPa (which approximately corresponds to a proofstrain $\sim 0.5\%$).
- The dwell-time t_d shall be 1 sec. A shorter alternate dwell-time t_a may be chosen; then a larger alternate proofstress σ_a must be chosen according to

the following equation:

$$\sigma_a = \sigma_p \left[\frac{t_d}{t_a} \right]^{1/n_d}$$

- The value of the dynamic fatigue parameter n_d is determined by a dynamic fatigue test method.
- For some applications, such as local networks or submarine systems, higher values of proofstress (or proofstrain) may be desired. Values such as 0.7 GPa or 1.4 GPa (or $\sim$ 1% and $\sim$ 2%) are for further study.

Refractive Index Profile. The refractive index profile of the fiber does not generally need to be known; if it must be measured, the reference test method given in ITU-T Rec. G.651 may be used.

9-5.11.3. Factory Length Specifications

Because the geometrical and optical characteristics of fibers given in Section 9-5.11.2 are barely affected by the cabling process, this section will give recommendations mainly relevant to transmission characteristics of cabled factory lengths.

Environmental and test conditions are paramount and are described in the guidelines for test methods (Rec. G.651).

Attenuation Coefficient. Optical fiber cables covered by this section generally have attenuation coefficients in the 1550-nm region below 0.5 dB/km. When they are intended for use in the 1300-nm region, their attenuation coefficient in that region is generally below 1 dB/km.

Note: The lowest values depend on the fabrication process, fiber composition and design, and cable design. Values in the range of 0.19–0.25 dB/km in the 1550-nm region have been achieved.

Chromatic Dispersion Coefficient. The following equation specifies the chromatic dispersion $D(\lambda)$, in psec/(nm · km), as

$$D(\lambda) = (\lambda - \lambda_0) S_0$$

where λ is the wavelength of interest (in nm), λ_0 is the zero-dispersion wavelength (in nm), and S_0 is the zero-dispersion slope [in psec/(nm^2 · km)]. The slope, S_0, is specified by its maximum value: $S_0 < S_{0\,\max}$. The zero-dispersion wavelength, λ_0, is specified by the nominal value of 1550 and its maximum tolerance, $\Delta\lambda_{0\,\max}$, above and below 1550 nm (considered symmetrical):

$$1550 - \Delta\lambda_{0\,\max} < \lambda_0 < 1550 + \Delta\lambda_{0\,\max}$$

In addition, the maximum absolute value of the dispersion coefficient, $D_{\max}$ [in psec/(nm · km)], is specified over the specified window width, $\Delta\lambda_w$ (in nm), above and below 1550 nm. Then

$$|D(\lambda)| < D_{\max} \qquad \text{for } 1550 - \Delta\lambda_w < \lambda < 1550 + \Delta\lambda_w$$

Users operating with a transmitter central wavelength separated from 1550 nm (either above or below) by $\Delta\lambda_t$ (in nm) may calculate the maximum absolute value

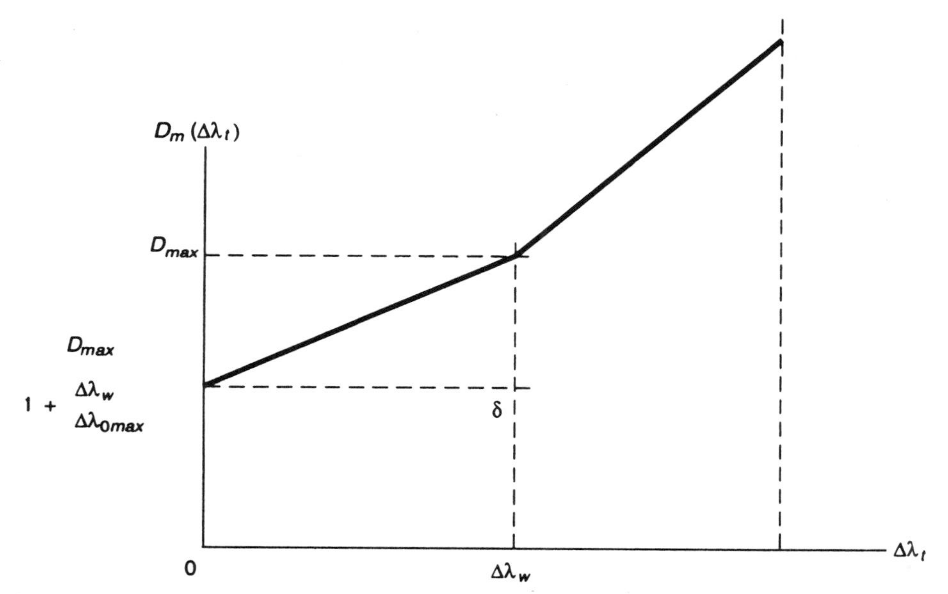

FIGURE 9-U.1. Maximum absolute value of the dispersion coefficient. (From Figure 1 / G.653, ITU-T Rec. G.653, 3 / 93, page 6.)

of the dispersion coefficient as

$$D_m(\Delta\lambda_t) = D_{\max}(\Delta\lambda_t + \Delta\lambda_{0\max})/\Delta\lambda_w + \Delta\lambda_{0\max} \qquad \text{for } 0 \le \Delta\lambda_t \le \Delta\lambda_w$$

$$D_m(\Delta\lambda_t) = D_{\max} + S_{0\max}(\Delta\lambda_t - \Delta\lambda_w) \qquad \text{for } \Delta\lambda_w \le \Delta\lambda_t \le 50 \text{ nm}$$

where $D_{\max} = D_m(\Delta\lambda_w)$. Figure 9-U.1 schematically illustrates the specification.
The specification of the dispersion coefficient is as follows:

$$\Delta\lambda_{0\max} \le 50 \text{ nm}$$

$$S_{0\max} \le 0.085 \text{ psec}/(\text{nm}^2 \cdot \text{km})$$

$$D_{0\max} = 3.5 \text{ psec}/(\text{nm} \cdot \text{km}) \text{ between 1525 and 1575 nm}$$

$$\Delta\lambda_w = 25 \text{ nm}$$

Note 1: The values above are provisionally specified in order to give guidance to fiber and system designers. Further study and tradeoffs between $\Delta\lambda_{0\max}$ and $S_{0\max}$ may be needed in the future to improve the fiber dispersion performances in the working wavelength window.

Note 2: It is not necessary to measure the chromatic dispersion coefficient on a routine basis.

9-5.11.4. Elementary Cable Sections

An elementary cable section usually includes a number of spliced factory lengths. The requirements for factory lengths are given in Section 9-5.11.4. The transmission parameters for elementary cable sections must take into account not only the performance of the individual cable lengths, but also, amongst other factors, such things as splice losses and connector losses (if applicable).

In addition, the transmission characteristics of the factory length fibers as well as such items as splices and connectors, and so on, will all have a certain probability distribution which often needs to be taken into account if the most economic designs are to be obtained. The following subsections should be read with this statistical nature of the various parameters in mind.

Section 9-5.11 was based on ITU-T Rec. G.653, Helsinki, 3/93.

11

Electromagnetic Wave Propagation

Equation (11-55) on page 767 of the *Manual* should be changed to:

$$r = \frac{1}{1 + d/d_0} \tag{11-55}$$

where, for $R_{0.01} \leq 100$ mm/hr,

$$d_0 = 35e^{-0.015R_{0.01}}$$

For $R_{0.01} > 100$ mm/hr, use the value 100 mm/hr in place of $R_{0.01}$.Remove any reference to L as path length and replace with d. Equation (11-56) should be changed to:

$$A_{\text{eff}} = A \times d \times r \tag{11-56}$$

where A = specific attenuation

d = actual path length

r = distance factor

From CCIR Rec. 530-4, Section 2.4.1, CCIR 8 Oct. 1992.

This new section follows Section 11-7.2.3, which ends on page 781 of the Manual.

11-7.2.4. Excess Attenuation Due to Rainfall, CCIR Method for Satellite Paths

Calculation of Long-Term Attenuation Statistics from Point Rainfall Rate. The following procedure provides estimates of the long-term statistics of the slant-path rain attenuation at a given location for frequencies up to 30 GHz. The following parameters are required:

$R_{0.01}$: Point rainfall rate for the location for 0.01% of an average year (mm/hr)
h_s: Height above mean sea level of the earth station (km)
θ: Elevation angle
φ: Absolute value of latitude of the earth station (degrees)
f: Frequency (GHz)

The geometry is illustrated in Figure 11-U.1.

Step 1: Calculate the effective height, h_R, for the latitude of the station φ:

$$h_R(\text{km}) = \begin{cases} 3.0 + 0.028\varphi, & 0 \leq \varphi < 36° \\ 4.0 - 0.075(\varphi - 36), & \varphi \geq 36° \end{cases} \tag{1}$$

Step 2: For $\theta \geq 5°$ compute the slant-path length, L_s, below the rain height from

$$L_s = \frac{(h_R - h_s)}{\sin\theta} \quad \text{km} \tag{2}$$

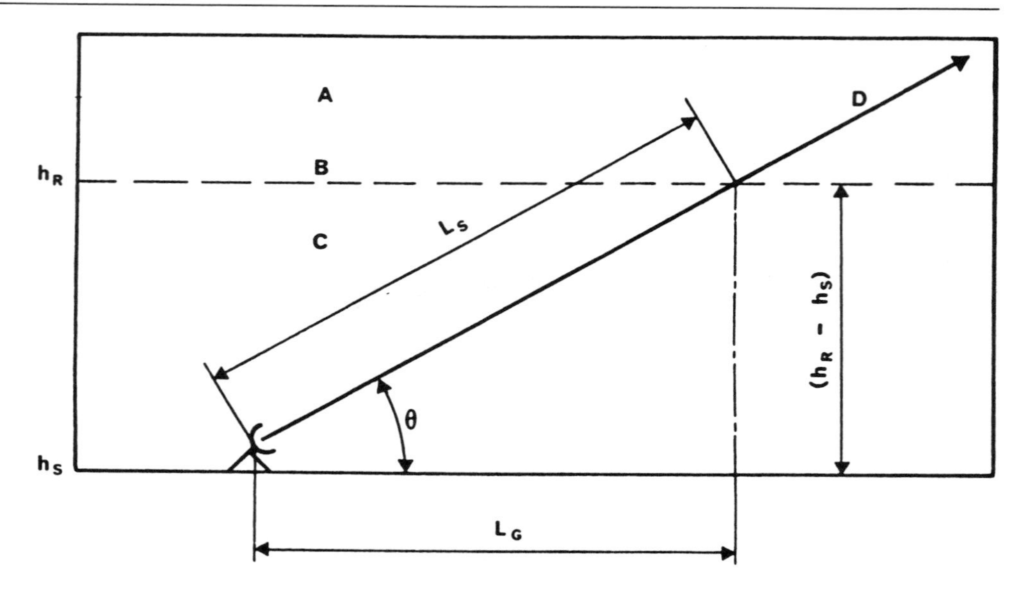

A: frozen precipitation

B: rain height

C: liquid precipitation

D: Earth-space path

FIGURE 11-U.1. Schematic presentation of an earth–space path giving the input parameters for the attenuation prediction process. (From Figure 1, ITU-R Rec. 618-2, ITU-R, Geneva, 10/92.)

For $\theta < 5°$ the following formula is used:

$$L_s = \frac{2\,(h_R - h_s)}{\left[\sin^2\theta + \dfrac{2\,(h_R - h_s)}{R_e}\right]^{1/2} + \sin\theta} \quad \text{km} \tag{3}$$

Step 3: Calculate the horizontal projection, L_G, of the slant-path length from

$$L_G = L_s \cos\theta \qquad \text{km} \tag{4}$$

Step 4: Obtain the rain intensity, $R_{0.01}$, exceeded for 0.01% of an average year (with an integration time of 1 min). If this information cannot be obtained from local data sources, an estimate can be obtained from the maps of rain climates given in Figures 11-40, 11-41, and 11-42.

Step 5: Calculate the reduction factor, $r_{0.01}$, for 0.01% of the time for $R_{0.01} \leq 100$ mm/hr:

$$r_{0.01} = \frac{1}{1 + L_G/L_0} \tag{5}$$

where $L_0 = 35\ \exp(-0.015\ R_{0.01})$. For $R_{0.01} > 100$ mm/hr, use the value 100 mm/hr in place of $R_{0.01}$.

Step 6: Obtain the specific attenuation, γ_R, using the frequency-dependent coefficients given in Table 11-8 and the rainfall rate, $R_{0.01}$, determined from Step 4, by using

$$\gamma_R = k(R_{0.01})^{\alpha} \qquad \text{dB/km} \tag{6}$$

Step 7: The predicted attenuation exceeded for 0.01% of an average year is obtained from

$$A_{0.01} = \gamma_R L_s r_{0.01} \qquad \text{dB} \tag{7}$$

Step 8: The estimated attenuation to be exceeded for other percentages of an average year, in the range 0.001–1%, is determined from the attenuation to be exceeded for 0.01% for an average year by using

$$\frac{A_p}{A_{0.01}} = 0.12 p^{-(0.546 + 0.043 \log p)} \tag{8}$$

This interpolation formula gives factors of 0.12, 0.38, 1, and 2.14 for 1%, 0.1%, 0.01%, and 0.001%, respectively.

Step 9: If desired, the value of p corresponding to a given value of A_p may be computed from the inverted form of equation (8):

$$p_R = 10^{11.628\left(-0.546 + \sqrt{0.298 + 0.172 \log(0.12 \cdot A_{0.01}/A_p)}\right)} \tag{9}$$

with the constraint that

$$\frac{A_{0.01}}{A_p} \geq 0.15 \tag{10}$$

When the complete prediction method above was tested using the procedure set out in Annex 1 to ITU-R Rec. 311, the results differed between high and low latitudes. For latitudes above 30°, the prediction was found to be in good agreement with available measurement data in the range 0.001–0.1%, with a standard deviation of some 35%, when used with concurrent rain rate measurements.

This method provides an estimate of the long-term statistics of attenuation due to rain. When comparing measured statistics with the prediction, allowance should be given for the rather large year-to-year variability in rainfall rate statistics (see ITU-Rec. 678).

Attenuation due to water cloud or fog of known liquid water content can be calculated from Rec. 840. Except for clouds of high water content, total attenuations due to cloud will not be great at frequencies below 30 GHz and are likely in any case to be included in the measured statistics. The effects of ice cloud, dry hail, and dry snow can be generally neglected for these frequencies.

As the frequency increases, the effect of clouds becomes steadily more important. However, recent measurements show that in some types of cloud (e.g., strato-cumulus) the additional attenuation produced at vertical incidence is no more than 0.5–1 dB, even at 150 GHz. On the other hand, clouds with a high liquid-water content, such as cumulo-nimbus, can cause an additional attenuation of about 4–5 dB at 100 GHz and up to 8 dB at 150 GHz.

Long-Term Frequency and Polarization Scaling of Rain Attenuation Statistics. The method described in the previous subsection may be used to investigate the dependence of attenuation statistics on elevation angle, polarization, and frequency and is therefore a useful general tool for scaling of attenuation according to these parameters.

If reliable attenuation data measured at one frequency are available, the following empirical formula giving an attenuation ratio directly as a function of frequency and attenuation may be applied for frequency scaling on the same path in the frequency range 7–50 GHz:

$$A_2 = A_1(\varphi_2/\varphi_1)^{1-H(\varphi_1,\varphi_2,A_1)} \tag{11}$$

where

$$\varphi(f) = \frac{f^2}{1 + 10^{-4}f^2} \tag{12a}$$

$$H(\varphi_1, \varphi_2, A_1) = 1.12 \times 10^{-3}(\varphi_2/\varphi_1)^{0.5}(\varphi_1 A_1)^{0.55} \tag{12b}$$

A_1 and A_2 are the equiprobable values of the excess rain attenuation at frequencies f_1 and f_2 (GHz), respectively.

Frequency scaling from reliable attenuation data is preferred, when applicable, rather than the prediction methods starting from rain data.

The radiometeorological basis of frequency scaling is discussed in ITU-R Report 721, where another method that makes use of the k and α constants is presented that is more appropriate where polarization scaling is required.

Seasonal Variations, Worst Month. System planning often requires the attenuation value exceeded for a time percentage, p_w, of the worst month. The following procedure is used to estimate the attenuation exceeded for a specified percentage of the worst month.

Step 1: Obtain the annual time percentage, p, corresponding to the desired worst-month time percentage, p_w, by using the equation specified in ITU-R Rec. 581 and by applying any adjustments to p as prescribed therein.

Step 2: For the path in question obtain the attenuation, A (dB), exceeded for the resulting annual time percentage, p, from the method described in the first subsection in Section 11-7.2.4 or from measured or frequency-scaled attenuation statistics. This value of A is the estimated attenuation for p_w percent of the worst month.

Curves giving the variation of worst-month values from their mean are provided in Section 11-7.5.

Variability in Space and Time Statistics. Precipitation attenuation distributions measured on the same path at the same frequency and polarization may show marked year-to-year variations. In the range 0.001–0.1% of the year, the attenuation values at a fixed probability level are observed to vary by more than 20% rms. When the models for attenuation prediction or scaling described in the first subsection in Section 11-7.2.4 are used to scale observations at a location to estimate for another path at the same location, the variations increase to more than 25% rms.

Site Diversity. Intense rain cells that cause large attenuation values on an earth–space link often have horizontal dimensions of no more than a few kilometers. Diversity systems able to re-route traffic to alternate earth stations, or with access to a satellite with extra on-board resources available for temporary allocation, can improve the system reliability considerably.

Two concepts exist for characterizing diversity performance: The "diversity improvement factor" is defined as the ratio of the single-site time percentage and the diversity time percentage, at the same attenuation level. "Diversity gain" is the difference (dB) between the single-site and diversity attenuation values for the same time percentage. Both parameters are important, depending on the system design approach, and prediction procedures for both are given below.

The procedures have been tested at frequencies between 10 and 30 GHz, which is the recommended frequency range of applicability. The diversity prediction procedures are only recommended for time percentages less than 0.1%. At time percentages above 0.1%, the rainfall rate is generally small and the corresponding site diversity performance is not significant.

Diversity Improvement Factor. The diversity improvement factor, I, is given by

$$I = \frac{p_1}{p_2} = \frac{1}{(1+\beta^2)}\left[1 + \frac{100\,\beta^2}{p_1}\right] \approx 1 + \frac{100\,\beta^2}{p_1} \tag{13}$$

where p_1 and p_2 are the respective single-site and diversity time percentages, and β is a parameter depending on link characteristics. The approximation on the right-hand side of equation (13) is acceptable because β^2 is generally small.

From a large number of measures carried out in the 10-to-20-GHz band, and mainly between 11 and 13.6 GHz, it has been found that the value of β^2 depends basically on the distance between the stations, and only slightly on the angle of elevation and the frequency. It is found that β^2 can be expressed by the following empirical relationship:

$$\beta^2 = 10^{-4} d^{1.33} \tag{14}$$

Figure 11-U.2 shows p_2 versus p_1 on the basis of equations (13) and (14).

Diversity Gain. The diversity gain, G (dB), between pairs of sites is calculated with the empirical expression given below. Parameters required for the calculation of diversity gain are:

d: Separation (km) between the two sites
A: Path rain attenuation (dB) for a single site
f: Frequency (GHz)
θ: Path elevation angle (degrees)
Ψ: Angle (degrees) made by the azimuth of the propagation path with respect to the baseline between sites, chosen such that $\Psi \leq 90°$

Step 1: Calculate the gain contributed by the spatial separation from

$$G_d = a(1 - e^{-bd}) \tag{15}$$

where

$$a = 0.78A - 1.94\,(1 - e^{-0.11A})$$

$$b = 0.59\,(1 - e^{-0.1A})$$

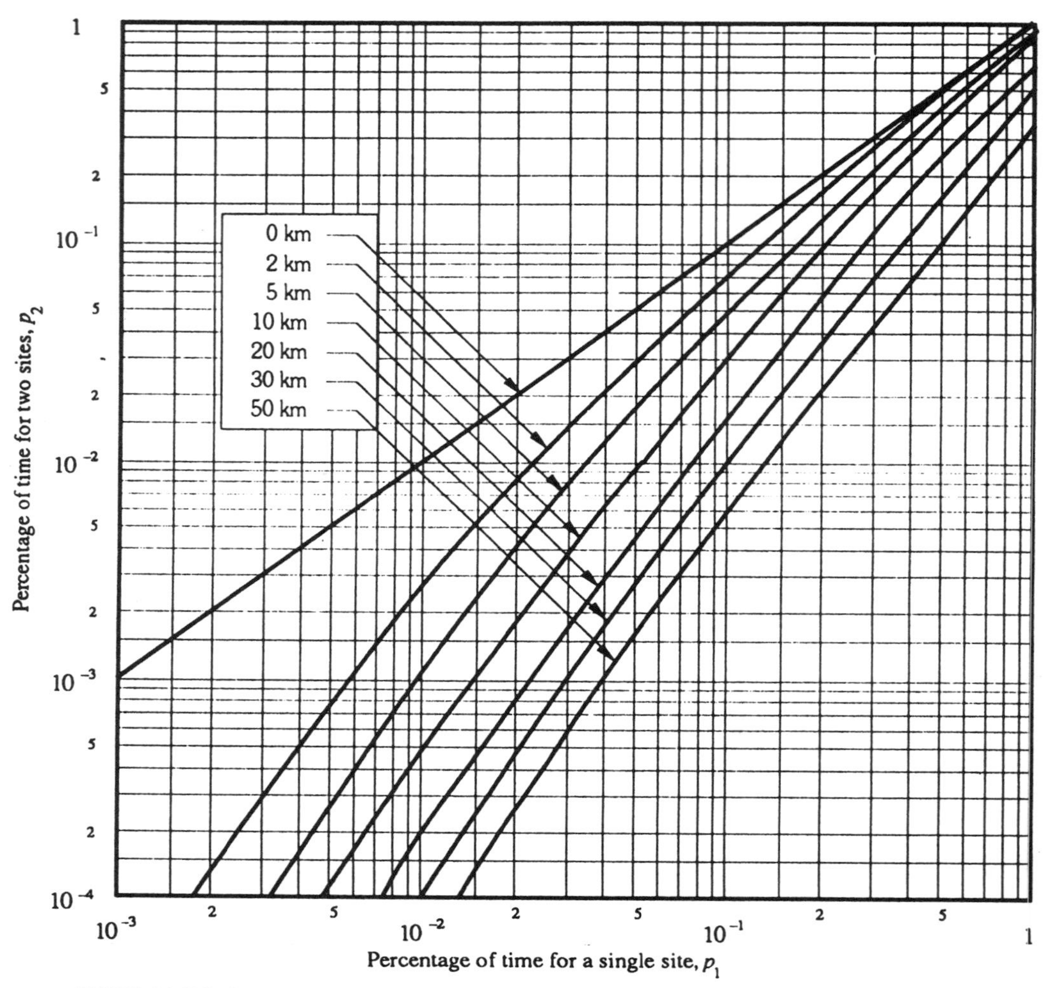

FIGURE 11-U.2. Relationship between percentages of time with and without diversity for the same attenuation (earth – satellite paths). (From Figure 2, ITU-R Rec. 618-2, 10/92, page 254.)

Step 2: Calculate the frequency-dependent gain from

$$G_f = e^{-0.025f} \tag{16}$$

Step 3: Calculate the term dependent gain on elevation angle from

$$G_\theta = 1 + 0.006\theta \tag{17}$$

Step 4 Calculate the baseline-dependent term from the expression

$$G_\Psi = 1 + 0.002\Psi \tag{18}$$

Step 5: Compute the net diversity gain as the product

$$G = G_d \cdot G_f \cdot G_\theta \cdot G_\Psi \quad \text{dB} \tag{19}$$

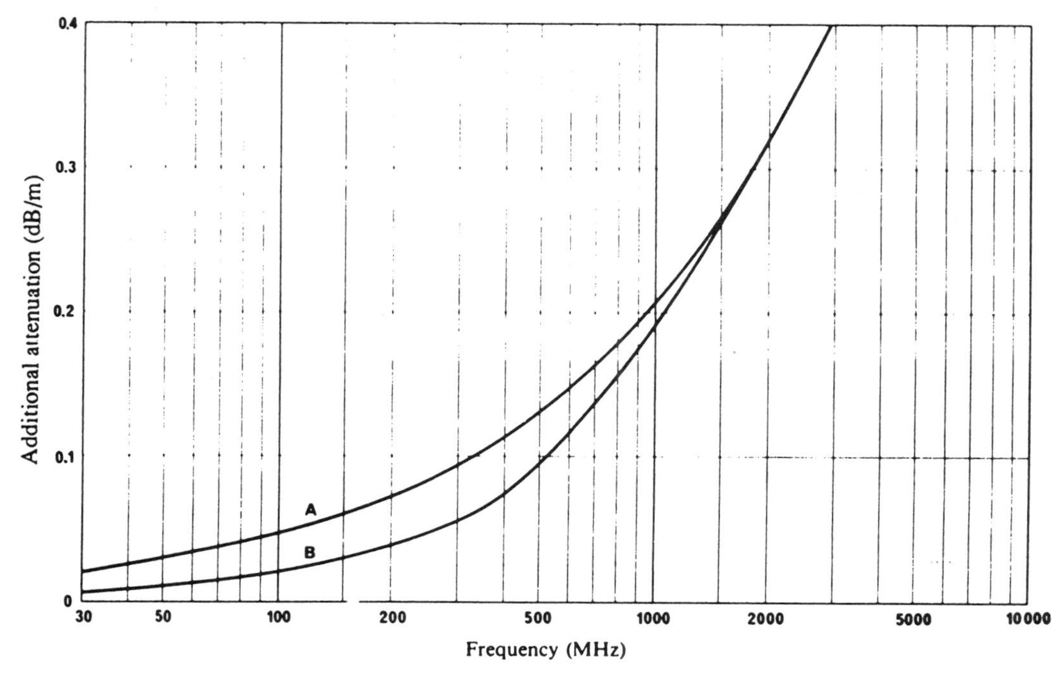

Note 1 – The data refer only to the additional attenuation caused by woodland to a ray passing through it and represent an approximate average for all types of woodland at frequencies up to about 3 000 MHz. When the attenuation inside the wood becomes large (for example more than 30 dB), the possibility of diffraction or surface-wave modes have to be considered.

FIGURE 11-U.3. Specific attenuation of woodland. (From Figure 1, ITU-R Rec. 833, Geneva, 10/92.)

When the above method was tested against the data in the CCIR site diversity data bank, the arithmetic mean and standard deviation were found to be 0.14 dB and 0.96 dB, respectively, with an rms error of 0.97 dB.

Section 11-7.2.4 is based on ITU-R Rec. 618-2, Geneva, 9/92.

11-7.3. Attenuation in Vegetation

Use Figure 11-U.3 for evaluating attenuation through woodland at VHF and UHF.

11-7.4. Signal Reception Inside Buildings

Representative UHF satellite signal attenuation observed within rooms located near an exterior wall in timber-framed private homes is summarized in Table 11-U.1. For interior rooms, 0.6 dB must be added to the tabulated values. For timber-frame buildings the attenuation shows little variation with weather or path elevation angle, but as the table illustrates, there is a systematic variation with frequency, polarization, construction materials, insulation, and position within the structure. Some aluminum-backed insulating and construction materials contribute up to 20 dB of loss.

TABLE 11-U.1
UHF Signal Attenuation (dB) Through Timber-Framed Buildings[a]

Building Condition		Frequency (MHz) and Polarization			
Exterior	Insulation (nonmetallic type)	860 H	860 V	1550 V	2569 V
Wood siding	Ceiling only	4.7	2.9	5.0	5.8
	Ceiling and wall	6.3	4.5	6.6	7.4
Brick veneer	Ceiling only	5.9	4.1	6.2	7.0
	Ceiling and wall	7.5	5.7	7.8	8.6

[a]The table is for rooms located near to the exterior wall; for interior rooms, 0.6 dB should be added.

Source: Table 2, ITU-R Rec. 679-1, Geneva, 9/92 (RPN Series).

11-7.5. Conversion of Annual Statistics to Worst-Month Statistics

The following is a model that provides for the conversion of the *annual* to the *worst-month* statistics:

1. The average annual worst-month time percentages in excess of p_w is calculated from the annual time percentage of excess p by use of the conversion factor Q.

$$p_w = Qp \qquad (1)$$

where $1 < Q < 12$, and both p and p_w refer to the same threshold levels.

2. Q is a two-parameter (Q_1, β) function of p (%):

$$Q(p) = \begin{cases} 12 & \text{for } p < \left(\dfrac{Q_1}{12}\right)^{1/\beta} \% \\ Q_1 p^{-\beta} & \text{for } \left(\dfrac{Q_1}{12}\right)^{1/\beta} < p < 3\% \\ Q_1 3^{-\beta} & \text{for } 3\% < p < 30\% \\ Q_1 \left(\dfrac{p}{30}\right)^{\log(Q_1 3^{-\beta})/\log(0,3)} & \text{for } 30\% < p \end{cases} \qquad (2)$$

3. The calculation of the average annual time percentage of excess from the given value of the average annual worst-month time percentage of excess is done through the inverse relationship:

$$p = p_w / Q \qquad (3)$$

and the dependence of Q on p_w can be easily derived from the above given dependence of Q on p. The resulting relationship for $12p_0 < p_w(\%) < Q_1 3^{(1-\beta)}$ is $(p_0 = (Q_1/12)^{1/\beta})$

$$Q = Q_1^{1/(1-\beta)} p_w^{-\beta(1-\beta)} \qquad (4)$$

4. For global planning purposes the following values for the parameters Q_1 and β should be used:

$$Q_1 = 2.85, \qquad \beta = 0.13$$

This leads to the following relationship between p and p_w:

$$p(\%) = 0.30\, p_w(\%)^{1.15} \tag{5}$$

for $1.9 \times 10^{-4} < p_w(\%) < 7.8$. Figure 11-U.4 provides an example of the dependence of Q on p.

5. For more precision the values of Q_1 and β for different climatic regions and various propagation effects given in Table 11-U.2 should be used where appropriate.
6. For trans-horizon mixed paths, β and Q_1 values are calculated from those values for sea and land given in Table 11-U.2, through linear interpolation using fractions of the link traversing sea and land, respectively, as weights.

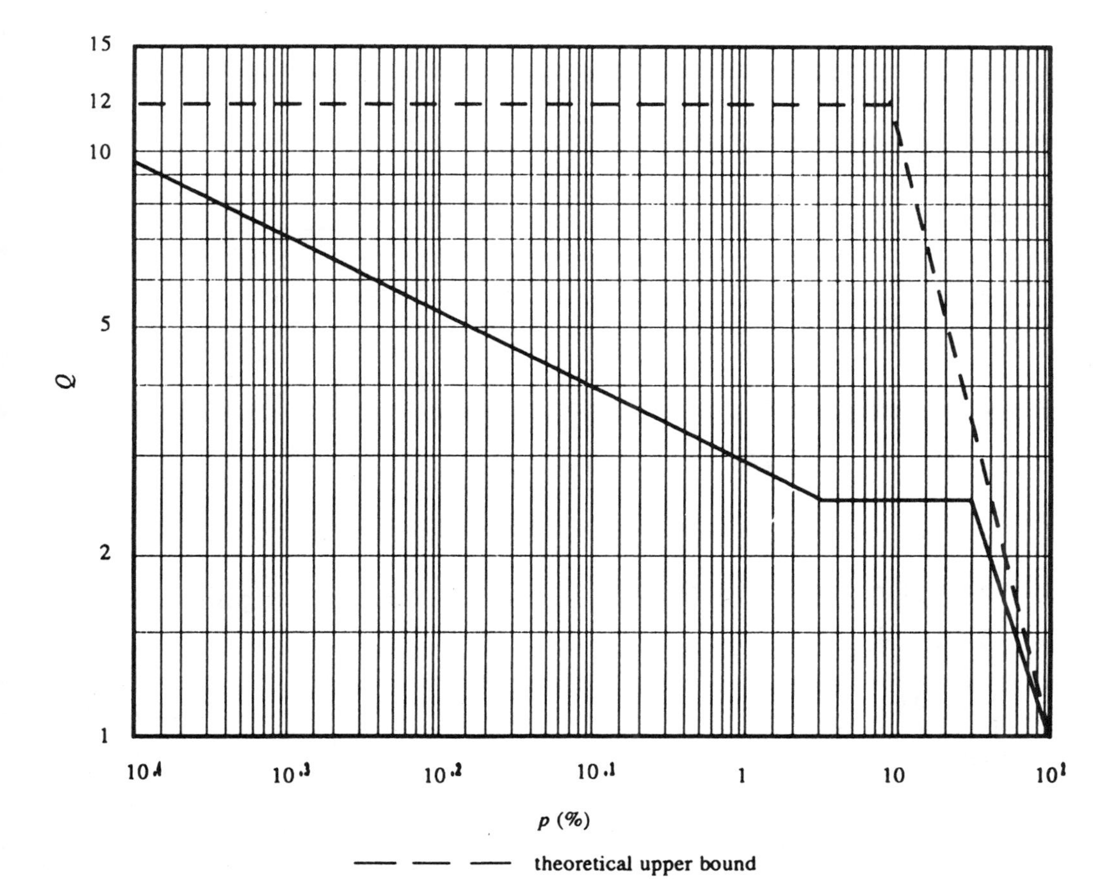

FIGURE 11-U.4. Example of the dependence of Q on p (solid line) with parameters $Q_1 = 2.85$ and $\beta = 0.13$. (From Figure 1, ITU-R Rec. 841, Geneva, 10/92, page 211.)

TABLE 11-U.2

β and Q_1 Values for Various Propagation Effects and Locations

Location	Rain Effect Terrestrial	Rain Effect Slant Path	Multipath	Trans-horizon Land	Trans-horizon Sea
Global	0.13; 2.85	0.13; 2.85	0.13; 2.85	0.13; 2.85	0.13; .85
CANADA Prairie & North	0.08; 4.3				
CANADA Coast & G. Lake	0.10; 2.7				
CANADA Cent. & Mount.	0.13; 3.0				
USA Virginia		0.15; 2.7			
JAPAN Tokyo	0.20; 3.0				
JAPAN Yamaguchi		0.15; 4.0			
JAPAN Kashima		0.15; 2.7			
CONGO	0.25; 1.5				
EUROPE Northwest	0.13; 3.0	0.16; 3.1	0.13; 4.0	0.18; 3.3	0.11; 5.0
EUROPE Mediterranean	0.14; 2.6	0.16; 3.1			
EUROPE Nordic	0.15; 3.0	0.16; 3.8	0.12; 5.0		
EUROPE Alpine	0.15; 3.0	0.16; 3.8			
EUROPE Poland	0.18; 2.6				
EUROPE Russia	0.14; 3.6				
INDONESIA	0.22; 1.7				

Source: Table 1, ITU-R Rec. 841, Geneva, 10/92, page 212.
This section is based on ITU-R Rec. 841, RPN Series, Geneva, 10/92.

11-7.6. Ionospheric Effects Influencing Radio Systems Involving Spacecraft

11-7.6.1. Ionospheric Effects

A signal carrier which penetrates the ionosphere is modified by the medium due to the presence of electrons and the earth's magnetic field. Large-scale changes due to the variation of electron density, as well as smaller-scale irregularities, affect the carrier. The effects include scintillation, absorption, variation in the direction of the arrival, propagation delay, dispersion, frequency change, and polarization rotation. These effects on transmission, at frequencies mainly above about 20 MHz, are treated in this Annex.

11-7.6.2. Scintillation

Introduction. Scintillations, as discussed in this Annex, are variations of amplitude, phase, polarization, and angle of arrival produced when radio waves pass through electron density irregularities in the ionosphere. Ionospheric scintillations present themselves as fast fluctuations of signal level with peak-to-peak amplitude fluctuations from 1 dB to over 10 dB and lasting for several minutes to several hours. The phenomena are caused by one of two types of ionospheric irregularities:

- Sufficiently high electron density fluctuations at scale comparable to the Fresnel zone dimension of the propagation path
- Sharp gradients of ambient electron density, especially in the direction transverse to the direction of propagation

Either type of irregularity is known to occur in the ionosphere under certain solar, geomagnetic, and upper atmospheric conditions, and the scintillations can become so severe that they represent a practical limitation for communication systems. Scintillations have been observed at frequencies from about 10 MHz to about 12 GHz.

For systems applications, scintillations can be characterized by the depth of fading and fading period. A useful index to quantify the severity of scintillation is the scintillation index, S_4, which is defined as the standard deviation of received power divided by the mean value of the received power—that is,

$$S_4^2 = \frac{< L^2 > - < L >^2}{< I >^2} \tag{1}$$

where I is the carrier intensity and $< >$ denotes ensemble average.

The fading period of scintillation varies over quite a large range from less than one-tenth of 1 sec to several minutes, because the fading period depends upon the apparent motion of the irregularities relative to the ray path and, in the case of strong scintillation, on its severity. The fading period of gigahertz scintillation ranges from approximately 1 to 10 sec. Long-period (on the order of tens of seconds) components of saturated scintillation (S_4 approaches 1) at VHF and UHF bands have also been observed.

Modeling/Scaling Rules for System Applications. Ionospheric scintillations exhibit a wide range of variations in frequency dependence, morphology patterns, and diurnal, seasonal, and solar cycle dependence. Different signal statistics have been found in different observations. An enormous amount of literature is available, and new findings based on refined measurement techniques and modeling methodology appear each year. System engineers are advised to use reliable/relevant published data for applications. If direct data and/or findings are not available or applicable, the modeling/scaling rules in the following sections should be used.

11-7.6.3. Frequency Dependence of Scintillation

If results from direct measurement are not available, an $f^{-1.5}$ frequency dependence on S_4 is recommended for engineering applications.

11-7.6.4. Instantaneous Statistics and Spectrum Behavior

Instantaneous Statistics. During an ionospheric scintillation event, the Nakagami density function is believed to be adequately close for describing the statistics of the instantaneous variation of amplitude. The density for the intensity of the signal is given by

$$p(I) = \frac{m^m}{\Gamma(m)} I^{m-1} \exp(-mI) \tag{2}$$

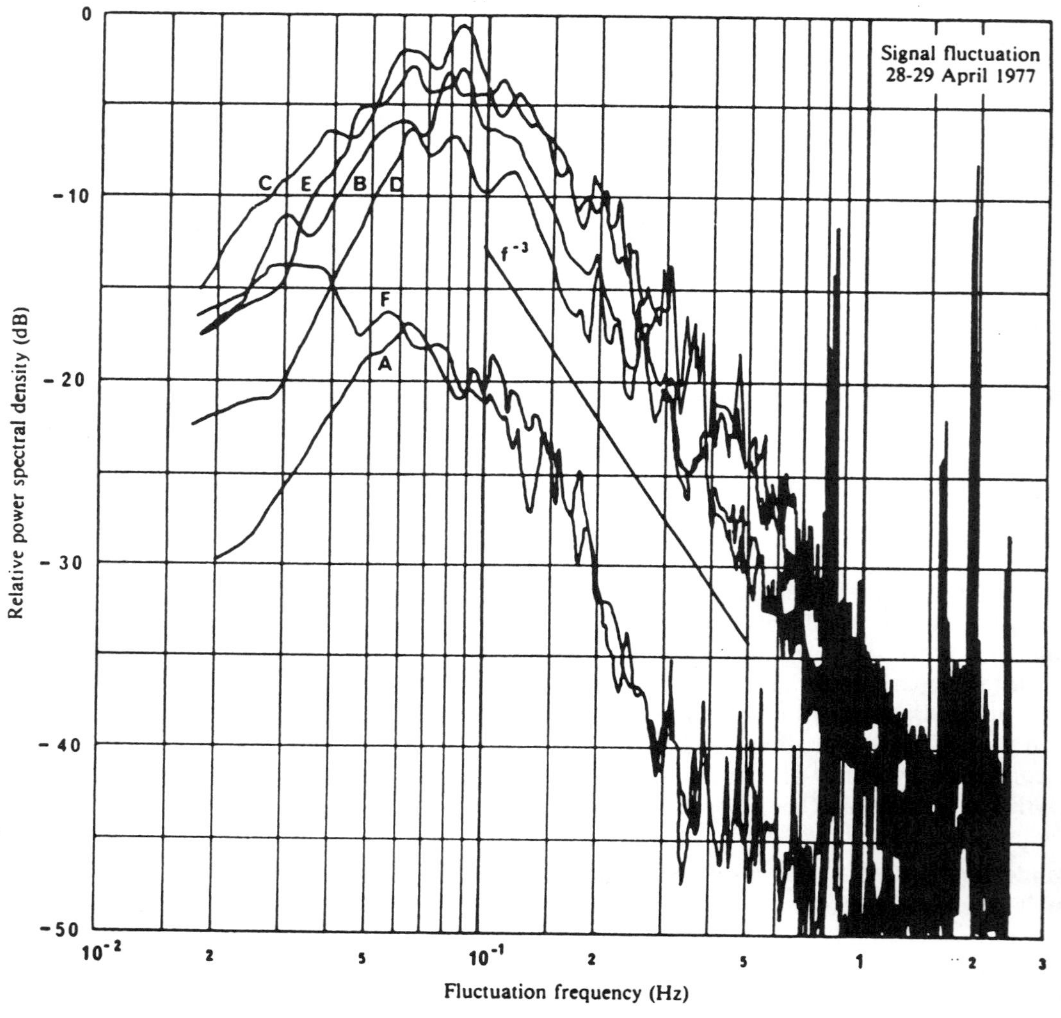

The scintillation event was observed during the evenings of 28-29 April 1977 at Taipei earth station.

A: 30 min before event onset
B: at the beginning
C: 1 h after
D: 2 h after
E: 3 h after
F: 4 h after

FIGURE 11-U.5. Power spectral density estimates for a geostationary satellite (INTELSAT IV) at 4 GHz. (From Figure 1, ITU-R Rec. 531-2, RPN Series, Geneva, 10/92, page 79.)

where the Nakagami "m-coefficient" is related to the scintillation index, S_4, by

$$m = 1/S_4^2 \tag{3}$$

In formulating equation (2) the average intensity level of I is normalized to be 1.0. The calculation of the fraction of time that the signal is above or below a given threshold is greatly facilitated by the fact that the distribution function corresponding to the Nakagami density has a closed form expression which is given

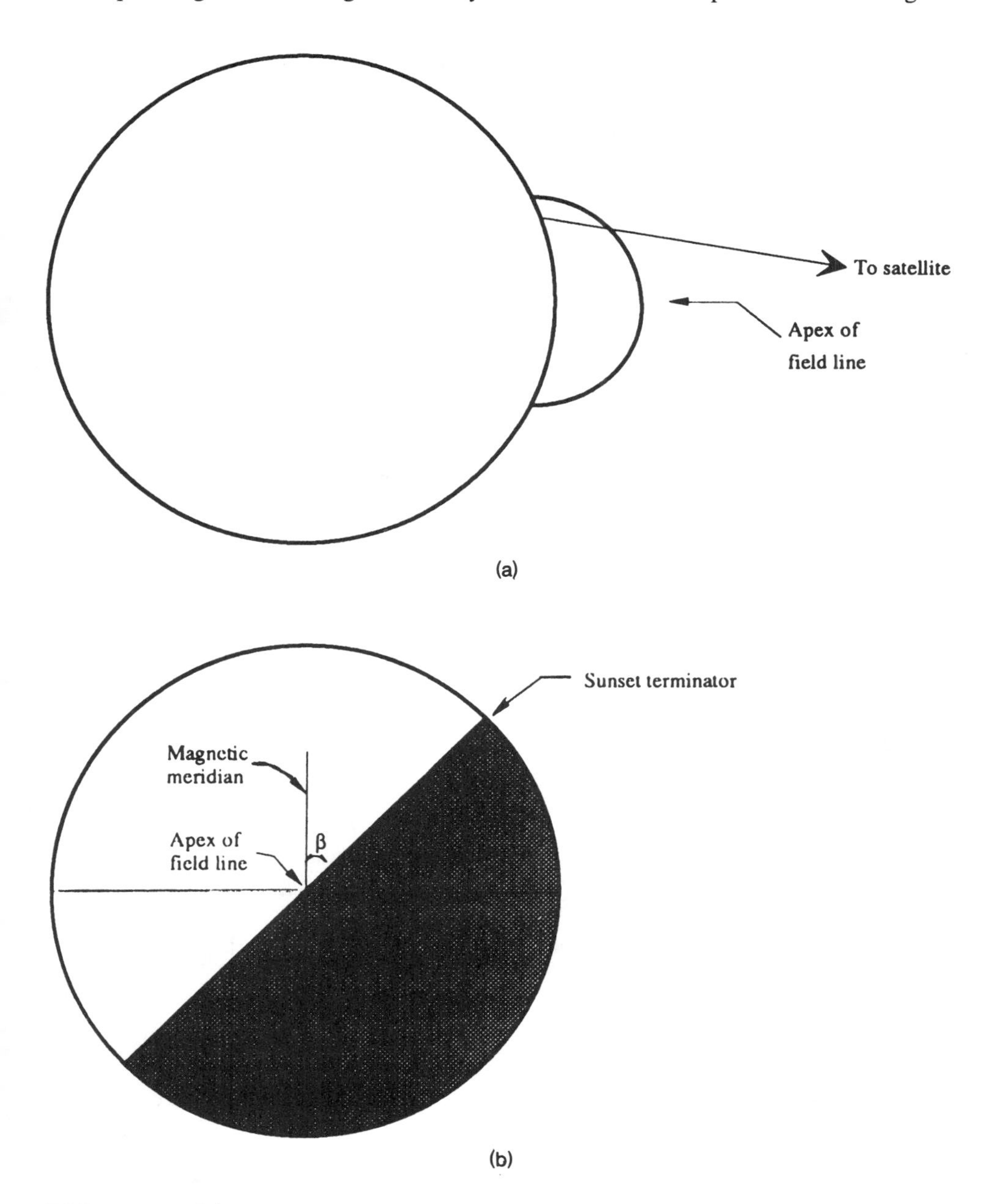

FIGURE 11-U.6. (a) The intersection of the propagation path with a magnetic field line at the F-region height. (b) The angle between the local magnetic meridian at the apex of the field line shown in part a and the sunset terminator. (Figures a and b are taken from Figures 2a and 2b of ITU-R Rec. 531-2, RPI Series, Geneva, 10/92.)

by

$$P(I) = \int_0^I p(x)\, dx = \frac{\Gamma(m, mI)}{\Gamma(m)} \tag{4}$$

where $\Gamma(m, mI)$ and $\Gamma(m)$ are the incomplete gamma function and gamma function, respectively. Using equation (4), it is possible to compute the fraction of time that the signal is above or below a given threshold during an ionospheric event. For example, the fraction of time that the signal is more than X dB below the mean is given by $P(10^{-X/10})$ and the fraction of time that the signal is more than X dB above the mean is given by $1 - P(10^{-X/10})$.

Spectrum Behavior. Because ionospheric scintillations are believed to be caused by relatively stationary refractive-index irregularities moving horizontally past the radio wave path, the spatial and temporal power spectra are related by the drift velocity. The actual relationship depends on the irregularity composition (power spectra) and a number of other physical factors. As a result, the power spectra exhibit a wide range of slopes, from f^{-1} to f^{-6} as have been reported from different observations.

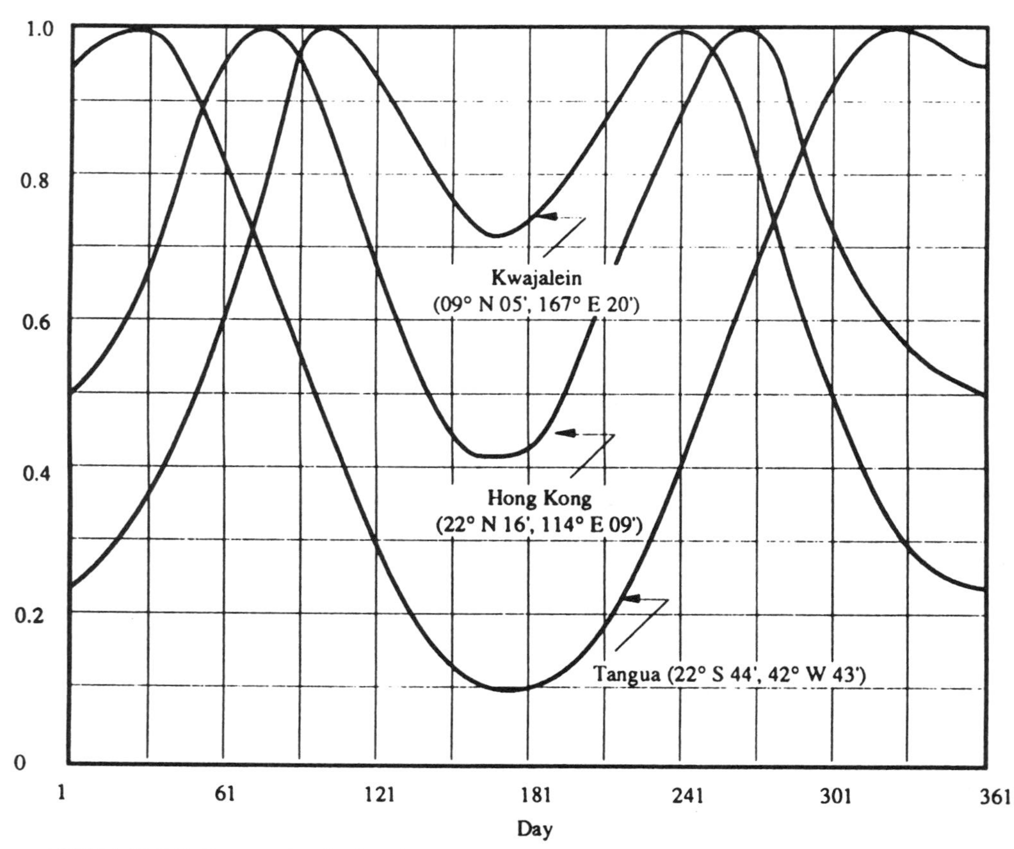

FIGURE 11-U.7. Seasonal weighting functions for stations in different longitude sectors. (From Figure 3, ITU-R Rec. 531-2, RPI Series, Geneva, 10/92, page 82.)

A typical spectrum behavior is shown in Figure 11-U.5. The f^{-3} slope as shown is recommended for system applications if direct measurement results are not available.

11-7.6.5. Geometric Considerations

Zenith Angle Dependence. In most models, S_4^2 is shown to be proportional to the secant of the zenith angle, i, of the propagation path. This relationship is believed to be valid up to $i \approx 70°$. At greater zenith angles, a dependence ranging between 1/2 and first power of sec i should be used.

Seasonal-Longitudinal Dependence. The occurrence of scintillations and magnitude of S_4 have a longitudinal as well as seasonal dependence that can be parameterized by the angle, β, shown in Figure 11-U.6. It is the angle between the sunset terminator and local magnetic meridian at the apex of the field line passing through the line-of-sight at the height of the irregularity slab. The weighting function for seasonal-longitudinal dependence is given by

$$S_4 \; \alpha \; \exp\left[-\frac{\beta}{W}\right] \tag{5}$$

where W is a weighting constant depending on location as well as calendar day of the year. As an example, using the data available from Tangua, Hong Kong, and Kwajalein, the numerical value of the weighting constant can be modeled as shown in Figure 11-U.7.

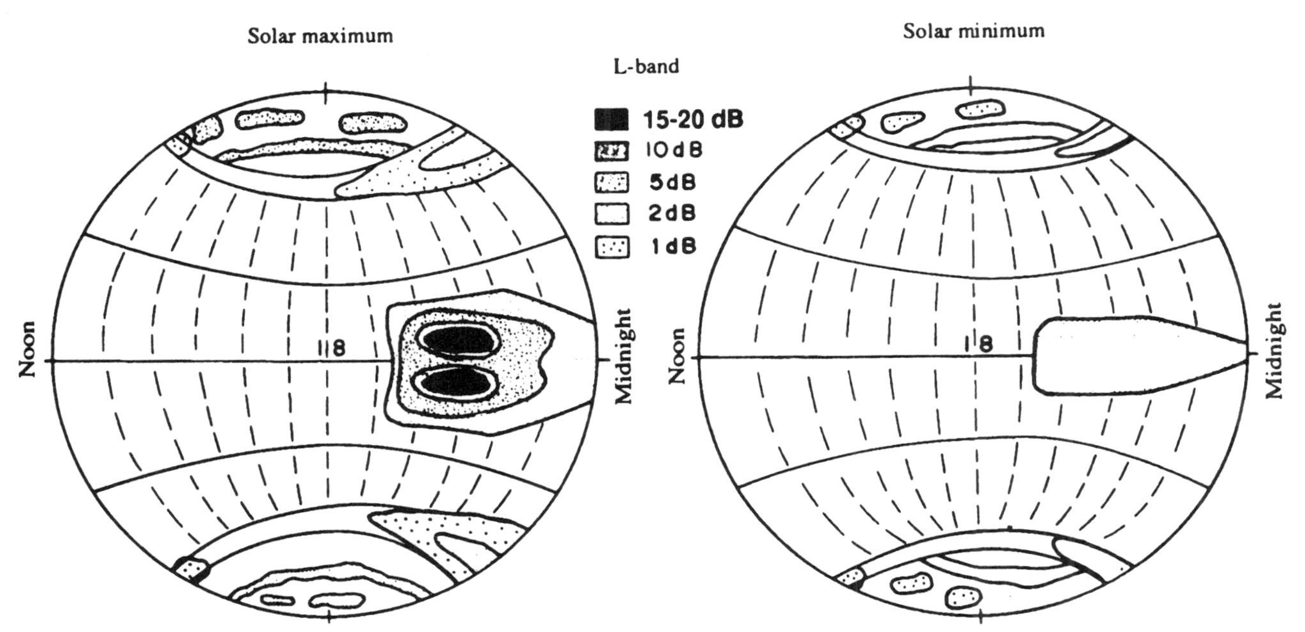

FIGURE 11-U.8. Illustrative picture of scintillation occurrence based on observations at L-band (1.6 GHz). (From Figure 4, ITU-R Rec. 531-2, RPI Series, Geneva, 10/92, page 82.)

11-7.6.6. Morphology

Scintillation can be categorized according to geomagnetic latitude, longitude, and local time. The latitudinal categorization consists of:

- Equatorial scintillations, which are caused by plasma bubbles in the equatorial anomaly regions
- Mid-latitude scintillations, which are correlated with the occurrence of ionospheric spread, F

TABLE 11-U.3
Solar-Geographical and Temporal Dependence of Ionospheric Scintillation

Parameter	Equatorial	Mid-latitude	High Latitude	
			Auroral	Polar
Activity level	Exhibits greatest extremes	Generally very quiet to moderately active	Generally moderately active to very active	Great intensity in high sunspot years
Diurnal	Maximum—nighttime Minimum—daytime	Maximum—nighttime Sporadic—daytime	Maximum—nighttime Maximum—daytime (not within polar cap)	Maximum—nighttime
Seasonal	Longitude-dependent Peaks in equinoxes Accra, Ghana Maximum—November and March Minimum—solstices Huancayo, Peru Maximum—October to March Minimum—May to July Kwajalein Islands Maximum—May Minimum—November and December	Maximum—spring Minimum—winter Tokyo, Japan Maximum—summer Minimum—winter	Pattern a function of longitude sector	Pattern a function of longitude sector
Solar cycle	Occurrence and intensity increase strongly with sunspot number	Tokyo, Japan Occurrence in night-time decreases with sunspot number; occurrence in daytime has little dependence	Occurrence and intensity increase strongly with sunspot number	Occurrence and intensity increase strongly with sunspot number
Magnetic activity	Longitudinal dependent Accra (Ghana) Occurrence decreases with K_p Huancayo, Peru March equinox Occurrence decreases with K_p June solstice Occurrence increases with K_p September equinox 0000–0400 hr (local time) Occurrence increases with K_p	Independent of K_p	Occurrence increases with K_p	Occurrence increases only slightly with K_p

Source: Table 1, ITU-R Rec. 531-2, RPI Series, Geneva, 10/92, page 83.

- High-latitude scintillations with effects that differ between the auroral and polar regions, though both are closely related to auroral activity and geomagnetic activities

A general sketch is provided in Figure 11-U.8 and a summary of scintillation characteristics is given in Table 11-U.3.

11-7.6.7. Cumulative Statistics

In design of ISDN and other radio systems, communications engineers have concerns not only with the system degradation during an event such as those described in sections 11-7.6.3, 11-7.6.4, and 11-7.6.5, but also with the long-term

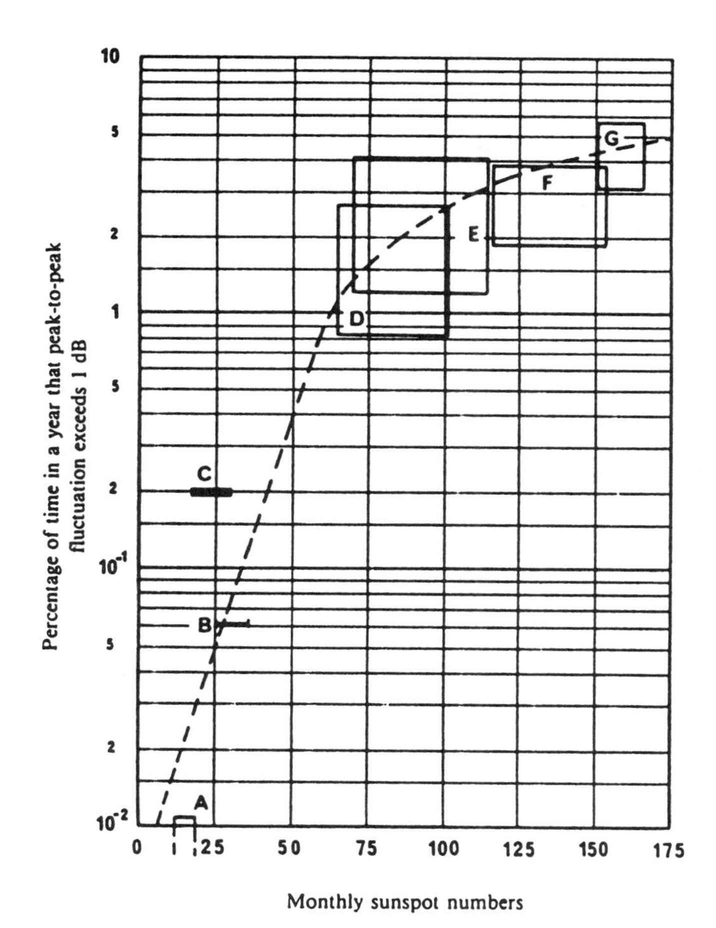

The squares are the ranges of variations over a year for different carriers

A: 1975-1976, Hong Kong and Bahrein, 15 carriers
B: 1974, Longovilo, 1 carrier
C: 1976-1977, Taipei, 2 carriers
D: 1970-1971, 12 stations, > 50 carriers
E: 1977-1978, Hong Kong, 12 carriers
F: 1978-1979, Hong Kong, 10 carriers
G: 1979-1980, Hong Kong, 6 carriers

FIGURE 11-U.9. Dependence of 4-GHz equatorial ionospheric scintillation on monthly mean sunspot number. (From Figure 5, ITU-R Rec. 531-2, RPI Series, Geneva, 10/92, page 84.)

cumulative occurrence statistics. For communications systems involving a geostationary satellite, which is the simplest radio system configuration, Figure 11-U.9 is recommended for the assessment and scaling of occurrence statistics. The sunspot numbers (SSN) cited are the 6-month averaged sunspot numbers available from the NTIA/ITS of the United States of America.

11-7.6.8. Simultaneous Occurrence of Ionospheric Scintillation and Rain Fading

Ionospheric scintillation and rain fading are two impairments of completely different physical origin. However, in equatorial regions at years of high sunspot number, the simultaneous occurrence of the two effects may have an annual percentage time that is significant to system design. The cumulative simultaneous occurrence time was about 0.06% annually as noted at 4 GHz at Djutiluhar earth station in Indonesia. This value is unacceptably high for ISDN type of applications.

The simultaneous events have signatures that are often vastly different from those when only a single impairment, either scintillation or rain alone, is present. While ionospheric scintillation alone is not a depolarization phenomenon, and rain fading alone is not a signal fluctuation phenomenon, the simultaneous events produce a significant amount of signal fluctuations in the cross-polarization channel. Recognition of these simultaneous events is needed for applications to satellite–earth radio systems which require high availability.

This section is based on ITU-R Rec. 531-2, RPI Series, Geneva 10/92.

13

Radio Systems

Add the following text to Section 13-2.1.1 (page 885) just above the subsection "Channel Arrangements for High- and Medium-to-High Capacity Digital Systems."

ITU-R Rec. 746 gives the following guidelines for radio-frequency channel arrangements for radio-relay (LOS microwave) systems.

Homogeneous patterns are preferred as the basis for radio-frequency channel arrangements. The preferred radio-frequency channel arrangements should be developed from the homogeneous pattern in accordance with the co-channel and alternated radio-frequency channel arrangements as shown in Figure 13-U.1.

The choice of radio-frequency channel arrangement depends on the values of cross-polar discrimination XPD, where XPD is defined as

$$\text{XPD}_{H(V)} = \frac{\text{Power received in polarization } H(V) \text{ transmitted in polarization } H(V)}{\text{Power received in polarization } H(V) \text{ transmitted in polarization } V(H)}$$

where the transmitted powers in polarizations $H(V)$ and $V(H)$ are equal.

The XPD parameter contributes to the value of carrier-to-interference ratio.

If $\text{XPD}_{\min}$ is the minimum value reached for the percentage time required, this value must be compared with the minimum value of carrier-to-interference $(C/I)_{\min}$ acceptable to the modulation adopted.

Co-channel arrangements can be used if

$$\text{XPD}_{\min} \geq (C/I)_{\min} \qquad \text{dB}$$

For alternated channel arrangements, an additional decoupling between adjacent radio-frequency channels NFD (net filter discrimination) can be obtained by using filter selectivity.

The condition to be satisfied is

$$\text{XPD}_{\min} + \text{NFD} \geq (C/I)_{\min} \qquad \text{dB}$$

The main parameters affecting the choice of radio-frequency channel arrangements are as follows:

- XS is defined as the radio-frequency separation between the center frequencies of adjacent radio-frequency channels on the same plane of polarization and in the same direction of transmission.

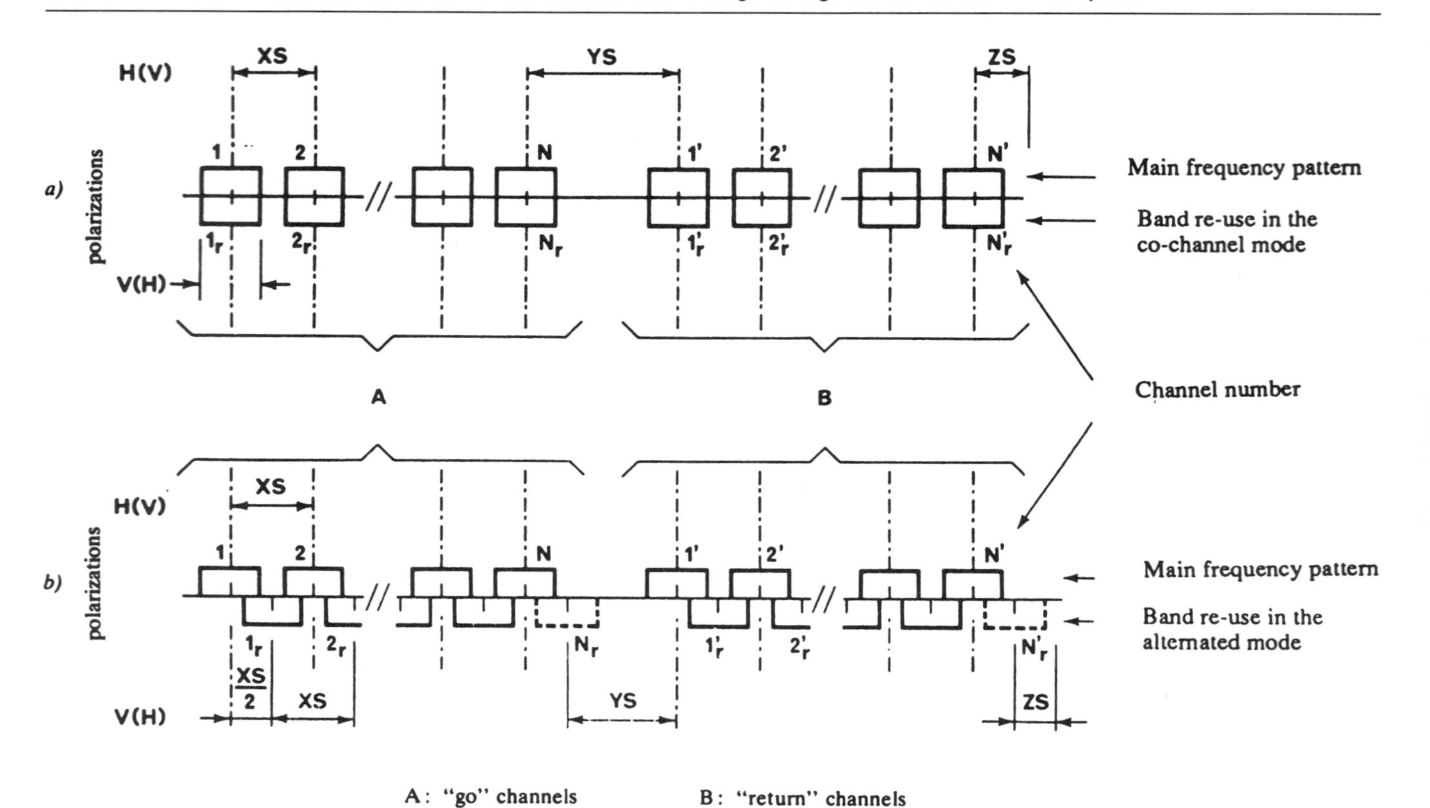

A: "go" channels B: "return" channels

FIGURE 13-U.1. Channel arrangements for the two possible schemes of frequency re-use considered in this subsection. (From Figure 1, ITU-R Rec. 746, RF Series, Geneva, 3/92, page 35.)

- YS is defined as the radio-frequency separation between the center frequencies of the go and return radio-frequency channels which are nearest to each other.
- ZS is defined as the radio-frequency separation between the center frequencies of the outermost radio-frequency channels and the edge of the frequency band; in the case where the lower and upper separations differ in value, Z_1S refers to the lower separation and Z_2S refers to the upper separation.

This section is based on ITU-R Rec. 746, RF Series, Geneva, 3/92.

This new section follows Section 13-2.1.1, which ends on page 889 of the Manual.

Radio-Frequency Channel Arrangements for Radio-Relay (Microwave) Systems Operating in the 18-GHz Band

Introduction. The preferred radio-frequency channel arrangement for digital radio-relay (LOS microwave) systems with a capacity on the order of 280 Mbps, 140 Mbps, and 34 Mbps operating in the 17.7- to 19.7-GHz band should be derived as follows: (1) Let f_0 be the frequency of the center of the band of frequencies occupied (MHz), (2) let f_n be the center of a radio-frequency channel in the lower half of the band (MHz), and (3) let f'_n be the center frequency of the radio-frequency channel in the upper half of the band. The frequencies (MHz) of individual channels are expressed by the following relationships:

Co-channel Arrangement. For systems with a capacity on the order of 280 Mbps we have:

Lower half of the band: $f_n = f_0 - 1110 + 220n$ MHz

Upper half of the band: $f'_n = f_0 + 10 + 220n$ MHz

where $n = 1, 2, 3,$ or 4. The frequency arrangement is illustrated in Figure 13-U.2a.

For systems with a capacity on the order of 140 Mbps we have

Lower half of the band: $f_n = f_0 - 1000 + 110n$ MHz

Upper half of the band: $f'_n = f_0 + 10 + 110n$ MHz

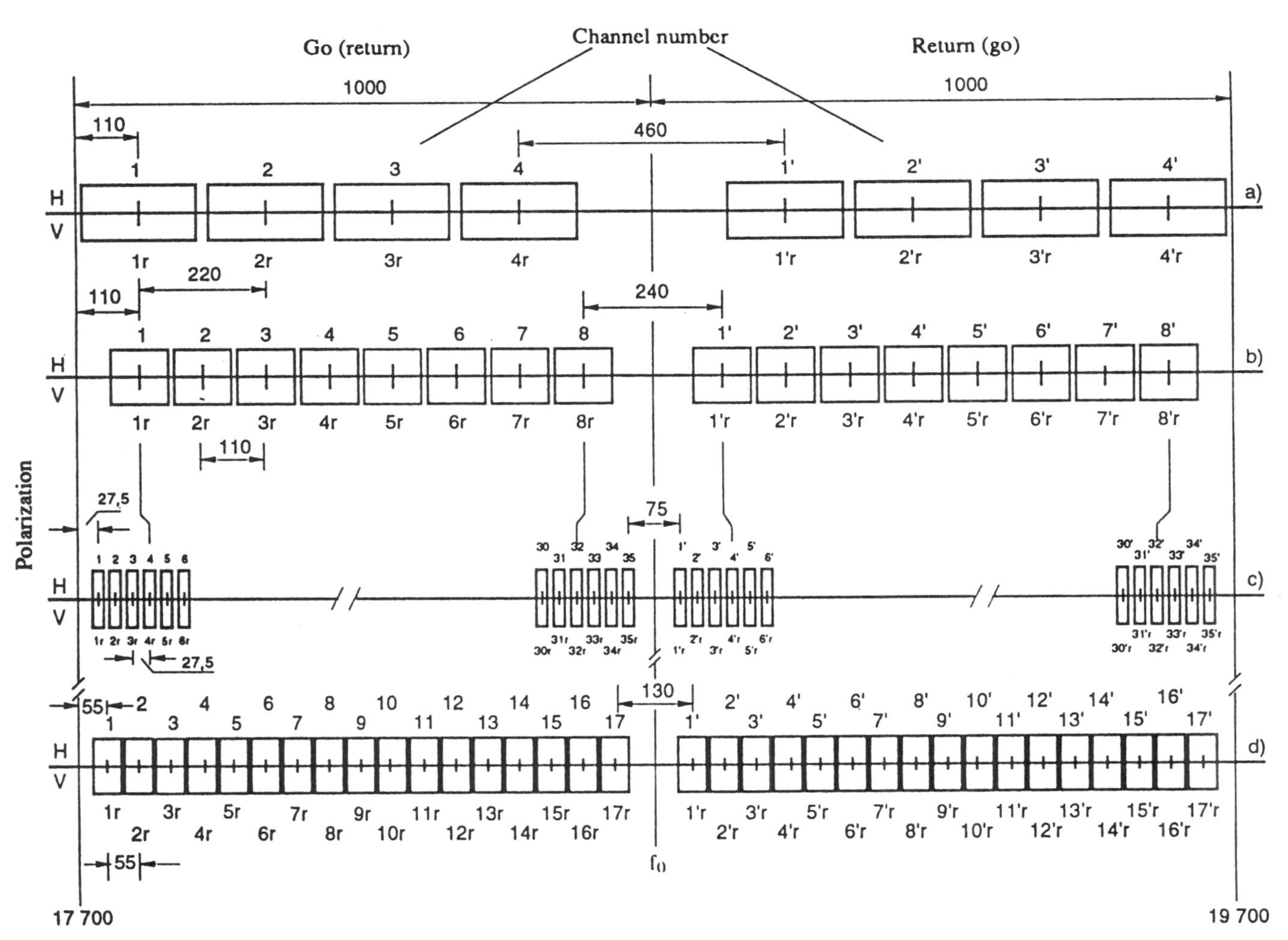

FIGURE 13-U.2. Radio-frequency channel arrangements for radio-relay systems operating in the 17.7- to 19.7-GHz band, co-channel arrangement. (From Figure 1, ITU-R Rec. 595-3, Geneva, 3/92, page 92.)

where: $n = 1, 2, 3, 4, 5, 6, 7$, or 8. The frequency arrangement is illustrated in Figure 13-U.2b.

For systems with a capacity of 34 Mbps we have

$$\text{Lower half of the band:} \quad f_n = f_0 - 1000 + 27.5n \quad \text{MHz}$$

$$\text{Upper half of the band:} \quad f'_n = f_0 + 10 + 27.5n \quad \text{MHz}$$

where $n = 1, 2, 3, \ldots, 35$. The frequency arrangement is illustrated in Figure 13-U.2c.

Interleaved Arrangement. For systems with a capacity on the order of 280 Mbps we have

$$\text{Lower half of the band:} \quad f_n = f_0 - 1000 + 110n \quad \text{MHz}$$

$$\text{Upper half of the band:} \quad f'_n = f_0 + 120 + 110\text{n} \quad \text{MHz}$$

where $n = 1, 2, 3, 4, 5, 6$, or 7. The frequency arrangement is illustrated in Figure 13-U.3a.

For systems with a capacity on the order of 140 Mbps we have

$$\text{Lower half of the band:} \quad f_n = f_0 - 945 + 55n \quad \text{MHz}$$

$$\text{Upper half of the band:} \quad f'_n = f_0 + 65 + 55n \quad \text{MHz}$$

where $n = 1, 2, 3, \ldots, 15$. The frequency arrangement is illustrated in Figure 13-U.3b.

The preferred radio-frequency channel arrangement for digital radio-relay systems with a capacity of 155 Mbps for use in the synchronous digital hierarchy is given above for the co-channel and interleaved channel arrangements for systems using 4-PSK-like modulation. For systems using 16-QAM-like modulation, the

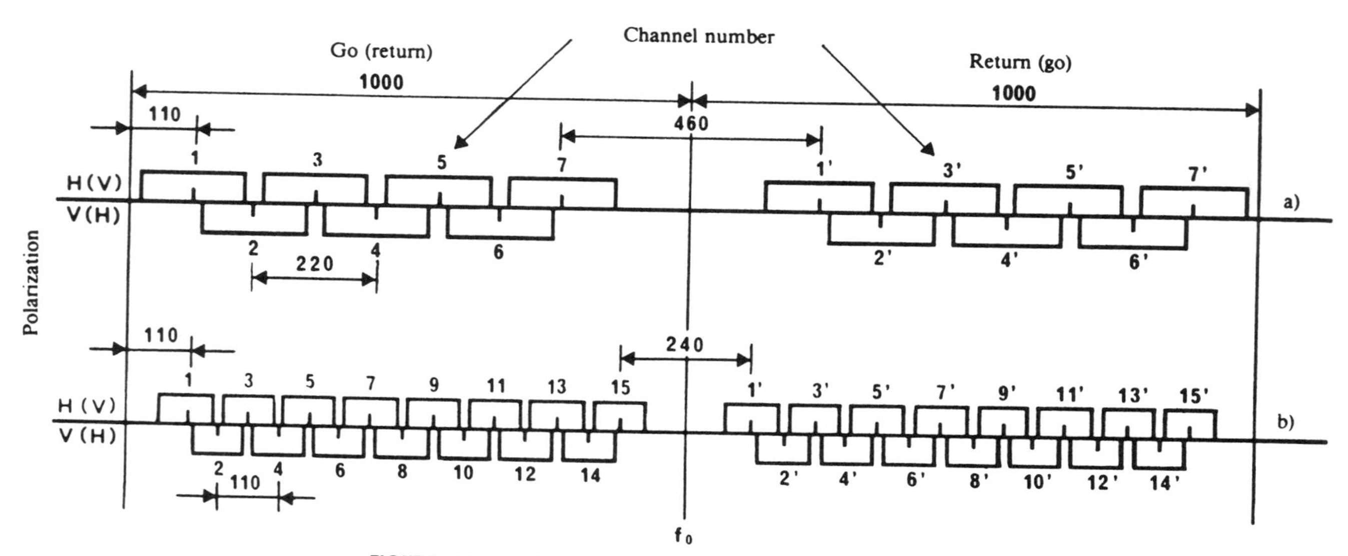

FIGURE 13-U.3. Radio-frequency channel arrangement for radio-relay systems operating in the 17.7- to 19.7-GHz band, interleaved arrangement. (From Figure 2, ITU-R Rec. 595-3, Geneva, 3/92, page 93.)

radio-frequency channel arrangement shown in Figure 13-U.2d is preferred for co-channel arrangement.

The frequencies of channels 2, 3, 4, . . . , 16 in Figure 13-U.2d are the same as the center frequencies for the interleaved arrangement (above) for channels 1, 2, 3, . . . , 15, respectively. Channel 1 and channel 17 in Figure 13-U.2d are allocated 55 MHz below channel 2 and above channel 16, respectively.

Guidelines for go and return channels for an international connection are as follows:

1. In the section through which an international connection is arranged to pass, all the go channels should be in one half of the band and all the return channels should be in the other half of the band.
2. Both horizontal and vertical polarizations should be used for each radio-frequency channel in the co-channel arrangement.
3. The center frequency f_0 should be 18,700 MHz.
4. For digital systems of small capacity (i.e., below about 10 Mbps), frequency allocations may be accommodated within any of the high-capacity channels or guard bands. Channels 1, 1′ and 8, 8′ and the guard bands of Figure 13-U.1b are suitable sub-band allocations for small-capacity utilizations. The selection of alternative allocations should not prevent the pairing of the go and return channels in the manner described in Figures 13-U.2 and 13-U.3.
5. For medium-capacity systems with bit rates different from that given above and for small-capacity systems, administrations may adopt other radio-frequency channel arrangements in conformity with the recommended pattern for high-capacity systems.
6. Due regard should be taken of the fact that in some countries another arrangement of the go and return channels which incorporates a mid-band allocation for small-capacity systems may be used, as shown in Figure 13-U.4.

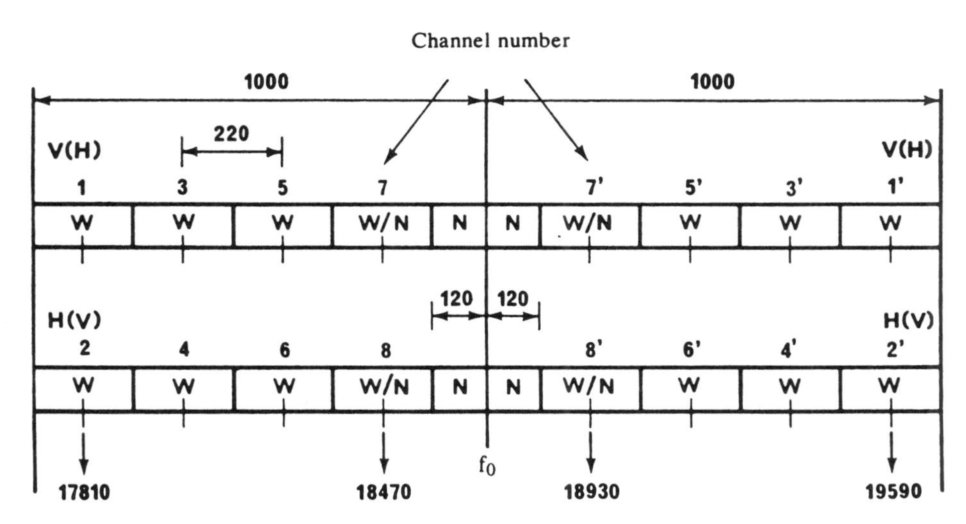

FIGURE 13-U.4. Co-channel radio-frequency arrangement for radio-relay system operating in the 18-GHz band referred to above. (From Figure 3, ITU-R Rec. 595-3, Geneva, 3/92, page 94.)

7. Due regard should be taken of the fact that in the countries where the band 17.7 to 21.2 GHz is available for the fixed service, other channel arrangements may be used (Annex 1).
8. Due regard should be taken of the fact that in some countries the band 17.7–19.7 GHz is subdivided to serve different applications in separate parts of the band.

Note 1: In establishing these systems, account should be taken of the need of passive sensors for earth exploration by satellite and space research in the band 18.6 to 18.8 GHz—particularly in Region 2, where these services have primary status in conformity with Rec. No. 706 and other relevant provisions (see No. 871) of the Radio Regulations (see Rec. 515, as well as Question 113/9 of CCIR).

Note 2: Actual gross bit rates may be as much as 5% or more higher than net transmission rates.

Description of a Radio-Frequency Channel Arrangement in the Band 17.7 to 19.7 GHz for North America. In North America this band was initially used for high-capacity digital transmission and then for small-capacity digital transmission systems. Usage has been extended to intermediate transmission capacities. More recently, new requirements have been put forward for this band by other services that originally used lower frequencies. The extent and variety of existing and identified future operational requirements have resulted in subdivision of the 17.7- to 19.7-GHz band.

The multiservice requirements are met by assigning separate bands to the major different service categories and by allowing for various radio-frequency channel widths to simultaneously increase usage versatility and spectral efficiency. The resulting composite radio-frequency channel arrangement is illustrated in Figure 13-U.5.

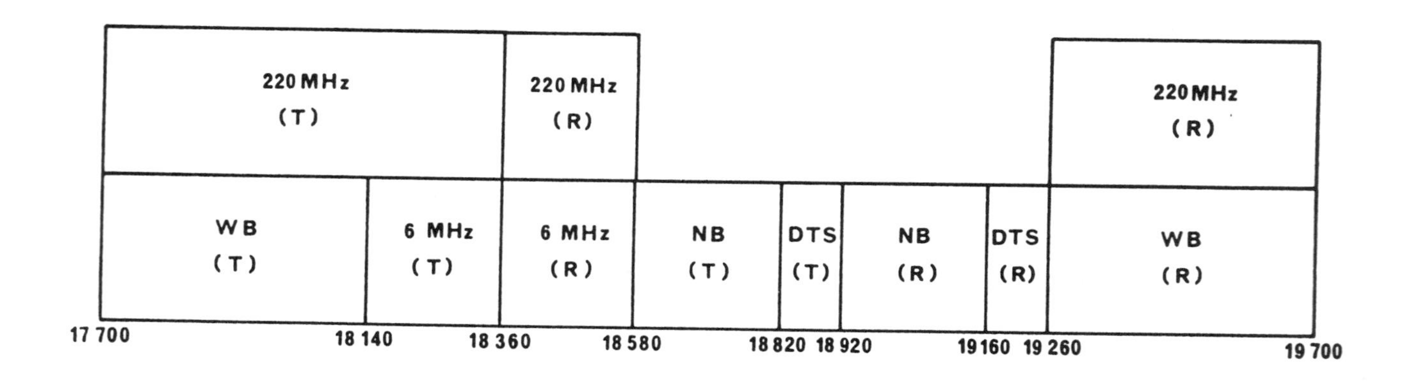

220 MHz: 220 MHz channels
WB: 10, 20, 40 and 80 MHz "wideband" channels
NB: 5, 10 and 20 MHz "narrowband" channels
DTS: 10 MHz digital termination system channels which may be sub-divided
6 MHz: 6 MHz channels for cable TV radio-relay systems
(T): transmit frequencies: go (return)
(R): receive frequencies: return (go)

FIGURE 13-U.5. Radio-frequency channel arrangements for digital and analog radio-relay systems operating in the 17.7- to 19.7-GHz band (North America). (From Figure 5, ITU-R Rec. 595-3, Geneva, 3/92, page 97.)

The versatility of this radio-frequency channel arrangement is exemplified by the overlap between the various WB, NB, and DTS channel bandwidths, as well as by the common transmit–receive frequency spacings for the adjacent NB and DTS services.

This subsection is based on ITU-R Rec. 595-3, Geneva, 3/92.

Radio-Frequency Channel Arrangements for Radio-Relay (LOS Microwave) Systems Operating in the 23-GHz Band. The ITU-R organization recommends that the radio-frequency channel arrangement for the 21.2- to 23.6-GHz band should be based on a homogeneous pattern. This homogeneous pattern with a preferred 3.5-MHz interval is defined by the relation

$$f_n = f_r + 3.5 + 3.5n$$

where $1 \leq n \leq 685$ and f_r is the reference frequency of the homogeneous pattern.

Another homogeneous pattern is based on a 2.5-MHz interval and is defined by the relation

$$f_n = f_r + 4 + 2.5n$$

where $1 \leq n \leq 959$ and again f_r is the reference frequency of the homogeneous pattern.

For international connections, the reference frequency of the homogeneous pattern is $f_r = 21{,}196$ MHz. The ITU-R organization states that other reference frequencies may be agreed upon by the administrations involved.

The following are further guidelines for channelization:

1. All go channels should be in one half of any bidirectional band, and all return channels should be in the other half.
2. The channel spacings, XS, the center gap, YS, and the distance to the lower and upper band limits, Z_1S and Z_2S, should be agreed upon by the administrations involved, dependent on the application and channel capacity envisaged. See ITU-R Rec. 746 for definitions of XS, YS, and ZS, and also see Section 13-2.1.1.

In the United States and Canada, the ITU-R organization reports that the most widespread use of the 21.2- to 23.6-GHz band is in the 21.8- to 22.4-GHz and 23.0- to 23.6-GHz portions for which a frequency pattern with 50-MHz channels has been adopted. The same pattern is being used in the remainder of the 21.2- to 23.6-GHz band because usage is spreading. Accordingly, a homogeneous pattern is in use based on a 2.5-MHz interval described above and is given by

$$f_n = f_r - 21 + 50n$$

where $n = 1, 2, 3, \ldots, 48$ and f_r (reference frequency) $= 21{,}196$ MHz.

For two-way operation, the go–return separation is about 1200 MHz. The typical systems in use include (a) digital transmission at data rates between about 1.5 and 8 Mbps and (b) a variety of analog video systems.

This subsection is based on ITU-R Rec. 637-1, Geneva, 3/92.

Radio-Frequency Channel Arrangements for Radio-Relay Systems in the 25.25- to 27.5-GHz and 27.5- to 29.5-GHz Bands. The preferred radio-frequency channel arrangement for the 25.25- to 27.5- and 27.5- to 29.5-GHz bands should be based on homogeneous patterns. This homogeneous pattern should have a 3.5-MHz interval defined by the relation

$$f_n = f_r + 3.5n$$

where $1 \leq n \leq 642$ for the band 25.25–27.5 GHz, $644 \leq n \leq 1214$ for the band 27.5–29.5 GHz, and f_r is the reference frequency for the homogeneous pattern.

There is also an arrangement based on a preferred 2.5-MHz interval which is defined by the relation

$$f_n = f_r + 1 + 2.5n$$

where $1 \leq n \leq 899$ for the band 25.25–27.5 GHz, $901 \leq n \leq 1699$ for the band 27.5–29.5 GHz, and f_r is the reference frequency of the homogeneous pattern.

For international connections, the reference frequency of the homogeneous pattern should be f_r = 25,249 MHz.

All go channels should be in one half of any bidirectional band, and all return channels should be in the other half.

The channel spacings, XS, center gap, YS, and the lower and upper band limits, Z_1S and Z_2S, should be agreed upon by the administration involved, dependent on the application and channel capacity envisaged. ITU-R Rec. 746 and Section 13-2.1.1 provide definitions of XS, YS, and ZS.

This subsection is based on ITU-R Rec. 748, Geneva, 3/92.

In Section 13-2.2.2 (page 901), replace head and text in section entitled "Allowable Bit Errors at the Output of the Hypothetical Reference Digital Path for Radio Links That Form Part of an ISDN."

Allowable Bit Error Ratios at the Output of the Hypothetical Digital Path for Radio-Relay (LOS Microwave) Systems Which May Form Part of an Integrated Services Digital Network. CCIR recommends that the following performance objectives are stated in each direction of the 64-kbps hypothetical reference digital path (HRDP) specified above. In these recommended performance criteria, fading, interference, and all other sources of performance degradation are taken into account in establishing the values given below.

The bit error ratio should not exceed the following values:

- 1×10^{-6} during more than 0.4% of any month; integration time 1 minute (minutes of degraded performance; see Notes 10 and 11).
- 1×10^{-3} during more than 0.054% of any month; integration time 1 second (severely errored seconds).
- The total errored seconds should not exceed 0.32% of any month (see Notes 8 and 9).

Note 1: The limits proposed are based on the best knowledge currently available but are subject to review in the future in light of further studies.

Note 2: The output bit stream from a digital radio-relay system suffers from jitter. The subject requires further study and is also being considered by the CCITT.

Note 3: This Recommendation relates to the hypothetical reference digital path (HRDP). The values given are for use by the system designer, and it is not intended that they should be quoted in specifications of equipment or used for acceptance tests.

Note 4: Contributions from multiplex equipment are not included.

Note 5: The Recommendation applies only when the system is considered to be available in accordance with CCIR Rec. 557 and includes periods of high bit error ratio exceeding 10^{-3} which persist for periods of less than 10 consecutive seconds. Periods of high bit error ratio which persist for 10 consecutive seconds duration or longer are taken into account by Rec. 557.

Note 6: The limits given for severely errored seconds above are based upon the 10-sec nonavailability criteria given in CCIR Rec. 557, and therefore will not necessarily include all forms of degradations in performance due to adverse propagation. Degradations due to adverse propagation which persist for 10 sec or longer will be limited by the requirements of Rec. 557.

Note 7: Adverse propagation conditions can result in a decrease of the wanted signal and/or an increase in the level of interfering signals.

Note 8: The relationship between the errored seconds of a 64-kbps channel and the corresponding parameters which may be measured directly at the bit rate of the radio-relay system is still under study. For the time being, the errored seconds should be measured only at the 64-kbps interface.

Note 9: The errored second objective described above could usually be satisfied when the objectives stated above and the residual bit error ratio (RBER) objective for 2500 km (see CCIR Rec. 634) are satisfied, taking into account typical cumulative error probability distributions.

Note 10: Measurements of BER are normally made at a much higher bit rate than 64 kbps—for example, at the bit rate of the radio-relay system.

Note 11: The seconds during which the bit error ratio exceeds 1×10^{-3} should not be taken into account in the integration time.

Note 12: The requirements are intended to meet the relevant performance objectives of CCITT Recs. G.821 and G.921 under all normally envisaged operating conditions. Rec. G.821 remains the overriding performance objective of the network.

Annex 1: Factors to Be Taken into Account When Determining Performance Requirements for Digital Radio-Relay (LOS Microwave) Systems Applied to the HRDP

Bit Error Ratio (BER) Objectives, DM and SES Criteria. ITU-T Rec. 821 states the following:

- The bit error ratio of the 25,000-km high-grade portion of the longest hypothetical reference connection should not exceed 1×10^{-6} for more than 4% of minutes within any month (DM).
- For a 27,500-km hypothetical reference connection (HRX), the bit error ratio should not exceed 1×10^{-3} for more than 0.2% of seconds within any month (SES).
- The bit error ratio of the 25,000-km high-grade portion of the hypothetical reference connection should not exceed 1×10^{-3} for more than 0.04% of seconds within any month. In addition, an allowance of 0.05% is given to a 2500-km hypothetical reference digital path (HRDP) for radio-relay systems to take account of adverse propagation conditions (SES).

Therefore, for a 2500-km HRDP the BER should not exceed 1×10^{-3} for more than 0.05% + 0.004%, where the 0.004% has been obtained by linearly scaling the 0.04% of the whole 25,000-km high-grade portion of the HRX.

With advances in high-speed data transmission services, some users may require higher transmission quality than specified above. In particular, users may become concerned with the frequency and duration of circuit interruptions.

Although it is difficult to predict the frequency of such breaks from a knowledge of the design of radio-relay systems, it may be appropriate to reduce the SES objective by considering whether the additional CCITT Rec. G.821 allowance for adverse propagation may be discarded. If this is the case, it will be necessary to include advanced technology in the design of radio-relay systems to achieve such high performance without compromising the economics of the system.

Considering that radio-relay systems have been designed to respect the existing SES objectives and that in some cases this already requires the use of the most advanced technologies and fading countermeasures, any reduction of the SES objective should be carefully examined.

Errored Seconds (ES) Objective. The bit error ratio (BER) parameter does not give an accurate estimate of the data circuit performance except when the error distribution is known. The more desirable alternative block error-ratio criterion unfortunately requires a knowledge of data block size and data rate which vary from user to user. The ES-based objective is one way in which the above limitations can be resolved. The CCITT has recommended that for ISDN services, including data services, a 3.2% ES objective should be attained on the 25,000-km high-grade portion in the longest hypothetical reference connection. A value of 0.32% ES for the hypothetical reference digital path is consistent with Rec. G.821 of the CCITT and is specified as such herein.

In general, there is no simple relationship between bit error ratio performance and the ES performance during periods of multipath fading.

Possible Error Burst Objective. With many commonly used digital demultiplex equipments, the occurrence of error bursts on large-capacity digital radio-relay systems will cause undesirable effects in the telephony network, such as multiple call dropout. Consideration may therefore have to be given to placing a limit on these error bursts. Area 1 of Figure 13-U.6 illustrates a possible limit which would serve to control the incidence of multiple call dropout, and is based upon proposals from the United Kingdom.

Error bursts in digital radio-relay systems which could cause multiple call dropout occur mainly because of fading. Analysis of field experiment results by one administration has shown that as fading events decrease in severity, or as fading countermeasures are applied, the number and length of error bursts will be reduced.

The above objective, which might incur inordinate economic penalties for digital radio-relay systems, is critically dependent upon the performance of demultiplex and signaling equipment. This would appear to be a matter for resolution with CCITT Study Group XVIII.

Other Factors. Timing jitter has an influence on digital performance, particularly on long paths, and requires further investigation.

It should be noted that the addition or loss of a bit, due for instance to switching without prior alignment to ensure coincidence or to a spurious impulse affecting the timing, may desynchronize the downstream transmission chain and cannot be considered an isolated error. This phenomenon affects system performance and should be studied further.

This section is based on ITU-R Rec. 594-3, RF Series, Geneva, 1991.

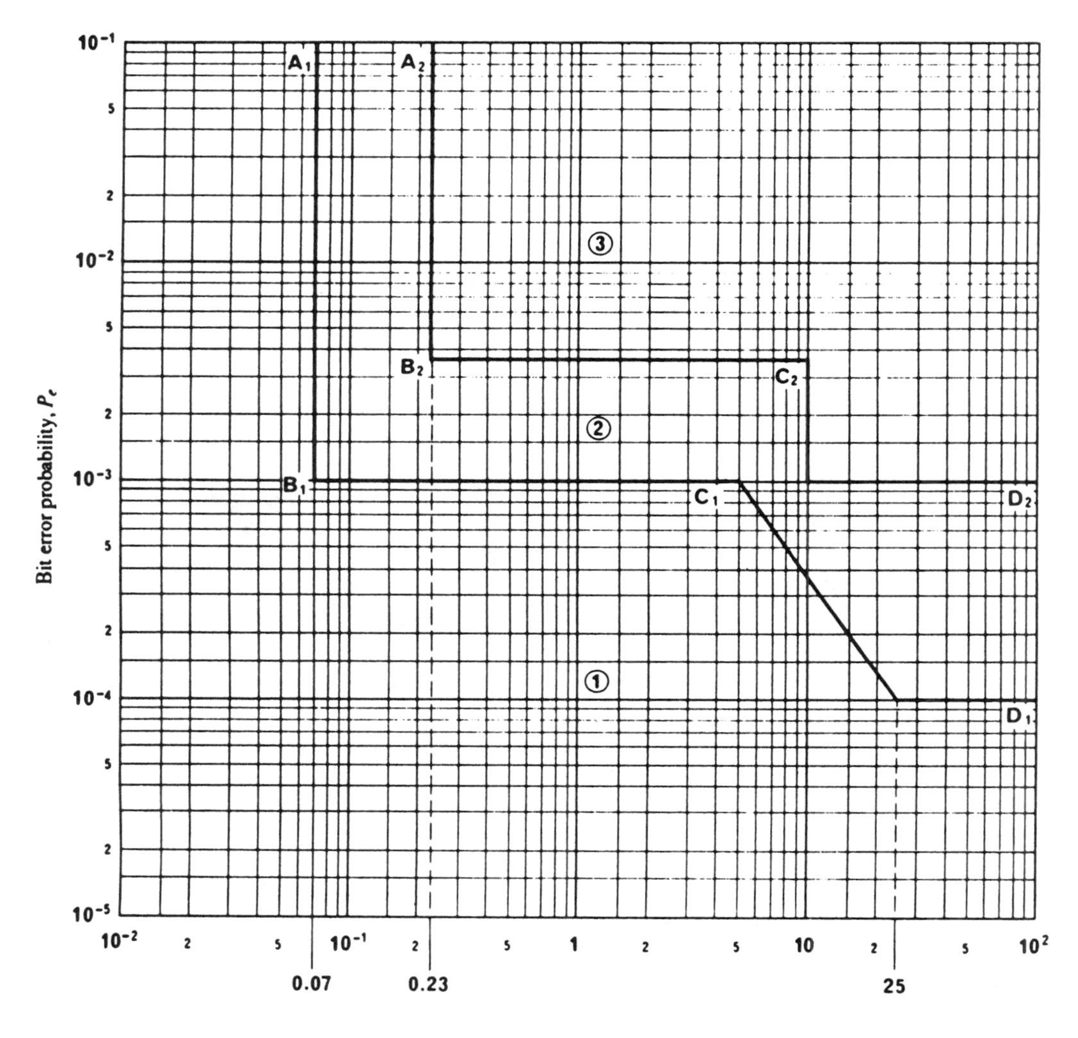

Area ①: error bursts confined within this area will have a negligible probability of causing multiple call dropout.

Area ②: error bursts extending into this area will have an increased probability of causing multiple call dropout and would probably be unacceptable.

Area ③: error bursts extending into this area are likely to have an unacceptably high probability of causing multiple call dropout.

A_1 B_1 C_1, A_2 B_2 C_2 } boundaries related to signalling performance criteria.

C_1 D_1, C_2 D_2 } boundaries related to subjective performance criteria.

FIGURE 13-U.6. Possible error burst objectives. (From Figure 1, ITU-R Rec. 594-3, RF Series, Geneva, 1991, page 8.)

This new section follows Section 13.2.11, which ends on page 954 of the Manual.

13-2.12. Typical Link Parameters: Selected Examples from the Fixed Service

Tables 13-U.1, 13-U.2, 13-U.3, and 13-U.4 give typical examples of link parameters for line-of-sight microwave systems.

Table 13-U.1 is taken from Table 1, ITU-R Rec. 759, RF Series, Geneva, 1991, page 227. Table 13-U.2 is taken from Table 8, ITU-R Rec. 758, RF Series, page 222. Table 13-U.3 is taken from Table 9, page 223 of the same recommendation. Table 13-U.4 is taken from Table 10, Rec. 758, page 224.

This new section follows Section 13.4.7.7, which ends on page 1040 of the Manual.

13-4.7.8. Contributions to Noise Temperature of an Earth-Station Receiving Antenna

Introduction. The noise temperature of an earth-station antenna is one of the factors contributing to the system noise temperature of a receiving system, and it may include contributions associated with atmospheric constituents such as water vapor, clouds, and precipitation, in addition to noise originating from extraterrestrial sources such as solar and cosmic noise. The ground and other features of the antenna environment, man-made noise and unwanted signals, and thermal noise generated by the receiving system, which may be referred back to the antenna terminals, could also make a contribution to the noise temperature of the earth-station antenna. Numerous factors contributing to antenna noise, particularly those governed by meteorological conditions, are not stable and the resulting noise will therefore exhibit some form of statistical distribution with time. A knowledge of these factors and their predicted variation would be a valuable aid to

TABLE 13-U.1

Selected Examples of Fixed Service Sharing Parameters for Bands Below 3 GHz

	Point-to-Point					Point-to-Multipoint				Trans-horizon
	Analog		Digital			Digital		Analog		Analog
Capacity	8 ch	960 ch/ 1 TV	64 kbps	2 Mbps	45 Mbps	Central Station	Out Station	Central Station	Out Station	72–312 ch
Modulation	FM-FDM	FM-FDM	4-PSK	8-PSK	64-QAM	4-PSK		94 ch FM-FDM		FM-FDM
Antenna gain (dB)	17–33	34	17–33	33	33	10–17	17–27	10	19	49
Transmitter power (dBW)	7	7	7	7	1	7	7	4	4	28
e.i.r.p. (dBW)	19	36	19	19	34	24	34	12	21	75
Bandwidth (MHz)	0.3	40	0.032	0.7	10	3.5	3.5	2	2	6
Receiver noise figure (dB)	8	10	4	4.5	4	3.5	3.5	9	9	2
Normal signal level (dBW)	−93	−64	−112	−90	−65	−90	−90	−97	−97	−65
Receiver input level for BER of 10^{-3}	NA[a]	NA	−137	−120	−106	−122	−122	NA	NA	NA
Maximum long-term interference[b]										
Total power (dBW)	−151	−129	−165	−151	−136	−141	−141	−142	−142	−138
Power spectral density [dB(W/4 kHz)]	−170	−169	−174	−173	−170	−170	−170	−169	−169	−172

[a]NA, not applicable.

[b]Interference values correspond to a degradation in receiver threshold of 1 dB or less.

Typically, the carrier level corresponding to the BER of 10^{-3} is around 4 dB lower than that for BER of 10^{-6}, and the carrier level difference between 10^{-6} and 10^{-10} BER points is also about 4 dB.

TABLE 13-U.2

Fixed Service Systems Parameters for Frequency Sharing Above 10 GHz

Frequency band (GHz)	10.7–11.7			12.2–12.44		13/14					14.4–15.35			17.7–19.7						
Modulation	4-PSK	FM-FDM	FM-TV	4-PSK	16-QAM	4-PSK	4-PSK	4-PSK	4-PSK	FM	64-QAM	FM-FDM	8-PSK	4-PSK	4-QAM	4-FSK	4-PSK	2-PSK	4-PSK	0-QPSK
Capacity	140 Mbps	960 ch	625-line PAL	13.9 Mbps	50.4 Mbps	2 Mbps	8 Mbps	16 Mbps	34 Mbps	1 Video	140 Mbps	2700 ch	156 Mbps	140 Mbps	140 Mbps	8 Mbps	8 Mbps	8 Mbps	34 Mbps	44.7 Mbps
Channel spacing (MHz)	67	40	40	20	20	3.5	7	14	28	28	28	40	40	110	55	20	20	20	27.5	40
Antenna gain (max.) (dB)	49	47	47	50	50	49	49	49	49	49	49	52	52	48	48	45	45	45	45	45
Feeder/multiplexer loss (min.) (dB)	5	5	5	1	1	0	0	0	0	0	2	5	5	7	7	0	0	0	0	3
Antenna type	3.7-m Dish	2.5-m Dish	2.5-m Dish	Dish	Dish	Dish	Dish	Dish	Dish	Dish	Dish	Dish	Dish	Dish	Dish	Dish	Dish	Dish	Dish	Dish
Max Tx output power (dBW)	10	10	10	−5	−5	10	10	10	10	10	5	3	0	−10	−4	−16	−6	−9	−8	−9
e.i.r.p. (max.) (dBW)	54	52	52	40	40	45	45	45	45	45	47	50	47	31	37	29	39	27	37	33
Receiver IF bandwidth (MHz)	68	29	29	12.3	17.2	1	2	4	17	24	40	56	50	68	68	8	4	8	18	40
Receiver noise figure (dB)	7	7	8	7	5	10	10	10	10	10	4	10	5	7	8	13	7	7	7	5
Receiver thermal noise (dBW)	−119	−121	−121	—		−134	−131	−128	−122	−120	−124			−119	−118	−122	−131	−128	−124	−125
Nominal Rx input level (dBW)	−62	−65	−65	−59 + M[a]	−59 + M	−74	−71	−68	−65	−65	−66	−48	−44	−64	−64	−65	−65	−65	−65	−70
Rx input level for 10^{-3} BER (dBW)	−104	NA[b]	NA			−116	−113	−111	−109	NA	−101			−103	−104	−106	−116	−116	−113	−106
Nominal short-term interference (dBW) (% time)																				
Nominal long-term interference (dBW)	−129	−131	−131			−144	−141	−138	−132	−130	−134			−129	−131	−132	−141	−138	−143	−131
Equivalent power [dB(W/4 kHz)]	—	−170	—			—	—	—	—	—	—			—	—	—	—	—		−171
Spectral density [dB(W/MHz)]	−147	—	−146			−144	−144	−144	−144	−144	−150			−147	−149	−141	−147	−147		
Refer to notes	c, d	c, e	c, d	f	f	d, g	d, g	d, g	d, g	d, g	d, g			c, d	c, d	c, d	c, d	c, d	c, d	g

[a] M, fade margin.

[b] NA, not applicable.

[c] Specified interference will reduce system C/N by 0.5 dB (interference 10 dB below receiver thermal noise floor).

[d] The specified interference level is total power within the receiver bandwidth.

[e] The specified interference level should be divided by the receiver bandwidth to obtain an average spectral density. The interference spectral density, averaged over any 4 kHz within the receiver bandwidth, must not exceed this value.

[f] Specified interference will have a relative contribution of no more than 10% of total noise.

[g] Specified interference will reduce system C/N by 1 dB (interference 6 dB below receiver thermal noise floor).

TABLE 13-U.3
Fixed Service System Parameters for Frequency Sharing Above 10 GHz (continued)

Frequency band (GHz)	21.12–23.6											25.25–27		
Modulation	2-FSK	2-FSK	2-FSK	4-PSK	4-PSK	FM	4-PSK	ASK	ASK	2-FSK	64-QAM	FSK	DFSK	FSK
Capacity	2 Mbps	4 Mbps	8 Mbps	34 Mbps	140 Mbps	1 Video	34 Mbps	2 Mbps	4 × 2 Mbps	2 Mbps	140 Mbps	6 Mbps		8 Mbps
Channel spacing (MHz)	7	7	14	28	112	28	28	28	28	5	40	40		20
												CS	OS	CS
Antenna gain (max.) (dB)	47	47	47	47	47	47	47	35	50	47	38.5	20	47	47
Feeder/multiplexer loss (min.) (dB) Antenna type	0 Dish	0 Dish	0 Dish	0 Dish	0 Dish	0 Dish	0 Dish	4 Dish	4 Dish	0 Dish	3 Dish	0 90° section	0 Dish	0 Dish
Max Tx output power (dBW)	0	0	0	0	0	0	0	−16	−14	−10	−4	−8	−10	−10
e.i.r.p. (max.) (dBW)	50	50	50	50	50	50	47	15	32	37	31.5	10	37	37
Receiver IF bandwidth (MHz)	2	4	8	17	70	24	18	5	14	2	40	16.4	16.4	16.4
Receiver noise figure (dB)	9	9	9	9	9	9	12	4	4	11	5	10	8	10
Receiver thermal noise (dBW)	−132	−129	−126	−123	−116	−121	−119	−133	−128		−123			
Nominal Rx input level (dBW)	−105 + M[a]	−104 + M	−103 + M	−100 + M	−94 + M	−84 + M	−87	−108 + M	−109 + M	−115	−73	−99 + M	−123 + M	−99 + M
Rx input level for 10^{-3} BER (dBW)	−108	—	−106	−103	−97	NA	−103	−112	−113		−96			
Nominal short-term interference (dBW) (% time)														
Nominal long-term interference (dBW)	−142	−139	−136	−133	−126	−131	−129	−139	−136		−131			
Equivalent power [dB(W/4 kHz)]	—	−170	—	—	—	—	—	—	—		−171			
Spectral density [dB(W/MHz)]	−143	−143	−143	−143	−143	−143	−141	−146	−148		−147			
Refer to notes	c,d	c,d	c,d	c,d	c,d	c,d	c,d	c,d	c,d	d,e		d,e		d,e

[a]M, fade margin.
[b]NA, not applicable.
[c]Specified interference will reduce system C/N by 1 dB (interference 6 dB below receiver thermal noise floor).
[d]The specified interference level is total power within the receiver bandwidth.
[e]Specified interference will have a relative contribution of no more than 10% of total noise.

TABLE 13-U.4
Fixed Service System Parameters for Frequency Sharing at 30- to 60-GHz

Frequency band (GHz)	37–39.5						54.25–57.2					
Modulation	2-FSK	2-FSK	4-PSK	4-PSK	FM	FM	2-FSK	2-FSK	4-PSK	4-PSK	FM	FM
Capacity	2 Mbps	8 Mbps	34 Mbps	140 Mbps	1 Video	1 Video	2 Mbps	8 Mbps	34 Mbps	140 Mbps	1 Video	1 Video
Channel spacing (MHz)	7	14	28	140	28	56	14	14	28	140	28	56
Antenna gain (max.) (dB)	47	47	47	47	47	47	47	47	47	47	47	47
Feeder/multiplexer loss (min.) (dB)	0	0	0	0	0	0	0	0	0	0	0	0
Antenna type	Dish	Dish	Dish	Dish	Dish	Dish	Dish	Dish	Dish	Dish	Dish	Dish
Max Tx output power (dBW)	0	0	0	0	0	0	0	0	0	0	0	0
e.i.r.p. (max.) (dBW)	50	50	50	50	50	50	50	50	50	50	50	50
Receiver IF bandwidth (MHz)	2	8	17	70	16	40	2	8	17	70	16	40
Receiver noise figure (dB)	11	11	11	11	12	12	11	11	11	11	12	12
Receiver thermal noise (dBW)	−130	−124	−121	−114	−120	−116	−130	−124	−121	−114	−120	−116
Nominal Rx input level (dBW)	−108 + M[a]	−102 + M	−99 + M	−93 + M	−98 + M	−85 + M	−108 + M	−102 + M	−99 + M	−93 + M	−98 + M	−85 + M
Rx input level for 10^{-3} BER (dBW)	−111	−105	−102	−95	NA	NA	−111	−105	−102	−95	NA	NA
Nominal short-term interference (dBW) (% time)												
Nominal long-term interference (dBW)	−140	−134	−131	−124	−130	−126	−140	−134	−131	−124	−130	−126
Equivalent power [dB(W/4 kHz)]	—	—	—		—	—	—	—	—		—	—
Spectral density [dB(W/MHz)]	−143	−143	−143	−143	−142	−142	−143	−143	−143	−143	−142	−142
Refer to notes	c, d	c, d	c, d		c, d	c, d	c, d	c, d	c, d		c, d	c, d

[a]M, fade margin.
[b]NA, not applicable.
[c]Specified interference will reduce system C/N by 1 dB (interference 6 dB below receiver thermal noise floor).
[d]The specified interference level is total power within the receiver bandwidth.

earth-station designers, and there is therefore the need to gather information on the antenna noise characteristics of existing earth stations in a form which can best be interpreted for future use.

This section presents results of antenna noise measurements made at 11.45 GHz, 11.75 GHz, 17.6 GHz, 18.4 GHz, 18.75 GHz, and 31.65 GHz. From the results measured at 17.6 GHz and 11.75 GHz, cumulative distributions of temperatures have been derived together with the dependence of the clear-sky noise temperature on the elevation angle.

Measuring Equipment. The antenna noise temperature measurements have been performed in the Netherlands using a series of radiometers equipped with a 10-m Cassegrain antenna fed by a corrugated horn. These measurements have also been performed in Japan using noise-adding-type and Dicke-type radiometers equipped with 13-m and 10-m Cassegrain antennas, and an 11.5-m offset Cassegrain antenna.

Noise measurements made in the Federal Republic of Germany were carried out on a 18.3-m-diameter antenna using the *y*-factor method, under clear-sky conditions.

Results of Measurements. Figure 13-U.7 shows the cumulative time distribution of the measured antenna noise temperature at 11.75 GHz and 17.6 GHz. The noise temperature shown in Figure 13-U.7 is the value measured at the output flange of the feedhorn.

The main contribution to noise temperature is caused by atmospheric attenuation. Other contributors are cosmic effects and noise from the ground.

The measurements presented in Figure 13-U.7 have been performed at an antenna elevation angle of 30°. The measurement period was between August 1975 and June 1977. The conditions during the measuring period can be considered as being typical for local rain conditions.

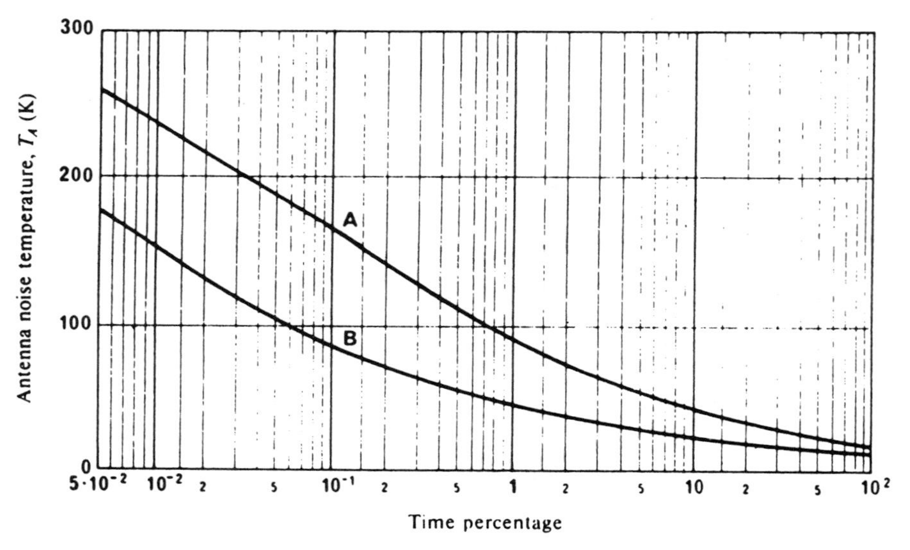

FIGURE 13-U.7. Measured antenna temperature as a function of the percentage of time each level was exceeded. (From Figure 3, ITU-R Rec. 733, RS Series, Geneva, 3/92, page 118.)

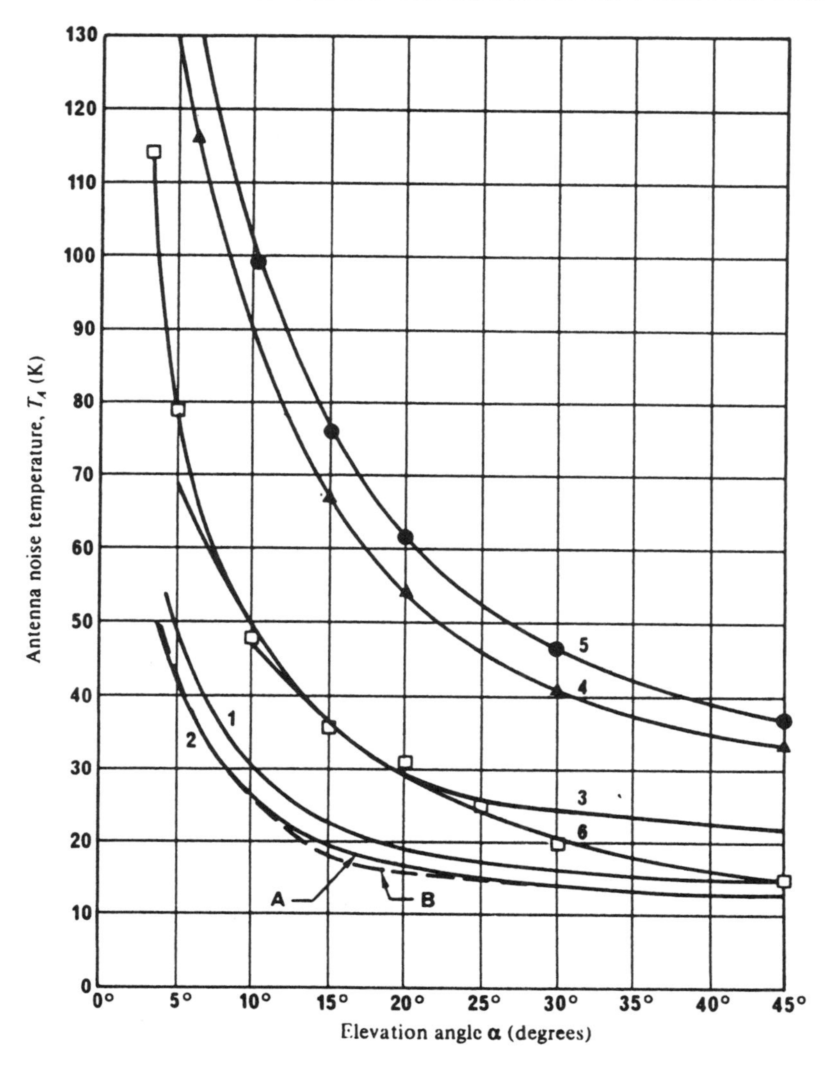

Note 1: Curves 1 to 6 are identified by reference to Table 13-U.5
Note 2: Measurement conditions were as follows:

Characteristics	Temperature (K)	Relative humidity (%)	Absolute humidity (g/m³)	Barometric pressure (mbar)
Curves 1 and 3	279	82	6	1016
Curve 2 A: calculated B: measured	294	51	10	1018
Curve 4 ▲: measured	296	50	10	1006
Curve 5 ●: measured	290	49	7	1013
Curve 6 □: measured	281.5	66	6	1017

FIGURE 13-U.8. Antenna noise temperature, T_A, as a function of elevation angle, α, of the antenna under clear-sky conditions. (From Figure 4, ITU-R Rec. 733, RS Series, Geneva, 3/92, page 119.)

Figure 13-U.8 shows the elevation dependence of antenna noise temperature with clear-sky conditions. The value of antenna noise temperature of Figure 13-U.8 corresponds to those of Figure 13-U.7 at 50% time percentage. An analysis of the measurement results given in Figure 13-U.8 showed that the antenna noise temperature consists of an elevation-dependent part and a component which is roughly constant. This constant part is formed by:

- Cosmic background microwave radiation having a value on the order of 2.8 K.
- Noise resulting from earth radiation. This contribution changes slightly with the angle of elevation of the antenna due to the side-lobe performance of the radiation diagram. A value on the order of 4–6 K is expected from this source.
- A noise contribution due to ohmic losses of the antenna system which is on the order of 0.04 dB. This component is expected to be 3–4 K.

The elevation-dependent part of the antenna noise temperature is caused by losses due to water and oxygen in the atmosphere; in order to estimate this elevation-dependent part, the curves of measured points in Figure 13-U.8 may be approximated by the following function and is accurate to 1% for elevation angles greater than 15°:

$$T_A = T_c + T_m(1 - \beta_0^{\text{cosec}\,\alpha}) \qquad \text{K} \tag{1}$$

where

T_A is the antenna noise temperature
T_c is the constant part of the noise temperature
T_m is the mean radiating temperature of the absorbing medium
β_0 is the transmission coefficient of the atmosphere in the zenith direction
α is the angle of elevation of the antenna.

In the range of angles of elevation between 5° and 90°, the constants of the function T_A are given in Table 13-U.5.

TABLE 13-U.5
Constants of the Function T_A

Reference No.	Frequency (GHz)	Antenna Diameter (m)	T_c (K)	β_0	Measuring Technique	Reference Station
1	11.75	10	8.3	0.9858	Radiometer	10-m OTS Netherlands
2	11.45	18.3	7.3	0.988	*y*-factor	18.3-m OTS/IS-V Federal Republic of Germany
3	17.6	10	8.3	0.9738	Radiometer	10-m OTS Netherlands
4	18.4	13	9.3	0.940	Radiometer	13-m CS Japan
5	31.65	10	11.5	0.934	Radiometer	10-m ECS Japan
6	18.75	11.5	4.5	0.970	Radiometer	11.5-m CS Japan

Source: Table 2, ITU-R Rec. 733, RS Series, Geneva, 3/92, page 120.

TABLE 13-U.6
Summary of Calculation Result and Measurement Value

Frequency (GHz)	Zenith Sky Temperature Calculation (K)	Zenith Sky Temperature Measurements (K)	Zenith Brightness Temperature Measurements (K)
11.75	3.2	3.9	6.7
17.6	7.8	7.2	10.0
18.4	14.7	16.7	19.5
31.65	14.3	18.3	21.1

Source: Table 3, ITU-R Rec. 733, RS Series, Geneva, 3/92, page 120.

Based on the constants given in Table 13-U.5 and for $\alpha = 90°$ in equation (1), the second term in this expression leads to the value of the zenith sky noise temperature caused by atmospheric attenuation. The zenith brightness temperature can be found by the addition of the zenith sky noise temperature and the cosmic microwave background radiation temperature. In this particular case, where atmospheric losses are very low, simple addition is allowed.

The zenith sky noise temperature can also be calculated using the humidity at the earth surface as an input parameter. The result of such calculation and the value found by measurements are summarized in Table 13-U.6.

Section 13-4.7.8 is based on ITU-R Rec. 733, Appendix 1 to Annex 1, RS Series, Geneva, 3/92.

13-4.7.9. Method of Determining Earth-Station Antenna Characteristics at Frequencies Above 10 GHz

Introduction. In communication-satellite systems operating at frequencies above 10 GHz, the specifications of the earth stations, in particular the figure of merit, must take account of G/T losses due to atmospheric effects and precipitation. These losses are generally specified for a percentage of time determined by the desired quality of the system.

The specification of the G/T must take account of losses:

- In the first place directly, because they lead to an increase in the required G/T
- In the second place indirectly, because they entail an increase in the noise temperature, T

The formulae given below are designed to standardize the methods used in determining the antenna characteristics from the standpoint of losses.

Specification of the Figure of Merit. The general formula used to specify the G/T of earth-station antennas at frequencies above 10 GHz is usually written as follows:

$$\frac{G}{T_i} - L_i \geq \left(K_i + 20 \log \frac{F}{F_0} \right) \quad \text{dB}\left(\text{K}^{-1}\right) \tag{2}$$

in the receiving band of the frequencies F for at least $(100 - P_i)\%$ of the time. L_i, expressed in decibels, is the additional loss on the downlink caused by the climatic conditions specific to the site of the earth station concerned, referred to nominal clear-sky conditions. T_i is the receive system noise temperature, including noise contribution due to L_i and referred to the input of the receiving low-noise amplifier.

The following is an example of dual specification for 11- to 12/14-GHz band TDMA/TV earth stations belonging to the European network (EUTELSAT):

$$\frac{G}{T_1} - L_1 \geq \left(37 + 20 \log \frac{F}{11.2}\right) \quad \text{dB}(\text{K}^{-1}) \qquad \text{under clear-sky conditions}$$

$$\frac{G}{T_2} - L_2 \geq \left(26.5 + 20 \log \frac{F}{11.2}\right) \quad \text{dB}(\text{K}^{-1}) \qquad \text{for at least 99.99\% of the year}$$

Calculation of Model. It is proposed to establish a relation $D = f(L_i, K_i, T_R)$ which may be used to determine the circular aperture diameter D for the antenna of an earth station with a (G/T_i) specified according to formula (2) and taking account of the receiving equipment noise temperature T_R.

Taking into account the expression for antenna gain G:

$$G = 10 \log\left[\eta\left(\frac{\pi DF}{c}\right)^2\right]$$

formula (2) may be expressed as follows:

$$20 \log D \geq (L_i + K_i)\ \text{dB} + 10 \log T_i - 10 \log \eta + 20 \log \frac{c}{\pi F_0} \tag{3}$$

where

D is the antenna diameter (m)
c is the speed of light: 3×10^8 m/sec
F_0 is the frequency (GHz)
η is the antenna efficiency at receiving port at frequency F_0
L_i is the atmospheric attenuation factor (referred to clear-sky conditions) expressed in decibels
K_i is the value specified for clear-sky figure of merit at frequency F_0, expressed in dB(K^{-1})
T_i is the noise temperature of the earth station, referred to the receiving port (K)

The earth-station noise temperature T_i is fairly accurately represented by the formula

$$T_i = \frac{L'_i - 1}{\alpha L'_i}(T_{\text{atm}} - T_c) + \frac{1}{\alpha}\left[T_c + T_s + (\alpha - 1)T_{\text{phys}}\right] + T_R \quad \text{K} \tag{4}$$

where

T_c is the antenna noise temperature due to clear sky
T_s is the antenna noise temperature due to ground
T_{atm} is the physical temperature of atmosphere and precipitations
T_{phys} is the physical temperature of the nonradiating elements of the antenna feed
T_R is the receiving equipment noise temperature
$\alpha \geq 1$ is the resistive losses due to nonradiating elements of the antenna feed
$L'_i \geq 1$ is the losses due to atmospheric effects and precipitation ratio
$L'_i = 10^{L_i/10}$, where L_i is expressed in decibels

Formula (4) may conveniently be expressed as follows:

$$T_i = T_A + (\Delta T_A) + T_R \tag{5}$$

where

T_A is the antenna noise temperature in clear-sky conditions ($L_i = 0$ dB):

$$T_A = \frac{T_c + T_s}{\alpha} + \frac{\alpha - 1}{\alpha} T_{\text{phys}} \tag{6}$$

ΔT_A is the additional antenna noise temperature caused by atmospheric and precipitation losses:

$$\Delta T_A = \frac{L'_i - 1}{\alpha L'_i}(T_{\text{atm}} - T_c) \tag{7}$$

Inserting relation (4) or relation (5) into relation (3), one can solve

$$D = f(L_i, K_i, T_R)$$

using additional data relating to the typical characteristics of earth-station antennas operating in the frequency band considered.

Sample Calculation. In the following example the diameter D of an EUTELSAT 11- to 12/14-GHz band TDMA/TV station antenna meeting the dual specification following equation (2) is calculated.

Assumptions

- The calculations are made at $F_0 = 11.2$ GHz for an elevation angle of about 30° above the horizon.
- The antenna performances at receiving port at the frequency F_0 are:

$\eta = 0.67$

$\left.\begin{array}{l} T_c = 15 \text{ K} \\ T_s = 10 \text{ K} \end{array}\right\}$ (typical values of contribution to antenna noise temperature at an elevation angle of 30°) at $F_0 = 11.2$ GHz

$T_{\text{atm}} = 270$ K

$T_{\text{phys}} = 290$ K

$\alpha = 1.122$ (resistive losses = 0.5 dB)

- The specifications are:

$$K_1 = 37 \text{ dB}$$

$$K_2 = 26.5 \text{ dB}$$

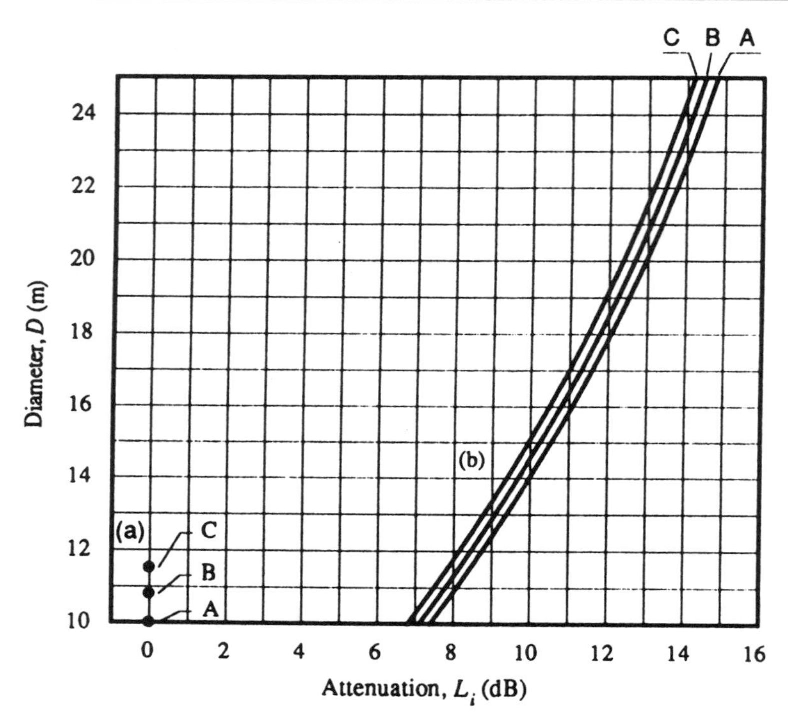

For two figures of merit at 11.2 GHz: and for three receiving equipment noise temperature values T_R :

(a): $G/T_1 = 37$ dB(K^{-1})
(b): $G/T_2 = 26.5$ dB(K^{-1})

A: $T_R = 130$ K
B: $T_R = 160$ K
C: $T_R = 190$ K

FIGURE 13-U.9. Variation of antenna diameter D as a function of attenuation L_i. (From Figure 9, ITU-R Rec. 733, RS Series, Geneva, 3/92, page 130.)

Calculation Results. Figure 13-U.9 shows two series of curves:

$$D = f(L_i)$$

parametered according to the dual specification for the clear-sky figure of merit, K_i, and according to three receiving equipment noise temperature values T_R (130 K, 160 K, and 190 K).

In the case of the example given above, if $T_R = 160$ K and if it is wished to install a station at a site where the propagation data are such that

$L_1 = 0$ dB under clear-sky conditions

$L_2 \leq 8$ dB for 99.99% of the year

the following two values for the antenna diameter are obtained:

$$D_1 = 10.70 \text{ m}$$

$$D_2 = 11.40 \text{ m}$$

so that $D \geq$ 11.40 m must be selected.

Section 13-4.7.9 has been extracted form ITU-R Rec. 733, RS Series, 3/92.

This new section follows Section 13-4.12.7, which ends on page 1188 of the Manual.

13-4.12.8. Maximum Permissible Level of Spurious Emissions from Very-Small-Aperture Terminals (VSATs)

ITU-R Rec. 726 recommends the following:

1. VSAT earth stations should satisfy the limits for radiated interference field strength specified in CISPR Publication 22 (1985) applicable to the Class A equipment over the frequency range from 30 MHz to 960 MHz.
2. Outside the band allocated to the fixed-satellite service (earth-to-space) within which the VSAT operating frequency is assigned the off-axis spurious e.i.r.p. from VSATs in any 100-kHz band shall be below the following limits for off-axis angles greater than 7°:
2.1. Carrier-off case (including receive-only VSAT earth stations)
960 MHz to 10.7 GHz 48 dBpW
10.7 GHz to 21.2 GHz 54 dBpW
21.2 GHz to 40 GHz 60 dBpW
2.2. Carrier-on case for transmit/receive VSAT earth stations except for the frequency range which is within ±400 MHz of the carrier frequency
960 MHz to 3.4 GHz 49 dBpW
3.4 GHz to 10.7 GHz 55 dBpW
10.7 GHz to 21.2 GHz 61 dBpW
21.2 GHz to 40 GHz 67 dBpW
3. The on-axis spurious e.i.r.p. shall not exceed 4 dBW in any 4-kHz band between 960 MHz and 40 GHz.
4. These values in this Recommendation should be revised to ensure the compatibility with the other services.
5. The following notes should be regarded as part of this Recommendation:

Note 1: In § 1, outdoor unit, indoor unit and the connecting cable shall be included. For transmit–receive VSAT earth stations, the antenna may be removed for this measurement and the transmitter output port shall be terminated by a terminating circuit.

Note 2: According to CISPR Publication 22 (1985) "in some countries the Class A equipment may be subject to restrictions on its sale or use" because "the Class A limits may be too liberal for domestic establishments and some residential areas." In order to avoid such restrictions to be applied to the sale or use of VSATs, the limits applicable to the Class B equipment should be satisfied.

Note 3: The worst-case off-axis spurious e.i.r.p. can be estimated from the measurement of the input power to the antenna transmission line assuming the antenna gain is 8 dBi or using measured data at off-axis angles greater than 7°.

Note 4: Intermodulation and spectral regrowth limits inside the band allocated to the fixed-satellite service (earth-to-space) within which the VSAT operating frequency is assigned are to be determined by the system design, subject to the satellite operator specifications and not covered by this Recommendation.

Note 5: For systems in which the VSAT earth stations are expected to transmit simultaneously at the same frequency (e.g., for systems employing CDMA), the limits given in § 3 should be reduced by $10 \log N$ (dB), where N is the maximum number of VSAT earth stations which are expected to transmit simultaneously at the same frequency within the overlapping band.

Note 6: The limit in § 3 applies to the 14-GHz band. The limit for VSATs operating in the 6-GHz and other frequency bands is under study.

Note 7: The exception of the frequency range of ±400 MHz in § 2.2 is considered appropriate for the present state-of-the-art technology. Further study is urgently needed to assess the feasibility of reducing this frequency range.

Section 13-4.12.8 is based on ITU-R Rec. 726, RS Series, Geneva, 3/92.

13-4.12.9. Maximum Permissible Levels of Interference in a Geostationary-Satellite Network for an HRDP When Forming Part of the ISDN in the Fixed-Satellite Service Caused by Other Networks of This Service Below 15 GHz

ITU-Rec. 735-1 recommends the following:

1. A geostationary network in the FSS operating in the frequency bands below 15 GHz should be designed and operated in such a manner that in any satellite hypothetical reference digital path which forms part of a 64-kbps ISDN connection the provisions of § 1 of Rec. ITU-R S.614 can be met when the aggregate interfering power from the earth and space station emissions of all other networks operating in the same frequency band or bands, assuming clear-sky conditions on the interference paths, does not exceed at the input to the demodulator:
   1.1 Twenty-five percent of the total system noise power under clear-sky conditions when the network does not practice frequency re-use.
   1.2 Twenty percent of the total system noise power under clear-sky conditions when the network does practice frequency re-use.
2. For a geostationary network in the FSS as mentioned in § 1, the internetwork interference caused by the earth and space station emissions of any one other network operating in the same frequency band or bands should be limited to 6% of the total system noise power under clear-sky conditions.
3. The maximum level of interference noise power caused to that network should be calculated on the basis of the following values for the receiving earth station antenna gain, in a direction at an angle φ (in degrees) referred to the main beam direction:

$$G = 32 - 25 \log \varphi \quad \text{dBi} \quad \text{for } 1^\circ \leq \varphi < 48^\circ$$

$$G = -10 \quad \text{dBi} \quad \text{for } 48^\circ \leq \varphi \leq 180^\circ$$

4. The following notes should be regarded as part of this Recommendation:

Note 1: For the calculation of the limits quoted in § 1.1, 1.2, and 2, it should be assumed that the total system noise power at the input to the demodulator is of thermal nature and includes all intersystem noise contributions as well as interference noise from other systems.

For interference not of a thermal nature, the permissible level of interference into a digital carrier should be based upon the degradation of the long-term performance objective as given in Annex 1 to Rec. ITU-R S.614.

Note 2: For this interference calculation, as applied to satellite networks operating in a fading environment, it should be assumed that the carrier power level is reduced, until the system performance coincides with the above BER and percentage of month (see Annex 1 for clarification).

Note 3: It is assumed in this Recommendation that the interference from other satellite networks is of a continuous nature at frequencies below 10 GHz; further study is required with respect to cases where interference is not of a continuous nature above 10 GHz.

Note 4: When interference is characterized by a nonuniform spectral distribution there may be cases where, for design purposes, a greater interference allocation of total system noise may be made to narrow-bandwidth carriers by the system designer. One model developed to address this is presented in detail in Annex 2.

Note 5: For networks using 8-bit PCM-encoded telephony, see Rec. ITU-R S.523.

Note 6: In some cases it may be necessary to limit the single-entry interference value to less than the value quoted in § 2 above in order that the total value recommended in § 1 may not be exceeded. In other cases, particularly in congested arcs of the geostationary-satellite orbit, administrations may agree bilaterally to use higher single-entry interference values than those quoted in § 2 above, but any interference noise power in excess of the value recommended in § 2 should be disregarded in calculating whether the total value recommended in § 1 is exceeded.

Note 7: There is an urgent need for study of the acceptability of an increase in the maximum total interference noise values recommended in § 1.

Note 8: Although this Recommendation has an upper frequency limit of 15 GHz, in the frequency range from 10 to 15 GHz short-term propagation data are not available uniformly throughout the world and there is a continuing need to examine such data to confirm an appropriate interference allowance to meet § 1.2 and 1.3 of Rec. ITU-R S.614.

Note 9: There is a need for urgent study to be given to the interference noise allowances appropriate to systems operating at frequencies above 15 GHz.

Note 10: The interference criteria of this Recommendation apply only to the transmission of digital services that fall under the provisions of Rec. ITU-T G.821 and Rec. ITU-R S.614. Further study by radiocommunication Study Group 4 is required regarding the performance objectives and appropriate interference criteria for other than 64-kbps digital transmissions within an ISDN connection as information on the performance requirements for such services becomes available to it.

Note 11: The principles of this Recommendation may also be applied to digital-satellite networks providing long-term performance objectives different from those in Rec. ITU-R S.614. This is a subject for further study.

Note 12: Special attention may have to be given to digital carriers with narrow bandwidths when they are being interfered with by analog TV transmissions. For such cases with artificial energy dispersal at the frame rate, the protection ratios given in Rec. ITU-R S.671 apply. This is a subject for further study.

Note 13: Attention should also be drawn to the interference from TDMA systems when, due to burst overlapping at the transponder input of an interfered-with system, the BER is increased relative to that of a synchronous burst allocation.

Note 14: In order to promote orbit efficiency, satellite networks operating in heavy rain environments are encouraged to use some form of fade compensation.

Note 15: This Recommendation is closely related to Rec. ITU-R S.614, a fact that needs to be considered in any future revisions to either of these two Recommendations.

Section 13-4.12.9 is based on ITU-R Rec. 726, RS Series, Geneva, 3/92.

This new text replaces Section 13-6.2, which starts on page 1245 of the Manual and ends on page 1249.

13-6.2. Communication by Meteor-Burst Propagation: ITU-R Rec. 843

ITU-R Rec. 843 recommends that the following information be used in the design and planning of meteor-burst communication systems.

13-6.2.1. Temporal Variations in Meteor Flux

At certain times of the year, meteors occur in the form of showers and may be prolific over durations of a few hours. There is, however, a general background of meteors incident upon the earth from all directions, and it is appropriate to consider only these sporadic meteors for communication-planning purposes.

For sporadic meteors at mid-latitudes there is a roughly sinusoidal diurnal variation of incidence with a maximum at 0600 hr and a minimum at 1800 hr local time. The ratio of maximum to minimum averages about four. In the Northern Hemisphere there is a seasonal variation of similar magnitude, with a minimum in February and maximum in July. Considerable day-to-day variability exists in the incidence of both sporadic and shower meteors.

The annual average flux of meteors incident per unit area and producing electron-line densities q exceeding a threshold q_0 per meter, $I\,(q > q_0)$, is given as

$$I\,(q > q_0) = \frac{160}{q_0} \qquad \mathrm{m^{-2}\,sec^{-1}} \tag{1}$$

By combining this overall meteor rate with a representative sinusoidal diurnal variation and the seasonal factor, M, from Figure 13-U.10 the average

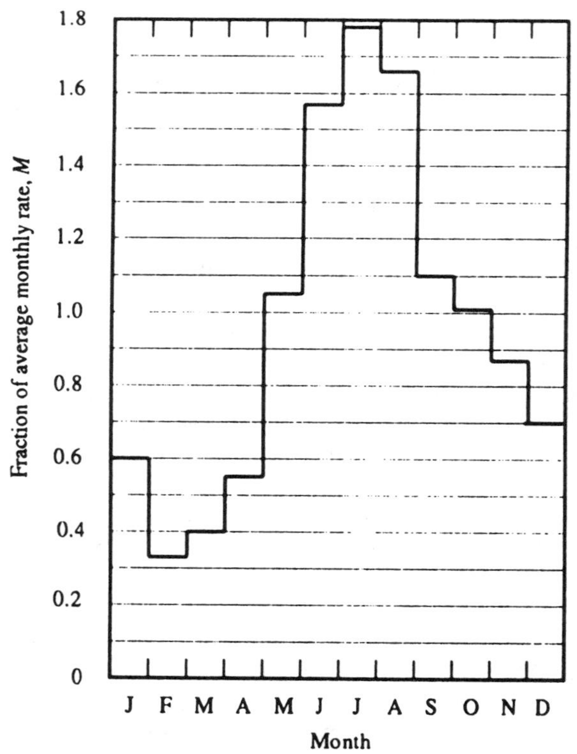

FIGURE 13-U.10. Month-to-month variation in sporadic meteor flux rate relative to the average value. (From Figure 1, ITU-R Rec. 843, RPI Series, Geneva, 10/92, page 92.)

temporal changes in meteor flux can be estimated:

$$\frac{160}{q_0} M\left[1 + 0.6\left(\sin \frac{\pi T}{12}\right)\right] \qquad \text{m}^{-2}\,\text{sec}^{-1} \tag{2}$$

where T is local time (hr).

For planning purposes it may only be necessary to consider the worst combination of month and local time.

13-6.2.2. *Spatial Variation in Meteor Flux*

Meteors occur in all parts of the world at all hours, but statistical information is incomplete on their geographical distribution and trail directions.

Until such times as spatial variations are quantified, it is recommended that flux estimates based on the method given in Section 13-6.2.1 be used at all latitudes.

13-6.2.3. *Underdense and Overdense Trails*

The ionized trails caused by meteors are classified as underdense or overdense according to the intensity of the ionization. The division between the two cases occurs for line densities of approximately 2×10^{14} electrons per meter. The amplitude of signals scattered from underdense trails may be calculated by summing the scattered field arising from each individual electron. Overdense trails are those for which the coupling between electrons cannot be ignored, in which case the reflecting properties are calculated as if the trail were a long metallic cylinder. At frequencies used in practice, the echoes from underdense trails show an abrupt start followed by an exponential decay, whereas those from overdense trails have more rounded envelopes and are of longer duration. The relative proportions of underdense and overdense echoes will depend on the system sensitivity.

The relation between number of trails and peak amplitude, A, can be approximated by

$$\text{Number of trails} \;\alpha\; (A)^{-\psi}$$

where ψ varies from 1.0 at low signal levels to greater than 2.0 at larger signal levels where the majority of trails are overdense. For many links, the index ψ is on the order of 1.1–1.4.

Results in the systems used so far indicate that echoes are predominantly from underdense trails. On this basis it is recommended that planning for a typical system should proceed on the basis that all meteor trails are of the underdense type.

13-6.2.4. *Effective Length and Radius of Meteor Trails*

Effective Length. The ray geometry for a meteor-burst propagation path is shown in Figure 13-U.11 between transmitter T and receiver R. P represents the tangent point and P′ a point further along the trail such that $(R_1' + R_2')$ exceeds $(R_1 + R_2)$ by half a wavelength. Thus PP′ (of length L) lies within the principal Fresnel zone, and the total length of the trail within this zone is $2L$. Provided that R_1 and R_2 are much greater than L, it follows that for practical cases:

$$L = \left[\frac{\lambda R_1 R_2}{(R_1 + R_2)(1 - \sin^2 \varphi \cos^2 \beta)}\right]^{1/2} \tag{3}$$

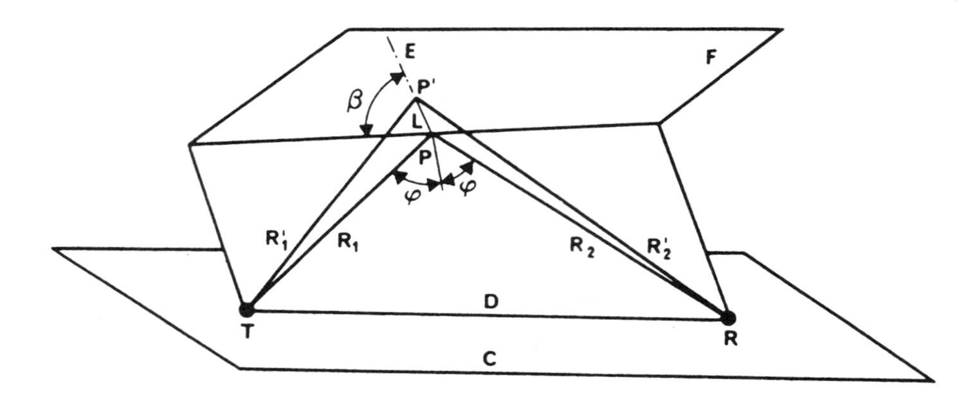

C: Earth's surface
D: plane of propagation
E: trail
F: tangent plane
β: angle between the trail axis and the plane of propagation
T: transmitter
R: receiver

FIGURE 13-U.11. Ray geometry for a meteor-burst path. (From Figure 2, ITU-R Rec. 843, RPI Series, Geneva, 10/92, page 93.)

where

φ is the angle of incidence
β is the angle between the trail axis and the plane of propagation
λ is the wavelength

Trail Radius. In order to evaluate the scattering cross section of the trail, it is usual to assume that ambipolar diffusion causes the radial density of electrons to have a Gaussian distribution and that the volume density is reduced while the line density remains constant.

The ionization trail immediately behind a meteor is formed near-instantaneously with a finite width. This is called the initial trail radius, r_0. An empirical relationship between r_0 and the meteor height is

$$\log r_0 = 0.035h - 3.45 \tag{4}$$

where

h is the trail height (km)
r_0 is the initial trail radius (m)

13-6.2.5. Received Power and Basic Transmission Loss

Received Power. Because any practical meteor-burst communication system will rely mainly on underdense trails, the overdense formulae are of less importance. Satisfactory performance estimates can be made using formulae for the underdense case with assumed values of q in the range 10^{13} to 10^{14} electrons per meter according to the prevailing system parameters.

The received power $p_R(t)$ after scattering from underdense trails at frequencies used in practice is as follows:

$$p_R(t) = \frac{p_T g_T g_R \lambda^2 \sigma a_1 a_2(t) a_2(t_0) a_3}{64\pi^3 R_1^2 R_2^2} \tag{5}$$

where

λ is the wavelength (m)
σ is the echoing area of the trail (m^2)
a_1 is the loss factor due to finite initial trail radius
$a_2(t)$ is the loss factor due to trail diffusion
a_3 is the loss factor due to ionospheric absorption
t is the time in seconds measured from the instant of complete formation of the first Fresnel zone
t_0 is half the time taken for the meteor to traverse the first Fresnel zone
p_T is the transmitter power (W)
$p_R(t)$ is the power available from the receiving antenna (W)
g_T is the transmit antenna gain relative to an isotropic antenna in free space
g_R is the receive antenna gain relative to an isotropic antenna in free space (lossless transmitting and receiving antennas are assumed)
R_1 and R_2 are distances (m); see Figure 13-U.11

The echoing area σ is given as

$$\sigma = 4\pi r_e^2 q^2 L^2 \sin^2 \alpha \tag{6}$$

where

r_e is the effective radius of the electron = 2.8×10^{-15} m
α is the angle between the incident electric vector at the trail and the direction of the receiver from that point

Because L^2 is directly proportional to λ, the echoing area, σ, is also proportional to λ and hence the received power for underdense trails varies as λ^3. Horizontal polarization normally is used at both terminals. The $\sin^2 \alpha$ term in equation (6) is then nearly unity for trails at the two hot spots.

The loss factor a_1 is given by

$$a_1 = \exp\left[-\frac{8\pi^2 r_0^2}{\lambda^2 \sec^2 \varphi}\right] \tag{7}$$

It represents losses arising from interference between the re-radiation from the electrons wherever the thickness of the trial at formation is comparable with the wavelength.

The factor $a_2(t)$ allows for the increase in radius of the trail by ambipolar diffusion. It may be expressed as

$$a_2(t) = \exp\left[-\frac{32\pi^2 Dt}{\lambda^2 \sec^2 \varphi}\right] \tag{8}$$

where D is the ambipolar diffusion constant in m^2 sec^{-1} given by

$$\log D = 0.067h - 5.6 \tag{9}$$

The increase in radius due to ambipolar diffusion can be appreciable even for as short a period as is required for the formation of the trail. The overall effect with regard to the reflected power is equal to that which would arise if the whole trail within the first Fresnel zone had expanded to the same extent as at its midpoint. Because this portion of trail is of length $2L$, the midpoint radius is that arising after a time lapse of L/V (sec), where V is the velocity of the meteor in msec^{-1}. For trails near the path midpoint ($R_1 \approx R_2 = R$), calling the time lapse t_0 gives the following:

- For trails at right angles to the plane of propagation ($\beta = 90°$):

$$t_0 \simeq \left(\frac{\lambda R}{2V^2}\right)^{1/2} \tag{10}$$

- For trails in the plane of propagation ($\beta = 0°$):

$$t_0 \simeq \left(\frac{\lambda R}{2V^2}\right)^{1/2} \cdot \sec \varphi \tag{11}$$

Substituting t_0 from equation (10) into equation (8) gives for the $\beta = 90°$ case:

$$a_2(t_0) = \exp\left[-\frac{32\pi^2}{\lambda^{3/2}}\left(\frac{D}{V}\right)\left(\frac{R}{2}\right)^{1/2}\frac{1}{\sec^2 \varphi}\right] \tag{12}$$

For $\beta = 0°$ the exponent in this expression is $\sec \varphi$ times greater.

The ratio of the ambipolar diffusion constant D to the velocity of the meteor V (required in the evaluation of received power) can be approximated by

$$D/V = \left[0.0015h + 0.035 + 0.0013(h - 90)^2\right] \times 10^{-3} \tag{13}$$

$a_2(t)$ is the only time-dependent term and gives the decay time of the reflected signal power. Defining a time constant T_{un} for the received power to decay by a factor e^2 (i.e., 8.7 dB) leads to

$$T_{un} = \frac{\lambda^2 \sec^2 \varphi}{16\pi^2 D} \tag{14}$$

With reflection at grazing incidence, $\sec^2 \varphi$ will be large and hence so is the echo-time constant. The echo-time constant is also increased by the use of lower frequencies.

Basic Transmission Loss. Basic transmission loss curves derived from equation (5) with $q = 10^{14}$ electrons per meter are given in Figure 13-U.12. Because the angle β can take any value between 0° and 90°, only these two extreme cases are shown. The advantage of lower propagation loss at the lower frequencies is clearly seen. Average meteor heights given from equation (15) have been used in deriving the curves. It should be noted that the prediction of system performance depends critically on the heights assumed.

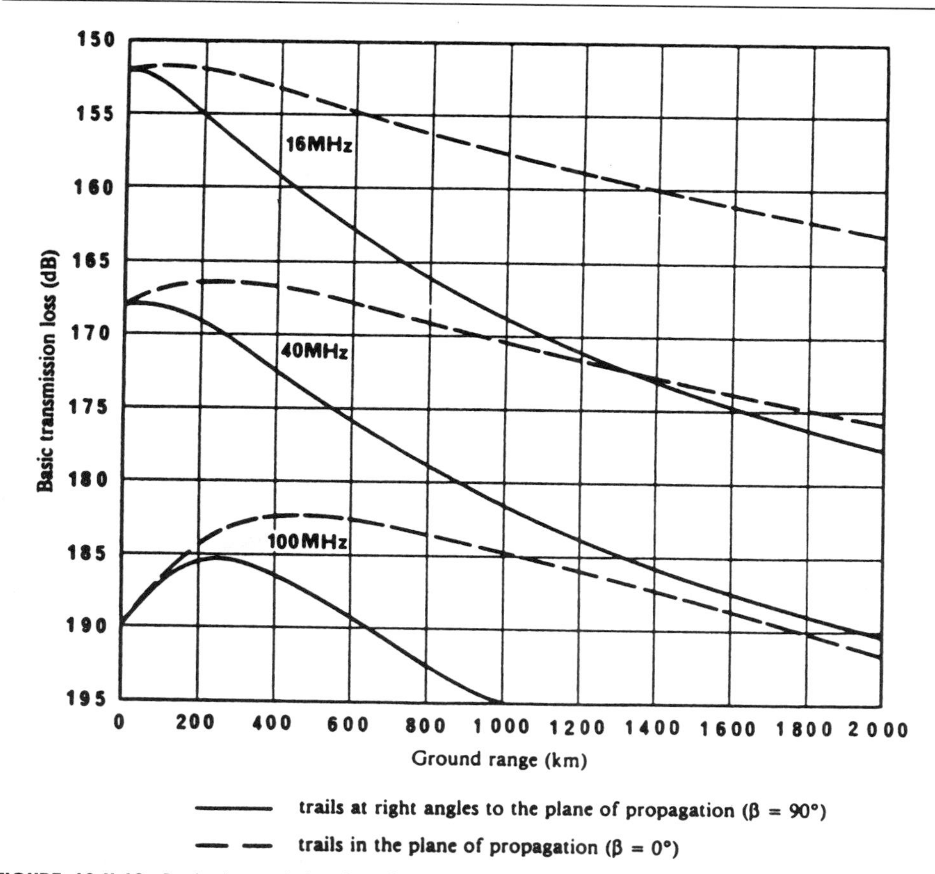

FIGURE 13-U.12. Basic transmission loss for underdense trails given by equation (5) with $q = 10^{14}$ electrons/m and horizontal polarization. (From Figure 3, ITU-R Rec. 843, RPI Series, Geneva, 10/92, page 96.)

13-6.2.6. The Underdense Echo Ceiling and Average Trail Height

Both the initial trail radius, r_0, and the ambipolar diffusion constant, D, increase with altitude. Consequently, the loss factors a_1 and $a_2(t_0)$ combine to reduce the number of underdense meteors occurring near the top of the meteor region which are useful for communication purposes. This effect is usually referred to as the underdense echo ceiling. Similar constraints have been observed to exist in the monostatic case. Figure 13-U.13 shows the measured height distribution of underdense echoes using various radar frequencies.

It can be seen that the lowest altitude at which underdense echoes occur is 85 km, and that the altitude distribution is approximately Gaussian at any frequency.

The average trail height h (km) at frequency f (MHz) is

$$h = -17 \log f + 124 \tag{15}$$

The average trail height is a function of other system parameters in addition to frequency. However, equation (15) is a good approximation.

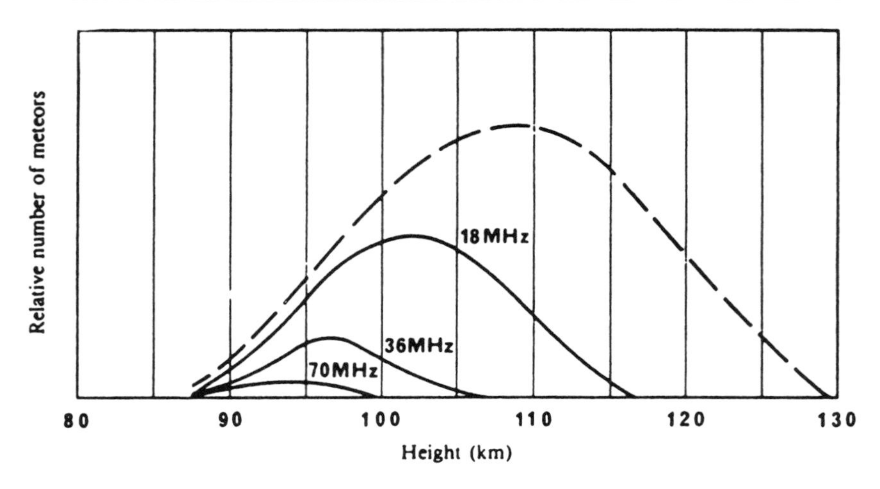

FIGURE 13-U.13. Height distribution of underdense meteors providing echoes at frequencies 18, 36, and 70 MHz. (From ITU-R Rec. 843, Fig. 4 RPI Series, Geneva, 10/92, page 97.)

13-6.2.7. Positions of Regions of Optimum Scatter

The scattering properties of straight meteor ionization trails are strongly aspect-sensitive. To be effective, it is necessary for the trails approximately to satisfy a specular reflection condition. This requires the ionized trail to be tangential to a prolate spheroid whose foci are at the transmitter and receiver terminals (see Fig. 13-U.11). The fraction of incident meteor trails which are expected to have usable orientations is approximately 5% in the area of the sky which is most effective. Figure 13-U.14 shows the estimated percentages of useful trails for a terminal separation of 1000 km. It may be seen that the optimum scattering regions are situated about 100 km to either side of the great circle, independent of path length.

The fraction of usable trails, p, for any path length, D', can be estimated using the following formula:

$$p = \frac{4L}{3\pi D'} \frac{\left[3(\xi^2 - \eta^2) - (1 - \eta^2)\right]\left[(\xi^2 - 1)(\xi^2 - \eta^2) - 4\xi^2 h^2/D'^2\right] - 4\eta^2(\xi^2 - 1)h^2/D'^2}{(\xi^2 - \eta^2)^2(\xi^2 - 1)\left[(\xi^2 - 1)(\xi^2 - \eta^2) - 4\xi^2 h^2/D'^2\right]^{1/2}} \tag{16}$$

where

$$\xi = (R_1 + R_2)/D'$$

$$\eta = (R_1 - R_2)/D'$$

13-6.2.8. Estimating the Useful Burst Rate

An assessment of the link power budget of a meteor-burst communication link can be made using the average trail height and other expressions presented above. Once a link appears viable, a more detailed analysis is required to estimate the rate at which useful meteor-burst signals will be passed.

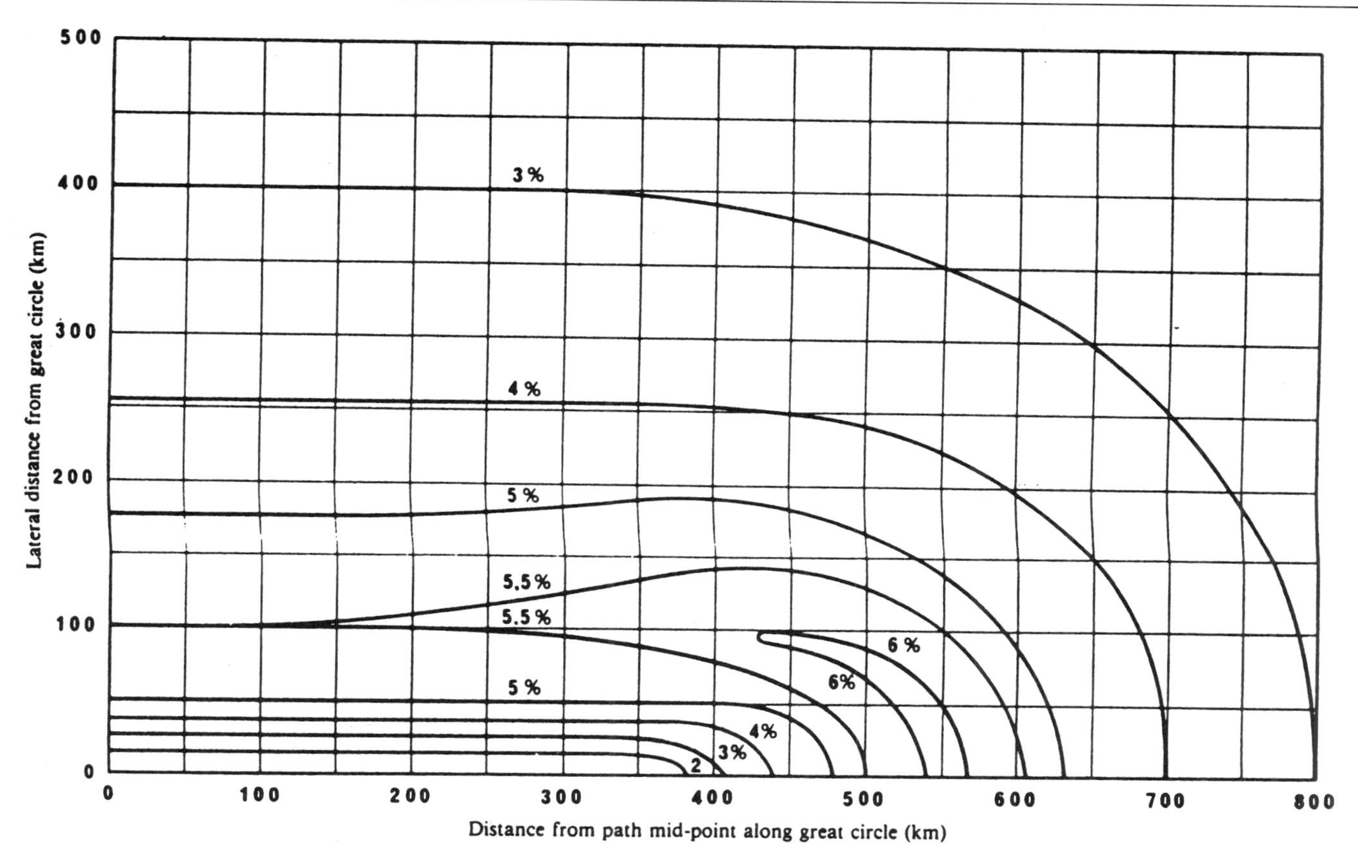

FIGURE 13-U.14. Estimated percentages of useful trails as a function of scattering position for a terminal separation of 1000 km. (From ITU-R Rec. 843, Fig. 5 RPI Series, Geneva, 10/92, page 97.)

The most rigorous methods of estimating useful burst rate typically involve the following stages:

1. Establish the minimum useful received signal power.
2. Utilize the equations of Section 13-6.2.5 to describe the variation of system parameters.
3. Compute the fraction of useful trails as a function of scattering position using equation (16).
4. Combine the estimated true height distribution of meteor trails with equation (2) to compute the volume density of meteor trails, as a function of q, in the atmosphere.
5. Integrate the product of stages 3 and 4 over the meteor region using at each point $q_{\min}$ derived at stage 2.

13-6.2.9. Antenna Considerations

The effect described in Section 13-6.2.7, together with the fact that the trails lie mainly in the height range 90–110 km, serves to establish the two "hot-spot" regions towards which both antennas should be directed. The two hot spots vary in relative importance according to time of day and path orientation. Generally, antennas used in practice should have beams broad enough to cover both hot spots. Thus the performance is not optimized, but on the other hand the need for beam swinging does not arise.

In general, horizontal polarization is preferred but vertical polarization could be useful for ranges in excess of approximately 1000 km where low angle cover is required from the antennas.

13-6.2.10. Considerations in the Choice of Frequency

The choice of frequency in a meteor-burst communication system is influenced by several factors.

Information Duty Cycle. The wavelength dependences of the maximum received power, $p_{R_{max}}$, and the duty cycle, D_C, as implied by equation (5) are such that for fixed transmitter power and antenna gains we have

$$p_{R_{max}} \sim \lambda^3 \tag{17}$$

This relationship holds for both underdense and overdense trails. The effect on the duty cycle will depend on the relative occurrence rate of these two types of trail. Assuming an intermediate occurrence rate, the duty cycle varies with wavelength as follows:

$$\begin{aligned} D_C &\sim p_{R_{max}} T_{un} \\ &\sim \lambda^5 \end{aligned} \tag{18}$$

In very quiet receiving locations the predominant noise at frequencies above 25 MHz is cosmic noise and the intensity of this varies as $\lambda^{2.3}$. Hence the information duty cycle, I_C, for a given bandwidth (i.e, the proportion of time a given signal/noise ratio is exceeded) varies with wavelength as follows:

$$I_C \sim \lambda^{2.7} \tag{19}$$

The relative frequency of occurrence of reflections as a function of signal amplitude depends on the sensitivity of the system. A common experimental result in the systems used so far may be expressed as

$$D_C \sim p_R^{-0.6} \tag{20}$$

where

D_C is the duty cycle (proportion of time exceeding the threshold A)
p_R is the received power corresponding to the threshold A

Because noise power is proportional to bandwidth B, the use of equation (20) leads to

$$I_C \sim B^{-0.6} \tag{21}$$

The average channel capacity, C, is the product of the signaling rate and the information duty cycle. The former term may be assumed to be proportional to the bandwidth. Therefore the average channel capacity is related to the bandwidth as follows:

$$C \sim B^{0.4} \tag{22}$$

For maximum information transfer, the bandwidth should be as large as possible.

Apart from questions of bandwidth availability, the increased noise in a wider bandwidth and hence the decreased availability of the required signal/noise ratio lead to a reduction in the information duty cycle; this in turn implies longer message delays. Moreover, a point is reached where the system has to rely on overdense trails, in which case equation (20) no longer holds. When the exponent of p_R to which D_C is proportional becomes less than -1 there is no advantage in speeding up the rate of signaling. It may be noted that the exponent is liable to fall below -1 for frequencies below 40 MHz on account of ionospheric scatter signals masking the weaker meteor-burst signals.

The usable bandwidth is not likely to be limited by the coherence bandwidth, because this is on the order of several megahertz for the main part of the burst. Even during the tails of the bursts, where there is fading on account of wind shears, the coherence bandwidth is some hundreds of kilohertz.

Interference. The high path loss associated with meteor-burst communication signals requires that the level of interfering signals be kept to a minimum. As a consequence, the operating frequency should be above that at which normal ionospheric modes propagate.

Ionospheric Absorption. Ionospheric absorption should be minimized, which requires the use of as high a frequency as possible. This is particularly of concern for systems operated at high latitudes, where auroral and polar cap absorption can attenuate and even totally absorb the signal if the operating frequency is too low.

Faraday Rotation. At certain times, Faraday rotation of meteor-burst communication signals will severely reduce the communication link capacity for frequencies below about 40 MHz.

The first influence is in conflict with the latter three, and when making a frequency choice the system designer must judge the appropriate weightings to be assigned to each.

13-6.2.11. Doppler Effects

Reflection from the head of a meteor gives rise to Doppler frequency shifts which may span the whole of the audio band. The Doppler frequency shift from a meteor trail is the result of ionospheric wind motions and could be on the order of 20 Hz at a frequency of 40 MHz. Propagation mechanisms, other than meteor scatter, may result in greater multipath and Doppler spreads.

Section 13-6.2 is taken from ITU-R Rec. 843, RPI Series, Geneva, 1992.

14

Antennas, Passive Repeaters, and Towers

The following text updates Section 14.8.3 (including Table 14-10 and Notes to Table 14-10), which starts on page 1332 and ends on page 1335 of the manual, with the following. Retain Table 14-9 on page 1334 of the manual, because it belongs to Section 14.8.2.

14-8.3. Design Wind Load on Typical Microwave Antennas/ Reflectors

Wind force data presented herein for parabolic antennas (including grid antennas) are described in the antenna axis system having the origin in the vertex of the reflector. The axial force (F_A) acts along the axis of the antenna. The side force (F_S) acts perpendicular to the antenna axis in the plane of the antenna axis and the wind vector. The twisting moment (M) acts in the plane containing F_A and F_S (see Figures 14-U.1, 14-U.2, and 14-U.3).

For horn antennas, the origin is at the intersection of the vertical antenna axis and a plane tangent to the bottom of the boresight cylinder. The axial force F_A acts parallel to the antenna boresight axis. The side force (F_S) acts perpendicular to F_A in the plane of F_A and the wind vector. The twisting moment M acts in the plane containing F_A and F_S (see Figure 14-U.4).

For flat-plate passive reflectors, the origin is at the centroid of the plate area. The axial force F_A acts along the normal to the plane. The side force (F_S) acts

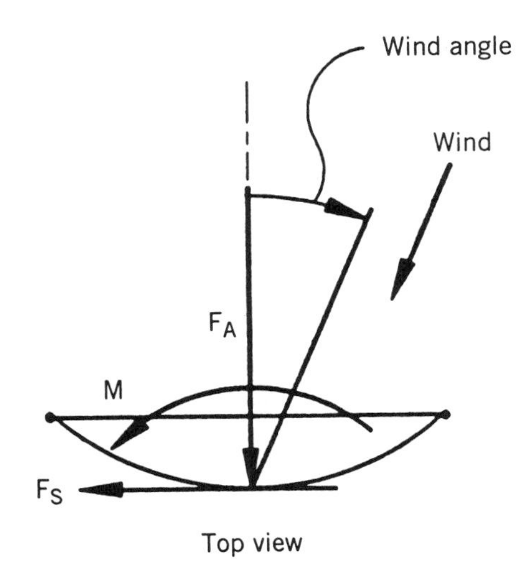

FIGURE 14-U.1. Wind forces on paraboloids and grids.

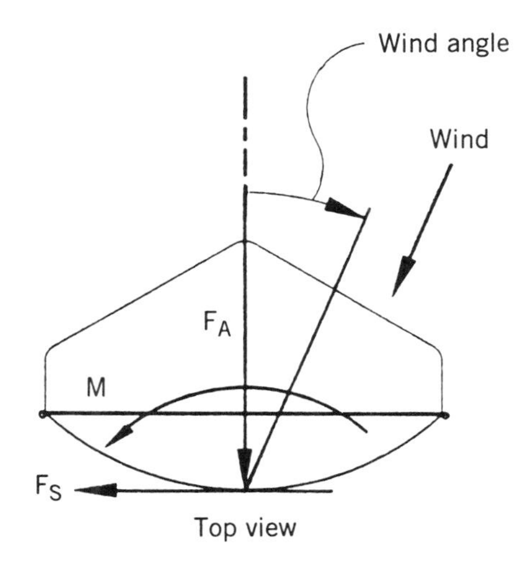

FIGURE 14-U.2. Wind forces on paraboloids with radomes. (Figures 14-U.1 and 14-U.2 are from EIA/TIA-222-E, Figures B1 and B2, page 68.)

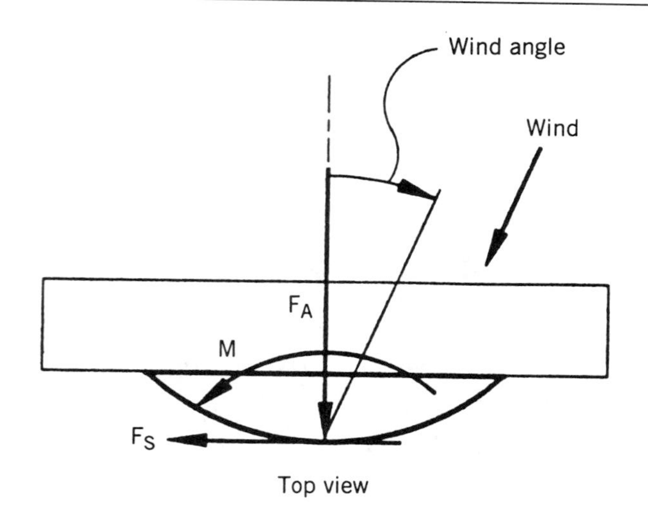

FIGURE 14-U.3. Wind forces on paraboloids with cylindrical shrouds.

perpendicular to F_A in the plane of F_A and the wind vector. The twisting moment M acts in the plane containing F_A and F_S (see Figure 14-U.5).

In all cases, the magnitudes of F_A, F_S, and M depend on the dynamic pressure of the wind, the projected frontal area of the antenna, and the aerodynamic characteristics of the antenna body. The aerodynamic characteristics vary with wind angle. The values of F_A, F_S, and M shall be calculated from the following equations:

$$F_A = C_A A K_Z G_H V^2 \quad \text{(lb)}, \qquad F_S = C_S A K_Z G_H V^2 \quad \text{(lb)},$$

$$M = C_M A D K_Z G_H V^2 \quad \text{(ft-lb)}$$

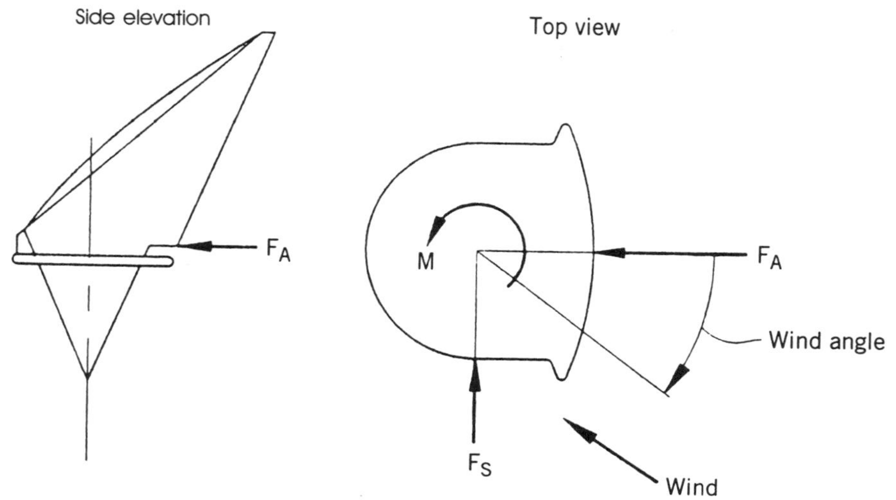

FIGURE 14-U.4. Wind forces on conical horn reflector antennas. (Figures 14-U.3 and 14-U.4 are from EIA/TIA-222-E, Figures B3 and B4, page 69.)

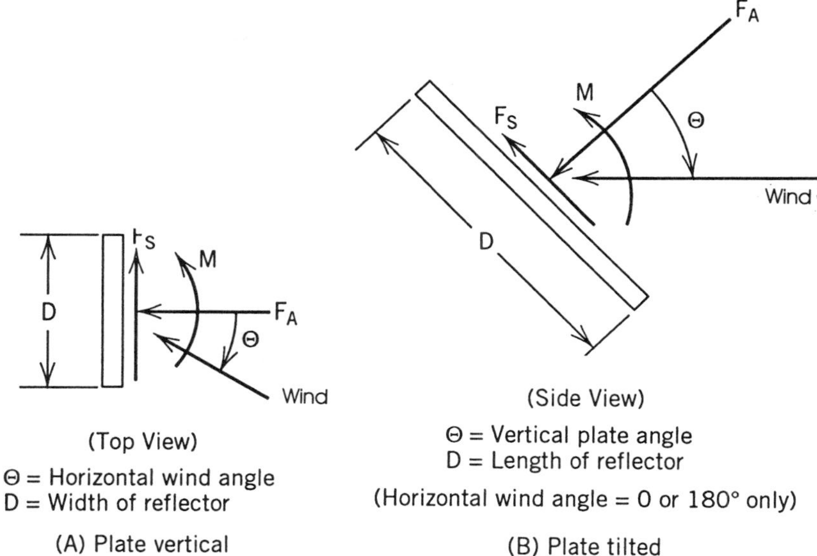

FIGURE 14-U.5. Wind forces on flat-plate passive reflectors. (From EIA/TIA-222-E, Figure B5, page 70.)

where

C_A, C_S, and C_M are the coefficients in Tables 14-U.1 through 14-U.6 as a function of wind angle θ

G_H is the gust factor from item 2 below (2.3.4 in EIA-222-E)

A is the outside aperture area (sq ft) of parabolic reflector, grid, or horn antenna, as well as plate area (sq ft) of passive reflector

D is the outside diameter (ft) of paraboloid reflector, grid, or horn antenna, as well as width or length (ft) of passive reflector (see Figure 14-U.5)

V is the basic wind speed (mph) from item 1 below (2.3.3 in EIA-222-E)

K_Z is the exposure coefficient from item 1 below (2.3.3 in EIA-222-E) with Z equal to the height of the origin of the axis system

θ is the wind angle (degrees) (see Figures 14-U.1 through 14-U.5 for positive sign conventions)

(*Note:* The coefficients used in Tables 14-U.1 through 14-U.6 are presented in the customary system of units used in North America. If SI units are desired, the results of the above equations may be converted using the conversion factors given in Chapter 30.)

1. The velocity pressure (q_Z) and the exposure coefficient (K_Z) shall be calculated from the equations

$$q_Z = 0.00256 K_Z V^2 \ (\text{lb/ft}^2) \text{ for } V \text{ in mi/hr or}$$

$$q_Z = 0.613 K_Z V^2 \ [\text{Pa}] \text{ for } V \text{ in m/sec}$$

$$K_Z = [z/33]^{2/7} \text{ for } z \text{ in ft or}$$

$$K_Z = [z/10]^{2/7} \text{ for } z \text{ in meters}$$

$$1.00 \le K_Z \le 2.58$$

V = Basic wind speed for the structure location (mi/hr) [m/sec]

z = Height above average ground level to midpoint of the section (ft) [m]

TABLE 14-U.1
Wind Force Coefficients for Typical Paraboloid without Radome

Wind Angle θ (degrees)	C_A	C_S	C_M
0	0.00397	0.00000	0.000000
10	0.00394	−0.00012	−0.000065
20	0.00396	−0.00013	−0.000097
30	0.00398	−0.00008	−0.000108
40	0.00408	0.00002	−0.000137
50	0.00426	0.00023	−0.000177
60	0.00422	0.00062	−0.000223
70	0.00350	0.00117	−0.000020
80	0.00195	0.00097	0.000256
90	−0.00003	0.00088	0.000336
100	−0.00103	0.00098	0.000338
110	−0.00118	0.00106	0.000343
120	−0.00117	0.00117	0.000366
130	−0.00120	0.00120	0.000374
140	−0.00147	0.00114	0.000338
150	−0.00198	0.00100	0.000278
160	−0.00222	0.00075	0.000214
170	−0.00242	0.00037	0.000130
180	−0.00270	0.00000	0.000000
190	−0.00242	−0.00037	−0.000130
200	−0.00222	−0.00075	−0.000214
210	−0.00198	−0.00100	−0.000278
220	−0.00147	−0.00114	−0.000338
230	−0.00120	−0.00120	−0.000374
240	−0.00117	−0.00117	−0.000366
250	−0.00118	−0.00106	−0.000343
260	−0.00103	−0.00098	−0.000338
270	−0.00003	−0.00088	−0.000336
280	0.00195	−0.00097	−0.000256
290	0.00350	−0.00117	0.000020
300	0.00422	−0.00062	0.000223
310	0.00426	−0.00023	0.000177
320	0.00408	−0.00002	0.000137
330	0.00398	0.00008	0.000108
340	0.00396	0.00013	0.000097
350	0.00394	0.00012	0.000065

Source: EIA/TIA-222-E, Table B1, page 62.

A. Unless otherwise specified, the basic wind speed (V) for the structure location shall be determined from Section 16 in EIA-222-E.

2. Gust Response Factors

A. For latticed structures, the gust response factor (G_H) shall be calculated from the equation

$$G_H = 0.65 + 0.60/(\mathrm{h}/33)^{1/7} \text{ for hr in ft or}$$

$$G_H = 0.65 + 0.60/(\mathrm{h}/10)^{1/7} \text{ for hr in meters}$$

$$1.00 \leq G_H \leq 1.25$$

B. For tubular pole structures, the gust response factor (G_H) shall be 1.69.

TABLE 14-U.2
Wind Force Coefficients for Typical Paraboloid with Radome

Wind Angle θ (degrees)	C_A	C_S	C_M
0	0.00221	0.00000	0.00000
10	0.00220	0.00038	−0.000204
20	0.00210	0.00076	−0.000285
30	0.00195	0.00105	−0.000277
40	0.00170	0.00125	−0.000205
50	0.00140	0.00136	−0.000114
60	0.00107	0.00128	−0.000002
70	0.00080	0.00118	0.000130
80	0.00058	0.00112	0.000268
90	0.00034	0.00104	0.000390
100	0.00008	0.00100	0.000434
110	−0.00017	0.00095	0.000422
120	−0.00042	0.00089	0.000404
130	−0.00075	0.00082	0.000357
140	−0.00105	0.00078	0.000232
150	−0.00133	0.00070	0.000132
160	−0.00154	0.00058	0.000063
170	−0.00168	0.00038	0.000022
180	−0.00177	0.00000	0.000000
190	−0.00168	−0.00038	−0.000022
200	−0.00154	−0.00058	−0.000063
210	−0.00133	−0.00070	−0.000132
220	−0.00105	−0.00078	−0.000232
230	−0.00075	−0.00082	−0.000357
240	−0.00042	−0.00089	−0.000404
250	−0.00017	−0.00095	−0.000422
260	0.00008	−0.00100	−0.000434
270	0.00034	−0.00104	−0.000390
280	0.00058	−0.00112	−0.000268
290	0.00080	−0.00118	−0.000130
300	0.00107	−0.00128	0.000002
310	0.00140	−0.00136	0.000114
320	0.00170	−0.00125	0.000205
330	0.00195	−0.00105	0.000277
340	0.00210	−0.00076	0.000285
350	0.00220	−0.00038	0.000204

Source: EIA/TIA-222-E, Table B2, page 63.

TABLE 14-U.3
Wind Force Coefficients for Paraboloid with Cylindrical Shroud

Wind Angle θ (degrees)	C_A	C_S	C_M
0	0.00323	0.00000	0.000000
10	0.00323	0.00025	−0.000072
20	0.00320	0.00045	−0.000116
30	0.00310	0.00060	−0.000133
40	0.00296	0.00072	−0.000125
50	0.00278	0.00078	−0.000083
60	0.00242	0.00094	−0.000022
70	0.00172	0.00122	0.000058
80	0.00070	0.00149	0.000178
90	−0.00028	0.00160	0.000251
100	−0.00088	0.00154	0.000288
110	−0.00138	0.00136	0.000292
120	−0.00182	0.00112	0.000266
130	−0.00220	0.00080	0.000237
140	−0.00239	0.00059	0.000199
150	−0.00245	0.00045	0.000158
160	−0.00249	0.00038	0.000112
170	−0.00255	0.00025	0.000059
180	−0.00260	0.00000	0.000000
190	−0.00255	−0.00025	−0.000059
200	−0.00249	−0.00038	−0.000112
210	−0.00245	−0.00045	−0.000158
220	−0.00239	−0.00059	−0.000199
230	−0.00220	−0.00080	−0.000237
240	−0.00182	−0.00112	−0.000266
250	−0.00138	−0.00136	−0.000292
260	−0.00088	−0.00154	−0.000288
270	−0.00028	−0.00160	−0.000251
280	0.00070	−0.00149	−0.000178
290	0.00172	−0.00122	−0.000058
300	0.00242	−0.00094	0.000022
310	0.00278	−0.00078	0.000083
320	0.00296	−0.00072	0.000125
330	0.00310	−0.00060	0.000133
340	0.00320	−0.00045	0.000116
350	0.00323	−0.00025	0.000072

Source: EIA/TIA-222-E, Figure B3, page 64.

TABLE 14-U.4
Wind Force Coefficients for Typical Grid Antenna without Ice

Wind Angle θ (degrees)	C_A	C_S	C_M
0	0.00137	0.00000	0.000000
10	0.00134	0.00026	0.000043
20	0.00130	0.00046	0.000074
30	0.00118	0.00059	0.000098
40	0.00104	0.00067	0.000115
50	0.00088	0.00070	0.000127
60	0.00060	0.00072	0.000135
70	0.00033	0.00070	0.000142
80	0.00010	0.00064	0.000126
90	−0.00013	0.00062	0.000111
100	−0.00030	0.00070	0.000120
110	−0.00048	0.00073	0.000129
120	−0.00068	0.00071	0.000131
130	−0.00086	0.00067	0.000127
140	−0.00104	0.00060	0.000114
150	−0.00122	0.00052	0.000095
160	−0.00140	0.00040	0.000070
170	−0.00150	0.00022	0.000038
180	−0.00152	0.00000	0.000000
190	−0.00150	−0.00022	−0.000038
200	−0.00140	−0.00040	−0.000070
210	−0.00122	−0.00052	−0.000095
220	−0.00104	−0.00060	−0.000114
230	−0.00086	−0.00067	−0.000127
240	−0.00068	−0.00071	−0.000131
250	−0.00048	−0.00073	−0.000129
260	−0.00030	−0.00070	−0.000120
270	−0.00013	−0.00062	−0.000111
280	0.00010	−0.00064	−0.000126
290	0.00033	−0.00070	−0.000142
300	0.00060	−0.00072	−0.000135
310	0.00088	−0.00070	−0.000127
320	0.00104	−0.00067	−0.000115
330	0.00118	−0.00059	−0.000098
340	0.00130	−0.00046	−0.000074
350	0.00134	−0.00026	−0.000043

Note: In the absence of more accurate data for a grid antenna with ice, use wind force coefficients for typical paraboloid without radome from Table 14-U.1.
Source: EIA/TIA-222-E, Table B4, page 65.

TABLE 14-U.5
Wind Force Coefficients for Typical Conical Horn Reflector Antenna

Wind Angle θ (degrees)	C_A	C_S	C_M
0	0.00338	0.00000	0.00000
10	0.00355	0.00004	−0.00005
20	0.00354	0.00025	−0.00007
30	0.00345	0.00077	−0.00001
40	0.00335	0.00142	0.00009
50	0.00299	0.00181	0.00023
60	0.00235	0.00208	0.00035
70	0.00154	0.00237	0.00044
80	0.00059	0.00248	0.00046
90	−0.00020	0.00245	0.00040
100	−0.00062	0.00240	0.00032
110	−0.00088	0.00235	0.00030
120	−0.00147	0.00225	0.00032
130	−0.00225	0.00201	0.00027
140	−0.00289	0.00167	0.00021
150	−0.00323	0.00113	0.00014
160	−0.00367	0.00052	0.00007
170	−0.00375	0.00010	0.00003
180	−0.00356	0.00000	0.00000
190	−0.00375	−0.00010	−0.00003
200	−0.00367	−0.00052	−0.00007
210	−0.00323	−0.00113	−0.00014
220	−0.00289	−0.00167	−0.00021
230	−0.00225	−0.00201	−0.00027
240	−0.00147	−0.00225	−0.00032
250	−0.00088	−0.00235	−0.00030
260	−0.00062	−0.00240	−0.00032
270	−0.00020	−0.00245	−0.00040
280	0.00059	−0.00248	−0.00046
290	0.00154	−0.00237	−0.00044
300	0.00235	−0.00208	−0.00035
310	0.00299	−0.00181	−0.00023
320	0.00335	−0.00142	−0.00009
330	0.00345	−0.00077	0.00001
340	0.00354	−0.00025	0.00007
350	0.00355	−0.00004	0.00005

Source: EIA/TIA-222-E, Table B5, page 66.

TABLE 14-U.6
Wind Force Coefficients for Typical Passive Reflector

Wind Angle θ (degrees)	C_A	C_S	C_M
0	0.00351	0.00000	0.000000
10	0.00348	0.00003	−0.000077
20	0.00341	0.00008	−0.000134
30	0.00329	0.00010	−0.000180
40	0.00309	0.00013	−0.000198
50	0.00300	0.00018	−0.000208
60	0.00282	0.00021	−0.000262
70	0.00178	0.00023	−0.000225
80	0.00071	0.00027	−0.000129
90	−0.00010	0.00030	0.000030
100	−0.00108	0.00035	0.000180
110	−0.00235	0.00039	0.000225
120	−0.00348	0.00036	0.000210
130	−0.00348	0.00029	0.000148
140	−0.00360	0.00023	0.000126
150	−0.00376	0.00019	0.000109
160	−0.00390	0.00012	0.000080
170	−0.00400	0.00008	0.000042
180	−0.00403	0.00000	0.000000
190	−0.00400	−0.00008	−0.000042
200	−0.00390	−0.00012	−0.000080
210	−0.00376	−0.00019	−0.000109
220	−0.00360	−0.00023	−0.000126
230	−0.00348	−0.00029	−0.000148
240	−0.00348	−0.00036	−0.000210
250	−0.00235	−0.00039	−0.000225
260	−0.00108	−0.00035	−0.000180
270	−0.00010	−0.00030	−0.000030
280	0.00071	−0.00027	0.000129
290	0.00178	−0.00023	0.000225
300	0.00282	−0.00021	0.000262
310	0.00300	−0.00018	0.000208
320	0.00309	−0.00013	0.000198
330	0.00329	−0.00010	0.000180
340	0.00341	−0.00008	0.000134
350	0.00348	−0.00003	0.000077

Source: EIA/TIA-222-E, Table B6, page 67.

TABLE 14-U.7

Allowable Twist and Sway Values for Parabolic Antennas, Passive Reflectors, and Periscope System Reflectors

A	B	C	D	E	F	G	H	I
		Parabolic Antennas		Passive Reflectors		Periscope System Reflectors		
3-dB Beamwidth $2\theta_{HP}$ for Antenna Only (degrees) (Note 8)	Deflection Angle at 10-dB Points (degrees) (Notes 1, 7)	Limit of Antenna Movement with Respect to Structure (degrees)	Limit of Structure Movement Twist or Sway at Antenna Attachment Point (degrees)	Limit of Passive Reflector Sway (degrees) (Notes 4, 5)	Limit of Passive Reflector Twist (degrees) (Note 4)	Limit of Reflector Movement with Respect to Structure (degrees)	Limit of Structure Twist at Reflector Attachment Point (degrees)	Limit of Structure Sway at Reflector Attachment Point (degrees)
5.6	5.0	0.4	4.6	3.5	2.5	0.2	4.8	2.3
5.6	4.8	0.4	4.4	3.3	2.4	0.2	4.6	2.2
5.4	4.6	0.4	4.2	3.2	2.3	0.2	4.4	2.1
5.1	4.4	0.4	4.0	3.0	2.2	0.2	4.2	2.0
4.9	4.2	0.4	3.8	2.9	2.1	0.2	4.0	1.9
4.7	4.0	0.3	3.7	2.8	2.0	0.2	3.8	1.8
4.4	3.8	0.3	3.5	2.6	1.9	0.2	3.6	1.7
4.2	3.6	0.3	3.3	2.5	1.8	0.2	3.4	1.6
4.0	3.4	0.3	3.1	2.3	1.7	0.2	3.2	1.5
3.7	3.2	0.3	2.9	2.2	1.6	0.2	3.0	1.4
3.5	3.0	0.3	2.7	2.1	1.5	0.2	2.8	1.4
3.4	2.9	0.2	2.7	2.0	1.45	0.1	2.8	1.3
3.3	2.8	0.2	2.6	1.9	1.4	0.1	2.7	1.3
3.1	2.7	0.2	2.5	1.8	1.35	0.1	2.6	1.25
3.0	2.6	0.2	2.4	1.8	1.3	0.1	2.5	1.2
2.9	2.5	0.2	2.3	1.7	1.25	0.1	2.4	1.15
2.8	2.4	0.2	2.2	1.6	1.2	0.1	2.3	1.1
2.7	2.3	0.2	2.1	1.6	1.15	0.1	2.2	1.05
2.6	2.2	0.2	2.0	1.5	1.1	0.1	2.1	0.1
2.5	2.1	0.2	1.9	1.4	1.05	0.1	2.0	0.95
2.3	2.0	0.2	1.8	1.4	1.0	0.1	1.9	0.9
2.2	1.9	0.2	1.7	1.3	0.95	0.1	1.8	0.85
2.1	1.8	0.2	1.6	1.2	0.9	0.1	1.7	0.8
2.0	1.7	0.2	1.5	1.1	0.85	0.1	1.6	0.75
1.9	1.6	0.2	1.4	1.1	0.8	0.1	1.5	0.7
1.7	1.5	0.2	1.3	0.1	0.75	0.1	1.4	0.65
1.6	1.4	0.2	1.2	0.9	0.7	0.1	1.3	0.6
1.5	1.3	0.1	1.2	0.9	0.65	0.1	1.2	0.55
1.4	1.2	0.1	1.1	0.8	0.6	0.1	1.1	0.5
1.3	1.1	0.1	1.0	0.7	0.55	0.1	1.0	0.45
1.2	1.0	0.1	0.9	0.7	0.5	0.1	0.9	0.4
1.1	0.9	0.1	0.8	0.6	0.45	0.1	0.8	0.35
0.9	0.8	0.1	0.7	0.5	0.4	0.1	0.7	0.3
0.8	0.7	0.1	0.6	0.4	0.35	0.1	0.6	0.25
0.7	0.6	0.1	0.5	0.4	0.3	0.1	0.5	0.2
0.6	0.5	0.1	0.4	0.3	0.25	0.1	0.4	0.15
0.5	0.4	0.1	0.3	0.2	0.2	0.07	0.3	0.13
0.3	0.3	0.05	0.25	0.2	0.15	0.05	0.25	0.10
0.2	0.2			0.14	0.1			
0.1	0.1			0.07	0.05			
						Only for configuration where antenna is directly under the reflector.		

Source: Appendix C, EIA/TIA-222-E, pages 71 and 72.

C. One gust response factor shall apply for the entire structure.
D. When cantilevered tubular or latticed pole structures are mounted on latticed structures, the gust response factor for the pole and the latticed structure shall be based on the height of the latticed structure without the pole. The stresses calculated for pole structures and their connections to latticed structures shall be multiplied by 1.25 to compensate for the greater gust response for mounted pole structures.

Table 14-U.7 presents allowable twist and sway values for microwave tower antenna and reflector systems.

14.8.3.1. Determination of Allowable Beam Twist and Sway for Cross-Polarization Limited Systems

A dual polarized antenna has a pattern like that shown in Figures 14-U.9 and 14-U.10. For most offset antennas the cross-polarized null is deep, as shown in Figure 14-U.9; for the center-fed antennas, the cross-polarized null is shallow and the envelope is as shown in Figure 14-U.10. In either case, as soon as the antenna is deflected from its normal position, the cross-polarization discrimination, XPD (the difference between the co-polarized and the cross-polarized signal), decreases.

Where on-path cross-polarization discrimination is critical to system performance, allowable beam deflection θ would be determined as shown in Figures will determine twist only and the antenna beamwidth will determine sway. For center-fed antennas, θ will determine both twist and sway. Table 14-U.8 provides allowable twist and sway values for cross-polarization limited systems.

Sections 14.8.3 and 14.8.3.1 were extracted from Appendix B and Appendix D, respectively, in EIA/TIA-222-E.

NOTES TO TABLE 14-U.7

Note 1: If values for columns A and B are not available from the manufacturer(s) of the antenna system or from the user of the antenna system, then values shall be obtained from Figure 14-U.6, 14-U.7, or 14-U.8.

Note 2: Limits of beam movement for twist or sway (treated separately in most analyses) will be the sum of the appropriate figures in columns C & D, G & H, and G & I. Columns G, H & I apply to a vertical periscope configuration.

Note 3: It is not intended that the values in this table imply an accuracy of beam width determination or structural rigidity calculation beyond known practicable values and computational procedures. For most microwave structures it is not practicable to require a calculated structural rigidity of less than 1/4-degree twist or sway with a 50-mi/hr (22.4-m/sec) basic wind speed.

Note 4: For passive reflectors the allowable twist and sway values are assumed to include the effects of all members contributing to the rotation of the face under wind load. For passives not elevated far above ground [approximately 5–20 feet (1.5–6 m) clearance above ground], the structure and reflecting face supporting elements are considered an integral unit. Therefore, separating the structure portion of the deflection is only meaningful when passives are mounted on conventional microwave structures.

Note 5: The allowable sway for passive reflectors is considered to be 1.4 times the allowable twist to account for the amount of rotation of the face about a horizontal axis through the face center and parallel to the face compared to the amount of beam rotation along the direction of the path as it deviates from the plane of the incident and reflected beam axis.

Note 6: Linear horizontal movement of antennas and reflectors in the amount experienced for properly designed microwave antenna system support structures is not considered a problem (no significant signal degradation attributed to this movement).

Note 7: For systems using a frequency of 450 MHz, the half-power beamwidths may be nearly 2θ degrees for some antennas. However, structures designed for microwave relay systems will usually have an inherent rigidity less than the maximum 5 degree deflection angle shown on the chart.

Note 8: The 3-dB beamwidths, $2\theta_{HP}$ in column A are shown for convenient reference to manufacturers' published antenna information. The minimum deflection reference for this standard is the allowable total deflection angle θ at the 10-dB points.

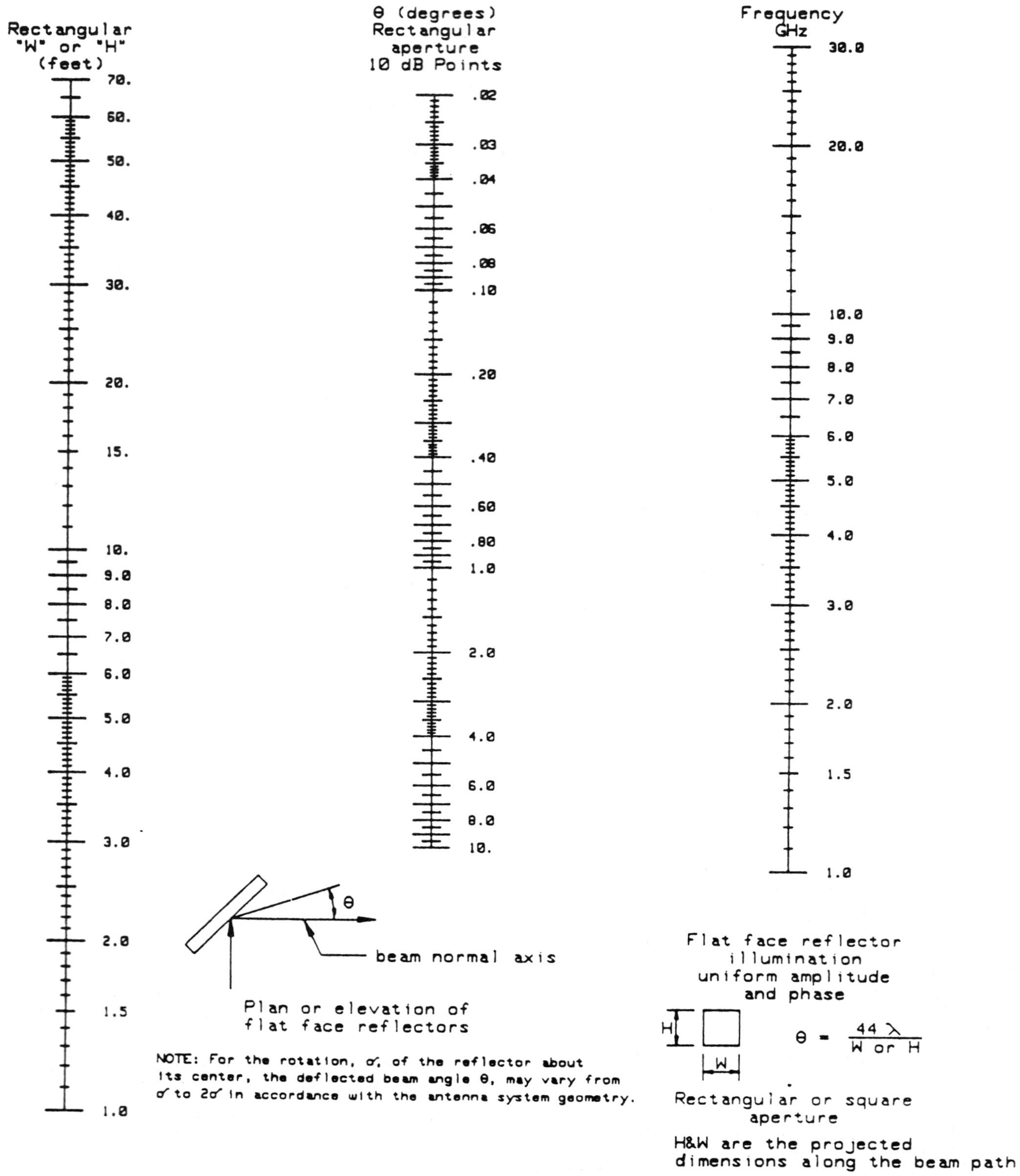

FIGURE 14-U.6. Nomogram showing deflection angle θ at 10-dB points for a rectangular aperture (flat face reflector). (From EIA/TIA-222-E, Figure C1, page 73.)

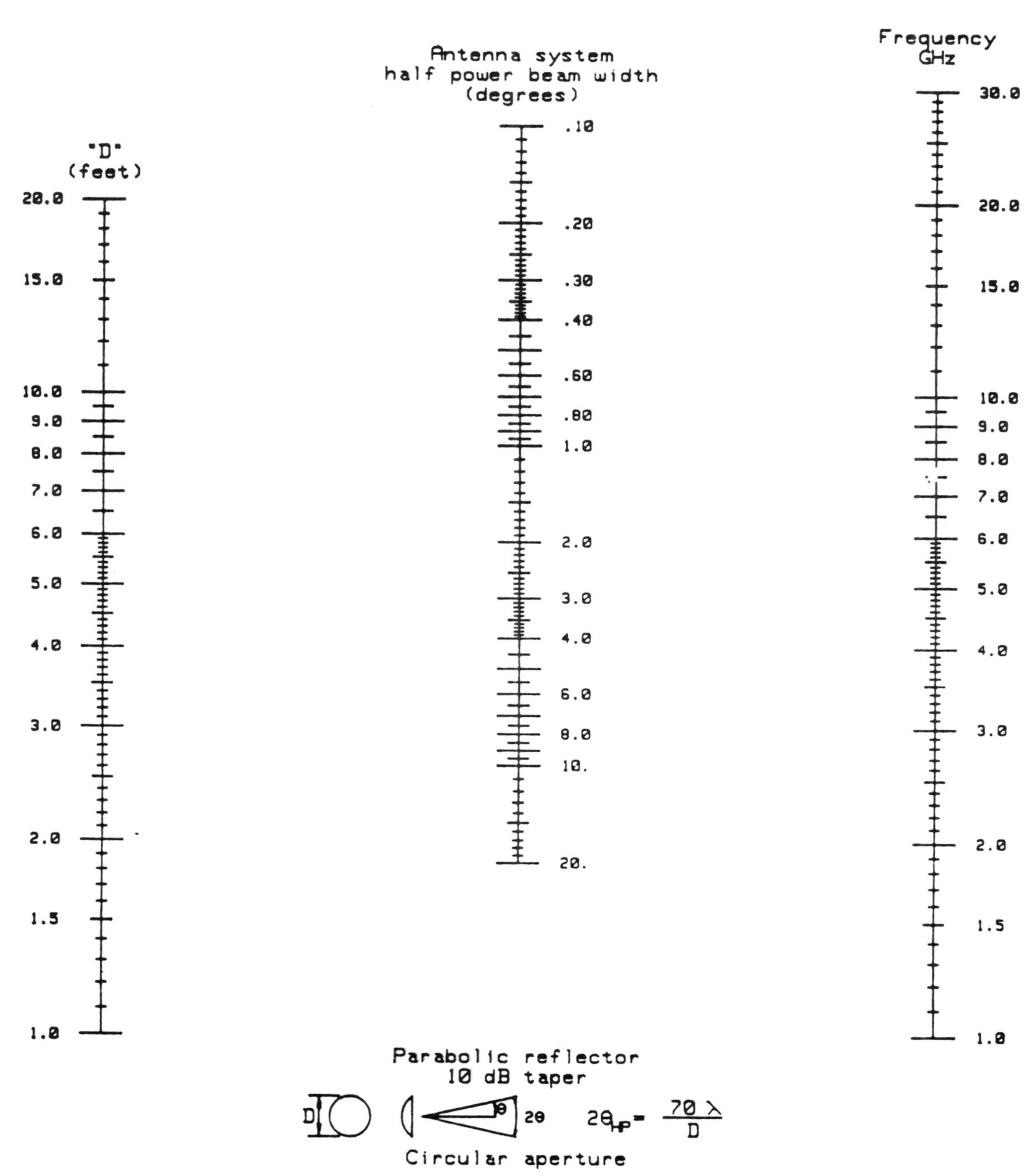

FIGURE 14-U.7. Nomogram showing nominal beamwidth at 3-dB points for a typical parabolic reflector. (From EIA/TIA-222-E, Figure C2, page 74.)

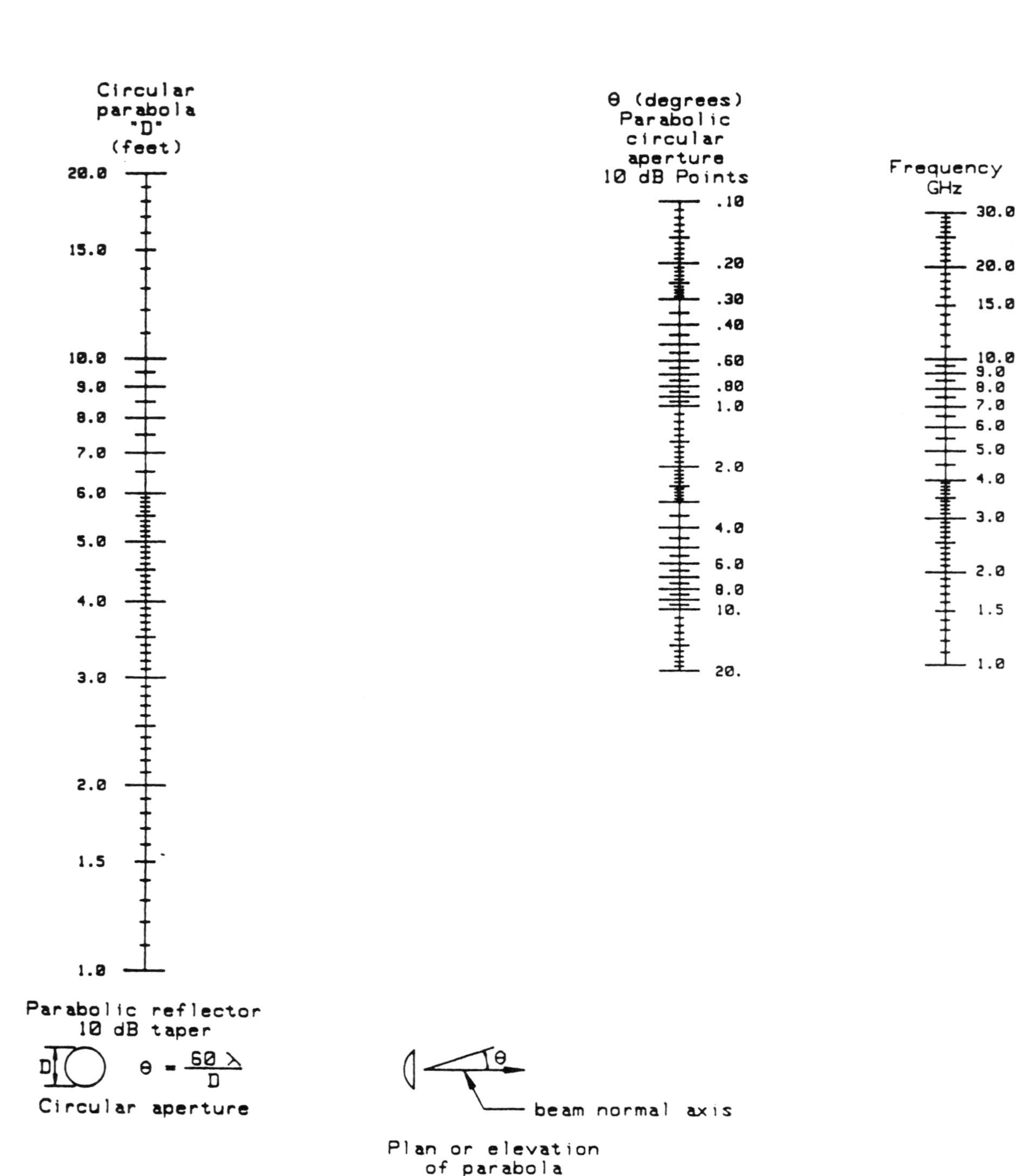

FIGURE 14-U.8. Nomogram showing deflection angle θ at 10-dB points for a circular aperture (parabolic surface contour). (From EIA/TIA-222-E, Figure C3, page 75.)

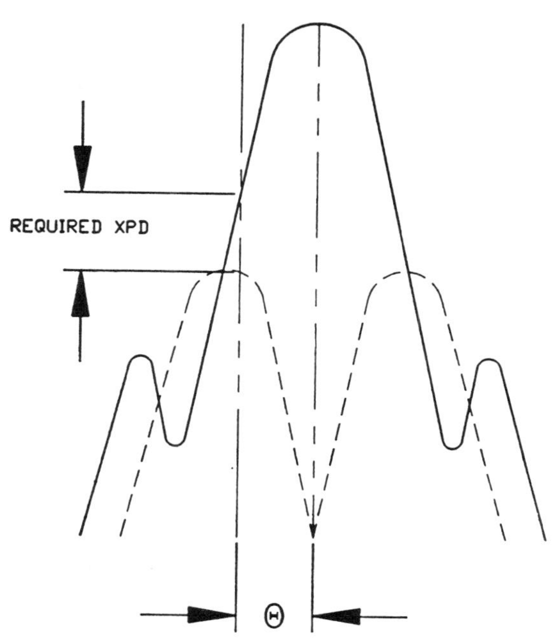

FIGURE 14-U.9. Offset fed antenna. (From EIA/TIA-222-E, Figure D1, page 78.)

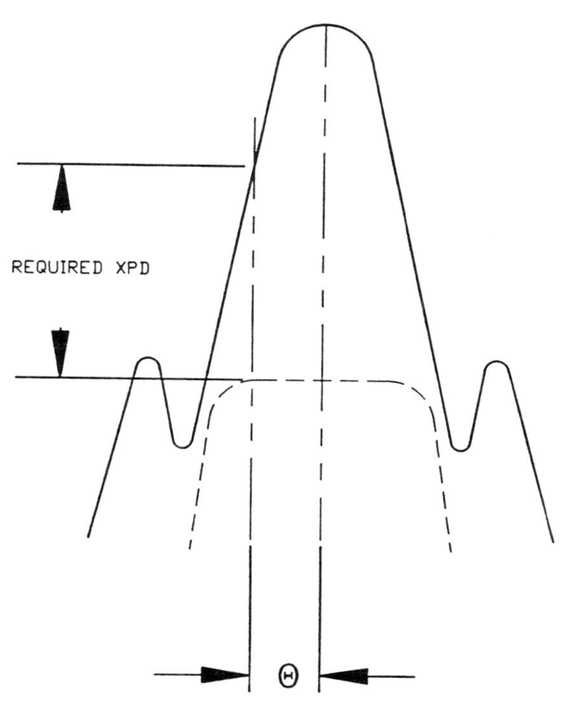

FIGURE 14-U.10. Center-fed antenna. (From EIA/TIA-222-E, Figure D2, page 79.)

TABLE 14-U.8
Allowable Twist and Sway for Cross-Polarization Limited Systems

Allowable Twist for Offset-Fed Antennas. Allowable Twist & Sway for Center-Fed Antennas			Allowable Sway for Offset-Fed Antennas			
A	B	C	D	E	F	G
Allowable Beam Twist or Sway for Cross-Polarization Limited Systems (θ) (degrees)	Limit of Antenna Movement with Respect to Structure (degrees)	Limit of Structure Movement at Antenna Attachment Point (degrees)	3-dB Beam-width $2\theta_{HP}$ for Antenna Only (degrees)	Deflection Angle at 10-dB Points (degrees)	Limit of Antenna Movement with Respect to Structure (degrees)	Limit of Structure Sway at Attachment Point (degrees)
5.0	0.5	4.5	5.8	5.0	0.4	4.6
4.0	0.4	3.6	5.6	4.8	0.4	4.4
3.0	0.3	2.7	5.4	4.6	0.4	4.2
2.0	0.2	1.8	5.1	4.4	0.4	4.0
1.0	0.1	0.9	4.9	4.2	0.4	3.8
0.9	0.09	0.81	4.7	4.0	0.3	3.7
0.8	0.08	0.72	4.4	3.8	0.3	3.5
0.7	0.07	0.63	4.2	3.6	0.3	3.3
0.6	0.06	0.54	4.0	3.4	0.3	3.1
0.5	0.05	0.45	3.7	3.2	0.3	2.9
0.4	0.04	0.36	3.5	3.0	0.3	2.7
0.3	0.03	0.27	3.4	2.9	0.2	2.7
0.2	0.02	0.18	3.3	2.8	0.2	2.6
0.1	0.01	0.09	3.1	2.7	0.2	2.5
			3.0	2.6	0.2	2.4
			2.9	2.5	0.2	2.3
			2.8	2.4	0.2	2.2
			2.7	2.3	0.2	2.1
			2.6	2.2	0.2	2.0
			2.5	2.1	0.2	1.9
			2.3	2.0	0.2	1.8
			2.2	1.9	0.2	1.7
			2.1	1.8	0.2	1.6
			2.0	1.7	0.2	1.5
			1.9	1.6	0.2	1.4
			1.7	1.5	0.2	1.3
			1.6	1.4	0.2	1.2
			1.5	1.3	0.1	1.2
			1.4	1.2	0.1	1.1
			1.3	1.1	0.1	1.0
			1.2	1.0	0.1	0.9
			1.1	0.9	0.1	0.8
			0.9	0.8	0.1	0.7
			0.8	0.7	0.1	0.6
			0.7	0.6	0.1	0.5
			0.6	0.5	0.1	0.4
			0.5	0.4	0.1	0.3
			0.3	0.3	0.05	0.25
			0.2	0.2		
			0.1	0.1		

Source: EIA/TIA-222-E, Table D1, page 80.

Due to the replacement of Section 14.8.3, Tables 14.11 and 14.12 should be renumbered to 14.10 and 14.11 as shown.

The values in Tables 14-10 and 14-11 apply in the horizontal plane. For restrictions on vertical plane radiation, see note in Section 2.1.8 of EIA RS-195-B [Ref. 8].

TABLE 14-10
FCC Antenna Radiation Standard A

Frequency Band (GHz)	3-dB Max. Beamwidth (degrees)	Gain Min. (dBi)	Angle (in Degrees) from Peak of Beam						
			5–10	10–15	15–20	20–30	30–100	100–140	140–180
			Minimum Radiation Suppression (dB)						
0.952–0.960	14	NA[a]	—	6	11	14	17	20	24
1.85–2.69	5	NA	12	18	22	25	29	33	39
3.7–4.2	NA	36.15	23	29	33	36	42	55	55
5.925–6.425	NA	38.15	25	29	33	36	42	55	55
6.525–6.875	1.5	NA	26	29	32	34	38	41	49
10.7–11.7	NA	38.15	25	29	33	36	42	55	55
12.2–12.7	1.0	NA	23	28	35	39	41	42	50
Above 12.7		To Be Specified in Authorization							

[a]NA, not applicable.

TABLE 14-11
FCC Antenna Radiation Standard B

Frequency Band (GHz)	3-dB Max. Beamwidth (degrees)	Gain Min. (dBi)	Angle (in Degrees) from Peak of Beam						
			5–10	10–15	15–20	20–30	30–100	100–140	140–180
			Minimum Radiation Suppression (dB)						
0.952–0.960	20.0	NA[a]	—	—	6	10	13	15	20
1.85–2.69	8.0	NA	5	18	20	20	25	28	36
3.7–4.2	NA	36.15	20	24	28	32	32	32	32
5.925–6.425	NA	38.15	20	24	28	32	35	36	36
6.525–6.875	2.0	NA	21	25	29	32	35	39	45
10.7–11.7	NA	38.15	20	24	28	32	35	36	36
12.2–12.7	2.0	NA	20	25	28	30	32	37	47
Above 12.7		To Be Specified in Authorization							

[a]NA, not applicable.

Section 14.9 was taken from Appendix III, EIA RS-195-B [Ref. 8]. Courtesy of Electronic Industries Association, Washington, D.C. Copyright Registration TX 297-832.

NOTES TO TABLE 14-U.8

Note 1: If values for columns D and E of the sway table and column A of the twist table are not available from the manufacturer(s) of the antenna system or from the user of the antenna system, then values shall be obtained from Figure 14-U.7 or 14-U.8.

Note 2: Limits of beam movement for twist or sway (treated separately in most analyses) are the sum of the appropriate figures in columns B and C of the twist table and the sum of the appropriate figures in columns F and G of the sway table.

Note 3: Linear horizontal movement of antennas and reflectors in the amount experienced for properly designed microwave antenna system support structures is not considered a problem (no significant signal degradation attributed to this movement).

Note 4: The 3-dB beamwidths, $2\theta_{HP}$ in column D are shown for convenient reference to manufacturers' standard published antenna information. The minimum deflection reference for this standard is the allowable total deflection angle θ at the 10-dB points.

Note 5: The values shown in this table depict angular deflections in two orthogonal planes normal to the boresight direction: vertical elevation (sway) and horizontal azimuth (twist). No allowance has been made for initial offsets due to mount skew, installation tolerances, paths not normal to the support structures, and so on. Special considerations will be required in those cases.

Note 6: It is not intended that the values in this table imply an accuracy of beamwidth determination or structural rigidity calculation beyond known practicable values and computational procedures. For most microwave structures it is not practicable to require a calculated structural rigidity of less than 1/4-degree twist or sway with a 50-mi/hr (22.4-m/sec) basic wind speed.

17

Data/Telegraph Transmission

This new section follows Section 17-10.1.7, which ends on page 1501 of the Manual.

A Duplex Modem Operating at Data Signaling Rates Up to 14,400 bps for Use on the General Switched Telephone Network or Leased Point-to-Point 2-Wire Telephone-Type Circuits

1. Introduction. This modem is intended for use on connections on general switched telephone networks (GSTNs) and on point-to-point 2-wire leased telephone-type circuits. The principal characteristics of the modem are as follows:

- **(a)** Duplex mode of operation on GSTN and point-to-point 2-wire leased circuits
- **(b)** Channel separation by echo cancellation techniques
- **(c)** Quadrature amplitude modulation for each channel with synchronous line transmission at 2400 symbols/sec
- **(d)** The following synchronous data signaling rates shall be implemented in the modem:
  - 14,400 bps trellis-coded
  - 12,000 bps trellis-coded
  - 9600 bps trellis-coded
  - 7200 bps trellis-coded
  - 4800 bps uncoded
- **(e)** Compatibility with Rec. V.32 modems at 9600 and 4800 bps
- **(f)** Exchange of rate sequences during start-up to establish the data signaling rate
- **(g)** A procedure to change the data signaling rate without retraining

Note 1: On international GSTN connections that utilize circuits that are in accord with Rec. G.235 (16-channel terminal equipments), it may be necessary to employ a greater degree of equalization within the modem than would be required for use on most national GSTN connections.

Note 2: The transmit and receive rates in each modem shall be the same. The possibility of asymmetric working remains for further study.

2. Line Signals

2.1. Carrier Frequency and Modulation Rate. The carrier frequency is to be 1800 $\pm$ 1 Hz. The receiver must be able to operate with a maximum received frequency offset of up to ± 7 Hz.

The modulation rate shall be 2400 symbols/sec $\pm 0.01\%$.

2.2. Transmitted Spectrum. The transmitted power level must conform to ITU-T Rec. V.2. With continuous binary ones applied to the input of the scrambler, the transmitted energy density at 600 Hz and 3000 Hz shall be attenuated 4.5 ± 2.5 dB with respect to the maximum energy density between 600 Hz and 3000 Hz.

2.3. Coding

2.3.1. Signal element coding for 14,400 bps. At 14,400 bps, the scrambled data stream to be transmitted is divided into groups of six consecutive data bits. The first two bits in time $Q1_n$ and $Q2_n$ in each group, where the subscript n designates the sequence number of the group, are first differentially encoded into $Y1_n$ and $Y2_n$ according to Table 17-U.1. The two differentially encoded bits $Y1_n$ and $Y2_n$ are used as input to a systematic convolutional encoder which generates a redundant bit $Y0_n$ (see Figure 17-U.1). This redundant bit and the six information-carrying bits $Y1_n$, $Y2_n$, $Q3_n$, $Q4_n$, $Q5_n$, and $Q6_n$ are then mapped into the coordinates of the signal element to be transmitted according to the signal space diagram shown in Figure 17-U.2.

2.3.2. Signal element coding for 12,000 bps. At 12,000 bps, the scrambled data stream to be transmitted is divided into groups of five consecutive data bits. The first two bits in time $Q1_n$ and $Q2_n$ in each group, where the subscript n designates the sequence number of the group, are first differentially encoded into $Y1_n$ and $Y2_n$ according to Table 17-U.1. The two differentially encoded bits $Y1_n$ and $Y2_n$ are used as input to a systematic convolutional encoder which generates a redundant bit $Y0_n$ (see Figure 17-U.1). This redundant bit and the five information-carrying bits $Y1_n$, $Y2_n$, $Q3_n$, $Q4_n$, and $Q5_n$ are then mapped into the coordinates of the signal element to be transmitted according to the signal space diagram shown in Figure 17-U.3.

TABLE 17-U.1
Differential Quadrant Coding with Trellis Coding

Inputs		Previous Outputs		Outputs	
$Q1_n$	$Q2_n$	$Y1_{n-1}$	$Y2_{n-1}$	$Y1_n$	$Y2_n$
0	0	0	0	0	0
0	0	0	1	0	1
0	0	1	0	1	0
0	0	1	1	1	1
0	1	0	0	0	1
0	1	0	1	0	0
0	1	1	0	1	1
0	1	1	1	1	0
1	0	0	0	1	0
1	0	0	1	1	1
1	0	1	0	0	1
1	0	1	1	0	0
1	1	0	0	1	1
1	1	0	1	1	0
1	1	1	0	0	0
1	1	1	1	0	1

Source: Table 1/V.32 bis, ITU-T Rec. V.32 bis, Geneva, 1992, page 2.

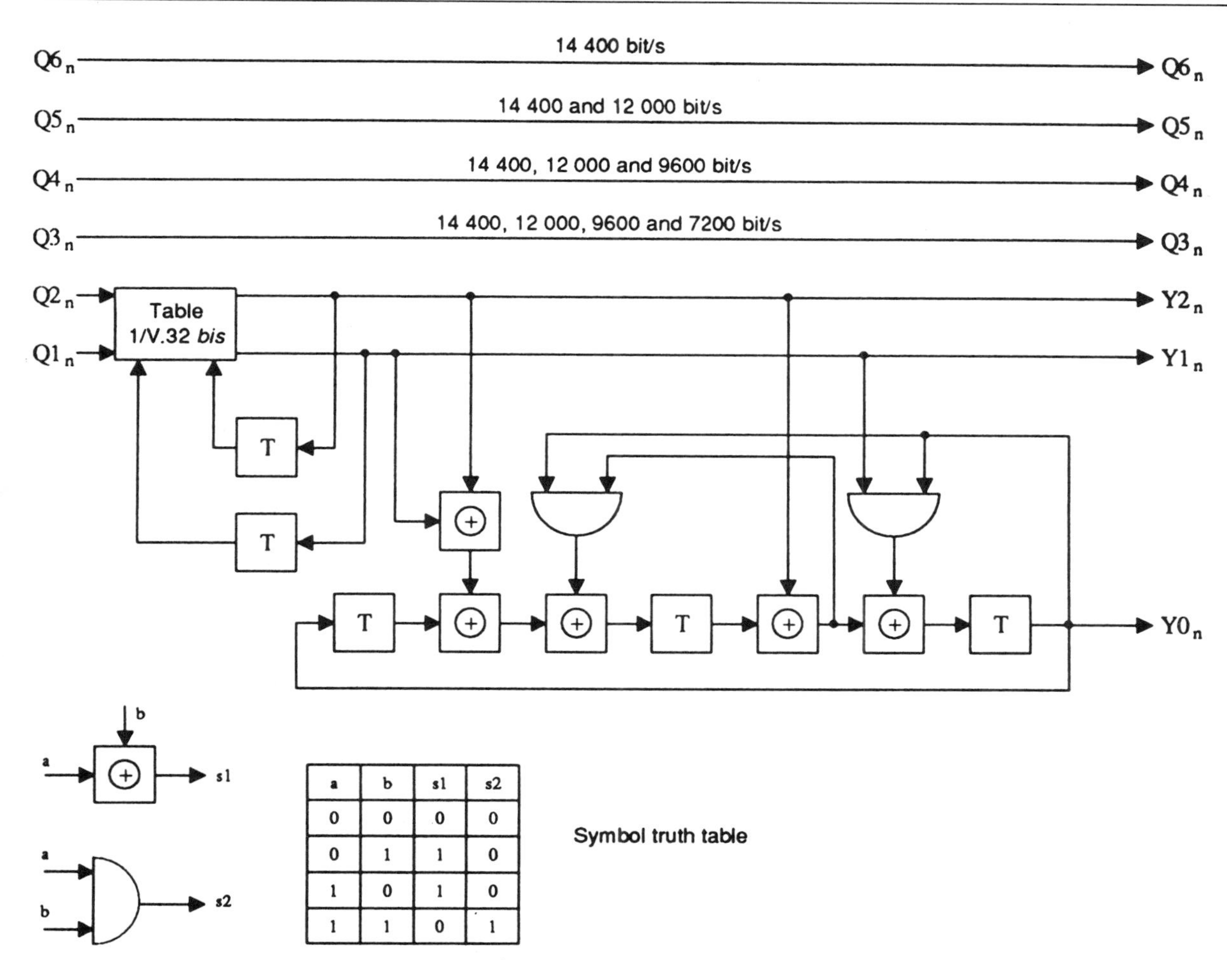

a	b	s1	s2
0	0	0	0
0	1	1	0
1	0	1	0
1	1	0	1

FIGURE 17-U.1. Trellis encoder. (From Figure 1/V.32 bis, ITU-T Rec. V.32 bis, Geneva, 1991, page 3.)

2.3.3. Signal element coding for 9600 bps. At 9600 bps, the scrambled data stream to be transmitted is divided into groups of four consecutive data bits. The first two bits in time $Q1_n$ and $Q2_n$ in each group, where the subscript n designates the sequence number of the group, are first differentially encoded into $Y1_n$ and $Y2_n$ according to Table 17-U.1. The two differentially encoded bits $Y1_n$ and $Y2_n$ are used as input to a systematic convolutional encoder which generates a redundant bit $Y0_n$ (see Figure 7-U.1). This redundant bit and the four information-carrying bits $Y1_n$, $Y2_n$, $Q3_n$, and $Q4_n$, are then mapped into the coordinates of the signal element to be transmitted according to the signal space diagram shown in Figure 17-U.4.

2.3.4. Signal element coding for 7200 bps. At 7200 bps, the scrambled data stream to be transmitted is divided into groups of three consecutive data bits. The first two bits in time $Q1_n$ and $Q2_n$ in each group, where the subscript n designates the sequence number of the group, are first differentially encoded into $Y1_n$ and $Y2_n$ according to Table 17-U.1. The two differentially encoded bits $Y1_n$ and $Y2_n$ are used as input to a systematic convolutional encoder which generates a redundant bit $Y0_n$ (see Figure 17-U.1). This redundant bit and the three information-carrying

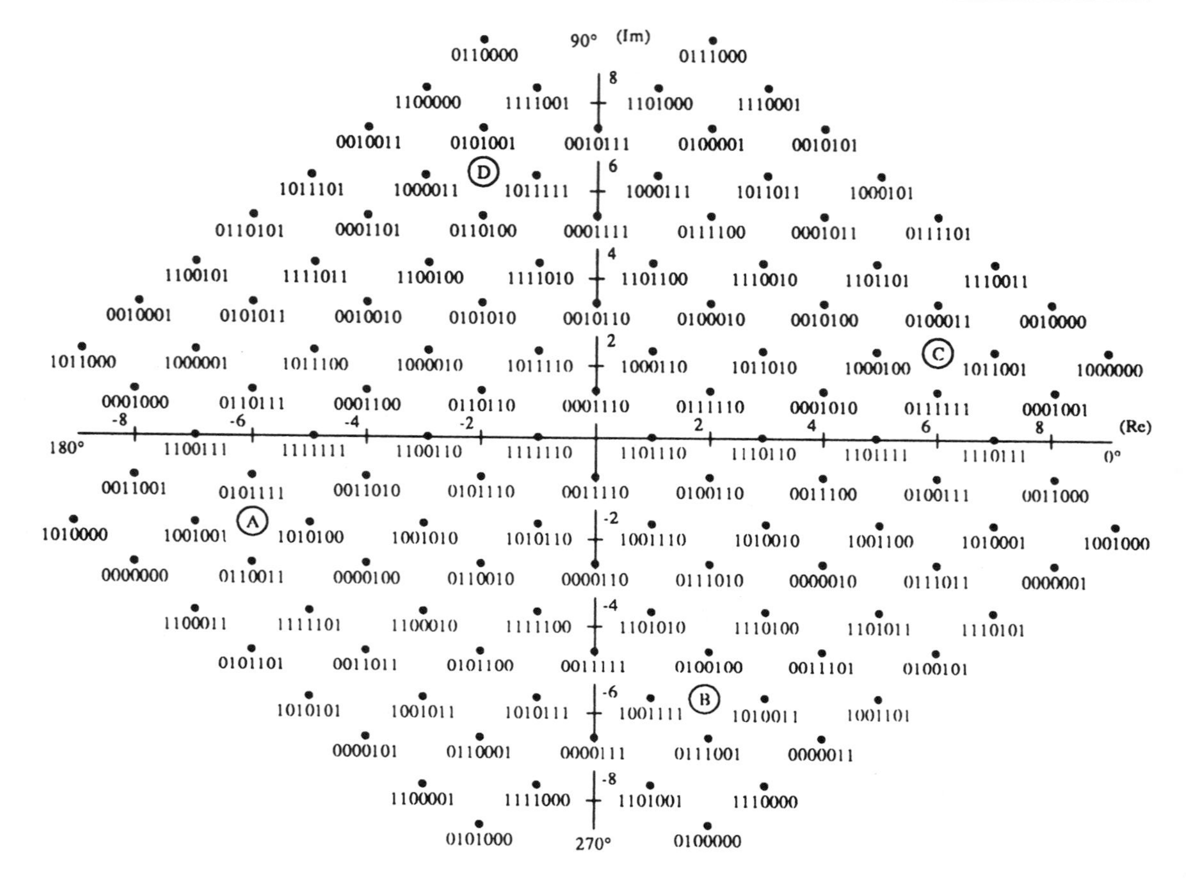

Note – Binary numbers refer to $Y0_n$, $Y1_n$, $Y2_n$, $Q3_n$, $Q4_n$, $Q5_n$, $Q6_n$.
A, B, C, D refer to synchronizing signal elements.

FIGURE 17-U.2. Signal space diagram and mapping for modulation at 14,400 bps. (From Figure 2-1/V.32 bis, ITU-T Rec. V.32 bis, Geneva, 1991, page 4.)

bits $Y1_n$, $Y2_n$, and $Q3_n$ are then mapped into the coordinates of the signal element to be transmitted according to the signal space diagram shown in Figure 17-U.5.

2.3.5. Signal element coding for 4800 bps. At 4800 bps, the scrambled data stream to be transmitted is divided into groups of two consecutive data bits. The two bits $Q1_n$ and $Q2_n$, where $Q1_n$ is first in time, where the subscript n designates the sequence number of the group, are first differentially encoded into $Y1_n$ and $Y2_n$ according to Table 17-U.2. The two differentially encoded bits $Y1_n$ and $Y2_n$ are then mapped into the coordinates of the signal element to be transmitted according to the signal space diagram shown in Figure 17-U.6.

3. DTE Interface. When a standardized physical interface for the interchange circuits is not present, the equivalent functionality of the circuits must still be provided (see Table 17-U.3).

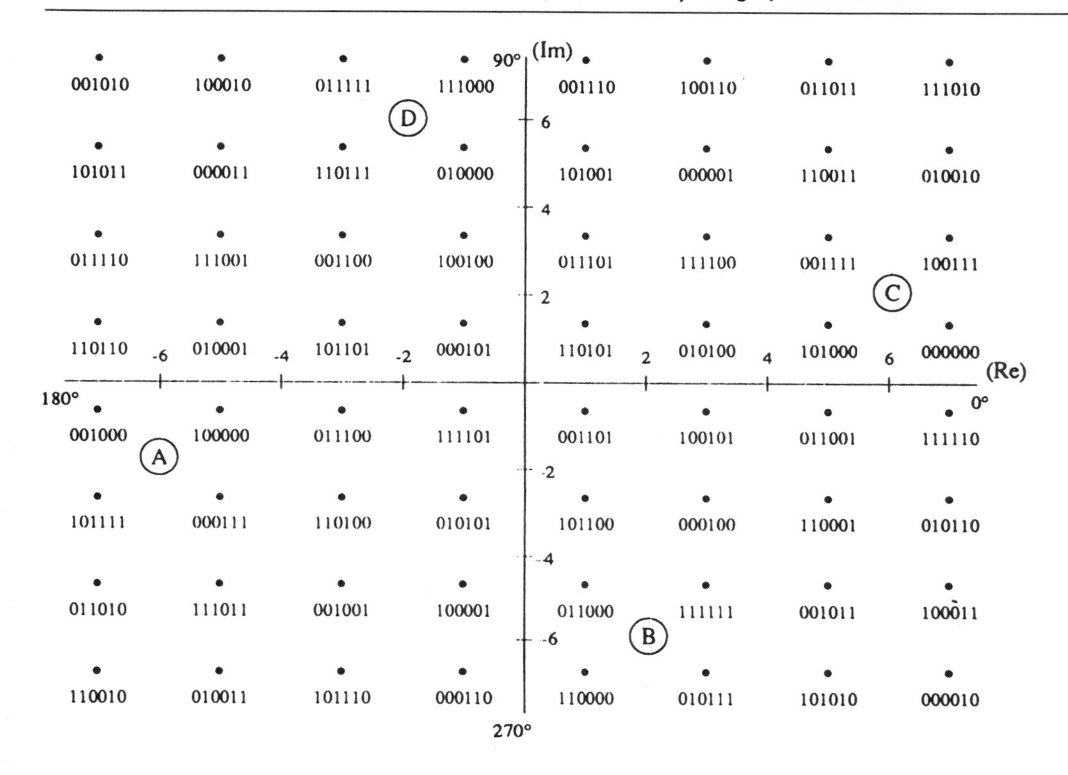

Note – Binary numbers refer to $Y0_n$, $Y1_n$, $Y2_n$, $Q3_n$, $Q4_n$, $Q5_n$.
A, B, C, D refer to synchronizing signal elements.

FIGURE 17-U.3. Signal space diagram and mapping for modulation at 12,000 bps. (From Figure 2-2/V.32 bis, ITU-T Rec. V.32 bis, Geneva, 1991, page 5.)

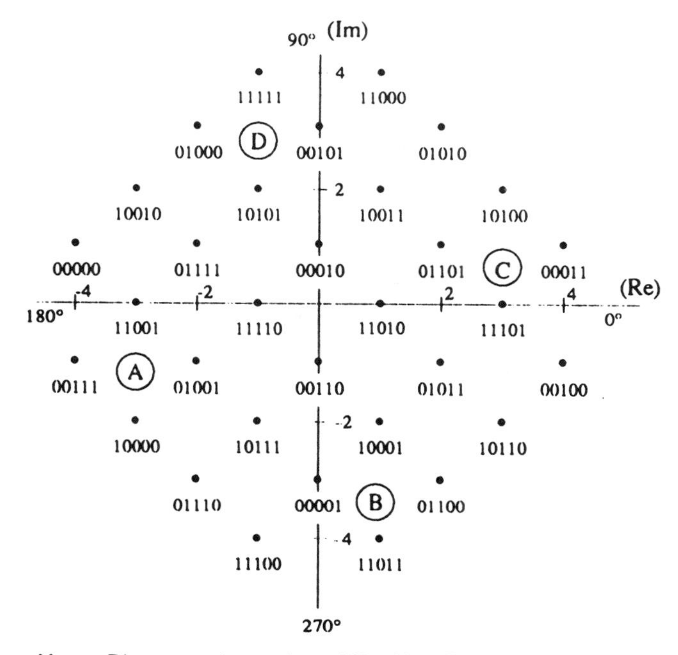

Note – Binary numbers refer to $Y0_n$, $Y1_n$, $Y2_n$, $Q3_n$, $Q4_n$.
A, B, C, D refer to synchronizing signal elements.

FIGURE 17-U.4. Signal space diagram and mapping for modulation at 9600 bps. (From Figure 2-3/V.32 bis, ITU-T Rec. V.32 bis, Geneva, 1991, page 6.)

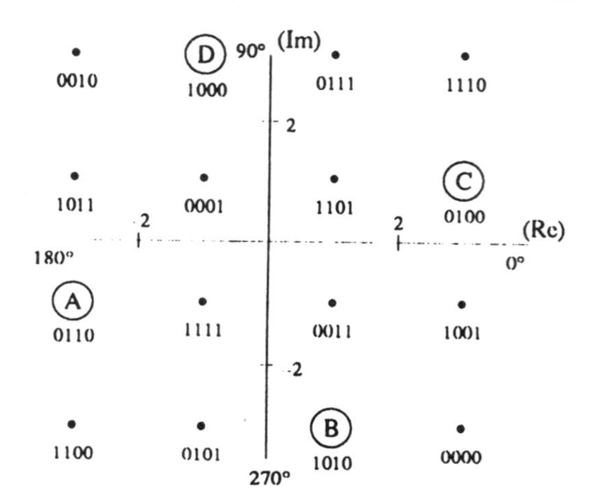

Note – Binary numbers refer to $Y0_n$, $Y1_n$, $Y2_n$.
A, B, C, D refer to synchronizing signal elements.

FIGURE 17-U.5. Signal space diagram and mapping for modulation at 7200 bps. (From Figure 2-4/V.32 bis, ITU-T Rec. V.32 bis, Geneva, 1991, page 6.)

3.1. Synchronous Interfacing. The modems shall accept synchronous data from the DTE on circuit 103 (see ITU-T Rec. V.24) under control of circuit 113 or 114. The modem shall pass synchronous data to the DTE on circuit 104 under the control of circuit 115. The modem shall provide the following to the DTE: (a) a clock on circuit 114 for transmit-data timing and (b) a clock on circuit 115 for receive-data timing. The transmit-data timing may, however, originate in the DTE and be

TABLE 17-U.2
Differential Quadrant Coding for 4800 bps

Inputs		Previous Outputs		Phase Quadrant Change	Outputs		Signal State for 4800 bps
$Q1_n$	$Q2_n$	$Y1_{n-1}$	$Y2_{n-1}$		$Y1_n$	$Y2_n$	
0	0	0	0	+90°	0	1	B
0	0	0	1		1	1	C
0	0	1	0		0	0	A
0	0	1	1		1	0	D
0	1	0	0	0°	0	0	A
0	1	0	1		0	1	B
0	1	1	0		1	0	D
0	1	1	1		1	1	C
1	0	0	0	+180°	1	1	C
1	0	0	1		1	0	D
1	0	1	0		0	1	B
1	0	1	1		0	0	A
1	1	0	0	+270°	1	0	D
1	1	0	1		0	0	A
1	1	1	0		1	1	C
1	1	1	1		0	1	B

Source: Table 2/V.32 bis, ITU-T Rec. V.32 bis, Geneva, 1991, page 7.

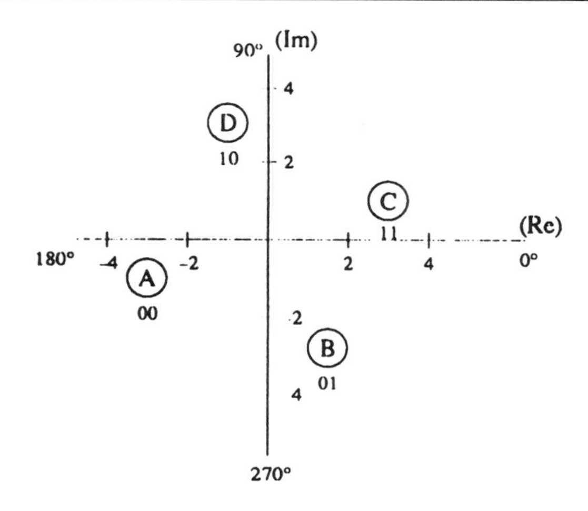

Note – Binary numbers refer to $Y1_n$ and $Y2_n$.
A, B, C, D refer to synchronizing signal elements.

FIGURE 17-U.6. Signal space diagram and mapping for modulation at 4800 bps. (From Figure 2-5/V.32 bis, ITU-T Rec. V.32 bis, Geneva, 1991, page 7.)

TABLE 17-U.3
Interchange Circuits for ITU-T Rec. V.32 Bis

Interchange Circuit		
No.	Description	
102	Signal ground or common return	
103	Transmitted data	
104	Received data	
105	Request to send	
106	Ready for sending	
107	Data set ready	
108/1 or	Connect data set to line	Note 1
108/2	Data terminal ready	Note 1
109	Data channel received line signal detector	
113	Transmitter signal element timing (DTE source)	Note 2
114	Transmitter signal element timing (DCE source)	Note 3
115	Receiver signal element timing (DCE source)	Note 3
125	Calling indicator	
140	Loopback/maintenance	
141	Local loopback	
142	Test indicator	

Note 1: This circuit shall be capable of operation as circuit 108/1 or circuit 108/2 depending on its use. Operation of circuits 107 and 108/1 shall be in accordance with § 4.4 of Rec. V.24.

Note 2: When the modem is not operating in a synchronous mode at the interface, any signals on this circuit shall be disregarded. Many DTEs operating in an asynchronous mode do not have a generator connected to this circuit.

Note 3: When the modem is not operating in a synchronous mode at the interface, this circuit shall be clamped to the OFF condition. Many DTEs operating in an asynchronous mode do not terminate this circuit.

Source: Table 3/X.32 bis, ITU-T Rec. V.32 bis, Geneva, 1991, page 8.

transferred to the modem via circuit 113. In some applications, it may be necessary to slave the transmitter timing to the receiver timing inside the modem.

After the start-up and retrain sequences, circuit 106 must follow the state of circuit 105 within 2 msec.

OFF to ON and ON to OFF transitions of circuit 109 should occur solely in accordance with the operating sequences defined in § 5 of ITU-T Rec. V.32 bis. Thresholds and response times are inapplicable because a line signal detector cannot be expected to distinguish wanted received signals from unwanted talker echos.

3.2. Asynchronous Character-Mode Interfacing. The modulation process operates synchronously. However, the modem may be associated with an asynchronous to synchronous conversion entity interfacing to the DTE in an asynchronous (or start–stop character) mode. The protocol for conversion shall be in accordance with Rec. V.14 or V.42. Other facilities such as data compression may also be employed.

3.3. Electrical Characteristics of Interchange Circuits. When a standardized physical interface is provided, the electrical characteristics conforming to Rec. V.28 will normally be used. Alternatively, the electrical characteristics conforming to Recs. V.10 and V.11 may be used. The connector and pole assignments specified by ISO 2110, corresponding to the electrical characteristics provided, shall be used.

3.4. Fault Condition on Interchange Circuits. The DTE shall interpret a fault condition on circuit 107 as an OFF condition using failure detection type 1.

The DCE shall interpret a fault condition on circuits 105 and 108 as an OFF condition using failure detection type 1.

All other circuits not referred to may use failure detection types 0 or 1.

Note: See § 7 of Rec. V.28 and § 11 of Rec. V.10.

4. Scrambler and Descrambler. A self-synchronizing scrambler/descrambler shall be included in the modem. Each transmission direction uses a different scrambler. The method of allocating the scramblers is described in "Scrambler/Descrambler Allocation," below. According to the direction of transmission, the generating polynomial is:

Call mode modem generating polynomial: (GPC) $= 1 + x^{-18} + x^{-23}$, or
Answer mode modem generating polynomial: (GPA) $= 1 + x^{-5} + x^{-23}$

At the transmitter, the scrambler shall effectively divide the message data sequence by the generating polynomial. The coefficients of the quotients of this division, taken in descending order, form the data sequence which shall appear at the output of the scrambler. At the receiver, the received data sequence shall be multiplied by the scrambler generating polynomial to recover the message sesequence.

4.1. Scrambler / Descrambler Allocation. On the general switched telephone network, the modem at the calling data station (call mode) shall use the scrambler with the GPC-generating polynomial and the descrambler with the GPA-generating polynomial. The modem at the answering data station (answer mode) shall use the scrambler with the GPA-generating polynomial and the descrambler with the GPC-generating polynomial. On point-to-point leased circuits or when calls are established on the GSTN by operators, call mode/answer mode designation will

be by bilateral agreement between administrations or users and the scrambler/descrambler allocation will be the same as used on the GSTN.
Extracted from ITU-T Rec. V.32 bis, Geneva, 1991.

This new section follows Section 17-12.2.5, which ends on page 1545 of the Manual.

17-12.2.6. Nine-Position Nonsynchronous Interface Between Data Terminal Equipment and Data Circuit-Terminating Equipment Employing Serial Binary Data Interchange.

1. Scope

1.1. General. The EIA/TIA-574 standard is applicable to the interconnection of data terminating equipment (DTE) and data circuit-terminating equipment (DCE) employing serial binary data interchange where a minimal number of control and information circuits are required. The standard defines interface mechanical characteristics and provides a functional description of the interchange circuits. The electrical characteristics applicable to this interface can be found in ANSA/EIA/TIA-562.

1.2. Connection. When used in conjunction with the above-mentioned electrical characteristics, this standard applies where equipment on one side of the DTE/DCE interface is intended for connection directly to the other side without additional technical considerations.

1.3. Bit Sequences. This standard applies directly to data communication systems where data are bit-serialized by the DTE and the DCE and places no restrictions on the arrangement or the sequence of bits.

1.4. Data Signaling Rates. This standard applies to data rates up to 64 kbps.

1.5. Nonsynchronous Communication. This standard applies to nonsynchronous serial binary data communications.

1.6. Classes of Service. This standard applies to switched, nonswitched, dedicated, leased, or private line service, either 2-wire or 4-wire. Consideration is given both to point-to-point and multipoint operation.

2. Interface Mechanical Characteristics

2.1. Point of Interface. The interface between the DTE and the DCE is located at a pluggable connector signal interface point between the two equipments as shown in Figure 17-U.7. A nine-position connector is specified for this interface. The DTE is provided with a male connector specified in Section 2.2. An interface cable

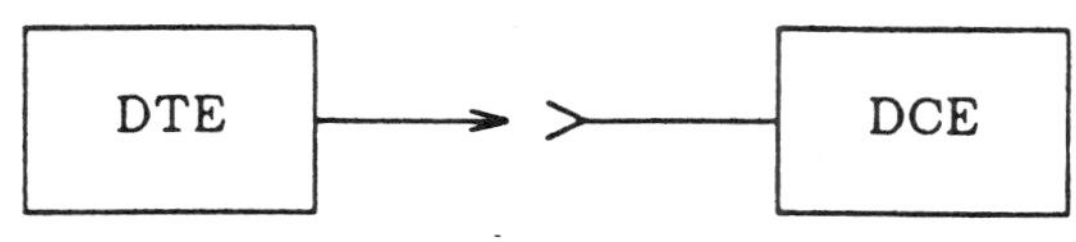

FIGURE 17-U.7. Point of interface.

Contact Number	Circuit	Description
1	109	Received Line Signal Detector
2	104	Received Data
3	103	Transmitted Data
4	108/2	DTE Ready
5	102	Signal Common
6	107	DCE Ready
7	105/133	Request to Send/Ready for Receiving
8	106	Clear to Send
9	125	Ring Indicator

FIGURE 17-U.8. Connector contact assignments.

terminated in a female connector should be provided for connection to the DCE. This connector is also specified in the following Section 2.2 of the reference specification.

2.2. Connector Contact Assignments. The connector contact assignments are shown in Figure 17-U.8.

Note: When hardware flow control is required, circuit 105 may take on the functionality of circuit 133.

3. Functional Descriptions of the Interchange Circuits. Each interchange circuit is characterized as one of the three following types:

- Signal common
- Data
- Control

A list of interchange circuits specifying the circuit number, name, type, and direction relative to the DCE is shown in Figure 17-U.9.

Circuit Number	Circuit Name	Circuit Direction	Circuit Type
102	Signal Common	-	Signal Common
103	Transmitted Data	To DCE	Data
104	Received Data	From DCE	Data
105	Request to Send	To DCE	Control
106	Clear To Send	From DCE	Control
107	DCE Ready	From DCE	Control
108/2	DTE Ready	To DCE	Control
109	Received Line Signal Detector	From DCE	Control
133	Ready for Receiving	To DCE	Control
125	Ring Indicator	From DCE	Control

FIGURE 17-U.9. Interchange circuits by category.

4. Signal Characteristics. The interchange circuits transferring data across the interface shall hold MARK (binary 1) or SPACE (binary 0) conditions for the total nominal duration of each signal element. Distortion tolerances are set forth in EIA-404-A, *Standard for Start–Stop Signal Quality for Non-Synchronous Data Terminal Equipment*.

The material in Section 17-12.2.6 has been derived from EIA/TIA-574, 10/90.

17-12.2.7. Simple Eight-Position Nonsynchronous Interface Between Data Terminal Equipment and Data Circuit-Terminating Equipment Employing Serial Binary Data Interchange

1. Technical Overview

1.1. General. EIA/TIA-561 applies to the interconnection of data terminating equipment (DTE) and data circuit-terminating equipment (DCE) employing serial binary data interchange where a minimal number of control and information circuits is required.

The electrical characteristics applicable to this interface can be found in ANSI/EIA/TIA-562.

1.2. Connection. When used in conjunction with the above-mentioned electrical characteristics standard, this standard applies where equipment on one side of the DTE/DCE interface is intended for connection directly to the other side without additional technical considerations.

1.3. Bit Sequences. This standard applies to data communication systems where data are bit-serialized by the DTE and the DCE and places no restrictions on the arrangement or the sequence of bits.

1.4. Data Signaling Rates. This standard is applicable for use at data signaling rates at or below 38.4 kbps. Equipments complying with this standard, however, need not operate over this entire signaling range. They may be designed to operate over a more narrow range as appropriate for the specific application.

1.5. Nonsynchronous Communication. This standard applies to nonsynchronous serial binary data communication systems.

1.6. Classes of Service. This standard applies to duplex switched, nonswitched, dedicated, leased, or private line service, either 2-wire or 4-wire.

2. Interface Mechanical Characteristics

2.1. Point of Interface. The interface between the DTE and the DCE is located at a pluggable connector signal interface point between the two equipments as shown in Figure 17-U.10. An eight-position connector is specified for this interface. The equipment is provided with either a female connector or a cable terminated in a male connector. When an interface cable is used, it incorporates a half-twist (i.e., pin 1 of the connector on one end is wired to pin 1 of the connector on the other end, etc.) and is terminated in male connectors. The configuration for connection of the interface cable at a point other than the signal interface point is not specified.

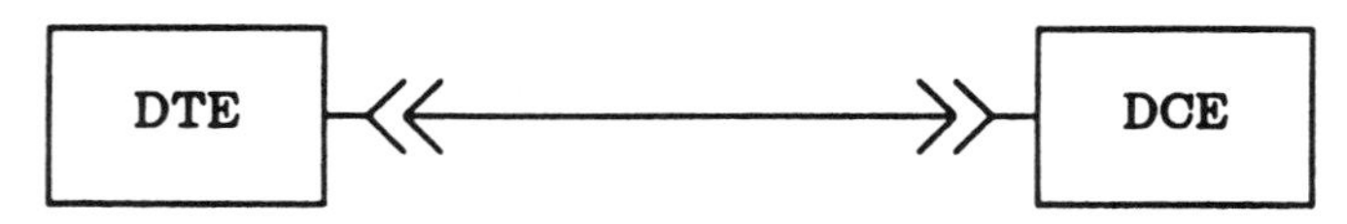

FIGURE 17-U.10. Point of interface.

Contact Number	Circuit	Description
1	125	Ring Indicator
2	109	Received Line Signal Detector
3	108/2	DTE Ready
4	102	Signal Common
5	104	Received Data
6	103	Transmitted Data
7	106	Clear to Send
8	105/133	Request to Send/Ready for Receiving

FIGURE 17-U.11. Connector contact assignments.

2.2. Connector Contact Assignments. Connector contact assignments are shown in Figure 17-U.11.

Note: When hardware flow control is required, circuit 105 may take on the functionality of circuit 133.

3. Functional Descriptions of the Interchange Circuits

3.1. Classification of Circuits. Each interchange circuit is characterized as one of the three following types:

- Signal common
- Data
- Control

A list of interchange circuits, specifying the number, name, direction (relative to the DCE), and type is shown in Figure 17-U.12.

Circuit Number	Circuit Name	Circuit Direction	Circuit Type
102	Signal Common	-	Signal Common
103	Transmitted Data	To DCE	Data
104	Received Data	From DCE	Data
105	Request to Send	To DCE	Control
106	Clear To Send	From DCE	Control
108/2	DTE Ready	To DCE	Control
109	Received Line Signal Detector	From DCE	Control
125	Ring Indicator	From DCE	Control
133	Ready for Receiving	To DCE	Control

FIGURE 17-U.12. Interchange circuits by category.

3.2. Signal Characteristics. Interchange circuits transferring data across the interface shall hold MARK (binary ONE) or SPACE (binary ZERO) conditions for the total nominal duration of each signal element.

Distortion tolerances are set forth in EIA-404-A, *Standard for Start–Stop Signal Quality for Non-Synchronous Data Terminal Equipment*.

Section 17-12.2.7 is based on EIA/TIA-561, 10/90.

17-12.2.8. High-Speed 25-Position Interface for Data Terminal Equipment and Data Circuit-Terminating Equipment Including Alternative 26-Position Connector

1. Applications. This standard applies where equipment on one side of the DTE/DCE interface is intended for connection directly to equipment on the other side without additional technical considerations. Applications where cable termination, signal waveshaping, interconnection cable distance, and mechanical configurations of the interface must be tailored to meet specific user needs are not precluded but are beyond the scope of this standard.

2. Serialization. This standard applies to data communication systems where the data are bit-serialized. There are no restrictions on the sequence of bits.

3. Data Rates. This standard is applicable for use at data signaling rates up to a nominal limit of 2.1 Mbps. Equipment complying with this standard, however, need not operate over this entire data signaling rate range. Equipment may be designed to operate over a narrower range as appropriate for the specific application.

4. Synchronous/Nonsynchronous Communication. EIA/TIA-530-A applies to both synchronous and nonsynchronous serial binary data communication systems.

5. Classes of Service. EIA/TIA-530-A applies to switched, nonswitched, dedicated, leased, or private line service, either 2-wire or 4-wire. It applies to both point-to-point and multipoint operation.

6. Signal Characteristics

6.1. Electrical Characteristics. The electrical characteristics of the interchange circuits are specified in the following standards:

1. EIA-422-A, *Electrical Characteristics of Balanced Voltage Digital Interface Circuits*
2. EIA-423-A, *Electrical Characteristics of Unbalanced Voltage Digital Interface Circuits*

For the purpose of assigning electrical characteristics, the interchange circuits defined in Section 4.3 of the reference publication are divided into two categories: Category I and Category II.

The meaning of circuit conditions "ON" and "OFF" for control and timing circuits, and "0," "1," "MARK," and "SPACE" for data circuits, shall be as defined in EIA-422-A for Category I circuits and as defined in EIA-423-A for Category II circuits.

Category I Circuits. The following interchange circuits are classified as Category I circuits:

Circuit BA (Transmitted Data)
Circuit BB (Received Data)
Circuit DA (Transmit Signal Element Timing, DTE Source)
Circuit DB (Transmit Signal Element Timing, DCE Source)
Circuit DD (Receiver Signal Element Timing, DCE Source)
Circuit CA (Request to Send)
Circuit CB (Clear to Send)
Circuit CF (Received Line Signal Detector)
Circuit CJ (Ready for Receiving)

The individual Category I circuits use the balanced electrical characteristics of EIA-422-A. Two leads are brought out to the interface connector for each Category I circuit, as shown in Figure 17-U.13a. Thus, each interchange circuit consists of a pair of wires interconnecting a balanced generator and a differential receiver.

(a) Category I Circuits

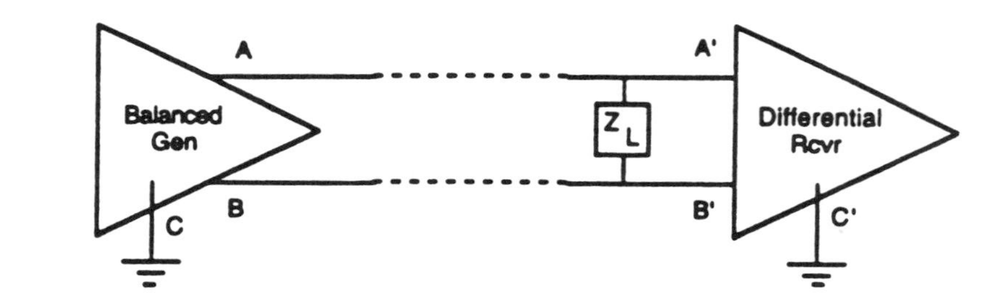

(b) Category II Circuits

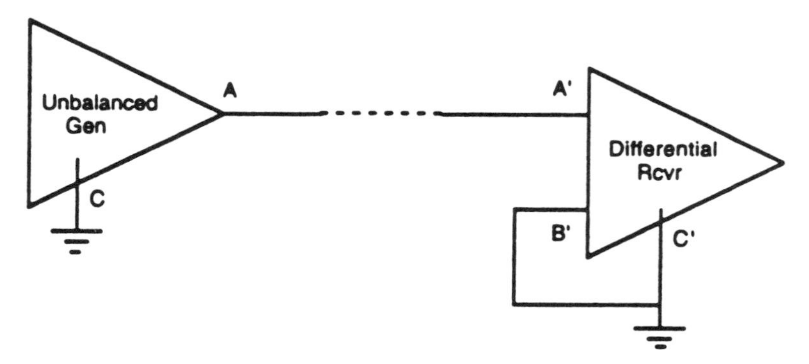

Note: The A, A', B, B', C, and C' designations are those specified in EIA-422-A and EIA-423-A

FIGURE 17-U.13. Generator and receiver connections at interface.

Category II Circuits. The following interchange circuits are classified as Category II circuits:

Circuit CC (DCE Ready)
Circuit CD (DTE Ready)
Circuit CE (Ring Indicator)
Circuit LL (Local Loopback)
Circuit RL (Remote Loopback)
Circuit TM (Test Mode)

Category II circuits use the unbalanced electrical characteristics of EIA-423-A. Each Category II interchange circuit consists of one wire interconnecting an unbalanced generator and a differential receiver, as shown in Figure 17-U.13b. Circuit AB (Signal Common) is the common return for Category II interchange circuits. EIA-423-A generators employ waveshaping suitable for operation over an interface cable length of at least 60 m (200 ft), the maximum cable length allowed for nontailored applications.

6.2. Protective Ground (Frame Ground). In DTEs and in DCEs, protective ground is a point that is electrically bonded to the equipment frame. It may also be connected to external grounds (e.g., through the third wire of the power cord).

It should be noted that protective ground (frame ground) is not an interchange circuit in this standard. If bonding of the equipment frames of the DCE and the DTE is necessary, a separate conductor should be used that conforms to the appropriate national or local electrical codes. Attention is called to the applicable Underwriters' Laboratories regulation applying to wire size and color coding.

6.3. Shield. In order to facilitate the use of shielded interconnecting cable, interface connector contact number 1 is assigned for this purpose. This will permit the cable associated with the DTE to be composed of tandem connectorized sections with continuity of the shield accomplished by connection through this contact in the connectors. Normally the DCE should make no connection to interface connector contact number 1. It is recognized that for certain electromagnetic interference (EMI) suppression situations, additional provisions may be necessary but are beyond the scope of this standard.

6.4. Circuit Grounding (Signal Common). Proper operation of the interchange circuits requires the presence of a path between the DTE circuit ground (circuit common) and the DCE circuit ground (circuit common). This path is obtained by means of interchange Circuits AB and AC, Signal Commons. Both the DCE and the DTE normally should have their circuit ground (circuit common) connected to their protective ground (frame ground) through a resistance of 100 Ω ($\pm 20\%$) having a power dissipation rating of 0.5 W.

Figure 17-U.14 illustrates the signal common and grounding arrangement.

6.5. "Fail-Safe" Operation

6.5.1. The receivers for the following interchange circuits:

Circuit CC (DCE Ready)
Circuit CA (Request to Send)
Circuit CD (DTE Ready)

shall be used to detect a power-off condition in the equipment connected across the interface and the disconnection of the interconnection cable. Detection of either of these conditions shall be interpreted as an OFF condition of the interchange circuit.

6.5.2. The receiver for each control circuit, except those control circuits specified in Section 6.5.1, shall interpret the situation where the conductor is not implemented in the interconnecting cable as an OFF condition.

6.6. General Signal Characteristics

6.6.1. Interchange circuits transferring data signals across the interface point shall hold the marking (binary ONE) and spacing (binary ZERO) conditions for the total nominal duration of each signal element.

Distortion tolerances for synchronous systems are set forth in EIA-334-A, *Signal Quality at Interface Between Data Terminal Equipment and Synchronous Data Circuit-Terminating Equipment for Serial Data Transmission*.

Standard nomenclature for specifying signal quality for nonsynchronous systems is set forth in EIA-363, *Standard for Specifying Signal Quality for Transmitting and Receiving Data Processing Terminal Equipments Using Serial Data Transmission at the Interface with Non-Synchronous Data Communication Equipment*.

Distortion tolerances for nonsynchronous systems are set forth in ANSI/EIA-404-A-1985, *Standard for Start–Stop Quality for Nonsynchronous Data Terminal Equipment*.

6.6.2. Interchange circuits transferring timing signals across the interface point shall hold ON and OFF conditions for nominally equal periods of time consistent with acceptable tolerances as specified in EIA-334-A.

Accuracy and stability of the timing information on Circuit DD (Receiver Signal Element Timing) are required only when Circuit CF (Received Line Signal Detector) is in the ON condition. Drift during the OFF condition of Circuit CF is acceptable; however, resynchronization of the timing information on Circuit DD must be accomplished as rapidly as possible following the OFF to ON transition of Circuit CF.

It is desirable that the transfer of timing information across the interface be provided during all times that the timing source is capable of generating this information (i.e., it should not be restricted only to periods when actual transmission of data is in progress). During periods when timing information is not provided on a timing interchange circuit, the interchange circuit shall be clamped in the OFF condition.

6.6.3. Tolerances on the relationship between data and associated timing signals shall be in accordance with EIA-334-A.

7. Interface Mechanical Characteristics

7.1. Definition of Mechanical Interface. The point of demarcation between the DTE and the DCE is located at a pluggable connector signal interface point between the two equipments which is less than 3 m (10 ft) from the DCE. A 25-position connector is the normal connector specified for all interchange circuits. An alternative 26-position connector (Alt A) is specified for use when a smaller physical connector is required. When the Alt A connector is used, the interface shall be referred to as "ANSI/TIA/EIA-530-A Alt A."

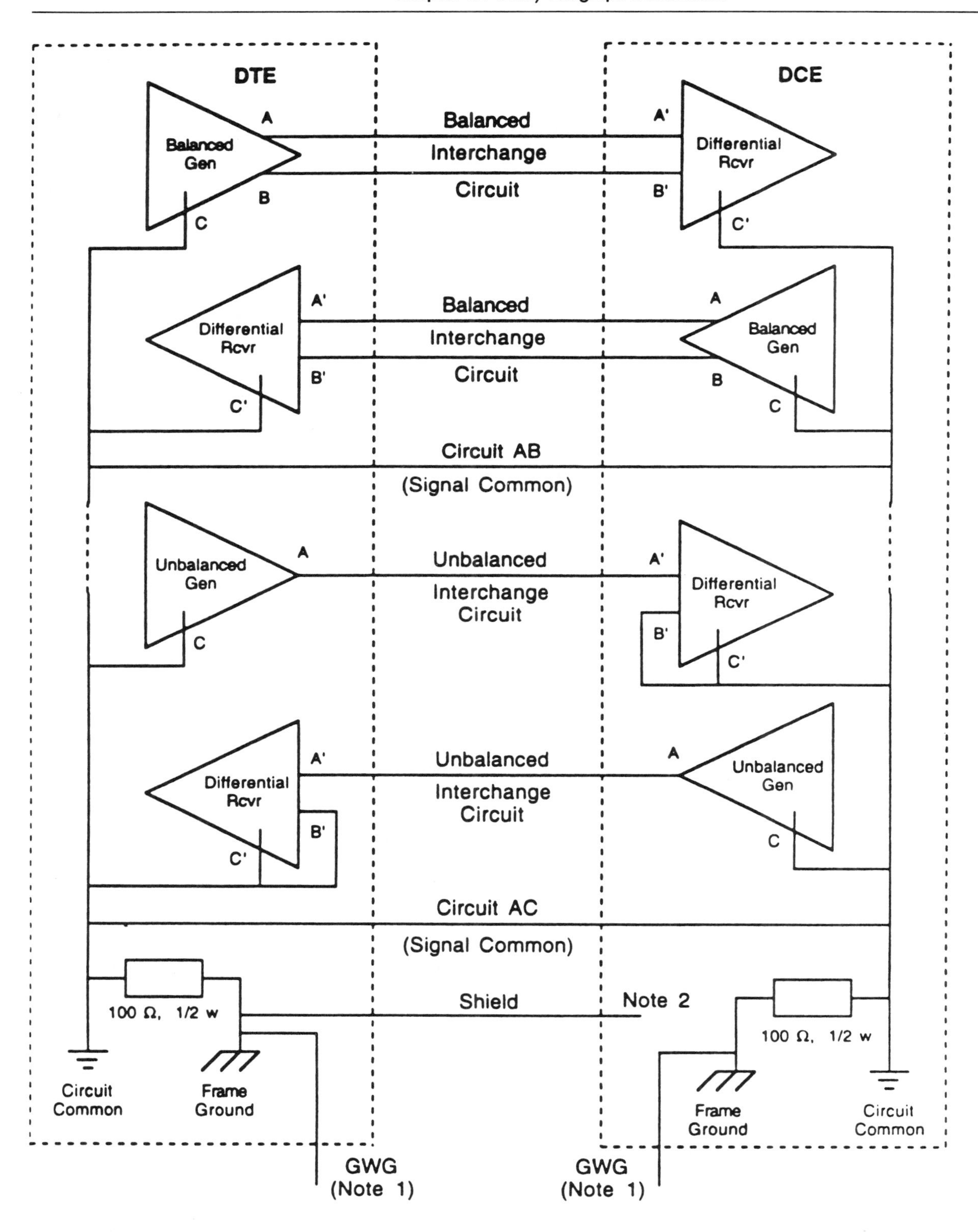

Note 1: GWG is green wire ground of power system.

Note 2: Normally no connection to shield on DCE (see Section 6.3).

FIGURE 17-U.14. Signal common and grounding arrangement.

The DCE shall be provided with the connector having female contacts and a male shell. The connector shall be either physically attached to the DCE or extended by means of a short cable (less than 3 m or 10 ft). The DTE shall be provided with a cable having the connector having male contacts and a female shell. The total length of the cable associated with the DTE shall not exceed 60 m (200 ft) for nontailored applications. The mechanical configuration for connections of the interface cable at points other than the point of demarcation is not specified.

7.2. Connector Contact Assignments. Contact assignments are listed in Figure 17-U.15. Interchange points A – A′ and B – B′ for each Category I circuit should be assigned twisted pairs in interconnecting cables to minimize crosstalk. CCITT numbers refer to circuits from CCITT Rec. V.24.

CONTACT NUMBER	CIRCUIT	CCITT NUMBER	INTERCHANGE POINTS	CIRCUIT CATEGORY	DIRECTION
1	Shield	-	-		
2	BA	103	A-A'	I	To DCE
3	BB	104	A-A'	I	From DCE
4	CA/CJ (Note 1)	105/133	A-A'	I	To DCE
5	CB	106	A-A'	I	From DCE
6	CC (Note 3)	107	A-A'	II	From DCE
7	AB	102A	C-C'	-	
8	CF	109	A-A'	I	From DCE
9	DD	115	B-B'	I	From DCE
10	CF	109	B-B'	I	From DCE
11	DA	113	B-B'	I	To DCE
12	DB	114	B-B'	I	From DCE
13	CB	106	B-B'	I	From DCE
14	BA	103	B-B'	I	To DCE
15	DB	114	A-A'	I	From DCE
16	BB	104	B-B'	I	From DCE
17	DD	115	A-A'	I	From DCE
18	LL	141	A-A'	II	To DCE
19	CA/CJ (Note 1)	105/133	B-B'	I	To DCE
20	CD (Note 3)	108/1, /2	A-A'	II	To DCE
21	RL	140	A-A'	II	To DCE
22	CE	125	A-A'	II	From DCE
23	AC	102B	C-C'		
24	DA	113	A-A'	I	To DCE
25	TM	142	A-A'	II	From DCE
26	(Note 2)	-	-	-	-

Note 1: When hardware flow control is required Circuit CA may take on the functionality of Circuit CJ.

Note 2: Contact 26 is contained on the Alt A connector only. No connection is to be made to this contact.

Note 3: In ANSI / EIA-530-87, Circuits CC and CD were Category I Circuits. Interoperation between Category I and II circuits is not possible without circuitry similar to that in Figure 17-U.17.

FIGURE 17-U.15. Connector contact assignments.

CIRCUIT MNEMONIC	CCITT NUMBER	CIRCUIT NAME	CIRCUIT DIRECTION	CIRCUIT TYPE
AB	102	Signal Common		COMMON
AC	102B	Signal Common		
BA	103	Transmitted Data	To DCE	DATA
BB	104	Received Data	From DCE	
CA	105	Request to Send	To DCE	CONTROL
CB	106	Clear to Send	From DCE	
CF	109	Received Line Signal Detector	From DCE	
CJ	133	Ready for Receiving	To DCE	
CE	125	Ring Indicator	From DCE	
CC	107	DCE Ready	From DCE	
CD	108/1, /2	DTE Ready	To DCE	
DA	113	Transmit Signal Element Timing (DTE Source)	To DCE	TIMING
DB	114	Transmit Signal Element Timing (DCE Source)	From DCE	
DD	115	Receiver Signal Element Timing (DCE Source)	From DCE	
LL	141	Local Loopback	To DCE	
RL	140	Remote Loopback	To DCE	
TM	142	Test Mode	From DCE	

FIGURE 17-U.16. Interchange circuits.

8. Functional Description of Interchange Circuits

8.1. Classification of Circuits. Interchange circuits fall into four general classifications:

- Signal common circuits
- Data circuits
- Control circuits
- Timing circuits

A list of interchange circuits showing circuit mnemonic, circuit name, circuit direction, and circuit type is given in Figure 17-U.16.

9. Interoperation of EIA/TIA-530-A with CCITT Rec. V.35. EIA/TIA-530-A may interoperate with equipment employing an interface which conforms to CCITT Rec. V.35 and utilizes the ISO 2593-1984 34-pin DTE/DCE Interface Connector and Pin Assignments. This may be via connecting cable or other connecting device. Table 17-U.4 lists the circuit name, mnemonic, and connector contact pin for each interface.

EIA/TIA-530-A provides balanced (EIA-422-A) generators and receivers on interchange circuits CA (Request to Send), CB (Clear to Send), and CF (Received Line Signal Detector). The corresponding interchange circuits in a V.35 interface utilize CCITT Rec. V.28 (EIA/TIA-232-E) electrical characteristics. In the case where an EIA-422-A balanced generator must interface with an EIA/TIA-232-E unbalanced receiver, a conversion must take place between the interfaces. Various

TABLE 17-U.4
Interconnection of EIA/TIA-530-A with CCITT Rec. V.35

ANSI/TIA/EIA-530-A			CCITT Rec. V.35			
Circuit Name and Mnemonic		Contact	Contact			Circuit Name and Mnemonic
Shield		1	A			Shield
Transmitted data	BA (A) BA (B)	2 14	P S	103 103	(A) (B)	Transmitted data
Received data	BB (A) BB (B)	3 16	R T	104 104	(A) (B)	Received data
Request to send	CA (A) CA (B)	4 19	C	105		Request to send (Note 1)
Clear to send	CB (A) CB (B)	5 13	D	106		Clear to send
DCE ready	CC	6	E	107		Data set ready
DTE ready	CD	20	H	108/1, /2		Data terminal ready (Note 2)
Signal common	AB	7	B	102		Signal common
Received line signal detector	CF (A) CF (B)	8 10	F	109		Data channel received line signal detector (Note 1)
Transmit signal element timing (DCE source)	DB (A) DB (B)	15 12	Y AA	114 114	(A) (B)	Transmitter signal element timing
Receiver signal element timing (DCE source)	DD (A) DD (B)	17 9	V X	115 115	(A) (B)	Receiver signal element timing
Local loopback	LL	18	L	141		Local loopback (Note 2)
Remote loopback	RL	21	N	140		Loopback/maintenance (Note 2)
Transmit signal element timing (DTE source)	DA (A) DA (B)	24 11	U W	113 113	(A) (B)	Transmitter signal element timing (Note 2)
Test mode	TM	25	NN	142		Test indicator
Ring indicator	CE	22	J	125		Calling indicator (Note 2)
Signal common	AC	23	B	102		Signal common

Note 1: A circuit similar to that shown in Figure 17-U.17 is required between these interfaces.
Note 2: These functions are not included in CCITT Rec. V.35 but are provided in ISO 2593 on an optional basis.

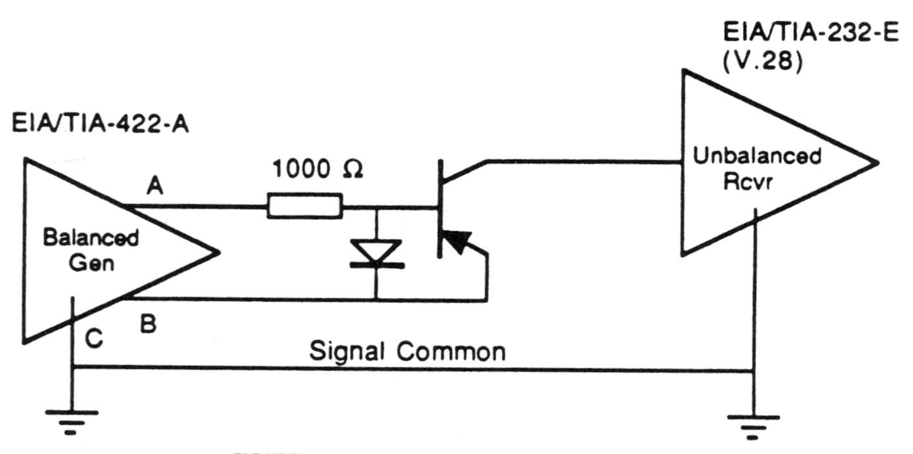

FIGURE 17-U.17. Balanced/unbalanced conversion.

methods may be used for this convertion. One such method is shown in Figure 17-U.17. The EIA/TIA-232-E receiver must be configured for "fail-safe" operation.

Section 17-12.2.8 was extracted from EIA/TIA-530-A and its Appendix B, 10/90.

17-12.2.9. Electrical Characteristics for an Unbalanced Digital Interface

1. Application. The provisions of EIA/TIA-562 may be applied to the circuits employed at the interface between equipments where the information being conveyed is in the form of binary signals. The electrical characteristics of the generator and receiver are specified as measured at the interface point specified in the referencing interface standard. The interconnecting cable is specified in terms of its electrical and physical characteristics. Typical points of applicability for this standard are shown in Figure 17-U.18.

The unbalanced voltage digital interface circuit will normally be specified for data, timing, or control circuits where the data signaling rate on these circuits is limited to a maximum of 64 kbps.

When interfacing with equipment designed to conform to EIA/TIA-232-E, the data signaling rate is restricted to the "nominal upper limit of 20 kbps" specified in that standard.

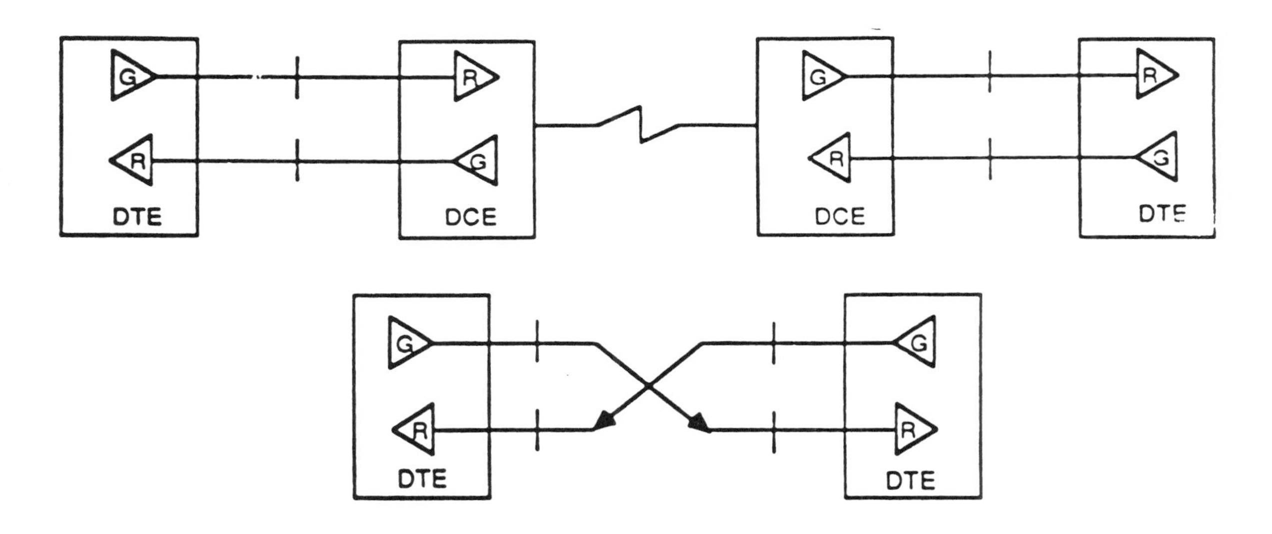

Legend :

DTE = Data Terminal Equipment

DCE = Data Circuit Terminating Equipment

G = Interface Generator

R = Interface Receiver (Load)

\+ = Unbalanced Interface Circuit

= Telecommunications Channel

FIGURE 17-U.18. Applications for an unbalanced digital interface.

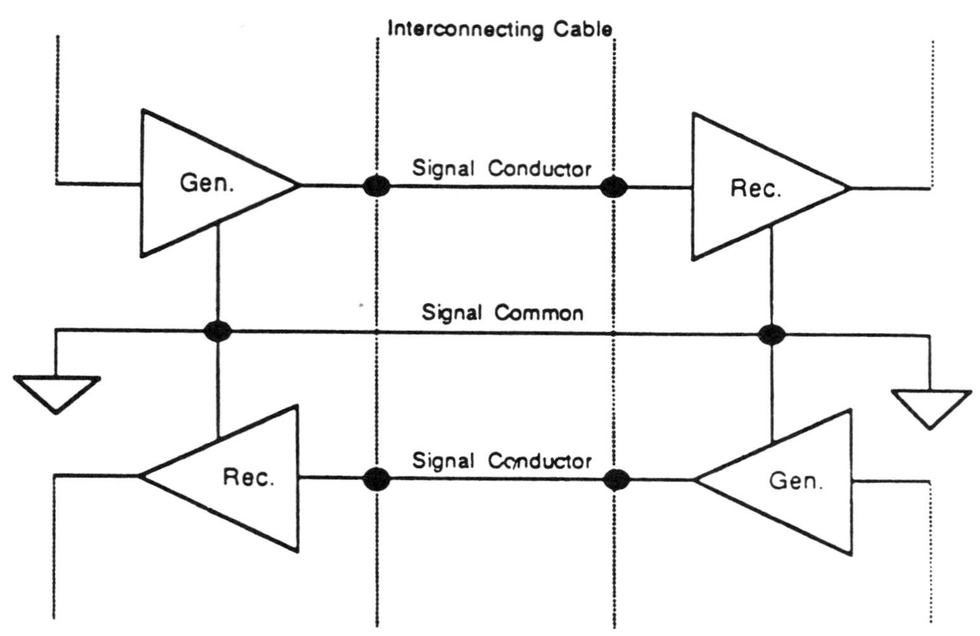

FIGURE 17-U.19. Unbalanced digital interface.

2. Electrical Characteristics. The unbalanced voltage digital interface circuit is shown in Figure 17-U.19. The circuit consists of three parts: the generator, the interconnecting cable, and the receiver. The electrical characteristics of the generator and receiver are specified as measured at the respective interface points.

The interconnecting cable between the generator and receiver interface points shown in Figure 17-U.19 consist of one or more signal conductors and at least one signal common. The signal common establishes a reference potential for the interchange circuits.

2.1. Generator Characteristics: Generator Interface Point (V_1). The generator electrical characteristics are specified in accordance with measurements illustrated in Figures 17-U.20 through 17-U.22. A generator circuit having these requirements results in an unbalanced voltage source that will produce a voltage (V_1) applied to the interconnecting cable as shown in Figure 17-U.20.

2.1.1. Generator Interface Point Interchange Equivalent Circuits. The signaling sense of the voltages appearing across the interconnecting cable is defined as follows: For timing and control interchange circuits, the function shall be considered ON when the voltage (V_1) on the interchange circuit is more positive than +3.3 V with respect to signal common and shall be considered OFF when the voltage (V_1) is more negative than −3.3 V with respect to signal common. The function is not uniquely defined for voltages in the generator interface point transition region between +3.3 V and −3.3 V.

2.1.2. Open Circuit Measurement. For either binary state, the magnitude of the voltage (V_1) measured between the generator output terminal and signal common with no load (open circuit) shall not exceed 13.2 V. For the opposite binary state, the polarity of the voltage (V_1) shall be reversed.

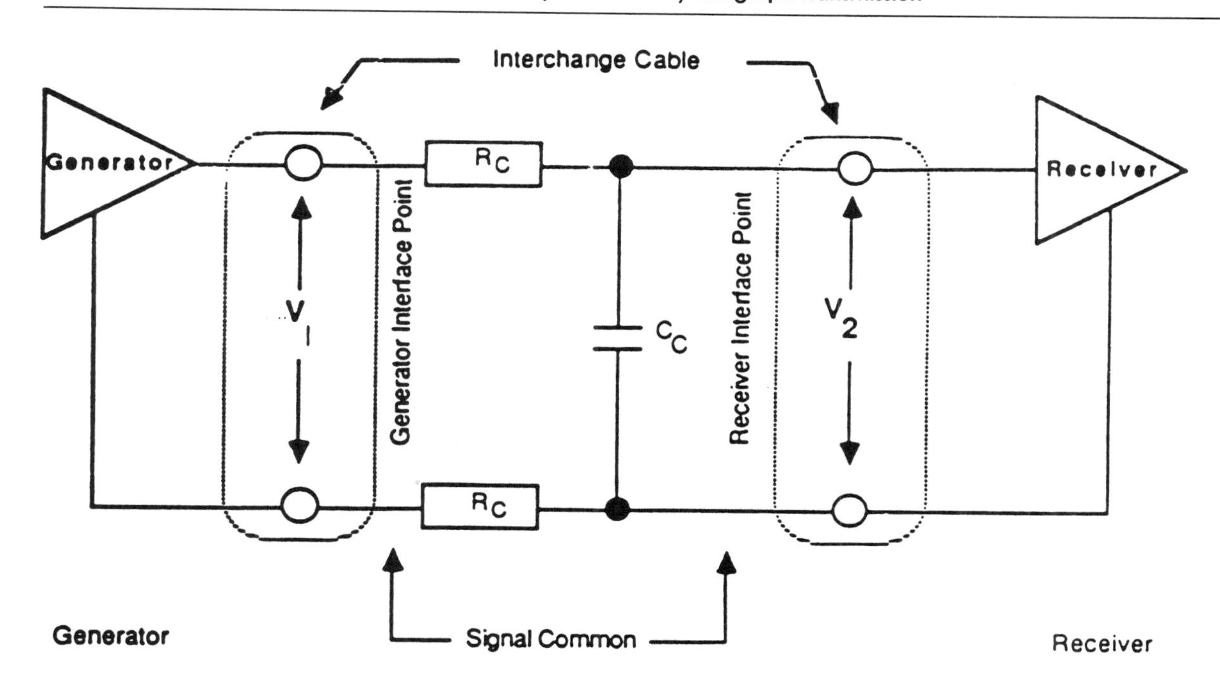

Interface Point = the physical connection between the DTE/ DCE and the interchange cable
C_C = cable capacitiance
R_C = total resistance of the cable per conductor
V_1 = voltage at the Generator Interface Point
V_2 = voltage at the Receiver Interface Point

FIGURE 17-U.20. Interchange circuit.

Notation	Interchange Voltage	
	Negative	Positive
Binary State	1	0
Signal Condition	Mark	Space
Function	OFF	ON

FIGURE 17-U.21. Interchange voltage.

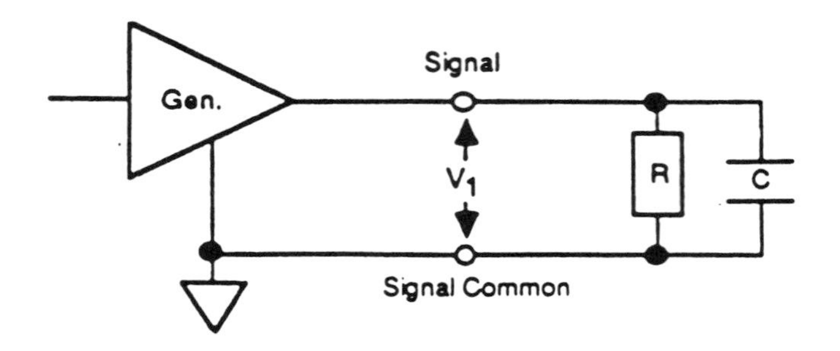

R = 3 kΩ ± 1%

C = 2500 pF ± 1% for Case 1
C = 1000 pF ± 1% for Case 2

FIGURE 17-U.22. Output signal applied waveform test set up.

2.1.3. Test Termination Measurement. With a test load (see Figure 17-U.22) connected between the generator output terminal and signal common, the magnitude of the voltage (V_1) measured between the generator output terminal and the signal common shall not be less than 3.7 V in either binary state.

2.1.4. Short-Circuit Measurement. With the generator output terminal short-circuited to signal common, the magnitude of the current flowing through the generator output terminal shall not exceed 60 mA for either binary state.

2.1.5. Power-Off Impedance. Under power-off conditions, the magnitude of the generator output leakage impedance (with voltages with a magnitude greater than 0 V and less than 2 V applied between the generator output terminal and signal common) shall not be less than 300 Ω.

2.1.6. Output Signal Applied Waveform Test. During transitions of the generator output between alternating binary states (i.e., one-zero-one-zero, etc.), the signal measured across the test load (Figure 17-U.22) connected between the generator output terminal and signal common shall be such that the voltage monotonically changes between 0.1 and 0.9 V_{ss}. Thereafter, the signal voltage shall not vary more than ±5% of V_{ss} as shown in Figure 17-U.23, until the next binary transition occurs. V_{ss} is defined as the voltage difference between the two steady-state values of the generator output.

The transition time between V_1 (+3.3 V) and V_1 (−3.3 V) shall meet the following requirements:

Case 1. If bit time or unit interval (tb) is greater than or equal to 50 μsec (up to 20 kbps), then transition time (tr) or (tf) shall be less than or equal to 3.1 μsec and greater than 220 nsec.

Case 2. If bit time or unit interval (tb) is less than 50 μsec but greater than or equal to 15.625 μsec (greater than 20 kbps), then transition time (tr) or (tf) shall be less than or equal to 2.1 μsec and greater than 220 nsec.

Maximum instantaneous rate of change (slew rate, +sr or −sr) shall not exceed 30 V per microsecond.

Minimum instantaneous rate of change shall be greater than or equal to 4 V per microsecond when measured at the generator interface point between +3.0 V and −3.0 V.

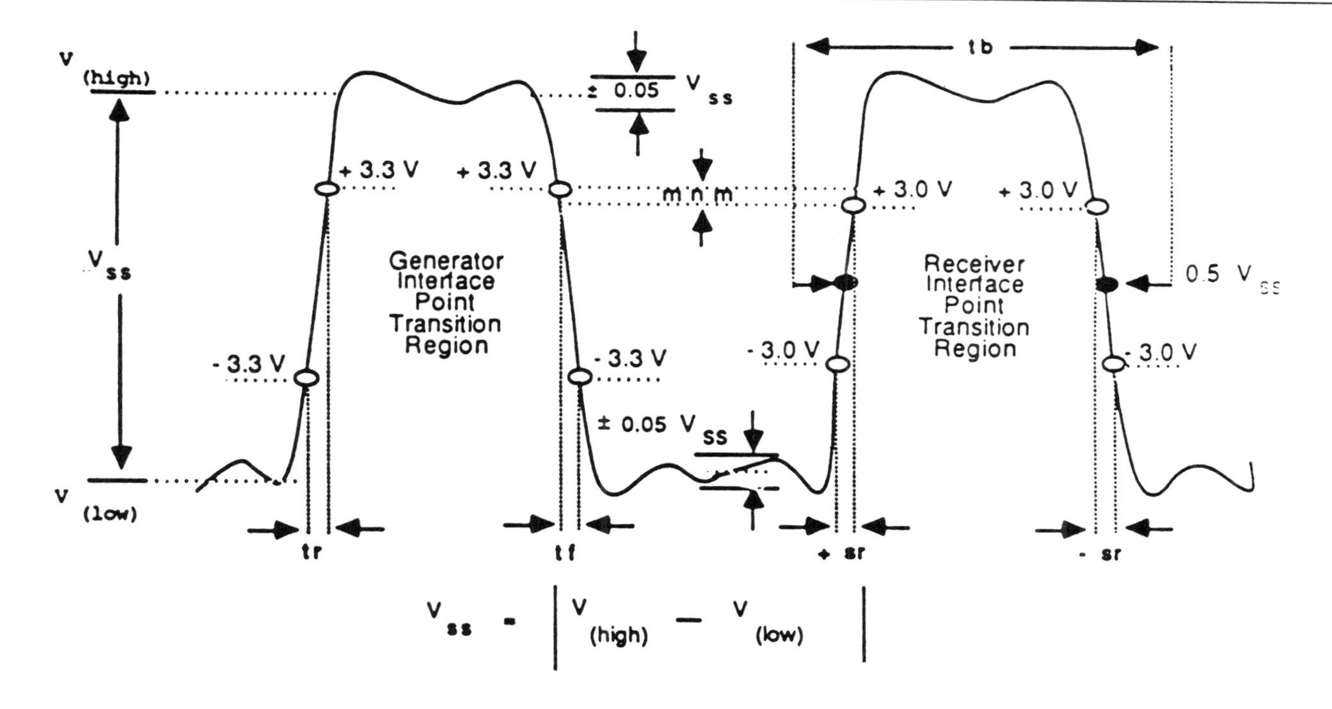

tf = Transition time (Fall time)
tb = Bit time (Unit Interval)
tr = Transition time (Rise time)
+ sr = Positive slew rate measurement points
- sr = Negative slew rate measurement points
m n m = Minimum noise margin

FIGURE 17-U.23. Generator output signal waveform template.

2.2. Receiver Characteristics: Receiver Interface Point (V_2)

2.2.1. Electrical Characteristics. The electrical load characteristics of the receiver are described below. See Figure 17-U.24.

2.2.1.1. Electrical Input Impedance. The input impedance of a receiver must have a DC resistance (R_L) part and a capacitive reactive part as follows:

1. R_L should be greater or equal to 3 kΩ and less than or equal to 7 kΩ.
2. C_L: It is desirable that the input capacitance be less than 100 pF; this includes the interface connector and circuit board capacitance as well as effective device input capacitances.

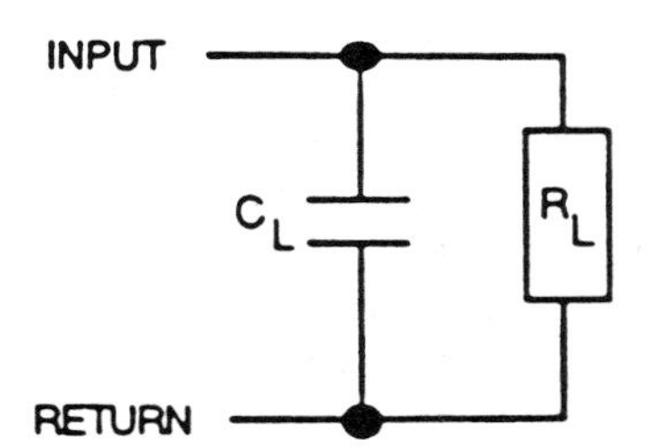

FIGURE 17-U.24. Receiver electrical equivalent.

The receiver must comply with the input impedance specified above with up to a maximum voltage of ± 15 V applied at the V_2 interface.

The receiver must not be damaged by voltages greater than ± 15 V but less than or equal to ± 25 V applied at the V_2 interface. The input impedance during this condition is not specified.

2.2.2. Receiver Interface Point Signal Definition

2.2.2.1. Timing and Control Interchange Circuits. For timing and control interchange circuits, the function shall be considered ON when the voltage (V_2) on the interchange circuit is more positive than $+3.0$ V with respect to signal common, and shall be considered OFF when the voltage (V_2) is more negative than -3.0 V with respect to signal common. The function is not uniquely defined for voltages in the transition region between $+3.0$ V and -3.0 V.

2.2.2.2. Data Interchange Circuits. For data interchange circuits, the signal shall be considered in the marking condition when the voltage (V_2) on the interchange circuit is more negative than -3.0 V with respect to signal common. The signal shall be considered in the spacing condition when the voltage (V_2) is more positive than $+3.0$ V with respect to signal common. The region between $+3.0$ V and -3.0 V is the transition region.

2.2.3. Receiver Fail-Safe Protection. Certain applications require detection of various fault conditions on interchange circuits. When this is required, the receiver circuitry shall detect a binary 1 (off condition) for the following fault conditions:

1. Generator in power-off condition
2. Open circuited interchange cable
3. Receiver not interconnected with a generator
4. Receiver input shorted to receiver interface point signal common

3. Interconnecting Cable Guidelines. An interchange cable may be composed of twisted or nontwisted pair (flat cable) or unpaired wires possessing the characteristics described below uniformly over its length. Cable commonly used in DCE/DTE applications should meet the specifications described below.

3.1. Conductor Resistance. It is desirable for proper operation over the interchange cable that the DC wire resistance not exceed 25 Ω per conductor.

3.2. Capacitive Cable Model. In order to assist the user in estimating cable capacitance, a model for capacitance is shown in Figure 17-U.25. It is important to note that the capacitance that exists from any conductor to a shield, in the case of shielded cable, or the stray capacitance that exists from any conductor to earth ground, in the case of nonshielded cable, is a significant factor in determining the maximum cable length and cannot be ignored. As a general rule, the capacitance from any conductor to shield is approximately twice the mutual capacitance, or capacitance that exists between two conductors. Therefore, shielded cable tends to reduce the maximum length by a factor of 3 over a calculated length based on strictly mutual capacitance. A typical error that is also made in calculating the maximum cable length for nonshielded cable is to again just factor in the mutual capacitance. The stray capacitance to earth ground is approximately 50% of the mutual capacitance and results in a reduction of approximately 33% of the original distance calculated based on only mutual capacitance.

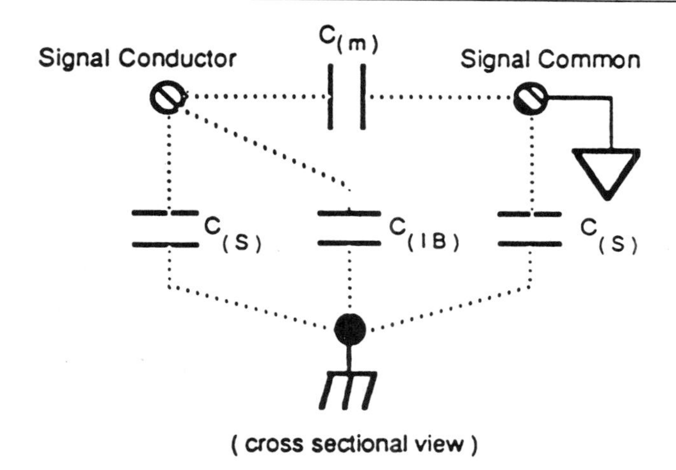

$C_{(m)}$ = mutual capacitance between conductors

$C_{(S)}$ = conductor to interchange cable shield capacitance (if shielded interchange cable is used) or stray capacitance to earth ground (if unshielded interchange cable is used)

$C_{(IB)}$ = capacitor imbalance between 2 conductors and shield

FIGURE 17-U.25. Interchange cable, capacitance model per unit length.

3.3. Interchange Capacitance Limits. The recommended maximum cable capacitance for speeds up to 20 kbps is 2500 pF, and for data rates above 20 kbps the recommended maximum cable capacitance should be limited to 1000 pF.

3.4. Simplified Electrical Equivalent Circuit. In order to assist the user, a simplified electrical equivalent circuit is given in Figure 17-U.26. The figure demonstrates the fact that the receiver signal common potential is never the same as the generator signal common potential. Additionally, the capacitance of either conductor to shield or conductor to earth ground becomes an integral part of the total capacitance load that the generator must drive in an unbalanced circuit.

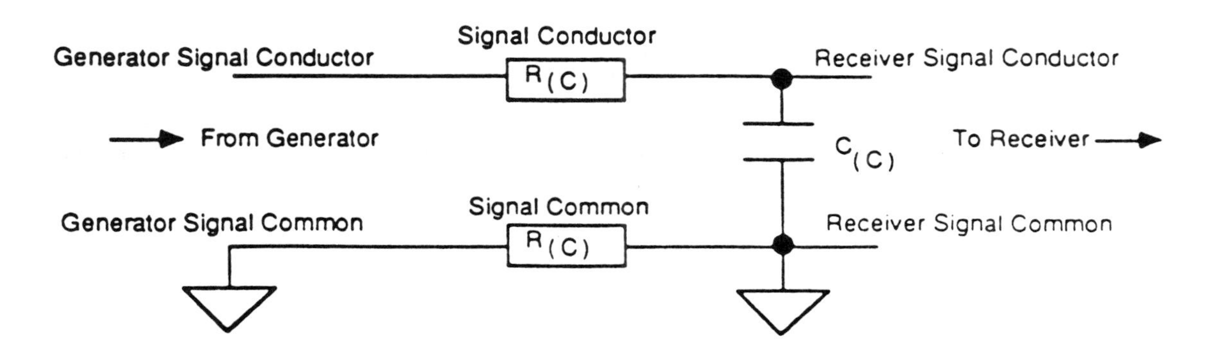

Where $C_{(C)} = C_{(m)} + C_{(S)}$

and $C_{(S)} = 2\ (C_{(m)})$ For Shielded Cable

or $C_{(S)} = 0.5\ (C_{(m)})$ For Non - Shielded Cable

and $R_{(C)}$ = Cable Resistance

FIGURE 17-U.26. Interchange cable capacitance limits.

3.5. Examples

3.5.1. 64-kbps Operation. The user needs to transmit 64 kbps reliably from point A to point B. The user has decided to follow the manufacturer's recommendation of using shielded cable when interconnecting his/her equipment. The user reads the shielded cable specifications and discovers that the cable contains 20 pF per foot of mutual capacitance (C_m).

The input capacitance of the receiver and the output capacitance for this example are both considered to be 100 pF each; this leaves the user 800 pF of capacitance for the interconnecting cable. This is because 64-kbps data transmission is classified as a Case 2 transmission (> 20 kbps), and therefore the upper capacitance limit is 1000 pF. It is very important that the user know how much capacitance the equipment interchange circuitry contains.

In this example the cable is shielded, and therefore the total capacitance per foot is equal to C_m (20 pF) + C_s (40 pF), or 60 pF per foot. Many cable manufacturers specify the capacitance to shield, in which case the above formula is not needed.

Dividing 800 pF by 60 pF reveals that the cable can only be 13.3 ft long. However, because most cables manufactured are cut to a standard length of 15 ft and because a small amount of additional capacitance (10% maximum) is permissible, the additional 100 pF created by the extra 1.7 ft of cable is acceptable. The final results reveal to the user that unless he/she reduces the cable capacitance, 15 ft is the maximum length for all data transmission rates above 20 kbps that requires the usage of shielded cable.

3.5.2. 19.2-kbps Operation. The user needs to transmit 19.2 kbps reliably from point A to point B. The user has decided to use nonshielded cable when interconnecting the equipment. The user reads the cable specifications and discovers that the cable contains 20 pF per foot of mutual capacitance (C_m).

As was the case above, the input capacitance for both the receiver and the generator interchange circuitry is considered to be 100 pF each. Because 19.2-kbps data transmission is classified as a Case 1 transmission (≤ 20 kbps), the upper limit for capacitance is 2500 pF. This leaves the user with 2300 pF for the interconnecting cable.

In this example the cable is nonshielded; therefore the total capacitance (C_c) equals C_m (20 pF) + C_s (10 pF), or 30 pF per foot.

Dividing 2300 pF by 30 pF reveals that the cable can only be 75 ft long. If the user requirements call for an interconnecting cable that is 190 ft long, then the user must select a cable that has only 8 pF of mutual capacitance (C_m).

Section 17-12.2.9 was extracted from EIA/TIA-562, 10/89.

Section 17-14 (pages 1552 and 1553 of the manual) has the following updates in the headings.

17-14. ADAPTING STANDARD DATA RATES TO THE 64-KBPS DIGITAL CHANNEL

17-14.1. Digital Data System (DDS)

17-14.1.1. Basic System. The Digital Data System (DDS) provides full-duplex point-to-point and multipoint private line digital data transmission at customer data rates of 2.4, 4.8, 9.6, and 56 kbit/sec. DDS provides data access to the national digital (PCM) network in North America, where the access data rates are submultiples of the DS1 rate (see Section 17-2.2).

DDS is based on the DS1 channel slot of 64 kbit/sec, utilizing the basic PCM 8-bit word. In DDS this 8-bit word is called a *byte*. However, 1 bit of the 8-bit byte is reserved to facilitate passing of network control information and to satisfy the minimum pulse density requirement for clock recovery on T1 lines. This bit is called the *C-bit*. Thus, the use of the 7 remaining bits in the byte results in a maximum service speed per displaced PCM voice channel of 56 kbit/sec (i.e., 7 bits/byte × 8000 bytes/sec = 56 kbit/sec). For the three lower service speeds of 2.4, 4.8, and 9.6 kbit/sec, requirements for multiplexing dictate the reservation of an additional bit per byte to establish synchronization patterns for routing each byte to the proper output port of the receiving submultiplexer. Thus, the maximum capacity of a subrate byte is 48 kbit/sec. Subsequently, this byte may be assigned in successive DS1 frames to five 9.6 kbit/sec, ten 4.8 bit/sec, or twenty 2.4 kbit/sec channels, respectively. The format of the DS0 signal is shown in Figure 17-67 with the makeup of the various subrates reviewed above. Figure 17-68 shows the DDS system block diagram.

17-14.1.2. Customer Data Service Unit Interface

Customer Data Service Unit Pin Assignments. Connector pin assignments for the customer Data Service Unit are shown in Tables 17-88 and 17-89. (This subsection continues on bottom of page 1553 of the *Manual*.)

This new section follows Section 17-14.1.2, which ends on page 1556 of the Manual.

17-14.2. Adaptation Based on CCITT Rec. V.110

17-14.2.1. Introduction

Standard data rates are based on the expression 75 × 2^n, where n is an integral number. The standard digital channel has a transmission rate (bit rate) of 64 kbps and n × 64 kbps. An incompatibility exists between the standard data rate and the standard digital channel bit rate. CCITT Rec. V.110 is one method that may be used to convert a standard data channel rate to a standard digital channel bit rate and vice versa. The discussion will use ISDN terminology. ISDN is covered in Chapter 19.

17-14.2.2. Conversion

1. Bit Rate Adaptation of Synchronous Data Rates Up to 19.2 kbps

1.1. General Approach. The bit rate adaptation functions within the TA are shown in Figure 17-U.27. The functions RA1 converts the user data signaling rate to an appropriate intermediate rate expressed by 2^k × 8 kbps (where k = 0, 1, or 2). RA2 performs the second conversion from the intermediate rates to 64 kbps. The data signaling rates of 48 and 56 kbps are converted directly into the 64-kbps B channel.

1.2. Adaptation of V-Series Data Signaling Rates to the Intermediate Rates. The intermediate rate used with each of the V-Series* data signaling rates is shown in Table 17-U.5.

*The V-Series are a group of modems and other data transmission devices and interfaces described in the ITU-T V Series of recommendations in this section. The data rates are again related to 75 × 2^n.

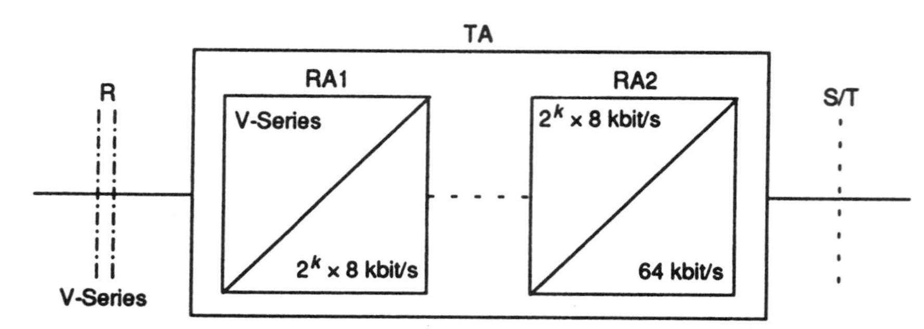

FIGURE 17-U.27. Two-step rate adaptation. (From ITU-T Rec. V.110, Figure 1/V.110, Geneva, 8/92, page 2.)

1.2.1. Frame Structure. The frame structure is shown in Table 17-U.6 and is described below. As shown in Table 17-U.6, the conversion of the V-Series rates to the intermediate rates uses an 80-bit frame. The octet zero contains all binary 0s, while octet 5 consists of a binary 1 followed by seven E bits (see Subsection 1.4 in this section). Octets 1–4 and 6–9 contain a binary 1 in bit number one (position), a status bit (S or X bit) in bit number 8 (position), and six data bits (D bits) in bit positions 2–7. The order of transmission is from left to right and top to bottom.

1.2.2. Frame Synchronization. The 17-bit frame alignment pattern consists of all 8 bits (set to binary 0) of octet zero and bit one (set to binary 1) of the following nine octets (see also Subsection 1.3 in this section).

1.2.3. Status Bits (S1, S3, S4, S6, S8, S9, and X). The bits S and X are used to convey channel control information associated with the data bits in the data transfer state, as shown in Table 17-U.7. The S bits are put into two groups, SA and SB, to carry the condition of two interchange circuits. The X bit is used to carry the condition of circuit 106, and it also signals the state of frame synchronization between TAs. The X bit can also be used optionally to carry flow control information between TAs supporting asynchronous terminal equipment. The usage is specified in Subsection 4.2 of this section.

The use of S and X bits for synchronization of entry to and exit from the data transfer state is specified in Section 17-14.2.4.

TABLE 17-U.5
First Step Rate Adaptation

Data Signaling Rate (bps)	Intermediate Rate		
	8 kbps	16 kbps	32 kbps
600	X		
1,200	X		
2,400	X		
4,800	X		
7,200		X	
9,600		X	
12,000			X
14,400			X
19,200			X

Source: Table 1/V.110, ITU-T Rec. V. 110, Geneva, 9/92, page 3.

TABLE 17-U.6
Frame Structure

Octet Number	Bit Number							
	1	2	3	4	5	6	7	8
0	0	0	0	0	0	0	0	0
1	1	D1	D2	D3	D4	D5	D6	S1
2	1	D7	D8	D9	D10	D11	D12	X
3	1	D13	D14	D15	D16	D17	D18	S3
4	1	D19	D20	D21	D22	D23	D24	S4
5	1	E1	E2	E3	E4	E5	E6	E7
6	1	D25	D26	D27	D28	D29	D30	S6
7	1	D31	D32	D33	D34	D35	D36	X
8	1	D37	D38	D39	D40	D41	D42	S8
9	1	D43	D44	D45	D46	D47	D48	S9

Source: Table 2/V.110, ITU-T Rec. V.110, Geneva, 9/92, page 3.

TABLE 17-U.7
General Mapping Scheme

108	S1, S3, S6, S8 = SA	107
105	S4, S9 = SB	109
Frame synch and 106/IWF	X	106

Source: Table 3/V.110, ITU-T Rec. V.110, Geneva, 9/92, page 4.

The mechanism for proper assignment of the control information from the transmitting signal rate adapter interface via these bits to the receiving signal rate adapter interface is shown in Table 17-U.7 and described in Section 17-14.2.4.

For the S and X bits, a ZERO corresponds with the ON condition whereas a ONE corresponds with the OFF condition.

Control information, conveyed by the S bits, and user data, conveyed by the D bits, should not have different transmission delays. The S bits should therefore

TABLE 17-U.8
Coordination Between S Bits and D Bits

S Bit	D Bit	
	Octet No.	Bit No.
S1	2	3(D8)
S3	3	5(D16)
S4	4	7(D24)
S6	7	3(D32)
S8	8	5(D40)
S9	9	7(D48)

Source: Table 4/V.110, ITU-T Rec. V.110, Geneva, 9/92, page 4.

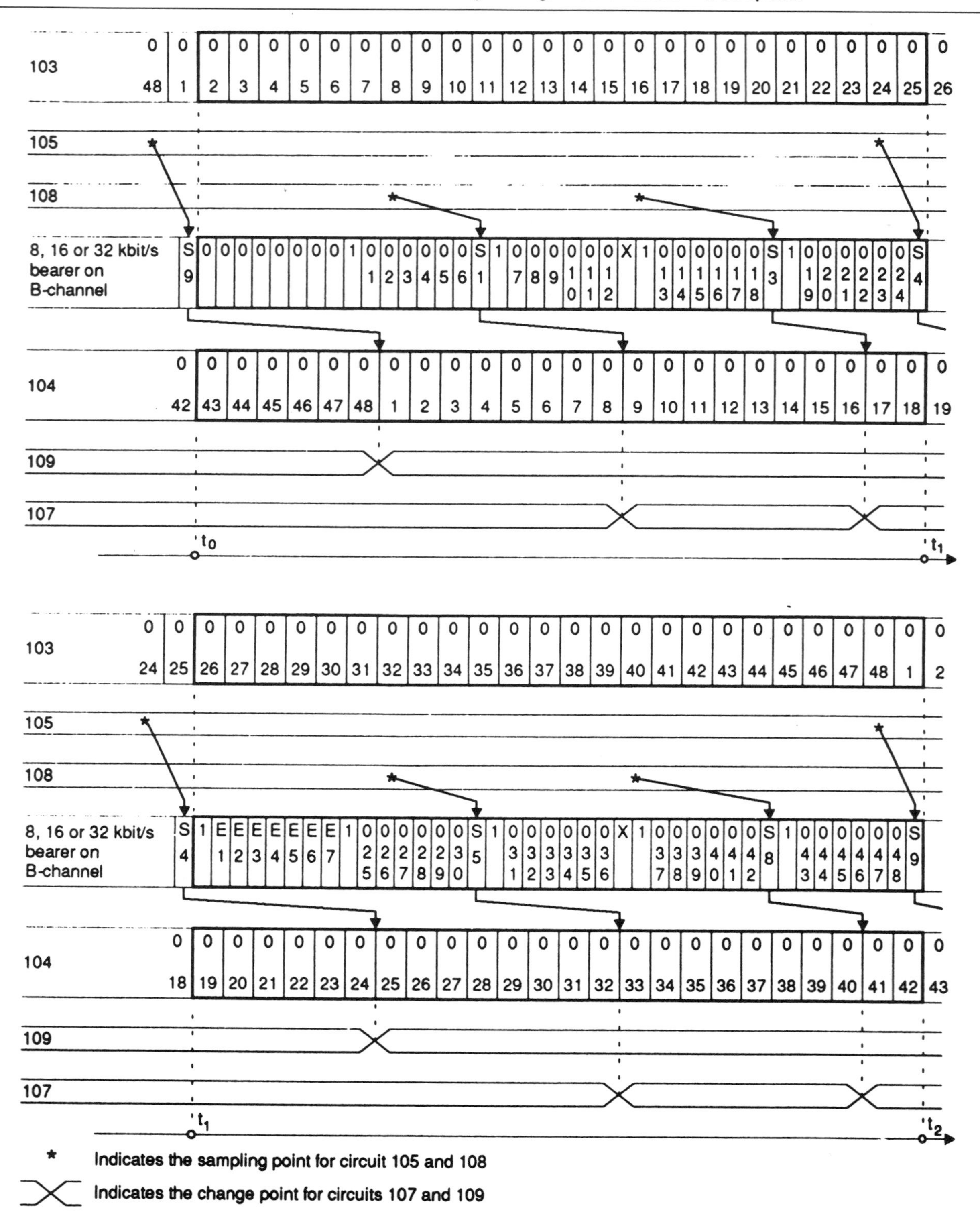

Note 1 – In order to maintain conformity with the bit rate adaptation of X.1 user classes of service described in Recommendation X.30 (I.461), the bits S1 and S6, S3 and S8, S4 and S9 are used to convey channel status information associated with the P-, Q- and R-bit groups respectively.

Refer to § 2.1.1.2.3 of Recommendation X.30 (I.461) for detailed information concerning the mapping of the information on circuit C of the X.21 interface to the S-bits and to the I-bits of the distant interface.

FIGURE 17-U.28. Coordination between D bits and S bits. (From Figure 2/V.110, ITU-T Rec. V.110, Geneva, 9/92, page 5.)

transmit control information sampled simultaneously with the D bits in the positions specified in Table 17-U.8 and as shown in Figure 17-U.28.

The X bit should be presented upon arrival to control circuit 106. Circuit 106 shall respond as defined in Subsection 3 of this section (X = ZERO, 106 = ON).

1.2.4. E-Bit Usage. The E bits are used to carry the following information:

1. Rate Repetition Information. Bits E1, E2, and E3, in conjunction with the intermediate rate (see Table 17-U.6), provide user data signaling rate (synchronous) identification. The coding of these bits is as shown in Table 17-U.9.
2. Network Independent Clock Information. Bits E4, E5, and E6 are used as specified in Section 17-14.2.5 to carry network-independent clock phase information.
3. Multiframe Information. Bit E7 is used as shown in Table 17-U.9.

1.2.5. Rate Negotiation. Negotiation of the synchronous rate may be appropriate in interworking situations involving interconnections with modems on the PSTN where the remote modem/DTE has the capability of operating at different rates depending upon the conditions. It may also be appropriate in interconnections for asynchronous transmission specified in Subsection 3 of this section and to accommodate split rate operation. The need for rate negotiation and the method is for further study.

1.2.6. Data Bits. Data are conveyed in D bits—that is, up to 48 bits per 80-bit frame. In this Recommendation the octet boundaries of the user's data stream are not defined.

TABLE 17-U.9
E-Bit Usage (Note 1)

Intermediate Rates (kbps)			E1	E2	E3	E4	E5	E6	E7
8 bps	16 bps	32 bps		(Note 2)			(Note 3)		
600			1	0	0	C	C	C	1 or 0 (Note 4)
1200			0	1	0	C	C	C	1
2400			1	1	0	C	C	C	1
		12,000	0	0	1	C	C	C	1
	7200	14,400	1	0	1	C	C	C	1
4800	9600	19,200	0	1	1	C	C	C	1

Note 1: The data signaling rates of 600, 2400, 4800, and 9600 bps are also Rec. X.1 user classes of service (see also Rec. X.30/I.461).

Note 2: Synchronous rate information is carried by bits E1, E2, and E3 as indicated. Asynchronous rate information must be provided with out-of-band signaling (layer 3 messages in the D channel) or with in-band parameter exchange as described in Appendix I of ITU-T Rec. V.110.

Note 3: C indicates the use of E4, E5, and E6 for the transport of network-independent blocking information (see Section 17-14.2.5). These bits shall be set to ONE when unused.

Note 4: In order to maintain compatibility with Rec. X.30 (I.461), the 600-bps user rate E7 is coded to enable the 4 × 80-bit multiframe synchronization. To this end, E7 in the fourth 80-bit frame is set to binary 0 (see Subsection 1.2.7 in this section; also see Table 17-U.10).

Source: Table 5/V.110, ITU-T Rec. V.110, Geneva, 9/92, page 6.

TABLE 17-U.10
600-bps to 8-kbps Intermediate Rate

0	0	0	0	0	0	0	0
1	D1	D1	D1	D1	D1	D1	S1
1	D1	D1	D2	D2	D2	D2	X
1	D2	D2	D2	D2	D3	D3	S3
1	D3	D3	D3	D3	D3	D3	S4
1	1	0	0	E4	E5	E6	E7[a]
1	D4	D4	D4	D4	D4	D4	S6
1	D4	D4	D5	D5	D5	D5	X
1	D5	D5	D5	D5	D6	D6	S8
1	D6	D6	D6	D6	D6	D6	S9

[a]See Note 4 to Table 17-U.9.
Source: Table 6a/V.110, ITU-T Rec. V.110, Geneva, 9/92, page 8.

TABLE 17-U.11
1200-bps to 8-kbps Intermediate Rate

0	0	0	0	0	0	0	0
1	D1	D1	D1	D1	D2	D2	S1
1	D2	D2	D3	D3	D3	D3	X
1	D4	D4	D4	D4	D5	D5	S3
1	D5	D5	D6	D6	D6	D6	S4
1	0	1	0	E4	E5	E6	E7
1	D7	D7	D7	D7	D8	D8	S6
1	D8	D8	D9	D9	D9	D9	X
1	D10	D10	D10	D10	D11	D11	S8
1	D11	D11	D12	D12	D12	D12	S9

Source: Table 6b/V.110, ITU-T Rec. V.110, Geneva, 9/92, page 8.

TABLE 17-U.12
2400-bps to 8-kbps Intermediate Rate

0	0	0	0	0	0	0	0
1	D1	D1	D2	D2	D3	D3	S1
1	D4	D4	D5	D5	D6	D6	X
1	D7	D7	D8	D8	D9	D9	S3
1	D10	D10	D11	D11	D12	D12	S4
1	1	1	0	E4	E5	E6	E7
1	D13	D13	D14	D14	D15	D15	S6
1	D16	D16	D17	D17	D18	D18	X
1	D19	D19	D20	D20	D21	D21	S8
1	D22	D22	D23	D23	D24	D24	S9

Source: Table 6c/V.110, ITU-T Rec. V.110, Geneva, 9/92, page 8.

1.2.7. Bit Assignment. The adaptation of 600-, 1200-, and 2400-bps data rates to the 8-kbps intermediate rate is shown in Tables 17-U.10, 17-U.11, and 17-U.12, respectively. The adaptation of 7200 and 14,400-bps data rates to the 16- and 32-kbps intermediate rates, respectively, uses the data bit assignments shown in Table 17-U.13. The adaptation of 4800-, 9600-, and 19,200-bps data rates to the 8-, 16-, and 32-kbps intermediate rates, respectively, uses the data bit assignments

TABLE 17-U.13
N^a × 3600 bps to the Intermediate Rate

0	0	0	0	0	0	0	0
1	D1	D2	D3	D4	D5	D6	S1
1	D7	D8	D9	D10	F[b]	F	X
1	D11	D12	F	F	D13	D14	S3
1	F	F	D15	D16	D17	D18	S4
1	1	0	1	E4	E5	E6	E7
1	D19	D20	D21	D22	D23	D24	S6
1	D25	D26	D27	D28	F	F	X
1	D29	D30	F	F	D31	D32	S8
1	F	F	D33	D34	D35	D36	S9

[a]$N = 2$ or 4 only.
[b]F = fill bit.
Source: Table 6d/V.110, ITU-T Rec. V110, Geneva, 9/92, page 8.

TABLE 17-U.14
N^a × 4800 bps to the Intermediate Rate

0	0	0	0	0	0	0	0
1	D1	D2	D3	D4	D5	D6	S1
1	D7	D8	D9	D10	D11	D12	X
1	D13	D14	D15	D16	D17	D18	S3
1	D19	D20	D21	D22	D23	D24	S4
1	0	1	1	E4	E5	E6	E7
1	D25	D26	D27	D28	D29	D30	S6
1	D31	D32	D33	D34	D35	D36	X
1	D37	D38	D39	D40	D41	D42	S8
1	D43	D44	D45	D46	D47	D48	S9

[a]$N = 1$, 2, or 4 only.
Source: Table 6e/V.110, ITU-T Rec. V. 110, Geneva, 9/92, page 8.

TABLE 17-U.15
12,000-bps to 32-kbps Intermediate Rate

0	0	0	0	0	0	0	0
1	D1	D2	D3	D4	D5	D6	S1
1	D7	D8	D9	D10	F[a]	F	X
1	D11	D12	F	F	D13	D14	S3
1	F	F	D15	F	F	F	S4
1	0	0	1	E4	E5	E6	E7
1	D16	D17	D18	D19	D20	D21	S6
1	D22	D23	D24	D25	F	F	X
1	D26	D27	F	F	D28	D29	S8
1	F	F	D30	F	F	F	S9

[a]*F*, fill.
Source: Table 6f/V.110, ITU-T Rec. V.110, Geneva, 9/92, page 8.

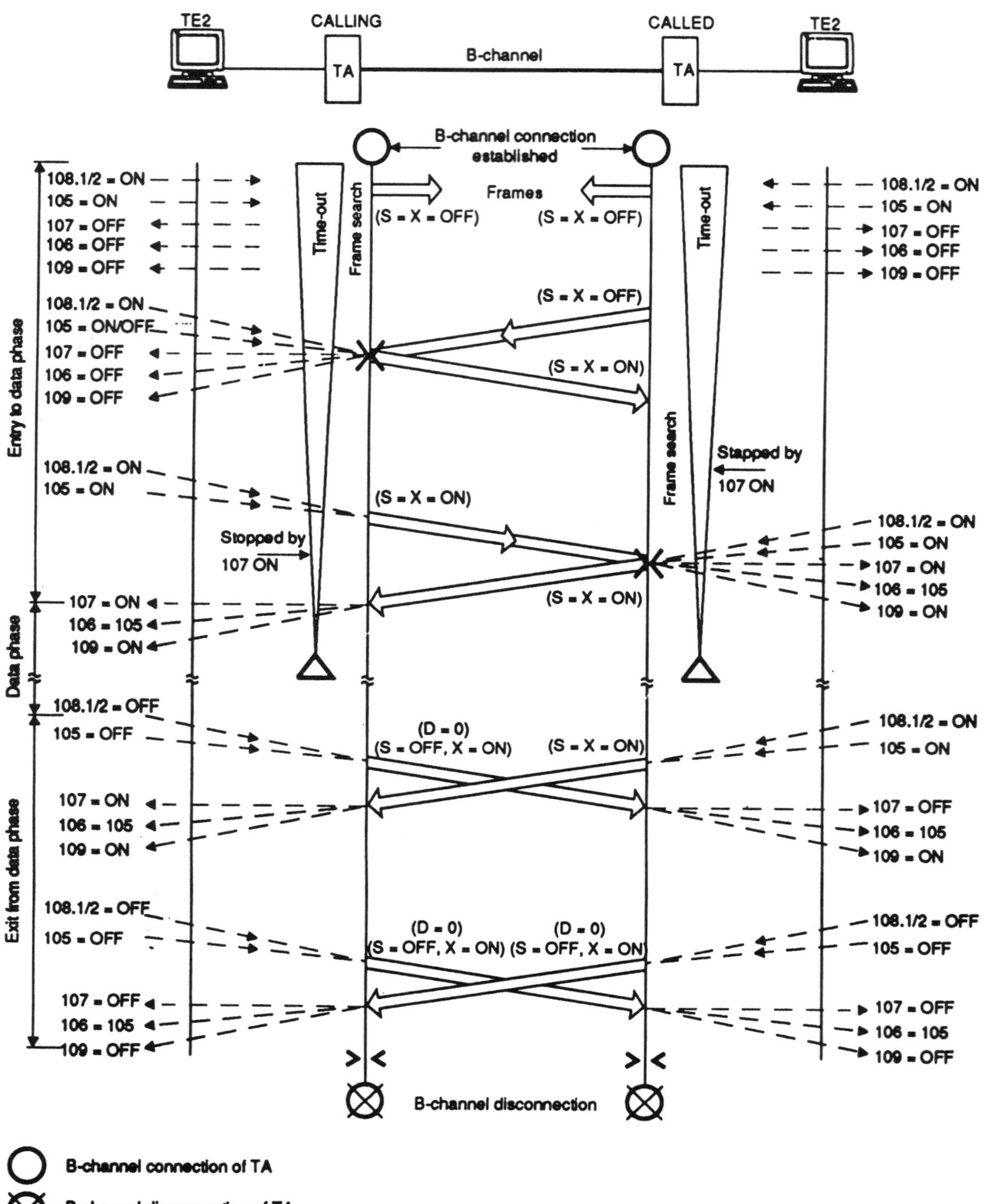

FIGURE 17-U.29. TA synchronization of entry to and exit from data transfer phase. *Note:* § 4.1.4 refers to Subsection 1.4 in Section 17.14.2.4. (From Figure 3/V.110, ITU-T Rec. V.110, page 9.)

shown in Table 17-U.14. The adaptation of 12,000-bps user data rate to the 32-kbps intermediate rate uses the data bit assignments shown in Table 17-U.15.

1.3. Frame Synchronization and Additional Signaling Capacity

1.3.1. Search for Frame Synchronization. The following 17-bit alignment pattern is used to achieve frame synchronization:

00000000	1XXXXXXX	1XXXXXXX	1XXXXXXX	1XXXXXXX
1XXXXXXX	1XXXXXXX	1XXXXXXX	1XXXXXXX	1XXXXXXX

To ensure a reliable synchronization, it is suggested that at least two 17-bit alignment patterns in consecutive frames be detected.

Once frame synchronization is achieved, it is suggested that a persistence check be made of the S = X = OFF condition of Subsection 1.2 (of Section 17-14.2.4) prior to proceeding to transparent data transfer with S = X = ON condition, as shown in Figure 17-U.29.

1.3.2. Frame Synchronization Monitoring and Recovery. Monitoring of the frame synchronization shall be a continuous process using the same procedures as for initial detection.

Loss of frame synchronization shall not be assumed unless at least three consecutive frames, each with at least one framing bit error, are detected.

Following loss of frame synchronization, the TA shall enter a recovery state as discussed in Subsection 1.5 of Section 17-14.2.4. If recovery is not successful, further maintenance procedures may be used.

1.4. Adaptation of Intermediate Rates to 64 kbps. Because rate adaptation of a single intermediate rate (e.g., 8, 16, or 32 kbps) to the 64-kbps B-channel rate and the possible multiplexing of several intermediate rate streams to the 64-kbps B-channel rate must be compatible to enable interworking, a common approach is needed for the second step rate adaptation and, possibly, for intermediate rate multiplexing. This second step rate adaptation method is described in ITU-T Rec. I.460.

2. Rate Adaptation of 48- and 56-kbps User Rates to 64 kbps. The 48- and 56-kbps user data rates are adapted to the 64-kbps B-channel rate in one step as shown in Tables 17-U.16 through 17-U.18, respectively.

TABLE 17-U.16
Adaptation of 48-kbps User Rate to 64 kbps

Octet Number	Bit Number							
	1	2	3	4	5	6	7	8
1	1	D1	D2	D3	D4	D5	D6	S1
2	0	D7	D8	D9	D10	D11	D12	X
3	1	D13	D14	D15	D16	D17	D18	S3
4	1	D19	D20	D21	D22	D23	D24	S4

Note 1: 48 kbps is also a Rec. X.1 user class of service (see also Rec. X.30/I.461, § 2.2.1).

Note 2: Refer to Subsection 1.2.3 in this section for the use of status bits and bit X; however, for international operation over restricted 64-kbps bearer capabilities, bit X must be set to binary 1.

Source: Table 7a/V.110, ITU-T Rec. V.110, Geneva, 9/92, page 10.

TABLE 17-U.17
Adaptation of 56-kbps User Rate to 64 kbps

Octet Number	Bit Number							
	1	2	3	4	5	6	7	8
1	D1	D2	D3	D4	D5	D6	D7	1
2	D8	D9	D10	D11	D12	D13	D14	1
3	D15	D16	D17	D18	D19	D20	D21	1
4	D22	D23	D24	D25	D26	D27	D28	1
5	D29	D30	D31	D32	D33	D34	D35	1
6	D36	D37	D38	D39	D40	D41	D42	1
7	D43	D44	D45	D46	D47	D48	D49	1
8	D50	D51	D52	D53	D54	D55	D56	1

Source: Table 7b/V.110, ITU-T Rec. V.110, Geneva, 9/92, page 10.

2.1. Frame Synchronization. At the user data rate of 48 kbps, the frame alignment pattern consists of 1011 in bit 1 of consecutive octets of one frame. To ensure reliable synchronization, it is suggested that at least five 4-bit alignment patterns in consecutive frames be detected.

At the user data rate of 56 kbps with the alternative frame structure based on Table 17-U.18, the frame alignment pattern consists of OYY1111 in bit 8 of consecutive octets of one frame. Bits marked with Y may be either "0" or "1." To ensure reliable synchronization, it is suggested that at least 5-bit (01111) alignment patterns in the 8-bit sequence of OYYY1111 in consecutive octets be detected.

Frame synchronization monitoring and recovery are described in Subsection 1.3.2 of this section.

TABLE 17-U.18
Alternative Frame Structure for the Adaptation of 56-kbps User Rate to 64 kbps

Octet Number	Bit Number							
	1	2	3	4	5	6	7	8
1	D1	D2	D3	D4	D5	D6	D7	0
2	D8	D9	D10	D11	D12	D13	D14	X
3	D15	D16	D17	D18	D19	D20	D21	S3
4	D22	D23	D24	D25	D26	D27	D28	S4
5	D29	D30	D31	D32	D33	D34	D35	1
6	D36	D37	D38	D39	D40	D41	D42	1
7	D43	D44	D45	D46	D47	D48	D49	1
8	D50	D51	D52	D53	D54	D55	D56	1

Note 1: Refer to Subsection 1.2.3 in this section for the use status bits and bit X.

Note 2: This table is a permitted option to provide for signaling to enter and to leave the data phase. However, the recommended approach shall be as in Table 17-U.17, and the responsibility shall be on the user of this table to ensure that interworking can be achieved.

Source: Table 7c/V.110, ITU-T Rec. V.110, Geneva, 9/92, page 11.

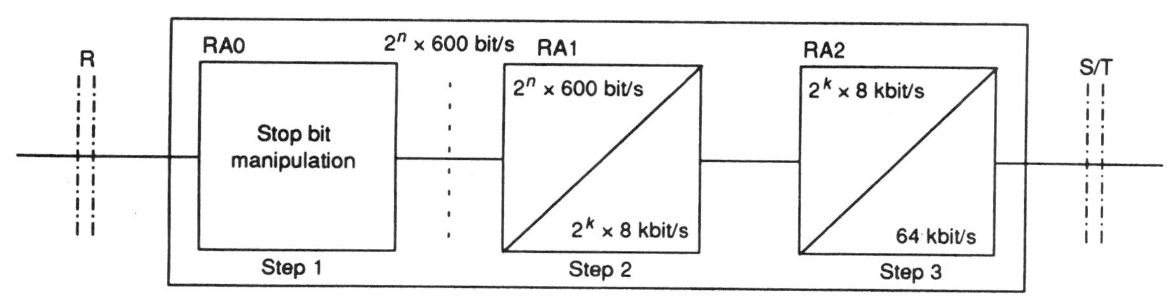

FIGURE 17-U.30. Three-step rate adaptation. (From Figure 4/V.110, ITU-T Rec. V.110, Geneva, 10/92, page 12.)

3. Adaptation for Asynchronous Rates Up to 19,200 bps

3.1. General Approach. The bit rate adaptation functions within the TA are shown in Figure 17-U.30. A three-step method is employed with the functional blocks RA0, RA1, and RA2. The RA0 function is an asynchronous-to-synchronous conversion step, for support of the rates specified in Table 17-U.19, using the same technique as defined in ITU-T Rec. V.14. It produces a synchronous bit stream defined by $2^n \times 600$ (where $n = 0$–5). The functions of RA1 and RA2 are the same as specified in Subsection 1 of this section. Function RA1 adapts the user rate to the next higher rate expressed by $2^k \times 8$ bits (where $k = 0$, 1, or 2). RA2 performs the second conversion to 64 kbps.

3.2. Supported Asynchronous User Rates. The asynchronous user rates to be supported, mandatory and optional, are specified in Table 17-U.19.

TABLE 17-U.19
Asynchronous User Rates

Data Rate (bps)	Rate Tolerance (%)	Number of Data Units	Number of Stop Elements	RA0/RA1 Rate (bps)	RA1 Rate (kbps)
50	±2.5	5	1, 5	600	8
75	±2.5	5.7 or 8	1 : 5 : 2	600	8
110	±2.5	7 or 8	1 or 2	600	8
150	±2.5	7 or 8	1 or 2	600	8
200	±2.5	7 or 8	1 or 2	600	8
300*	±2.5	7 or 8	1 or 2	600	8
600*	+1 − 2.5	7 or 8	1 or 2	600	8
1,200*	+1 − 2.5	7 or 8	1 or 2	1,200	8
2,400*	+1 − 2.5	7 or 8	1 or 2	2,400	8
3,600	+1 − 2.5	7 or 8	1 or 2	4,800	8
4,800*	+1 − 2.5	7 or 8	1 or 2	4,800	8
7,200	+1 − 2.5	7 or 8	1 or 2	9,600	16
9,600*	+1 − 2.5	7 or 8	1 or 2	9,600	16
12,000	+1 − 2.5	7 or 8	1 or 2	19,200	32
14,400	+1 − 2.5	7 or 8	1 or 2	19,200	32
19,200	+1 − 2.5	7 or 8	1 or 2	19,200	32

Note 1: Asterisk indicates rate whose support is mandatory for universal TA.
Note 2: Number of data bits includes possible parity bits.
Source: Table 8/V.110, ITU-T Rec. V.110, Geneva, 10/92, page 12.

3.3. Asynchronous-to-Synchronous Conversion (RA0). The RA0 function is only used with asynchronous V-Series interfaces. Incoming asynchronous data are padded by the addition of stop elements to fit the nearest channel rate defined by $2^n \times 600$ bps. Thus, a 7200-bps user data signaling rate shall be adapted to a synchronous 9600-bps stream, and a 110-bps user data signaling rate shall be adapted to a synchronous 600-bps stream. The resultant synchronous stream is fed to RA1. Padding with stop elements is inhibited during the transmission of the break signal as described in Subsection 3.5 of this section.

3.4. Overspeed/Underspeed. A terminal adaptor shall insert additional stop elements when its associated terminal is transmitting with a lower than nominal character rate. If the terminal is transmitting characters with an overspeed of up to 1% (or 2.5% in the case of nominal speeds lower than 600 bps), the asynchronous-to-synchronous converter may delete stop elements as often as is necessary to a maximum of one every eight characters at 1% overspeed. The converter on the receiving side shall detect deleted stop elements and reinsert them in the received data stream (circuit 104).

The nominal length of the start elements and data units shall be the same for all characters. The length of the stop element may be reduced as much as 12.5% by the receiving converter for nominal speeds exceeding 300 bps to allow for overspeed in the transmitting terminal. For nominal speeds less than or equal to 300 bps, a 25% reduction in stop element is allowed.

3.5. Break Signal. The terminal adaptor shall detect and transmit the break signal as follows:

If the converter detects M to $2M + 3$ bits, all of start polarity, where M is the number of bits per character in the selected format including start and stop elements, the converter shall transmit $2M + 3$ bits of start polarity.

If the converter detects more than $2M + 3$ bits, all of start polarity, the converter shall transmit all these bits as start polarity.

For the cases where the asynchronous rate is lower than the synchronous rate for the converter, the following rules shall apply:

- The converter shall transmit start polarity (to RA1) for a time period equal to $2M + 3$ bits at the asynchronous rate if the converter has detected M to $2M + 3$ bits of start polarity.
- The converter shall transmit (to RA1) start polarity for a time period as long as the received break condition if the converter has detected more than $2M + 3$ bits of start polarity.
- The $2M + 3$ or more bits of start polarity received from the transmitting side shall be output to the receiving DTE.
- The DTE must transmit on circuit 103 at least $2M$ bits of stop polarity after the start polarity break signal before sending further data characters. The converter shall then regain character synchronism from the following stop to start transition.

3.6. Parity Bits. Possible parity bits included in the user data are considered as data bits by the RA0 function.

4. Flow Control. A flow control option, for use with TAs supporting asynchronous DTEs, is described in this section. Flow control allows the connection of asynchronous DTEs operating at different user data rates by reducing the character output of the faster to that of the slower. Support of flow control will require the use of end-to-end (TA-to-TA) protocol defined in Subsection 4.2 (of this section) and an incoming line (from network) buffer in addition to a selected local protocol (see Subsection 2.4.1 in this section). Depending upon the local flow control protocol employed, there will also be a requirement for character buffering from the DTE interface. The size of this buffer is not defined in this Recommendation because it is dependent upon implementation.

Local flow control of the DTE interface is required where the DTE operates at a rate higher than the synchronous rate established between TAs. End-to-end flow control is required where the synchronous rate established between TAs is consistent with the operating rate of one DTE (or interworking unit) and higher than the synchronous rate consistent with the operating rate of the other DTE (or interworking unit). Both local and end-to-end flow control could be required in some applications.

4.1. Local Flow Control. Connection may be made between TAs connected to asynchronous DTEs operating at two different speeds. It is the responsibility of the TA connected to the faster DTE to execute a local flow control protocol to reduce the character rate to that of the slower DTE. This operation will require some buffer storage in the TA. A TA may support several different local flow control protocols, although only one will be selected at any one time. There are a number of such protocols in use, some of which are detailed in the following text.

4.1.1. 105/106 Operation. This is an out-of-band flow control mechanism, utilizing two of the interchange circuits specified in Rec. V.24. If a DTE requires to transmit a character, it turns ON circuit 105 (request to send). The DTE can only begin transmission when it receives in return circuit 106 ON (ready for sending). If during transmission of a block of characters circuit 106 goes OFF, the DTE must cease transmission (after completing the transmission of any character of which transmission has started) until circuit 106 turns ON again.

4.1.2. XON/XOFF Operation. This is an in-band flow control mechanism using two characters of the International Alphabet No. 5 (IA5) set for XON and XOFF operation. If a DTE receives an XOFF character, it must cease transmission. When it receives an XON character, it may resume transmission. The characters typically used for XON and XOFF are device control one (DC1) and device control three (DC3) (bit combination 1/1 and 1/3 in Rec. T.50), respectively, although alternative bit combinations can be used.

4.1.3. Other Methods. Alternative and nonstandard methods of flow control are in use, and these may be mapped on the TA flow control protocol.

4.2. End-to-End (TA-to-TA) Flow Control. Matching (by reduction) of the transmitted character rate of the DTE to the rate of the TA is not sufficient in all cases to guarantee correct operation, and end-to-end flow control may be required.

The X bit is used to carry flow control information. A TA will buffer incoming characters. When the number of buffered characters exceeds a threshold TH1, depending upon implementation, the TA will set the X bit of its outgoing frames to OFF.

Upon receipt of a frame containing an X bit set to OFF, a TA will execute its selected local flow control procedure indicating that the attached DTE must stop sending characters, and it will cease the transmission of data after completion of the character in progress by setting the data bits in the outgoing frames to ONE.

When the buffer contents of a TA which has initiated an end-to-end flow control drops below threshold TH2, the TA will reset the outgoing X bit to ON.

When the far-end TA receives a frame with the X bit set to ON, it will recommence data transmission and, by use of the local flow control procedure, indicate to the attached DTE that it may continue.

Note: There may be a delay between initiation of the end-to-end flow control protocol and termination of the incoming character stream. The characters arriving during this time must be buffered, and the total buffer size will depend upon the character rate, round trip delay, and the buffer threshold.

4.3. Use of Channel Capacity. Upon accepting a call from a TA supporting flow control and operating at a different user rate and/or intermediate rate, the called TA will adopt the identical intermediate rate and bit repetition factor. This will override the parameters normally selected. In such cases, the TA connected to the faster DTE will execute a local flow control procedure to reduce the character rate to that of the slower DTE.

Thus, if a faster DTE calls a slower DTE, the faster intermediate channel rate and bit repetition factor will be adopted by the TAs on both ends. To reduce the character rate received by the slower DTE, its TA will exercise end-to-end flow control and cause the TA on the calling side to utilize local flow control.

If a slower DTE calls a faster DTE, the slower intermediate channel rate and bit repetition factor will be adopted by the TAs on both ends. To reduce the character rate transmitted by the faster DTE, its TA will exercise local flow control.

If the called TA does not implement the intermediate rate and bit repetition factor used by the calling TA, the call shall be rejected.

4.4. Requirements of a TA Supporting Flow Control. The following are general requirements for a TA supporting flow control:

1. A TA supporting flow control shall be capable of operating with an intermediate rate and bit repetition factor that is independent of the asynchronous speed used at its DTE interface.
2. A TA supporting flow control shall, if possible, adapt to the intermediate rate and bit repetition factor required for an incoming call. User rate information will be obtained from signaling.
3. A TA supporting flow control shall be capable of executing a local flow control protocol to reduce the character rate to that of the far-end DTE.
4. A TA supporting flow control will support the use of end-to-end (TA-to-TA) flow control using the X bit and will contain a character buffer.

17-14.2.3. Interchange Circuits

1. Essential and Optional Interchange Circuits. The essential and optional interchange circuits are shown in Table 17-U.20.

2. Timing Arrangements. The TA shall derive ISDN timing from the received bit stream of the ISDN's basic user–network interface (see Sections 5 and 8 of Rec. I.430). This network timing shall be used by the TA to provide the DTE with transmitter signal element timing on circuit 114 and receiver signal element timing on circuit 115.

3. Circuit 106. After the start-up and retrain synchronization sequences, the ON state of circuit 106 shall be delayed relative to the ON state of circuit 105 (where implemented) by an interval of at least N bits (a value of N equal to 24 has been proposed, but the value is for further study). ON to OFF state transitions of circuit

TABLE 17-U.20
Interchange Circuits

Number	Description	Notes
102	Signal ground or common return	
102a	DTE common return	2
102b	DCE common return	2
103	Transmitted data	
104	Received data	
105	Request for sending	3
106	Ready for sending	
107	Data set ready	
108/1	Connect data set to line	4
108/2	Data terminal ready	4
109	Data channel received line signal detector	
111	Data signaling rate selector (DTE source)	5
112	Data signaling rate selector (DCE source)	5
113	Transmitter signal element timing (DTE source)	6
114	Transmitter signal element timing (DCE source)	
115	Receiver signal element timing (DCE source)	
125	Calling indicator	7
140	Loopback/maintenance test	8
141	Local loopback	8
142	Test indicator	8

Note 1: All essential circuits and any others which are provided shall comply with the functional and operational requirements of Rec. V.24. All interchange circuits provided shall be properly terminated in the data terminal equipment and in the data circuit-terminating equipment in accordance with the appropriate Recommendation for electrical characteristics (see Subsection 5 in this section).

Note 2: Interchange circuits 102a and 102b are required where the electrical characteristics defined in Rec. V.10 are used at data signaling rates above 20 kbps.

Note 3: Not required for DTEs that operate with DCEs in the continuous carrier mode.

Note 4: This circuit shall be capable of operating as circuit 108/1 or 108/2, depending on its use (by the associated TE).

Note 5: The use of this circuit is for further study.

Note 6: The use of circuit 113 is for further study, because its application is restricted by the synchronous nature of ISDN.

Note 7: This circuit is used with the automatic answering terminal adaptor function.

Note 8: The use for loopback testing is for further study.

Source: Table 9/V.110, ITU-T Rec. V.110, Geneva, 10/92, page 16.

106 shall follow ON to OFF state transitions of circuit 105 (when implemented) by less than 2 msec. Where circuit 105 is not implemented, the initial circuit 106 transition to the ON state shall be delayed by an interval greater than or equal to N bits relative to the corresponding transition in the state of circuit 109. Subsequent transitions in the state of circuit 106 should occur solely in accordance with the operating sequences defined in Section 17-14.2.4, or when used for the optional flow control in Subsection 4 of this section.

4. Circuit 109. OFF and ON and ON and OFF transitions of circuit 109 should occur solely in accordance with the operating sequence defined in Section 17-14.2.4.

5. Electrical / Mechanical Characteristics of Interchange Circuits

5.1. Basic ISDN User–Network Interface. The electrical and mechanical characteristics of the basic ISDN user–network interface are described in Sections 8 and 10 of ITU-T Rec. I.430.

5.2. TE2/TA (DTE/DCE) Interface

5.2.1. Rates Less Than or Equal to 19.2 kbps. Use of electrical characteristics conforming to ITU-T Rec. V.28 is recommended together with the connector and pin assignment plan specified by ISO 2110.

Note: Manufacturers may wish to note that the long-term objective is to replace electrical characteristics specified in ITU-T Rec. V.28, and that Study Group XVII has agreed that the work shall proceed to develop a more efficient, all-balanced interface for the V-Series application which minimizes the number of interchange circuits (ITU-T Rec. V.230).

5.2.2. Rates Greater Than 19.2 kbps. Use of electrical characteristics conforming to Rec. V.10 and/or V.11 is recommended together with the use of the connector and pin assignment plan specified by ISO 4902.

- **(i)** Concerning circuits 103, 104, 113, 114, and 115, both the generators and the receivers shall be in accordance with Rec. V.11.
- **(ii)** In the case of circuits 105, 106, 107, and 109, generators shall comply with Rec. V.10 or, alternatively, Rec. V.11. The receivers shall comply with Rec. V.10, category 1, or Rec. V.11 without termination.
- **(iii)** In the case of all other circuits, Rec. V.10 applies, with receivers configured as specified by Rec. V.10 for category 2.

Alternatively, the interface defined in Appendix II to Rec. V.35, together with connector and pin assignment plan specified by ISO 2593, may be used.

6. Fault Conditions on Interchange Circuits. See Section 7 of Rec. V.28 for association of the receiver failure detection types.

6.1. The DTE should interpret a fault condition on circuit 107 as an OFF condition using failure detection type 1.

6.2. The data circuit-terminating equipment (DCE) should interpret a fault condition on circuits 105 and 108 as an OFF condition using failure detection type 1.

6.3. All other circuits not referred to above may use failure detection type 0 or 1.

17-14.2.4. Operating Sequence

1. TA Duplex Operation. When using the TA to provide data transmission service within ISDN, the call is established over a 64-kbps connection using the procedures applicable to the particular network and/or terminal configuration.

The internal arrangement of the TA functional parts and the DTE (with a V-Series-type interface) is not within the scope of this Recommendation. It is assumed that means are provided to control the entry to and the exit from the data transfer mode. For example, it is assumed that the means are provided to control circuits 108/1 (Connect data set to line) or 108/2 (Data terminal ready) internally —that is, within the station at the customer premises. However, for the purpose of this Recommendation, circuit 108/2, as defined in Rec. V.24, is assumed.

1.1. Idle or Ready State

1.1.1. During the idle (or ready) state, the TA (DCE) will be receiving the following from the DTE:

Circuit 103 = Continuous binary 1
Circuit 105 = See Note
Circuit 108/1 = OFF; circuit 108/2 = ON

Note: In many duplex DTEs circuit 105 is either permanently in the ON condition or it is not present. If not present, the function must be set in an ON condition in the TA. See paragraph 1.2.4 (below) for the case where a duplex DTE can operate circuit 105.

1.1.2. During the idle (or ready) state, the TA will transmit continuous binary 1s into the B and D channels (i.e., all bits of Table 17-U.6 = binary 1).

1.1.3. During the idle (or ready) state, the TA (DCE) will transmit the following toward the DTE:

Circuit 104 = Continuous binary 1

Circuit 107 = OFF

Circuit 106 = OFF

Circuit 109 = OFF

1.2. Connect TA to Line State

1.2.1. When the TA is to be switched to the data mode, circuit 108 must be ON. Switching to the data mode causes the TA to transmit the following toward the ISDN (refer to Table 17-U.6):

Note 1: At this time, circuit 103 is not connected to the data channel (e.g., the binary 1 condition of the data bits is generated within the TA).

Note 2: In the following description only the interoperation between TE2/TA (DTE/DCE) interface and the intermediate rate frames (see Tables 17-U.10 to 17-U.15) and the 64-kbps frame of Tables 17-U.16 and 17-U.18 are discussed. The second step of rate adaptation encoding and decoding and the multiplexing and demultiplexing of the ISDN basic user–network interface are discussed in Recs. I.460 and I.430, respectively.

1.2.2. At this time (i.e., switching to data mode) the receiver in the TA will begin to search for the frame synchronization pattern in the received bit stream (see Subsections 1.3.1 and 2.1 in Section 17-14.2.2).

1.2.3. When the receiver recognizes the frame synchronization pattern, it causes the S and X bits in the transmitted frames to be turned ON (provided that circuit 108 is ON).

1.2.4. When the receiver recognizes that the status of bits S and X are in the ON condition, it will perform the following functions:

(a) Turn ON circuit 107 toward the DTE and stop timer T1.
Note: A duplex DTE that implements and is able to operate circuit 105 may be expected to turn this circuit ON at any time. However, if not previously turned ON, it must be turned ON in response to the ON condition on circuit 107.

(b) Then, circuit 103 may be connected to the data bits in the frame; however, the DTE must maintain a binary 1 condition until circuit 106 is turned ON in the next portion of the sequence.

(c) Turn ON circuit 109 and connect the data bits to circuit 104.
Note: Binary 1 is being received on circuit 104 at this time.

(d) After an interval of N bits (see Subsection 3 in Section 17-14.2.3), it will turn ON circuit 106.

(e) Circuit 106 transitioning from OFF to ON will cause the transmitted data to transition from binary 1 to the data mode.

If circuit 107 has not been turned ON, after expiring of timer T1 the TA shall be disconnected according to the procedures given in Subsection 1.4 of this section.

1.3. Data Transfer State

1.3.1. While in the data transfer state, the following circuit conditions exist:

(a) Circuits 105 (when implemented), 106, 107, 108/1 or 108/2, and 109 are in the ON condition.
(b) Data are being transmitted on circuit 103 and received on circuit 104.

1.4. Disconnect Mode

1.4.1. At the completion of the data transfer phase, the local DTE will indicate a disconnect request by turning OFF circuit 108. This will cause the following to occur:

(a) The status bits S in the frame toward ISDN will turn OFF, and status bits X are kept ON.
(b) Circuit 106 will be turned OFF.
(c) The data bits in the frame will be set to binary 0.

1.4.2. If circuit 108 is still ON at the remote TA, this TA will recognize the transition of the status bits from ON to OFF and the data bits from data to binary 0 as a disconnected signal and it will turn OFF circuits 107 and 109. This DTE should respond by turning OFF circuit 108 and transferring to disconnected mode. The disconnection will be signaled via the ISDN D-channel signaling protocol. At this time, the DTE–DCE interface should be placed in the idle (or ready) state.

1.4.3. The TA at the station that originated the disconnect request will recognize reception of S = OFF or the loss of framing signals as a disconnect acknowledgment and turn OFF circuits 107 and 109 and transfer to disconnected mode. The disconnection will be signaled via the ISDN D-channel signaling protocol. At this time, the DTE–DCE interface should be placed in the idle (or ready) state.

1.5. Loss of Frame Synchronization. In the event of loss of frame synchronization, the TA should attempt to resynchronize as follows:

(a) Place circuit 104 in binary 1 condition (passes from the data mode).
(b) Turn OFF status bit X in the transmitted frame.
(c) The remote TA upon recognition of status bit X OFF will turn OFF circuit 106, which will cause the remote DTE to place circuit 103 in a binary 1 condition.
(d) The local TA should attempt to resynchronize on the incoming signal.
(e) If after an interval of three seconds the local TA cannot attain synchronization, it should send a disconnect request by turning OFF all of the status bits for several (at least three) frames with data bits set to binary 0 and then disconnect by turning OFF circuit 107 and transferring to the disconnected mode as discussed in paragraph 1.4.2 above.
Note: The values of three seconds and three frames are provisional and should be confirmed or amended after further study.
(f) If resynchronization is achieved, the TA should turn ON status bit X toward the distant station.
(g) If resynchronization is achieved, the TA (which has turned OFF circuit 106) should, after an interval of *N* bits (see Subsection 3 in Section

17-14.2.3), turn ON circuit 106. This will cause circuit 103 to change from binary 1 to the data mode.
Note: During a resynchronization attempt, circuits 107 and 109 should remain ON.

2. TA Half-Duplex Operation. The data call establishment for the interworking of half-duplex DTEs equipped with V-Series-type interfaces is the same as discussed in Subsection 1 above. The only difference between half-duplex operation is in the control of circuits 105, 106, and 109, as follows.

Note: This is a unique application; therefore, TA arranged for half-duplex operation will not be able to interwork with either a V-Series or an X-Series duplex DTE (TE2).

2.1. In a TA arranged to accommodate half-duplex DTEs, circuit 109 will be under the control of the status bits SB in the incoming frame, as follows:

(a) If at the local interface circuit 109 is OFF and circuit 104 is in the binary 1 state, the DTE may "request to send" by turning ON circuit 105.
(b) The TA will then turn ON status bits SB in the transmitted frame which will turn ON circuit 109 in the remote interface and connect circuit 104 to the data bit stream of the incoming frame.
(c) After an N-bit interval (see Subsection 3 in Section 17-14.2.3) the local TA will turn ON circuit 106, which will allow the local DTE to transmit data on circuit 103.
(d) Upon completion of the transmission the local DTE will turn OFF circuit 105. This will in turn:
- Turn OFF circuit 106 in the local interface and circuit 103 will revert to the binary 1 state
- Turn OFF status bits SB which will in turn at the remote TA turn OFF circuit 109 and place circuit 104 in a binary 1 condition

(e) At this time the remote DTE is able to reverse the sequence by turning ON circuit 105.

17-14.2.5. Network-Independent Clocks

In cases where synchronous data signals at user rates up to and including 19.2 kbps are received from outside the ISDN (e.g., through an interworking unit from a DTE/modem on the PSTN), the data may not be synchronized to the ISDN. The following method shall be used to enable transfer of those data signals and the corresponding bit timing information via the 80-bit frame to the receiving TA. Such a situation would exist where the signals are received through an interworking unit from voice-band data modems on the analog PSTN where the transmit data from the remote modem is synchronized to the modem clock (normal case for such applications). The frequency tolerance of such modems is 100 ppm.

1. Measurement of Phase Differences. The phase difference between the following two frequencies will be measured:

(i) R1 = 0.6 × the nominal intermediate rate (except where fill bits are used; see note), synchronized with the ISDN.
(ii) R2 = 0.6 × the nominal intermediate rate (except where fill bits are used; see note), derived from and synchronized with the bit timing received from the remote synchronous source (e.g., modem).
Note: Clocks R1 and R2 are nominally either 4800, 9600, or 19,200 Hz at 8 kbps, 16 kbps, and 32 kbps intermediate rate, respectively.

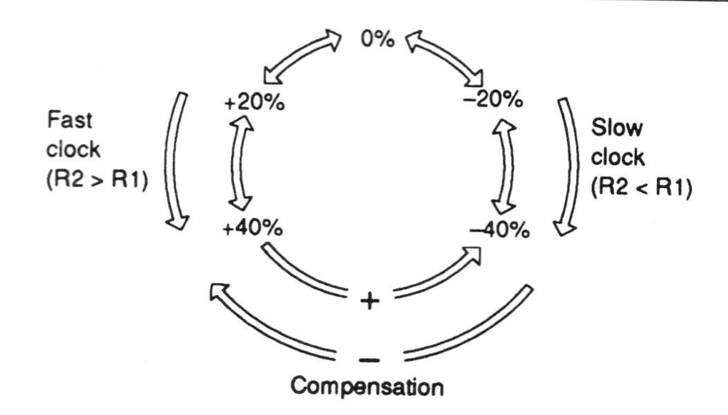

Note 1 – Phase measurements are given relative to R1 by the formula: Phase = phase(R2) – phase(R1).

Note 2 – Receipt of a bit combination requiring an illegal move of more than one state will cause a legal move of one state in the appropriate direction.

Note 3 – The initial state of both the receiving and transmitting sides of the TA will be 0%.

FIGURE 17-U.31. Network-independent clocking state diagram. (From Figure 5/V.110, ITU-T Rec. V.110, Geneva, 9/92, page 21.)

Where fill bits are used, in the cases of 7200 and 14,400 bit/s, R1 and R2 will have the same nominal rate as the user bit rate.

Compensation will affect one, one-half, one-quarter, or one-eighth of a user data bit, depending upon the bit repetition factor.

A state diagram for the transmitting TA showing the phase of R2 relative to R1 appears in Figure 17-U.31. Table 17-U.21 shows the related bit coding.

Comparison of R1 and R2 will give a phase difference relative to R1 which will be encoded as shown in Table 17-U.21. The resultant 3-bit code will be transmitted in bit positions E4, E5, and E6 and will be used for clock control at the receiving TA.

TABLE 17-U.21
Coding E Bits for Network-Independent Clocking

Displacement (in % of nominal R1 clock period at $n \times 4800$ bps, $n = 1, 2$, or 4)	Coding in the 80-Bit Frame		
	E4	E5	E6
Nominally 0	1	1	1
+20	0	0	0
+40	0	0	1
−40	0	1	0
−20	0	1	1
Compensation control			
Positive compensation of a one	1	0	1
Positive compensation of a zero	1	0	0
Negative compensation	1	1	0

Source: Table 10/V.110, ITU-T Rec. V.110, Geneva, 9/92, page 21.

To avoid continuous jitter between neighboring displacement positions, hysteresis shall be applied, as follows:

The displacement code shall be changed only when the measured phase difference between R1 and R2 is 15% (of the R1 clock period) more or less than the difference indicated by the existing displacement code.

Example: Bit combination 000 indicates a phase difference of nominally 20%. This bit combination will be changed into 001 when the measured phase difference is 35% or more and into 111 when the measured phase difference is 5% or less.

2. Positive / Negative Compensation. On transition from the +40% state to the −40% state, an extra user D bit has to be transmitted in the 80-bit frame, using bit E6 (positive compensation). At the receiving TA, this extra bit will be inserted between D24 and D25 as shown in Table 17-U.6, immediately following the E bits.

On transition from the −40% state to the +40% state, a bit combination is transmitted in the 80-bit frame (E4, E5, and E6 = 1, 1, 0, respectively), indicating to the receiving TA that bit D25 of the 80-bit frame, being set to ONE, does not contain user data and should be removed (negative compensation).

3. Encoding. The encoding of the measured phase difference for clock control and the positive/negative compensation control overrides and replaces the clock control coding.

Section 17-14.2 was extracted from ITU-T Rec. V.110, Geneva, 10/92.

17-14.3. Transmission of Start–Stop Characters Over Synchronous Bearer Channels

17-14.3.1. General

CCITT Rec. V.14 describes a method of conveying start–stop characters over synchronous bearer channels using an asynchronous-to-synchronous converter in the data rate range up to 19.2 kbps. Start–stop characters at data rates below or equal to 300 bps can be conveyed over synchronous bearer channels by oversampling at a rate of at least 1200 bps.

Note: This conversion method replaces the conversion method applied earlier in CCITT Recs. V.22, V.22 bis, V.26, and V.32.

This converter may be an intermediate device inserted into data lines of both circuit 103 in the transmitter and circuit 104 in the receiver inside a synchronous DCE (see Figure 17-U.32), or it may be a stand-alone unit in certain applications.

17-14.3.2. Data Rates

This conversion method is limited to data rates up to 19.2 kbps. The standard data rates of CCITT Rec. V.5 are preferred.

The nominal data rates for both the start–stop characters and the synchronous DCE shall be the same. The tolerance of the data rate of the synchronous transmission shall be ±0.01%.

17-14.3.3. Data Rate Ranges of the Start–Stop Characters at the Converter Input

The conversion method is capable of tolerating the signaling rates of the DTE in two ranges:

(a) Basic range: +1% to −2.5%
(b) Extended range: +2.3% to −2.5%

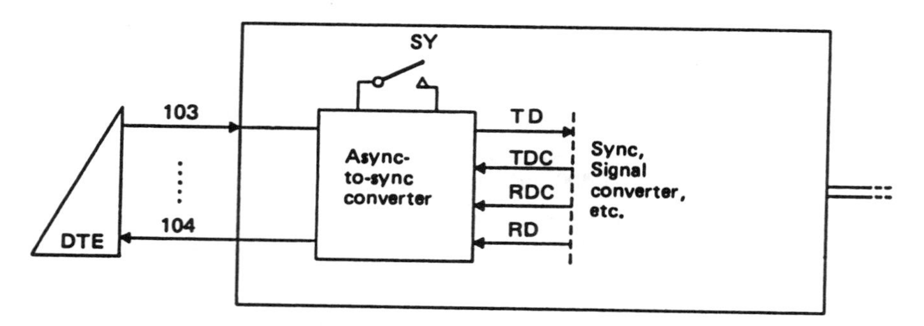

103 Transmitted data; data input to the DCE

TD Transmitted data; the synchronous output of the converter following the async-to-sync conversion of start-stop characters to be transmitted

TDC Transmitter signal element timing; internal timing information for the generation of synchronous transmitted data

RDC Receiver signal element timing; internal timing information associated with synchronous received data

RD Received data; the input of the converter for the restoration of start-stop characters

104 Received data; data output from the DCE

SY Synchronous mode; selection of the required mode of operation (asynchronous or synchronous)

FIGURE 17-U.32. Diagram showing inclusion of an async-to-sync converter into a synchronous DCE. (From Figure A-1/V.14, CCITT Rec. V.14, CCITT, Melbourne, 1988, page 48.)

The use of the basic signaling rate range is preferred because it results in lower distortion. The choice of range shall be made at the time of installation and shall be the same for both transmitter and receiver. It is not intended to be under customer control.

17-14.3.4. Start–Stop Character Format

It shall be possible to condition the converter to accept the following formats, namely

(a) A one-unit start element, followed by seven data units, and a stop element of one unit in length (9-bit characters)
(b) A one-unit start element, followed by eight data units, and a stop element of one unit in length (10-bit characters)
(c) A one-unit start element, followed by nine data units, and a stop element of one unit in length (11-bit characters)

The converter may also accept characters consisting of:

(d) A one-unit start element, followed by six data units, and a stop element of one unit in length (8-bit characters)

Note that character formats c and d do not conform to International Alphabet No. 5.

The character format selected shall be the same for both transmitter and receiver. The characters shall be in accordance with Rec. V.4 regardless of whether they conform to International Alphabet No. 5. It shall be possible to transmit characters continuously or with any additional continuous stop element of arbitrary length between characters.

Note: In each of the four formats, data units can be replaced by additional stop units. For example, format c will allow 11-bit characters consisting of a one-unit start element, followed by eight data units and a stop element of two units, to be handled.

17-14.3.5. Margin of the Converter Input

The effective net margin of the converter for transmitting start–stop characters applied to the input of the converter shall be at least 40%. CCITT states that this figure is for further study.

17-14.3.6. Selection of Synchronous or Asynchronous Modes of Operation

Selection of synchronous or asynchronous modes of operation shall be provided by a switch (or similar means) enabling the user to perform normal transmission and testing in each mode of operation, respectively. In the synchronous mode of operation the converter is totally bypassed in both directions.

17-14.3.7. Asynchronous-to-Synchronous Conversion Method

The general method to handle the speed differences between the intracharacter signaling rate of the start–stop characters and the data signaling rate of the synchronous bearer channel will be the insertion/deletion of stop elements at the transmitter and reinsertion of deleted stop elements at the receiver. Means are provided to transfer continuous start polarity (break signals) as well.

1. Transmitter. In the transmit direction the start–stop characters shall be adapted to the signaling rate of the synchronous bearer channel by:

- Deleting stop elements in case of overspeed of the start–stop characters
- Insertion of additional stop elements in case of underspeed of the start–stop characters

1.1. Basic Signaling Rate Range. No more than one stop element shall be deleted for any eight consecutive characters.

1.2. Extended Signaling Rate Range. No more than one stop element shall be deleted for any four consecutive characters.

2. Receiver. The intracharacter signaling rate provided by the converter shall be in the range of the nominal data rate to the limit of the specified overspeed tolerance —that is, +1% in the basic and +2.3% in the extended data signaling range. The length of the stop element shall not be reduced by more than 12.5% for the basic signaling rate range (or 25% for the optional extended signaling rate range) to allow for overspeed in the transmitting terminal. The nominal length of the start and data elements for all characters shall be the same.

Note: Equipments exist in the field which delete stop elements more frequently than specified in Subsections 1.1 and 1.2. However, in these equipments there will always be at least one additional inserted stop element between deleted stop elements.

3. Break Signal

3.1. Transmitter. If the converter detects M to $2M + 3$ bits all of "start" polarity, where M is the number of bits per character in the selected format, the converter shall transmit $2M + 3$ bits of "stop" polarity. If the converter detects more than

$2M + 3$ bits all of "start" polarity, the converter shall transmit all these bits as "start" polarity.

Note: The converter must receive at least $2M$ bits of "stop" polarity after the "start" polarity break signal in order to ensure that it regains the character synchronism.

3.2. Receiver. The $2M + 3$ or more bits of "start" polarity received from the transmitting modem shall be transferred to the output of the converter, and the character synchronism shall be regained from the following "stop" to "start" transition.

Note: In some earlier implementations an uninitiated NUL character may precede the break signal at the output of the converter when no measures have been taken to prevent this.

4. Tandem Operation. Tandem operation between two ends comprising async-to-sync conversions can be established only by using cascaded synchronous bearer channels.

5. Testing Facilities. All the tests recommended in the relevant CCITT Recommendations can be performed in asynchronous operation as well, where this converter is used, with the exception of self test end-to-end.

Note: Other interchange circuits which are provided are not involved in the operation of the async-to-sync converter but must comply with the requirements of the relevant DCE Recommendations, including the conditions of the timing circuits (i.e., 113, 114, and 115) during both the asynchronous and synchronous modes of operation.

Section 17-14.3 is based on CCITT Rec. V.14, Melbourne, 1988 and Helsinki, 3/93.

18

Data Networks

This new Section 18.5.2 updates Section 18.5.2 on pages 1728–1738 of the Manual.

18-5.2. Transmission Control Protocol/Internet Protocol TCP/IP

18-5.2.1. Background and Application

TCP/IP protocol family was developed by the U.S. Department of Defense (DoD) for the ARPANET (Advanced Research Projects Agency Networks). ARPANET was one of the first large advanced packet networks. It dates back to 1968 and was well into existence before CCITT and ISO took interest in layered protocols.

The TCP/IP suite of protocols [Ref. 1] has wide acceptance today. These protocols are used on both LANs and WANs. They are particularly attractive for their internetworking capabilities. The internet protocol (IP) competes with the CCITT Rec. X.75 protocol [Ref. 2].

The architectural model of the IP [Ref. 3] uses terminology that differs from the OSI reference model. (The IP predates OSI). Figure 18-U.1 shows the relationship between TCP/IP and related DoD protocols and the OSI model. It traces data traffic from an originating host, which runs an applications program, to another host in another network. This may be a LAN-to-WAN-to-LAN, a LAN-to-LAN, or a WAN-to-WAN connectivity. The host enters its own network by means of a network access protocol such as HDLC or an IEEE 802 series protocol. The common LAN–WAN–LAN connectivity is shown in Figure 18-U.2.

A LAN connects via a gateway (or router) to another network. Typically, a gateway has three protocols. Two of these protocols connect to each of the attached networks (e.g., LAN and WAN); the third protocol is the IP, which provides the network-to-network interface.

Hosts typically have four protocols. To communicate with routers (gateways), a network access protocol and IP are required. A transport layer protocol ensures reliable communication between the hosts because end-to-end capability is not provided in either the network access nor internet protocols. Hosts also must have application protocols such as E-mail or file transfer protocols.

18-5.2.2. TCP/IP and Data-Link Layers

TCP/IP is transparent to the type of data-link layer involved because it is also transparent whether it is operating in a LAN or WAN domain or among them. However, there is document support for Ethernet (IEEE 802 series), ARCNet for LANs, and X.25 for WANs.

Figure 18-U.3 shows how upper OSI layers are encapsulated with TCP and IP header information and then incorporated into the data-link-layer frame.

OSI	TCP / IP and Related Protocols		
Application	File transfer	Electronic mail	Terminal emulation
Presentation Session	File transfer protocol (FTP)	Simple mail transfer protocol (SMTP)	Telnet protocol
Transport	Transmission control protocol (TCP)		User datagram protocol (UDP)
Network	Address resolution protocol (ARP)	Internet protocol (IP)	Internet control message protocol (ICMP)
Data link	----------Network interface cards---------- CSMA/CD (Ethernet), Token Ring, ARCNet, StarLan		
Physical	----------Transmission media---------- Wire pair, fiber optics, coaxial cable, radio		

FIGURE 18-U.1. How TCP/IP and associated DoD protocols relate to OSI.

For the case of IEEE 802 series LAN protocols, advantage is taken of the LLC common to all 802 protocols. The LLC extended header contains the SNAP (subnetwork access protocol) such that we have three octets for the LLC header and five octets in the SNAP. The LLC header has its fields fixed as follows:

DSAP = 10101010 (destination service access point)
SSAP = 10101010 (source service access point)
Control = 00000011 for unnumbered information (UI frame)

The five octets in the SNAP have three assigned for protocol ID or organizational code and two for "EtherType." EtherType assignments are shown in Table 18-U.1. EtherType refers to the general class of LANs based on CSMA/CD.

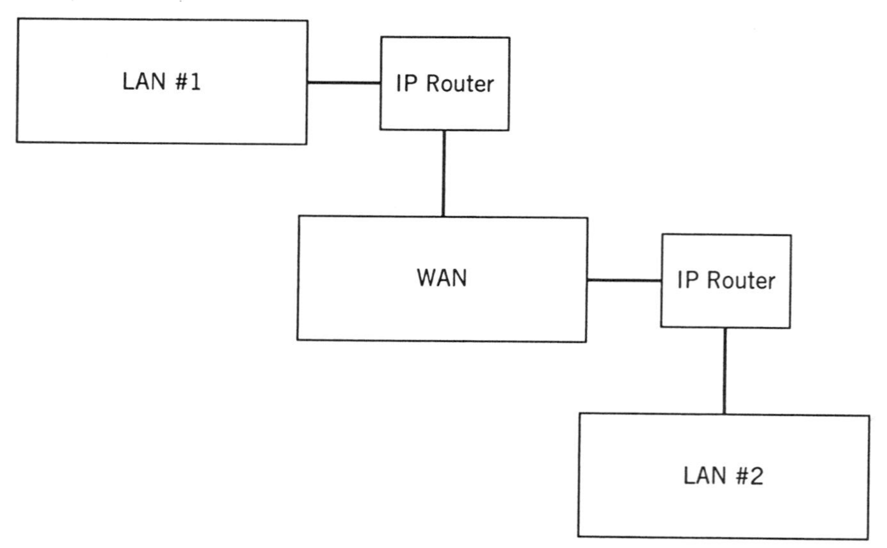

FIGURE 18-U.2. Internet from one LAN to another LAN via a WAN.

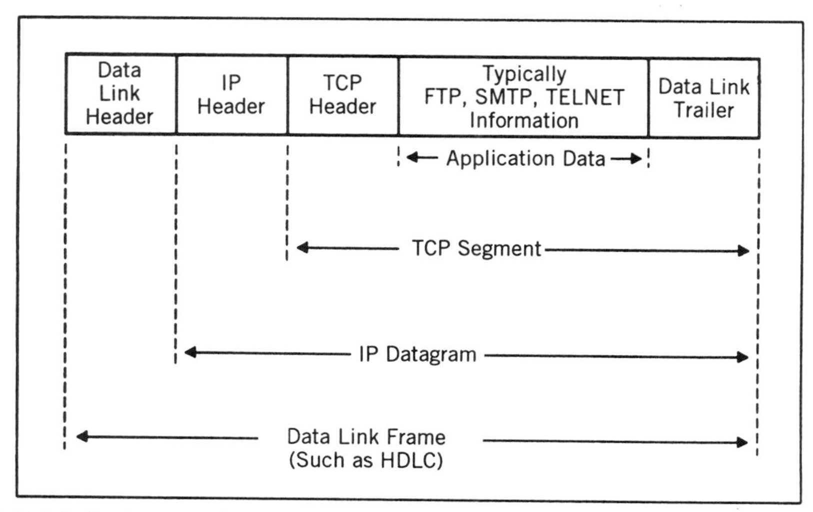

FIGURE 18-U.3. The incorporation of upper-layer PDUs into a data-link-layer frame showing relationship with TCP and IP.

Figure 18-U.4 shows how TCP/IP relates to OSI and IEEE 803, and Figure 18-U.5 illustrates an IEEE 802 frame with LLC data incorporating TCP and IP.

Addressing is resolved by a related TCP/IP protocol called ARP (address resolution protocol), which performs the mapping of the 32-bit internet address into a 48-bit IEEE 802 address.

TABLE 18-U.1
EtherType Assignments

Ethernet Decimal	Hex	Description
512	0200	XEROX PUP
513	0201	PUP address translation
1536	0600	XEROX NS IDP
2048	0800	DOD internet protocol (IP)
2049	0801	X.75 internet
2050	0802	NBS internet
2051	0803	ECMA internet
2052	0804	Chaosnet
2053	0805	X.25 level 3
2054	0806	Address resolution protocol (ARP)
2055	0807	XNS compatibility
4096	1000	Berkeley trailer
21000	5208	BBN Simnet
24577	6001	DEC MOP dump/load
24578	6002	DEC MOP remote console
24579	6003	DEC DECnet phase IV
24580	6004	DEC LAT
24582	6005	DEC
24583	6006	DEC
32773	8005	HP probe
32784	8010	Excelan
32821	8035	Reverse ARP
32824	8038	DEC LANBridge
32823	8098	Appletalk

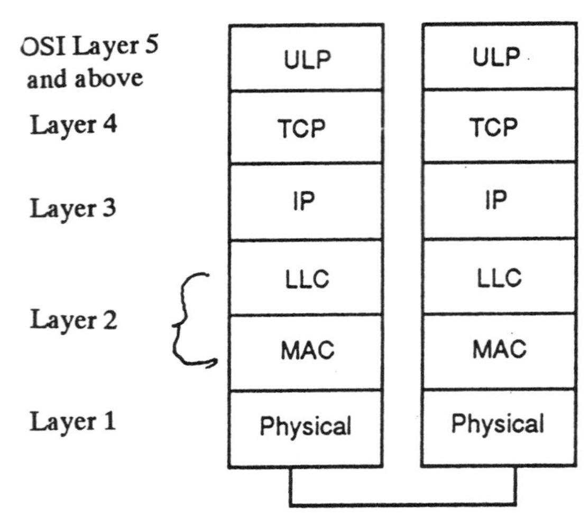

FIGURE 18-U.4. How TCP/ IP working with IEEE 802 relates to OSI. ULP = upper layer protocol.

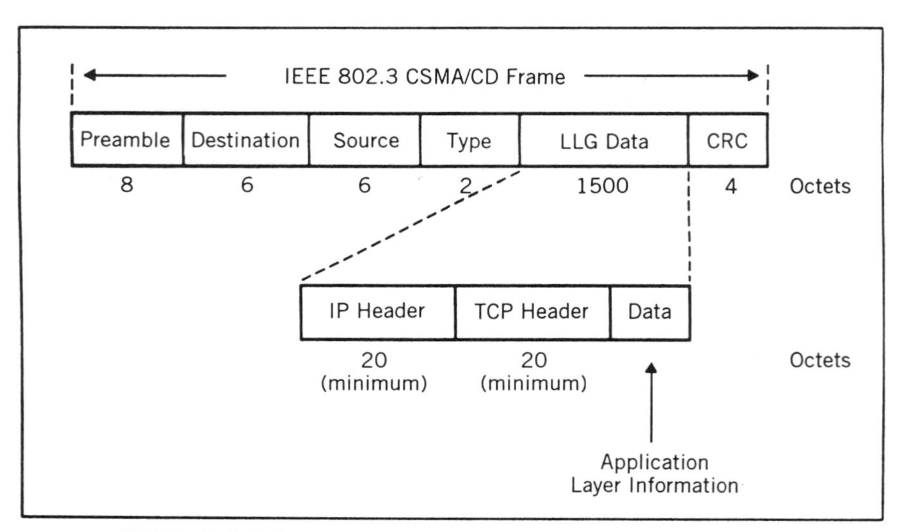

FIGURE 18-U.5. An IEEE 802 frame showing LLC and TCP/IP functions.

Another interface problem is the limited IP datagram length of 576 octets, where with the 802 series the frames have considerably larger length limits.

18-5.2.3. The IP Routing Function

In OSI the network layer functions include routing and switching of a datagram through the telecommunications subnetwork. The IP provides this essential function. It forwards the datagram based upon the network address contained within the IP header. Each datagram is independent and has no relationship with other datagrams. There is no guaranteed delivery of the datagram from the standpoint of the IP. However, the next higher layer, the TCP layer, provides for the reliability that the IP lacks. It also carries out segmentation and reassembly functions of a datagram to match frame sizes of data-link-layer protocols.

Addresses determine routing and, at the far end, equipment (hardware). Actual routing derives from the IP address, and equipment addresses derive from the data-link-layer header (typically the 48-bit Ethernet address).

User data from upper-layer protocols is passed to the IP layer. The IP layer examines the network address (IP address) for a particular datagram and determines if the destination node is on its own local area network or some other network. If it is on the same network, the datagram is forwarded directly to the destination host. If it is on some other network, it is forwarded to the local IP router (gateway). The router, in turn, examines the IP address and forwards the datagram as appropriate. Routing is based on a lookup table residing in each router or gateway.

18-5.2.4. Detailed IP Operation

The IP provides connectionless service, meaning that there is no call setup phase prior to exchange of traffic. There are no flow control nor error control capabilities incorporated in IP. These are left to the next higher layer, the transmission control protocol (TCP). The IP is transparent to subnetworks connecting at lower layers, and thus different types of networks can attach to an IP gateway.

Whereas prior to this discussion we have used the term *segmentation*, meaning to break up a data file into manageable segments, frames, packets, blocks, and so on, the IP specifications refer to this as *fragmentation*. IP messages are called *datagrams*. The minimum datagram length is 576 octets, and the maximum length is 65,535 octets. Fragmentation resolves PDU (protocol data unit) sizes of the different networks with which IP carries out an interface function. For example, X.25 packets typically have data fields of 128 octets, Ethernet limits the size of a PDU to 1500 octets, and so forth. Of course, IP does a reassembly of the "fragments" at the opposite end of its circuit.

Description of the IP Datagram. The IP datagram format is shown in Figure 18-U.6. The datagram format should be taken in context with Figures 18-U.4 and 18-U.5, showing how the IP datagram relates to TCP and the data-link layer.

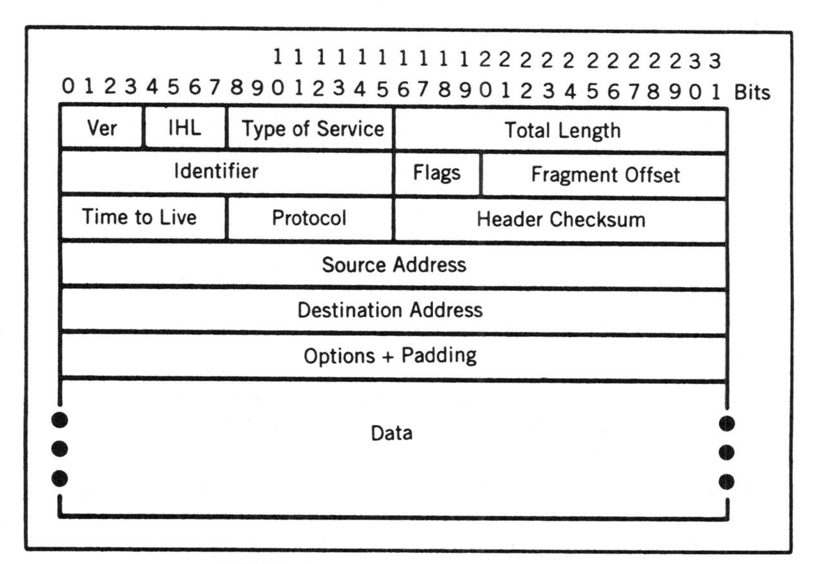

FIGURE 18-U.6. The IP datagram format.

In Figure 18-U.6, we move from left to right and top down in our description:

Version field (4 bits) gives the release number of the IP version for which a particular gateway or router is equipped.

Header length (4 bits) is measured in units of 32 bits. A header without certain options, such as QoS, typically has 20 octets. Thus its length is 5 (32 bits or 4 octets per unit; $5 \times 4 = 20$).

The *type of service* (TOS) field (8 bits) is used to identify several QoS parameters provided by IP. The field has eight bits broken down into four active groupings, and the last two bits are reserved. The first is *Precedence*, which consists of three bits as follows:

000	Routine	001	Priority
010	Immediate	011	Flash
100	Flash override	101	CEITIC/ECP
110	Internetwork control	111	Network control

Delay, 1 bit: 0 = normal, 1 = low
Throughput, 1 bit: 0 = normal, 1 = high
Reliability: 0 = normal, 1 = high

The *total length* field specifies the total length of an IP datagram in question. The unit of measure is the octet, and it includes the length of the header and data fields. The maximum possible length of a datagram is $2^{16} - 1$, and the minimum length is 576 octets.

Segmentation (fragmentation) and reassembly are controlled by three fields in the header. These are *identifier* (16 bits), *flags* (3 bits), and *fragment offset* (3 bits). The term *fragment* suggests part of a whole; thus the identifier field identifies a fragment as part of a complete datagram, along with the source address. The flag bits determine if a datagram can be fragmented. When it can be fragmented, one of the flag bits shows whether the fragment is the last fragment of a datagram. The fragmentation offset field gives the relative position of the fragment regarding the original datagram. It is initially set at 0 and then set to the proper number by the fragmenting gateway.

The *time-to-live* (TTL) field's basic purpose is to prevent routing loops. In this we mean a routing that eventually routes back on itself. In telephony, it is sometimes called *ring-around-the-rosy*. It is used to measure the time a datagram has been in the internet. Each internet gateway checks this number and will discard that datagram if the TTL equals zero. There are numerous ways to implement TTL, some being vendor-specific. It is sometimes used for diagnostics by network management features such as SNMP [Ref. 4].

The *protocol* field (8 bits) identifies the next-higher-layer protocol the datagram expects at the destination host. Table 18-U.2, taken from RFC 1060 [Ref. 5], shows the protocol decimal numbering scheme and the corresponding protocol for each number.

The next field is the *header checksum* field (16 bits). This provides error detection on the header.

Source address and *destination address* fields are each 32 bits long. Of course, the source address is the address of the originating host and the destination address is the address of the destination host.

The IP address structure is shown in Figure 18-U.7. It shows four address formats in which the lengths of the two component fields making up the address field change with each of the different formats. These component fields are the "network address" and the local address" fields. The first bits of the address field

TABLE 18-U.2
IP Protocol Field Numbering (Assigned Internet Protocol Numbers)

Decimal	Key Word	Protocol
0		Reserved
1	ICMP	Internet control message protocol
2	IGMP	Internet group management protocol
3	GGP	Gateway-to-gateway protocol
4		Unassigned
5	ST	Stream
6	TCP	Transmission control protocol
7	UCL	UCL
8	EGP	Exterior gateway protocol
9	IGP	Interior gateway protocol
10	BBN-MON	BBN-RCC monitoring
11	NVP-II	Network voice protocol
12	PUP	PUP
13	ARGUS	ARGUS
14	EMCON	EMCON
15	XNET	Cross net debugger
16	CHAOS	Chaos
17	UDP	User datagram protocol
18	MUX	Multiplexing
19	DCN-MEAS	DCN measurement subsystems
20	HMP	Host monitoring protocol
21	PRM	Packet radio monitoring
22	XNS-IDP	XEROX NS IDP
23	TRUNK-1	Trunk-1
24	TRUNK-2	Trunk-2
25	LEAF-1	Leaf-1
26	LEAF-2	Leaf-2
27	RDP	Reliable data protocol
28	IRTP	Internet reliable TP
29	ISO-TP4	ISO transport class 4
30	NETBLT	Bulk data transfer
31	MFE-NSP	MFE network services
32	MERIT-INP	MERIT internodal protocol
33	SEP	Sequential exchange
34–60		Unassigned
61		Any host internal protocol
62	CFTP	CFTP
63		Any local network
64	SAT EXPAK	SATNET and backroom EXPAK
65	MIT-SUBN	MIT subnet support
66	RVD	MIT remote virtual disk
67	IPPC	Internet plur. packet core
68		Any distributed file system
69	SAT-MON	SATNET monitoring
70		Unassigned
71	IPCV	Packet core utility
72–75		Unassigned
76	BRSAT-MON	Backroom SATNET monitoring
77		Unassigned
78	WB-MON	Wideband monitoring
79	WB-EXPAK	Wideband EXPAK
80–254		Unassigned
255		Reserved

Source: RFC 1060 [Ref. 5] and *Internet Protocol Transition Workbook* [Ref. 6].

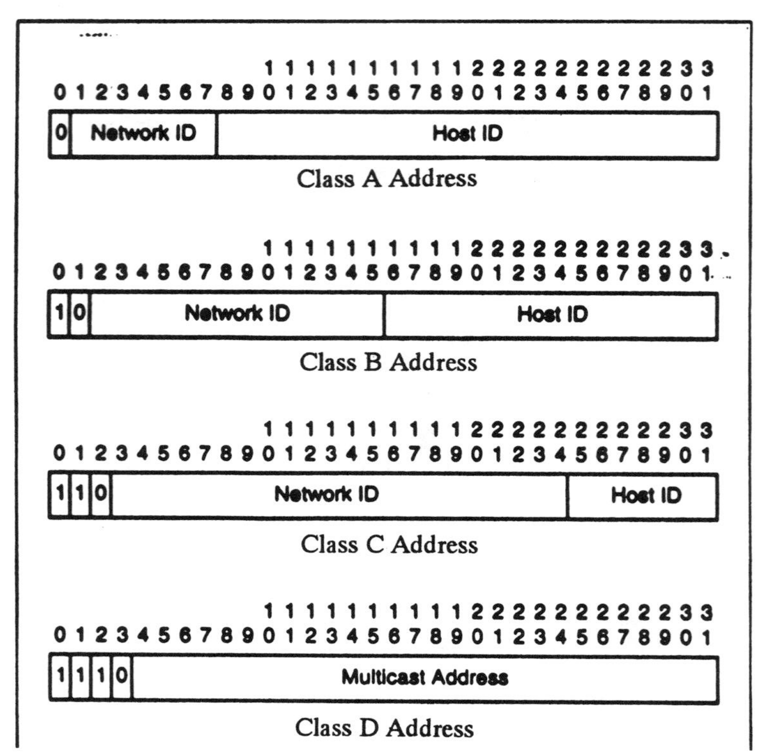

FIGURE 18-U.7. Internet protocol address formats.

specify the format or "class"—that is, Class A, B, C, or D. The "local address" is often called "host address."

Class A addressing is for very large networks, such as what was ARPANET. The field starts with binary 0, indicating that it is a Class A format. In this case the local or host address component field is 24 bits long and has an address capacity of 2^{24}. *Class B* addressing is for medium-sized networks, such as campus networks. The field begins with 10 to indicate that it is a Class B format; the network component field is 14 bits long, and the host or local address component field is 16 bits in length. The *Class C* format is for small networks with a very large network ID field with an addressing capacity of 2^{24} and a considerably small host ID field of only 8 bits (2^8 addressing capacity). The address field in this case starts with 110.

The *Class D* format is for multicasting, a form of broadcasting. Its first four bits are the sequence 1110.

IP Routing. A gateway (router) needs only the network ID portion of the address to perform its routing function. Each router or gateway has a routing table which consists of destination network addresses and specified next-hop gateway.

Three types of routing are performed by the routing table:

1. Direct routing to locally attached devices
2. Routing to networks that are reached via one or more gateways
3. Default routing to destination network in case the first types of routing are unsuccessful

Suppose a datagram (or datagrams) is (are) directed to a host which is not in the routing table resident in a particular gateway. Likewise, there is a possibility that the network address for that host is also unknown. These problems may be resolved with the *address resolution protocol* (ARP) [Ref. 7].

First the ARP searches a mapping table which relates IP addresses with corresponding physical addresses. If the address is found, it returns the correct address to the requester. If it cannot be found, the ARP broadcasts a *request* containing the IP target address in question. If a device recognizes the address, it will reply to the request where it will update its ARP *cache* with that information. The ARP *cache* contains the mapping tables maintained by the ARP module.

There is also a *reverse address resolution protocol* (RARP) [Ref. 8]. It works in a similar fashion as the ARP, but in reverse order. RARP provides an IP address to a device when the device returns its own hardware address. This is particularly useful when certain devices are booted and only know their own hardware address.

Routing with IP involves a term called *hop*. A hop is defined as a link connecting adjacent nodes (gateways) in a connectivity involving IP. A *hop count* indicates how many gateways (nodes) must be traversed between source and destination.

One part of an IP routing algorithm can be *source routing*. Here an upper-layer protocol (ULP) determines how an IP datagram is to be routed. One option is that the ULP passes a listing of internet addresses to the IP layer. Here information is provided on the intermediate nodes required for transit of a datagram in question to its final destination.

Each gateway makes its routing decision based on a resident routing list or routing table. If a destination resides in another network, a routing decision is required by the IP gateway to implement a route to that other network. In many cases, multiple hops are involved and each gateway must carry out routing decisions based on its own routing table.

A routing table can be static or dynamic. The table contains IP addressing information for each reachable network and closest gateway for the network, and it is based on the concept of shortest routing, thus routing through the closest gateway.

Involved in IP shortest routing is the *distance metric*, which is a value expressing minimum number of hops between a gateway and a datagram's destination. An IP gateway tries to match the destination network address contained in the header of a datagram with a network address entry contained in its routing table. If no match is found, the gateway discards the datagram and sends an ICMP message back to the datagram source.

Internet Control Message Protocol (ICMP). ICMP [Ref. 9] is used as an adjunct to IP when there is an error in datagram processing. ICMP uses the basic support of IP as if it were a higher-level protocol; however, ICMP is actually an integral part of IP and is implemented by every IP module.

ICMP messages are sent in several situations: for example, when a datagram cannot reach its destination, when a gateway does not have the buffering capacity to forward a datagram, and when the gateway can direct the host to send traffic on a shorter route.

ICMP messages typically report errors in the processing of datagrams. To void the possibility of infinite regress of messages about messages, and so on, no ICMP messages are sent about ICMP messages. Also ICMP messages are only sent about errors in handling fragment zero of fragmented datagrams. (*Note*: Fragment zero has the fragment offset equal to zero.)

Message Formats. ICMP messages are sent using the basic IP header (see Figure 18-U.6). The first octet of the data portion of the datagram is an ICMP-type field.

The "data portion" is the last field at the bottom of Figure 18-U.6. The ICMP-type field determines the format of the remaining data. Any field labeled *unused* is reserved for later extensions and is fixed at zero when sent, but receivers should not use these fields (except to include them in the checksum). Unless otherwise noted under individual format descriptions, the values of the internet header fields are as follows:

- Version: 4
- IHL (internet header length): length in 32-bit words
- Type of service: 0
- Total length: Length of internet header and data in octets
- Identification, flags, and fragment offset: Used in fragmentation, as in the basic IP protocol described above.
- Time to live (in seconds): Because this field is decremented at each machine in which the datagram is processed, the value in this field should be at least as great as the number of gateways which this datagram will traverse.
- Protocol: ICMP = 1
- Header checksum: The 16-bit one's complement of the one's complement sum of all 16-bit words in the header. For computing the checksum, the checksum field should be zero. The reference RFC (RFC 792) states that this checksum may be replaced in the future.
- Source address: The address of the gateway or host that composes the ICMP message. Unless otherwise noted, this is any of a gateway's addresses.
- Destination address: The address of the gateway or host to which the message should be sent.

There are eight distinct ICMP messages covered in RFC 792:

1. Destination unreachable message
2. Time exceeded message
3. Parameter problem message
4. Source quench message
5. Redirect message
6. Echo or echo reply message
7. Timestamp or timestamp reply message
8. Information request or information reply message

Example: Destination Unreachable Message. The ICMP fields in this case are shown in Figure 18-U.8.

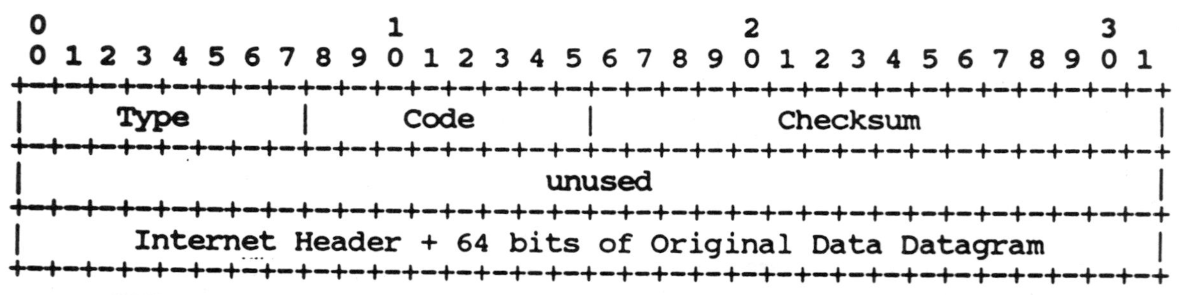

FIGURE 18-U.8. A typical ICMP message format; a destination unreachable message.

IP Fields
Destination address: The source network and address from the original datagram's data.

ICMP Fields

Type 3
Code: 0 = net unreachable
1 = host unreachable
2 = protocol unreachable
3 = port unreachable
4 = fragmentation needed and DF set
5 = source route failed
Checksum: as above
Internet header + 64 bits of data datagram

The internet header plus the first 64 bits of the original datagram's data are used by the host to match the message to the appropriate process. If a higher-level protocol uses port numbers, they are assumed to be in the first 64 data bits of the original datagram's data.

Description. If, according to the information in the gateway's routing tables, the network specified in the internet destination field of a datagram is unreachable (e.g,., the distance to the network is infinity), the gateway may send a destination unreachable message to the internet source host of the datagram. In addition, in some networks the gateway may be able to determine if the internet destination host is unreachable. Gateways in these networks may send destination unreachable messages to the source host when the destination host is unreachable.

If, in the destination host, the IP module cannot deliver the datagram because the indicated protocol module or process port is not active, the destination host may send a destination unreachable message to the source host.

Another case is when a datagram must be fragmented to be forwarded by a gateway yet the "Don't Fragment" flag is on. In this case the gateway must discard the datagram and may return a destination unreachable message.

It should be noted that Codes 0, 1, 4, and 5 may be received from a gateway; Codes 2 and 3 may be received from a host (RFC 792 [Ref. 9]).

18-5.2.5. The Transmission Control Protocol (TCP)

TCP Defined. TCP [Refs. 10 and 11] was designed to provide reliable communication between pairs of processes in logically distinct hosts on networks and sets of interconnected networks. TCP operates successfully in an environment where the loss, damage, duplication or misorder of data and network congestion can occur. This robustness in spite of unreliable communications media makes TCP well-suited to support commercial, military, and government applications. TCP appears at the transport layer of the protocol hierarchy. Here, TCP provides connection-oriented data transfer that is reliable, ordered, full-duplex, and flow-controlled. TCP is designed to support a wide range of upper-layer protocols (ULPs). The ULP can channel continuous streams of data through TCP for delivery to peer ULPs. The TCP breaks the streams into portions which are encapsulated together with appropriate addressing and control information to form a segment—the unit of exchange between TCPs. In turn, the TCP passes the segments to the network layer for transmission through the communication system to the peer TCP.

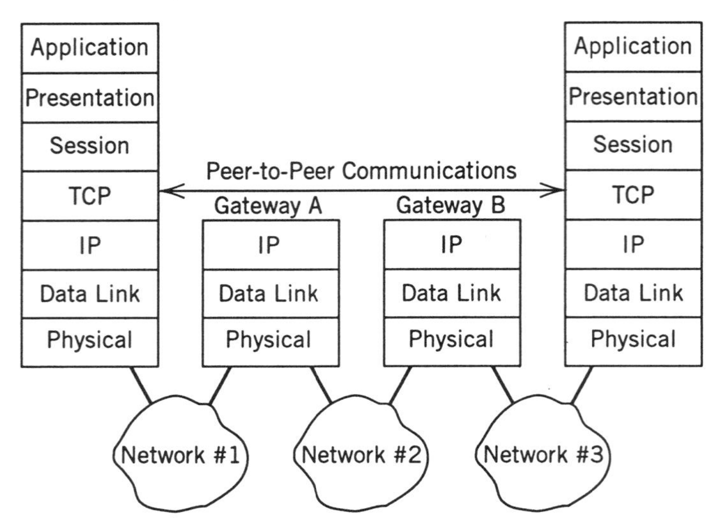

FIGURE 18-U.9. Protocol layers showing the relationship of TCP with other layered protocols.

As shown in Figure 18-U.9, the layer below the TCP in the protocol hierarchy is commonly the IP layer. The IP layer provides a way for the TCP to send and receive variable-length segments of information enclosed in internet datagram "envelopes." The internet datagram provides a means for addressing source and destination TCPs in different networks. The IP also deals with fragmentation or reassembly of TCP segments required to achieve transport and delivery through the multiple networks and interconnecting gateways. The IP also carries information on the precedence, security classification, and compartmentation of the TCP segments, so this information can be communicated end-to-end across multiple networks.

TCP Mechanisms. TCP builds its services on top of the network layer's potentially unreliable services with mechanisms such as error detection, positive acknowledgments, sequence numbers, and flow control. These mechanisms require certain addressing and control information to be initialized and maintained during data transfer. This collection of information is called a *TCP connection*. The following paragraphs describe the purpose and operation of the major TCP mechanisms.

PAR Mechanism. TCP uses a positive acknowledgment with retransmission (PAR) mechanism to recover from the loss of a segment by the lower layers. The strategy with PAR is for a sending TCP to retransmit a segment at timed intervals until a positive acknowledgment is returned. The choice of retransmission interval affects efficiency. An interval that is too long reduces data throughput, whereas one that is too short floods the transmission media with superfluous segments. In TCP, the timeout is expected to be dynamically adjusted to approximate the segment round-trip time plus a factor for internal processing; otherwise performance degradation may occur. TCP uses a simple checksum to detect segments damaged in transit. Such segments are discarded without being acknowledged. Hence, damaged segments are treated identically to lost segments and are compensated for by the PAR mechanism. TCP assigns sequence numbers to identify each octet of the data stream. These enable a receiving TCP to detect duplicate and out-of-order segments. Sequence numbers are also used to extend the PAR mechanism by allowing a single acknowledgment to cover many segments worth of data. Thus, a sending TCP can still send new data although previous data have not been acknowledged.

Flow Control Mechanism. TCP's flow control mechanism enables a receiving TCP to govern the amount of data dispatched by a sending TCP. The mechanism is based on a *window* which defines a contiguous interval of acceptable sequence-numbered data. As data are accepted, TCP slides the window upward in the sequence-number space. This window is carried in every segment, enabling peer TCPs to maintain up-to-date window information.

Multiplexing Mechanism. TCP employs a multiplexing mechanism to allow multiple ULPs within a single host and multiple processes in a ULP to use TCP simultaneously. This mechanism associates identifiers, called *ports*, to ULP processes accessing TCP services. A ULP connection is uniquely identified with a *socket*, the concatenation of a port and an internet address. Each connection is uniquely named with a socket pair. This naming scheme allows a single ULP to support connections to multiple remote ULPs. ULPs which provide popular resources are assigned permanent sockets, called *well-known sockets*.

ULP Synchronization. When two ULPs wish to communicate, they instruct their TCPs to initialize and synchronize the mechanism information on each to open the connection. However, the potentially unreliable network layer (i.e., internet protocol layer) can complicate the process of synchronization. Delayed or duplicate segments from previous connection attempts might be mistaken for new ones. A handshake procedure with clock-based sequence numbers is used in connection opening to reduce the possibility of such false connections. In the simplest handshake, the TCP pair synchronizes sequence numbers by exchanging three segments, thus the name *three-way handshake*.

ULP Modes. A ULP can open a connection in one of two modes, passive or active. With a passive open, a ULP instructs its TCP to be *receptive* to connections with other ULPs. With an active open, a ULP instructs its TCP to actually initiate a three-way handshake to connect to another ULP. Usually an active open is targeted to a passive open. This active/passive model supports server-oriented applications where a permanent resource, such as a database management process, can always be accessed by remote users. However, the three-way handshake also coordinates two simultaneous active opens to open a connection. Over an open connection, the ULP pair can exchange a continuous stream of data in both directions. Normally, TCP groups the data into TCP segments for transmission at its own convenience. However, a ULP can exercise a *push* service to force TCP to package and send data passed up to that point without waiting for additional data. This mechanism is intended to prevent possible deadlock situations where a ULP waits for data internally buffered by TCP. For example, an interactive editor might wait forever for a single input line from a terminal. A push will force data through the TCPs to the awaiting process. A TCP also provides the means for a sending ULP to indicate to a receiving ULP that "urgent" data appear in the upcoming data stream. This urgent mechanism can support, for example, interrupts or breaks. When a data exchange is complete, the connection can be closed by either ULP to free TCP resources for other connections. Connection closing can happen in two ways. The first, called a *graceful close*, is based on the three-way handshake procedure to complete data exchange and coordinate closure between the TCPs. The second, called an *abort*, does not allow coordination and may result in the loss of unacknowledged data.

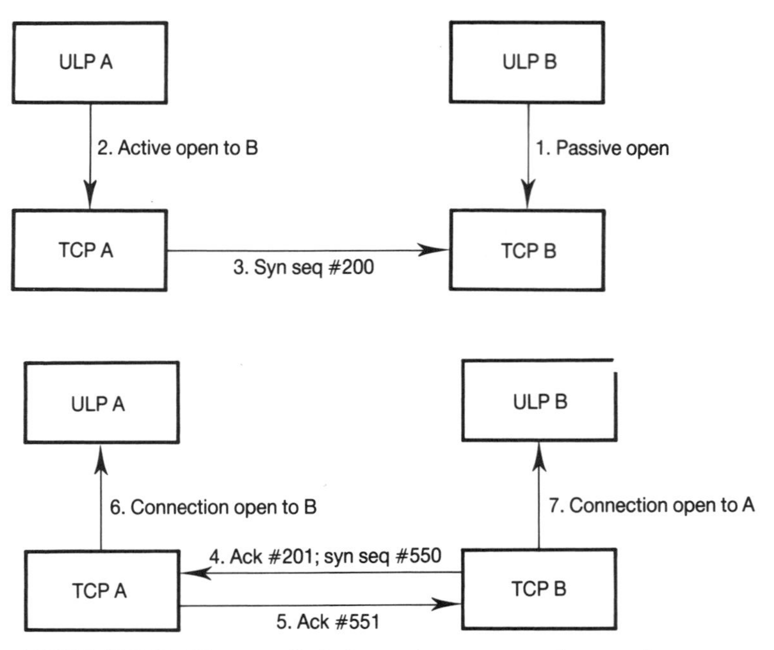

FIGURE 18-U.10. Diagrams illustrating a simple connection opening.

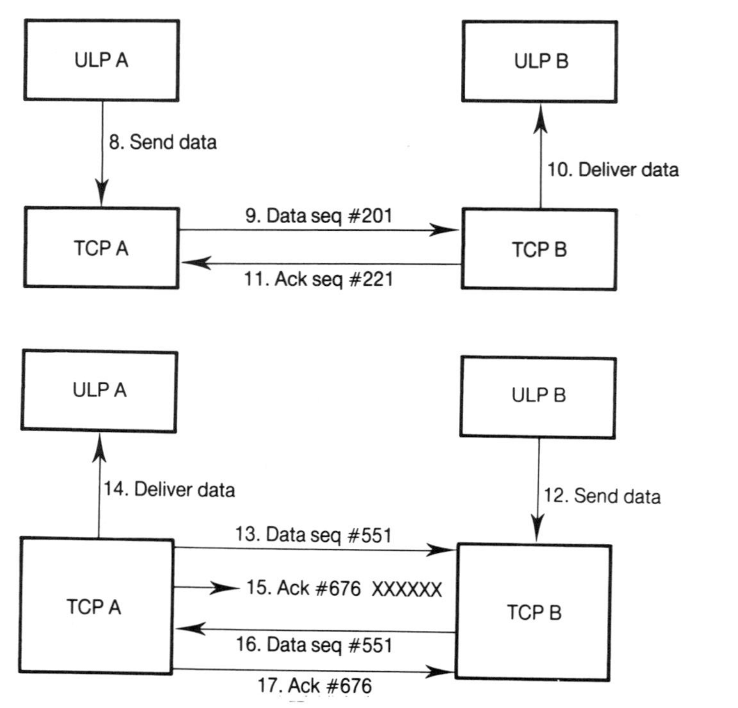

FIGURE 18-U.11. Diagrams illustrating two-way data transfer.

Scenario. The following scenario provides a walk-through of a connection opening, data exchange, and connection closing as might occur between the database management process and the user mentioned above (see Figure 18-U.10). The scenario focuses more on the three-way handshake mechanism in connection with opening and closing, as well as on the positive acknowledgment with retransmis-

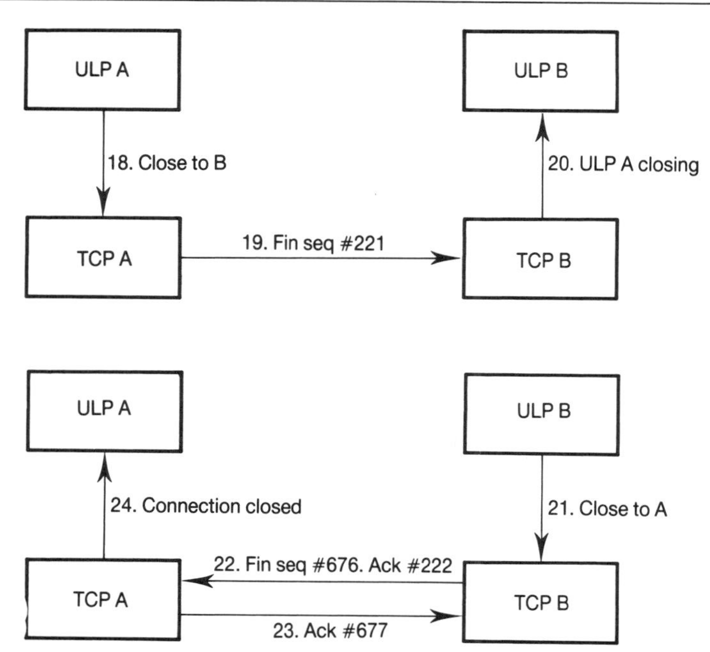

FIGURE 18-U.12. Diagrams illustrating a graceful connection close.

sion mechanism supporting reliable data transfer. Although not pictured, the network layer transfers the information between TCPs. For the purpose of this scenario, the network layer is assumed not to damage, lose, duplicate, or change the order of data unless explicitly noted. The scenario is organized into three parts:

a. A simple connection opening (steps 1–7) (Figure 18-U.10)
b. Two-way data transfer (steps 8–17) (Figure 18-U.11)
c. A graceful connection close (steps 18–24) (Figure 18-U.12)

Scenario Notation. The following notation is used in Figures 18-U.10–18-U.12

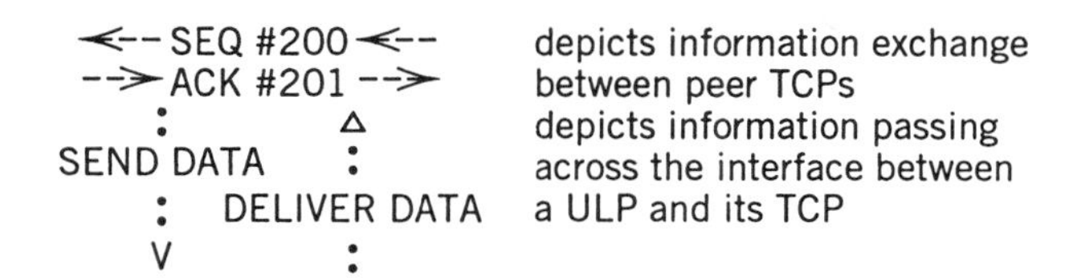

1. ULP B (the DB manager) issues a PASSIVE OPEN to TCP B to prepare for connection attempts from other ULPs in the system.
2. ULP A (the user) issues an ACTIVE OPEN to open a connection to ULP B.
3. TCP A sends a segment to TCP B with an OPEN control flag, called a SYN, carrying the first sequence number (shown as SEQ #200) it will use for data sent to B.
4. TCP B responds to the SYN by sending a positive acknowledgment, or ACK, marked with the next sequence number expected from TCP A. In the same segment, TCP B sends its own SYN with the first sequence number for its data (SEQ #550).
5. TCP A responds to TCP B's SYN with an ACK showing the next sequence number expected from B.

6. TCP A now informs ULP A that a connection is open to ULP B.
7. Upon receiving the ACK, TCP B informs ULP B that a connection has been opened to ULP A.
8. ULP A passes 20 octets of data to TCP A for transfer across the open connection to ULP B.
9. TCP A packages the data in a segment marked with the current "A" sequence number.
10. After validating the sequence number, TCP B accepts the data and delivers it to ULP B.
11. TCP B acknowledges all 20 octets of data with the ACK set to the sequence number of the next data octet expected.
12. ULP B passes 125 bytes of data to TCP B for transfer to ULP A.
13. TCP B packages the data in a segment marked with the "B" sequence number.
14. TCP A accepts the segment and delivers the data to ULP A.
15. TCP A returns an ACK of the received data marked with the sequence number of the next expected data octet. However, the segment is lost by the network and never arrives at TCP B.
16. TCP B times out waiting for the lost ACK and retransmits the segment. TCP A receives the retransmitted segment, but discards it because the data from the original segment have already been accepted. However, TCP A re-sends the ACK.
17. TCP B gets the second ACK.
18. ULP A closes its half of the connection by issuing a CLOSE to TCP A.
19. TCP A sends a segment marked with a CLOSE control flag, called a FIN, to inform TCP B that ULP A will send no more data.
20. TCP B gets the FIN and informs ULP B that ULP A is closing.
21. ULP B completes its data transfer and closes its half of the connection. TCP B sends an ACK of the first FIN and its own FIN to TCP A to show ULP B's closing. TCP A gets the FIN and the ACK, then responds with an ACK to TCP B. TCP A informs ULP A that the connection is closed (not pictured). TCP B receives the ACK from TCP A and informs ULP B that the connection is closed.

TCP Header Format. The TCP header format is shown in Figure 18-U.13. It should be noted that TCP works with 32-bit segments.

Source Port. The "port" represents the source ULP initiating the exchange. The field is 16 bits long.

Destination Port. This is the destination ULP at the other end of the connection. This field is also 16 bits long.

Sequence Number. Usually, this value represents the sequence number of the first octet of a segment. However, if an SYN is present, the sequence number is the initial sequence number (ISN) covering the SYN; the first data octet is then numbered SYN + 1. The "SYN" is the *synchronize control flag*. It is the opening segment of a TCP connection. SYNs are exchanged from either end. When a connection is to be closed, there is a similar "FIN" sequence exchange.

Acknowledgment Number. If the ACK control bit* is set (bit 2 of the 6-bit control field), this field contains the value of the next sequence number that the sender of the segment is expecting to receive. This field is 32 bits long.

*ACK: A control bit (acknowledge) occupying no sequence space, which indicates that the acknowledgment field of this segment specifies the next sequence number the sender of this segment is expecting to receive, hence acknowledging receipt of all previous sequence numbers.

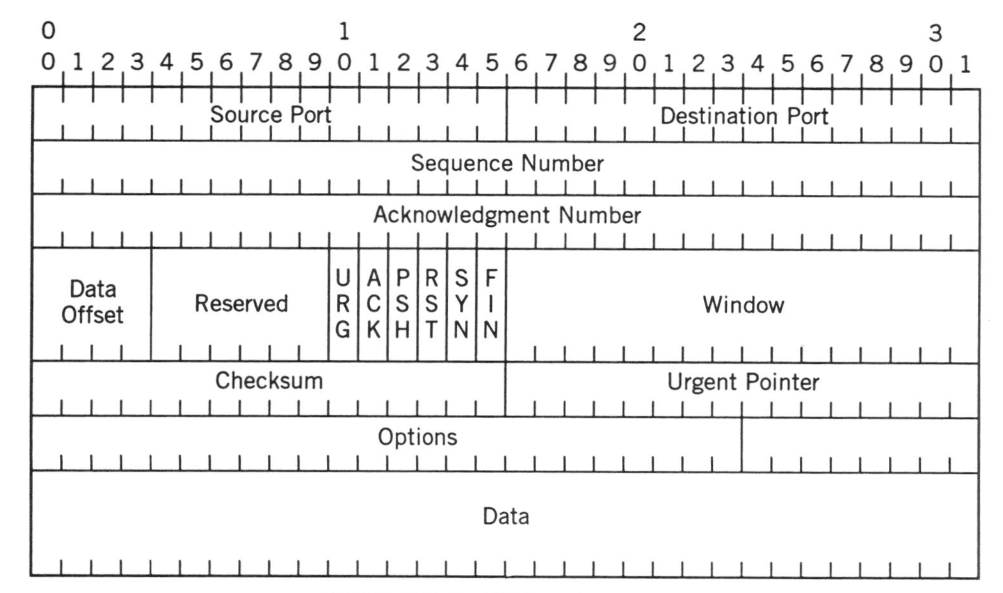

FIGURE 18-U.13. TCP header format.

Data Offset. This field indicates the number of 32-bit words in the TCP header. From this value the beginning of the data can be computed. The TCP header is an integral number of 32 bits long. The field size is 4 bits.

Reserved. This is a field of 6 bits set aside for future assignment. It is set to zero.

Control Flags. The field size is six bits covering six items (1 bit per item):

URG: Urgent pointer field significant
ACK: Acknowledgment field significant
PSH: Push function
RST: Reset the connection
SYN: Synchronize sequence numbers
FIN: No more data from sender

Window. The number of data octets beginning with the one indicated in the acknowledgment field which the sender of this segment is willing to accept. The field is two octets in length.

Checksum. The checksum is the 16-bit one's complement of the one's complement sum of all 16-bit words in the header and text. The checksum also covers a 96-bit pseudo-header conceptually prefixed to the TCP header. This pseudo-header contains the source address, the destination address, the protocol, and TCP segment length.

Urgent Pointer. This field indicates the current value of the urgent pointer as a positive offset from the sequence number in this segment. The urgent pointer points to the sequence number of the octet following the urgent data. This field is only to be interpreted in segments with the URG control bit set. The urgent point field is two octets long.

Options. This field is variable in size, and present options occupy space at the end of the TCP header and are a multiple of 8 bits in length. All options are included in the checksum. An option may begin on any octet boundary. There are two cases of an option:

a. Single octet of option-kind
b. An octet of option-kind, an octet of option length, and the actual option data octets

Options include "end of option list," "no-operation," and "maximum segment size."

Padding. The field size is variable, The padding is used to ensure that the TCP header ends and data begin on a 32-bit boundary. The padding is composed of zeros.

TCP Entity State Diagram. Figure 18-U.14 summarizes TCP operation with a TCP entity state diagram.

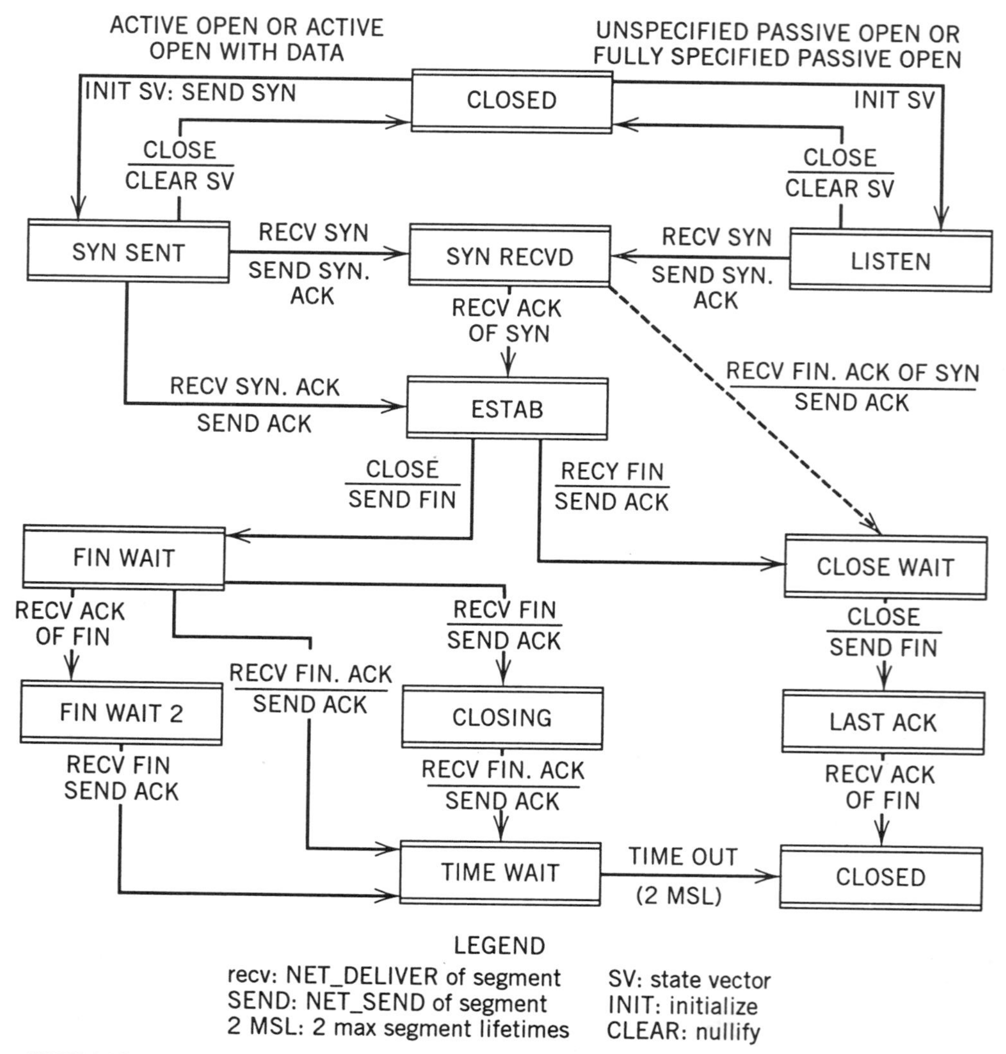

FIGURE 18-U.14. TCP entity state summary [Ref. 11].

Note: RFC is an abbreviation for *Request for Comment* and, when followed by a number, provides a form of specification number for TCP/IP and related DoD protocols.

Section 18-5.2 has been extracted from various RFCs and MIL-STD-1778.

This new section follows Section 18-6.3, which ends on page 1745 of the Manual.

18-7. THE ISO TRANSPORT PROTOCOL

The ISO Transport protocol [Ref. 12] has been designed to be positioned in the OSI layered model in the transport layer. The standard specifies the following:

a. Five classes of procedures for the connection-oriented transfer of data and control information from one transport entity to a peer transport entity:
Class 0—Simple class
Class 1—Basic error recovery class
Class 2—Multiplexing class
Class 3—Error recovery and multiplexing class
Class 4—Error detection and recovery class

b. The means of negotiating the class of procedures to be used by the transport entities

c. The structure and encoding of the transport protocol data units used for the transfer of data and control information

TABLE 18-U.3
Transport Service Primitives

Primitive		Parameter
T-CONNECT	Request Indication	Called address Calling address Expedited data option Quality of service TS user-data
T-CONNECT	Response Confirm	Responding address Quality of service Expedited data option TS user-data
T-DATA	Request Indication	TS user-data
T-EXPEDITED DATA	Request Indication	TS user-data
T-DISCONNECT	Request	TS user-data
T-DISCONNECT	Indication	Disconnect reason TS user-data

TABLE 18-U.4
Network Service Primitives

Primitive		X/Y	Parameter	X/Y/Z
N-CONNECT	Request Indication Response Confirm	X X X X	Called address Calling address NS user-data QoS parameter set Responding address Receipt confirmation selection	X X Z X Z Y
N-DATA	Request Indication	X X	NS user-data Confirmation request	X Y
N-DATA ACKNOWLEDGE	Request Indication	Y Y		
N-EXPEDITED DATA	Request Indication	Y Y	NS user-data	Y
N-RESET	Request Indication Response Confirm	X X X X	Originator Reason	Z Z
N-DISCONNECT	Request Indication	X X	NS user-data Originator Reason	Z Z Z

X: The transport protocol assumes that this facility is provided in all networks.
Y: The transport protocol assumes that this facility is provided in some networks and a mechanism is provided to optionally use the facility.
Z: The transport protocol does not use this parameter.
Note 1: The parameters listed in this table are those in the current network service (first DP 8348).
Note 2: The way the parameters are exchanged between the transport entity and the NS provider is a local matter.

18-7.1. Services Provided and Assumed

Services Provided by the Transport Layer. Information is transferred to and from the transport service user in the transport service primitives listed in Table 18-U.3.

Services Assumed from the Network Layer. Information is transferred to and from the network service provider in the network service primitives listed in Table 18-U.4.

Service Primitive. The IEEE [Ref. 13] defines a service primitive as "an abstract, implementation-independent interaction between a service user and a service provider."

18-7.2. Functions of the Transport Layer

The functions of the transport layer are those necessary to bridge the gap between the services available from the network layer and those offered to the transport service users. The following functions, depending on the selected class and options, are used at all times during a transport connection:

- **a.** Transmission of transport protocol data units (TPDUs)
- **b.** Multiplexing and demultiplexing, a function used to share a single network connection between two or more transport connections
- **c.** Error detection, a function used to detect the loss, corruption, duplication, misordering, or misdelivery of TPDUs
- **d.** Error recovery, a function used to recover from detected and signaled errors

Connection Establishment. The purpose of connection establishment is to establish a transport connection between two transport service users. The following functions of the transport layer during this phase must match the transport service users' requested quality of service (QoS) with the services offered by the network layer:

- **a.** Select the network service which best matches the requirement of the transport service user, taking into account charges for various services.
- **b.** Decide whether to multiplex multiple transport connections onto a single network connection.
- **c.** Establish the optimum TPDU size.
- **d.** Select the functions that will be operational upon entering the data transfer phase.
- **e.** Map transport addresses onto network addresses.
- **f.** Provide a means to distinguish between two different transport connections.
- **g.** Transport the transport service user data.

Data Transfer. The purpose of data transfer is to permit duplex transmission of PDUs between two transport service users connected by the transport connection. This purpose is achieved by means of two-way simultaneous communication and by the following functions, some of which are used or not used in accordance with the result of the selection performed in connection establishment:

- **a.** Concatenation and separation, a function used to collect several TPDUs into a single network service data unit (SDU) at the sending transport entity and to separate the TPDUs at the receiving transport entity.
- **b.** Segmenting and reassembly, a function used to segment a single data transport SDU into multiple TPDUs at the sending transport entity and to reassemble them into their original format at the receiving transport entity.
- **c.** Splitting and recombining, a function allowing the simultaneous use of two or more network connections to support the same transport connections.
- **d.** Flow control, a function used to regulate the flow of TPDUs between two transport entities on one transport connection.
- **e.** Transport connection identification, a means to uniquely identify a transport connection between the pair of transport entities supporting the connection during the lifetime of the transport connection.
- **f.** Expedited data, a function used to bypass the flow control of normal data PDU. Expedited data PDU flow is controlled by separate flow control.
- **g.** Transport SDU delimiting, a function used to determine the beginning and ending of a transport SDU.

Release. The purpose of release is to provide disconnection of the transport connection, regardless of the current activity.

18-7.3. Classes and Options

A class defines a set of functions. Options define those functions within a class which may or may not be used. The five classes were listed previously above. It should be noted that transport connections of Classes 2, 3, and 4 may be multiplexed together on the same network connection.

Negotiation. The use of classes and options is negotiated during connection establishment. The choice made by the transport entities will depend upon:

- **a.** The transport service user's requirements expressed via the T-CONNECT service primitives
- **b.** The quality of the available network services
- **c.** The user-required service versus cost ratio acceptable to the transport service user.

Choice of Network Connection. The following list classifies services in terms of quality with respect to error behavior in relation to user requirements; its main purpose is to provide a basis for the decision regarding which class of transport protocol should be used in conjunction with given network connection:

Type A. Network connection with acceptable residual error rate (for example, not signaled with reset or disconnect) and acceptable rate of signaled errors.
Type B. Network connections with acceptable residual error rate (for example, not signaled with disconnect or reset) but unacceptable rate of signaled errors.
Type C. Network connections with unacceptable residual error rate.

(*Note:* Interpret "signaled errors" as the underlying error rate and interpret "residual error rate" as the error rate after error recovery has taken place.)

With the above, it must be assumed that each transport entity is aware of the quality of service provided by particular network connections.

Characteristics of Class 0. Class 0 provides the simplest type of transport connection and is fully compatible with ITU-T Rec. S.70 for Teletex terminals. Class 0 has been designed to be used with Type A network connections.

Characteristics of Class 1. Class 1 provides a basic transport connection with minimum overhead. The main purpose of this class is to overcome network disconnect and reset. Selection of this class is usually based on reliability criteria. Class 1 has been designed for use with Type B network connections.

Characteristics of Class 2. Class 2 provides a way to multiplex several transport connections on a single network connection. This class has been designed to be used with Type A network connections.

Use of Explicit Flow Control with Class 2. The objective is to provide flow control to help avoid congestion at transport connection endpoints and on the network connection. Typical use is when traffic is heavy and continuous, or when there is intensive multiplexing. Use of flow control can optimize response times and resource utilization.

Nonuse of Explicit Flow Control with Class 2. Here the objective is to provide a basic transport connection with minimal overhead suitable when explicit disconnection of the transport connection is desirable. This option would typically be used for unsophisticated terminals and when no multiplexing on network connections is required. Expedited data are never available.

Characteristics of Class 3. Class 3 provides the characteristics of Class 2 plus the ability to recover from network disconnect or reset. Selection of this class is usually based upon reliability criteria. Class 3 has been designed to be used with Type B network connections.

Characteristics of Class 4. Class 4 provides the characteristics of Class 3 plus the capability to detect and recover from errors which occur as a result of the low grade of service available from the network service provider. The kinds of errors to be detected include: TPDU loss, TPDU delivery out of sequence, TPDU duplication, and TPDU corruption. These errors may affect control TPDUs as well as data TPDUs.

This class also provides for increased throughput capability and additional resilience against network failure. Class 4 has been designed to be used with Type C network connections.

18-7.4. Model of the Transport Layer

A transport entity communicates with its transport service users through one or more transport service access points (TSAPs) by means of service primitives. Service primitives will cause or be the result of TPDU exchanges between the peer transport entities supporting a transport connection. These protocol exchanges are effected using the services of the network layer through one or more network service access points (NSAPs).

Transport connection endpoints are identified in end systems by an internal implementation-dependent mechanism so that the transport service user and the transport entity can refer to each transport connection. Figure 18-U.15 illustrates a model of the transport layer.

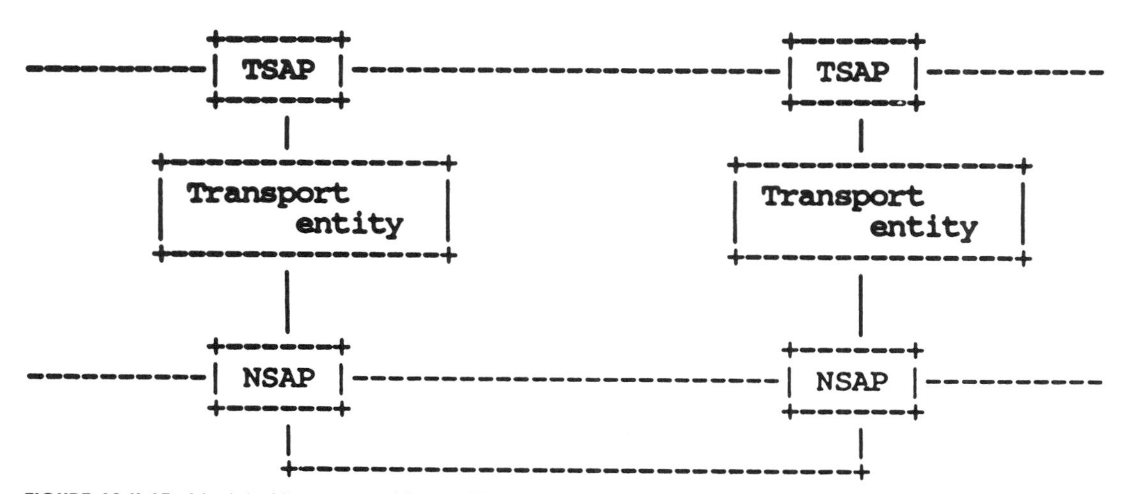

FIGURE 18-U.15. Model of the transport layer. *Note*: For purpose of illustration, this figure shows only one TSAP and one NSAP for each transport entity. In certain instances, more than one TSAP and/or more than one NSAP may be associated with a particular transport entity.

TABLE 18-U.5
Transport Protocol Data Unit (TPDU) Code

	Validity Within Classes					Code
	0	1	2	3	4	
CR: Connection request	x	x	x	x	x	1110 xxxx
CC: Connection confirm	x	x	x	x	x	1101 xxxx
DR: Disconnect request	x	x	x	x	x	1000 0000
DC: Disconnect confirm		x	x	x	x	1100 0000
DT: Data	x	x	x	x	x	1111 0000
ED: Expedited data		x	NF	x	x	0001 0000
AK: Data acknowledgment		NRC	NF	x	x	0110 zzzz
EA: Expedited data acknowledgment		x	NF	x	x	0010 0000
RJ: Reject		x		x		0101 zzzz
ER: TPDU error	x	x	x	x	x	0111 0000
Not available (see Key)						0000 0000
						0011 0000
						1001 xxxx 1010 xxxx

xxxx (bits 4-1): Used to signal the CDT (credit field) (set to 0000 in Classes 0 and 1)
zzzz (bits 4-1): Used to signal CDT in Classes 2, 3, and 4 (set to 1111 in Class 1)
NF: Not available when the nonexplicit flow control option is selected
NRC: Not available when the receipt confirmation option is selected

18-7.5. Structure and Coding of TPDUs

Table 18-U.5 specifies those TPDUs which are valid for each class and the code for each TPDU.

18-7.5.1. Structure

All of the TPDUs contain an integral number of octets. The octets in a TPDU are numbered starting from 1 and increasing in the order that they are put into a network service data unit (NSDU). The bits in an octet are numbered from 1 to 8, where bit 1 is the low-ordered bit. When consecutive octets are used to represent a binary number, the lower octet number has the least significant value.

Note: When the encoding of a TPDU is represented by a diagram, several of which are given below, the following representation is used:

a. Octets are shown with the lowest-numbered octet to the left and the high-numbered octets further to the right.
b. Within an octet, bits are shown with bit 8 to the left and bit 1 to the right.

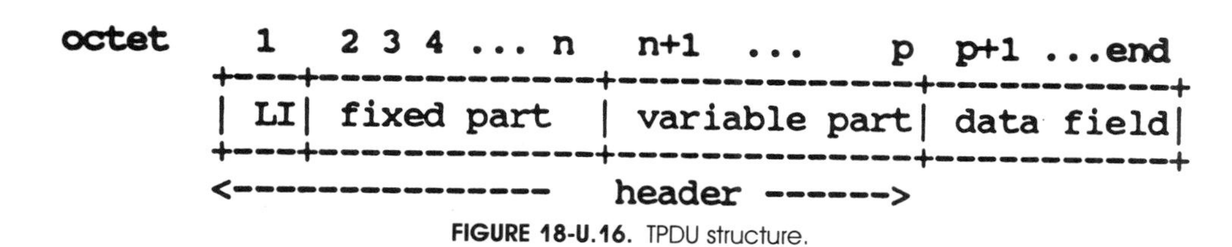

FIGURE 18-U.16. TPDU structure.

The TPDU structure is shown in Figure 18-U.16 The TPDU contains, in the following order:

1. The header, comprising:
   a. The length indicator (LI) field
   b. The fixed part
   c. The variable part, if present
2. The data field, if present

Length Indicator Field. This field is the first octet in the TPDU format. The length is indicated by a binary number, with the maximum value of 254 (1111 1110). The length indicated is the header length in octets, including parameters but excluding the length indicator field and user data, if any. The value 255 (1111 1111) is reserved for possible extensions. If the length indicated exceeds the size of the network service (NS) user data which are present, this is a protocol error.

Fixed Part. The fixed part contains frequently occurring parameters including the code of the TPDU. The length and the structure of the fixed part are defined by the TPDU code and in certain cases by the protocol class and the formats in use (normal or extended). If any of the parameters of the fixed part have an invalid value, or if the fixed part cannot be contained within the header (as defined by the LI), this is a protocol error.

Note: In general, the TPDU code defines the fixed part unambiguously. However, different variants may exist for the same TPDU code as in normal and extended formats.

TPDU Code. This field contains the TPDU code and is contained in octet 2 of the header. It is used to define the structure of the remaining header. This field is a full octet except in the following cases:

1110 xxxx	Connection request
1101 xxxx	Connection confirm
0101 xxxx	Reject
0110 xxxx	Data acknowledgment

where xxxx (bits 4–1) is used to signal the CDT (credit). Only those codes defined in Table 18-U.5 are valid.

Variable Part. The variable part is used to define less frequently used parameters. If the variable part is present, it will contain one or more parameters. Each parameter contained within the variable part is structured as shown in Figure 18-U.17.

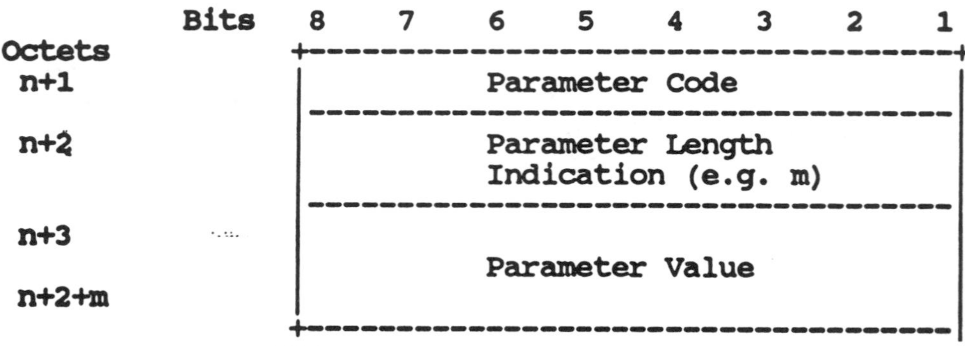

FIGURE 18-U.17. Format of the variable part.

The following discussion refers to Figure 18-U.17:

1. The parameter code field is coded in binary.

Note: Without extensions, it provides a maximum number of 255 different parameters. However, as noted below, bits 8 and 7 cannot take every possible value, so that the practical maximum number of different parameters is less. Parameter code 1111 1111 is reserved for possible extensions of the parameter code.

2. The parameter length indication indicates the length, in octets, of the parameter value field.

Note: The length is indicated by a binary number, m, with a theoretical maximum value of 255. The practical maximum value of m is lower, For example, in the case of a single parameter contained within the variable part, two octets are required for the parameter code and the parameter length indication itself. The value of m is limited to 258. For larger fixed parts of the header and for each succeeding parameter, the maximum value of m decreases.

3. The parameter codes use bits 8 and 7 with the value 00.

4. The parameters defined in the variable part may be in any order. If any parameter is duplicated, then the later value will be used. A parameter not defined in the ISO standard is treated as a protocol error in any received TPDU except a CR TPDU. In a CR TPDU it is ignored. If the responding transport entity selects a class for which a parameter of the CR TPDU is not defined, it may ignore this parameter, except the class and option, and alternative protocol class parameters which shall always be interpreted. A parameter defined by the ISO transport standard but having an invalid value will be treated as a protocol error in any received TPDU except a CR TPDU (CR = connection request). In a CR TPDU it will be treated as a protocol error if it is either the class and option parameter or the alternative class parameter or the additional option parameter; otherwise it will be either ignored or treated as a protocol error.

Checksum Parameter (Class 4 Only). All TPDU types may contain a 16-bit checksum parameter in their variable part. This parameter will be present in a CR TPDU and will also be present in all other TPDUs except when the nonuse of the checksum option is selected.

Parameter code: 1100 0011
Parameter length: 2
Parameter value: Result of checksum algorithm

```
 1   2          3          4        5    6     7       8      p   p+1...end
+---+-------+----------+----------+----+----+-------+--------+---------+
|LI|CR CDT|     DST - REF          |SRC-REF|CLASS  |VARIAB. |USER     |
|  |1110  |0000 0000|0000 0000     |   |   |OPTION |PART    |DATA     |
+---+-------+----------+----------+----+----+-------+--------+---------+
```

FIGURE 18-U.18. Structure of the CR TPDU.

Data Field. This field contains transparent user data. Restrictions on its size are given for each TPDU type.

18-7.5.2. Detailed TPDU Types

In this subsection we have selected the CR TPDU as an example of TPDU type. A brief description of the CR TPDU is provided below.

Connection Request TPDU. The CR TPDU has a maximum length of 128 octets. Its structure is shown in Figure 18-U.18. The length indicator (LI) field is as described in Section 18-7.5.1 above.

Fixed Part (Octets 2 to 7). The structure of this part contains the following:

- **a.** CR: Connection request code: 1110, bits 8–5 of octet 2.
- **b.** CDT: Initial credit allocation (set to 0000 in Classes 0 and 1 when specified as preferred class). Bits 4–1 of octet 2.
- **c.** DST-REF: Set to zero.
- **d.** SRC-REF: Reference selected by the transport entity initiating the CR TPDU to identify the requested transport connection.
- **e.** CLASS and OPTION: Bits 8–5 of octet 7 defines the preferred transport protocol class to be operated over the requested transport connection. This field will take on one of the following values:

0000	Class 0
0001	Class 1
0010	Class 2
0011	Class 3
0100	Class 4

The CR TPDU contains the first choice of class in the fixed part. Second and subsequent choices are listed in the variable part if required.

Bits 4–1 of octet 7 define options to be used on the requested transport connection and their coding is shown in Table 18-U.6.

Variable Part (Octets 8 through p). The following parameters are permitted in the variable part:

a. *Transport service access point identifier (TSAP-ID)*

Parameter code:	1100 0001 for the identifier of the calling TSAP.
	1100 0010 for the identifier of the called TSAP.
Parameter length:	Not defined.
Parameter value:	Identifier of the calling or called TSAP, respectively.

TABLE 18-U.6
Coding of Bits 4-1 of Octet 7

Bit		Option
4	0	Always
3	0	Always
2	= 0	Use of normal formats in all cases
	= 1	Use of extended formats in Classes 2, 3, and 4
1	= 0	Use of explicit flow control in Class 2
	= 1	No use of explicit flow control in Class 2

Note 1: The connection establishment procedure does not permit a given CR TPDU to request use of transport expedited data transfer service (additional option parameter) and no use of explicit flow control in Class 2 (bit 1 = 1).

Note 2: Bits 4 to 1 are always zero in Class 0 and have no meaning.

If a TSAP-ID is given in the request, it may be returned in the confirmation.

b. *TPDU size*. This parameter defines the maximum TPDU size (in octets including the header) to be used over the requested transport connection. The coding of this parameter is as follows:
Parameter Value:
0000 1101 8192 octets (not allowed in Class 0)
0000 1100 4096 octets (not allowed in Class 0)
0000 1011 2048 octets
0000 1010 1024 octets
0000 1001 512 octets
0000 1000 256 octets
0000 0111 128 octets
Default value is 0000 0111 (128 octets)

c. *Version number* (not used if Class 0 is the preferred class)
Parameter code: 1100 0100
Parameter length: one octet
Parameter value field: 0000 0001
Default value is 0000 0001 (not used in Class 0)

d. *Security parameters* (user defined)
Parameter code: 1100 0101

e. *Checksum*

The material in Section 18-7 was extracted from ISO Protocol Specification ISO DP 8073 as contained in RFC 905 [Ref. 12].

18-8. IBM SYSTEM NETWORK ARCHITECTURE (SNA)

18-8.1. Background

The first version of SNA was implemented by IBM in 1974 to bring order out of chaos in resource sharing data circuits replacing such file access systems as BTAM, RTAM, and TCAM. These were replaced by VTAM (virtual telecommunications access method).

The traditional SNA is a hierarchy of connected network resources: host processors such as the IBM 3081, communication controllers such as the IBM 3725, cluster controllers typically the IBM 3274, and terminals such as the IBM 3278.

VTAM runs in a host processor where all SNA application subsystems use VTAM as their telecommunications access method. The network control program (NCP) runs in a communications controller and provides network management for all remote network resources that are attached to the communications controller.

Some of the major evolutionary trends in SNA have been the following:

- A very large infrastructure of SNA products from IBM and other suppliers
- Increasing configuration flexibility, particularly to exploit advances in transmission technology
- Inclusion of network standards as these have become available
- Greater attention to network management services
- Widening support for non-SNA services
- Expansion of routing and transport services to keep pace with installation of ever larger networks
- Increasing the functions available to end users [Refs. 14 and 15]

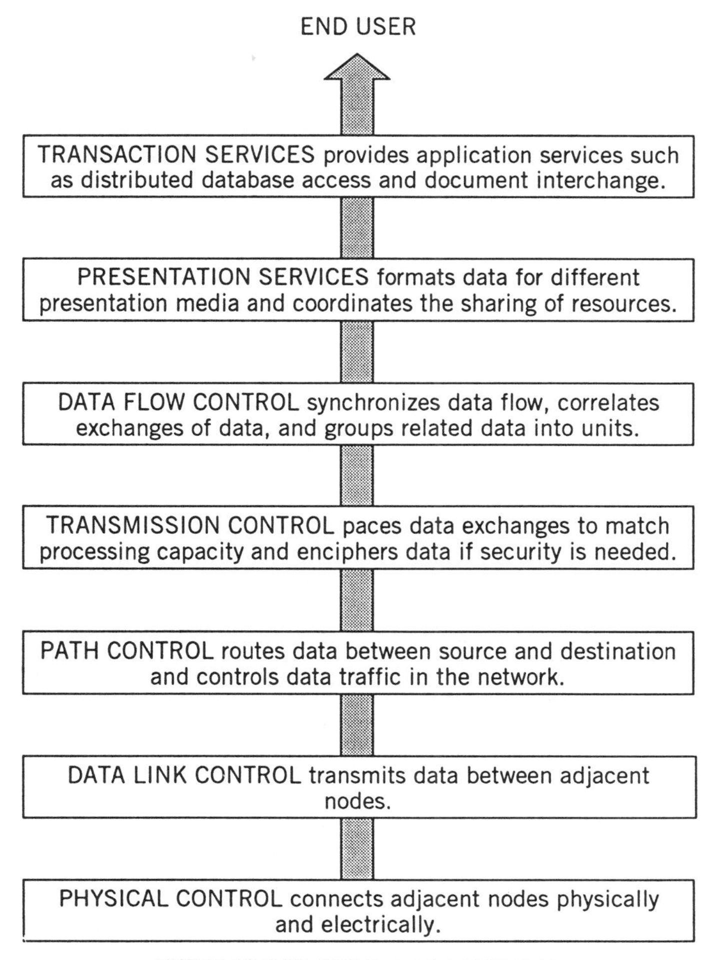

FIGURE 18-U.19. SNA layered architecture.

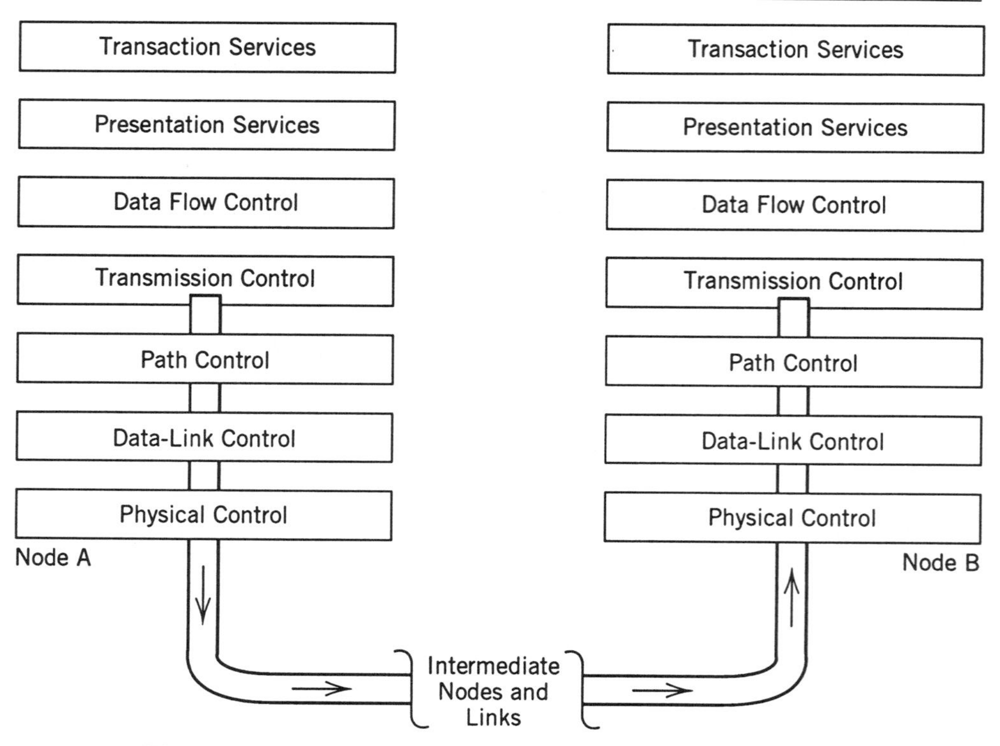

FIGURE 18-U.20. Communication between two transmission control layers.

18-8.2. The SNA Layered Architecture

SNA functions are divided into a hierarchical structure of seven well-defined layers. Each layer in the architecture performs a specific set of functions. Figure 18-U.19 identifies SNA's seven layers and describes their major functions.

SNA defines formats and protocols between layers that permit equivalent layers (layers at the same level within the hierarchy) to communicate with one another. Similar to OSI, each layer performs services for the next higher layer, requests services from the next lower layer, and communicates with equivalent layers. One example is the encryption of data as required by an end user. The procedure by which the two transmission control layers communicate is outlined in Figure 18-U.20.

The two transmission layers shown in Figure 18-U.20 encrypt and decrypt data independently of the functions of any other layer. The transmission control layer in the originating node encrypts the data it receives from the the data flow control layer. It then requests that the path control layer route the encrypted data to the destination node. The transmission control layer at the destination node decrypts the data that the path control layer delivered. It then requests that the data flow control layer give the encrypted data to the destination end user.

The interaction of hardware and software components of SNA is shown in Figure 18-U.21.

The SNA *networking blueprint* is shown in Figure 18-U.22. The networking blueprint supports the implementation of multiple protocols as well as integrates these protocols into a cohesive, modular structure. The networking blueprint defines layers of functions plus a systems management backplane. SNA *advanced peer-to-peer networking* (APPN) is part of the transport layer, and it is one of the protocols that can be used in this layer.

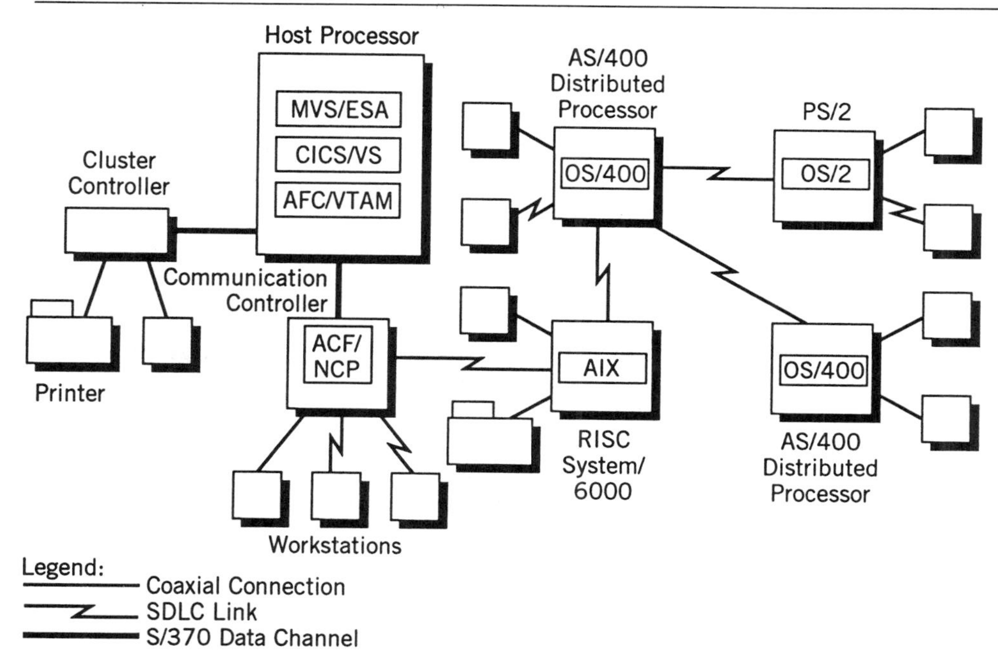

FIGURE 18-U.21. Hardware and software components of an SNA network.

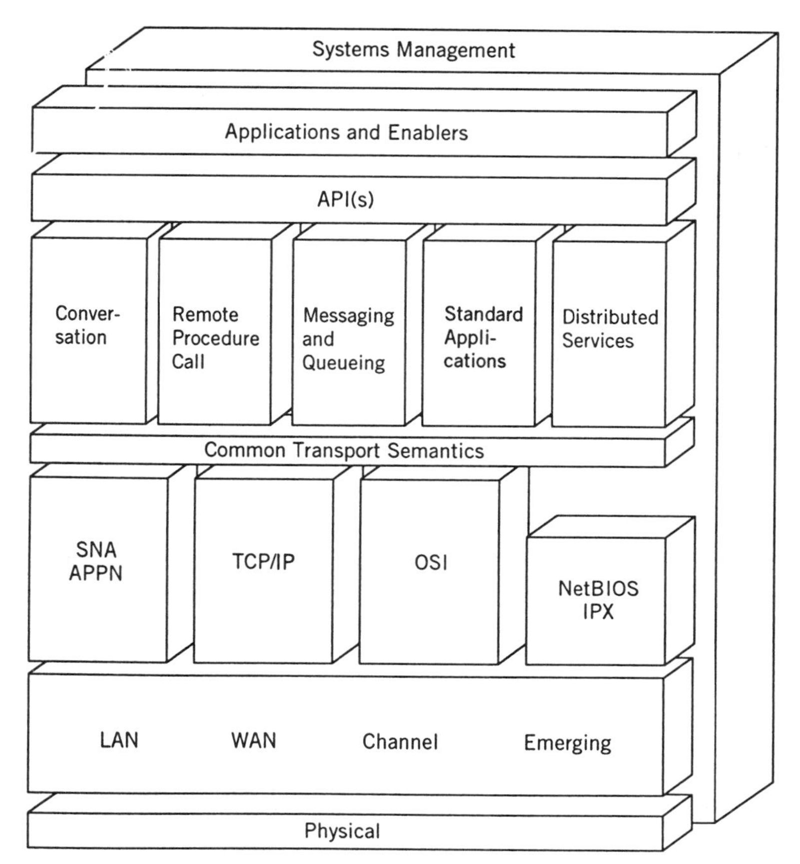

FIGURE 18-U.22. The SNA networking blueprint.

High-performance routing (HPR) is a small but powerful extension to APPN. It enhances data routing performance by decreasing intermediate node processing. HPR increases session reliability via *nondisruptive path switch*. The main components of HPR are *rapid-transport protocol* (RTP) and *automatic network routing* (ANR).

18-8.3. Advanced Peer-to-Peer Networking (APPN)

APPN ties together diverse platforms, topologies, and applications into a single network. APPN's any-to-any connectivity makes it possible for large and small networks alike to communicate over local- and wide-area networks, across slow and fast links.

APPN provides two basic functions: keeping track of the location of resources in the network, and selecting the best path to route data between resources. APPN nodes dynamically exchange information about each other, eliminating the need for customers having to deal with complicated system and path definitions. APPN nodes limit the information they exchange, enabling more efficient use of network resources.

18-8.4. Architectural Components of an SNA Network

A data network can be described as a configuration of nodes and links. Nodes are the network component that send data over, and receive data from, the network. A node may be a processor, controller, or workstation. Links are the network components that connect adjacent nodes. Of course, nodes and links work together in transferring data through a network.

A *node* is a set of hardware and associated software components that implement the functions of the seven architectural layers. Although all seven layers are implemented in a given node, nodes can differ based on their architectural components and the sets of functional capabilities they implement. Nodes with different architectural components represent different *node types*. Four types of nodes exist:

Type 4 (T4)
Type 5 (T5)
Type 2.0 (T2.0)
Type 2.1 (T2.1)

Nodes that perform different functions are said to act in different *network roles*. In some cases a given network node can act in different network roles. A T4 node, for example, can perform an interconnection role between nodes at different levels of the subnetwork hierarchy, or between nodes in different subarea networks. The functions performed in these two roles are referred to as *boundary function* and *gateway function*, respectively. T2.1 and T5 nodes can act in several different network roles. Node roles fall into two broad categories: *hierarchical roles* and *peer-oriented roles*.

18-8.4.1. Hierarchical Roles

Hierarchical roles are those in which certain nodes have a controlling or *mediating* function with respect to the actions of other nodes. SNA hierarchical networks are characterized by nodes of all four types acting in hierarchical roles. Within such networks, nodes are categorized as either *subarea nodes* (SNs) or *peripheral nodes* (PNs). Subarea nodes provide services for and control over peripheral nodes. Networks consisting of subarea and peripheral nodes are referred to as *subarea networks*.

Subarea Nodes. Type 5 (T5) and type 4 (T4) nodes can act as subarea nodes. T5 subarea nodes provide the SNA functions that control network resources, support transaction programs, support network operators, and provide end-user services. Because these functions are provided by host processors, T5 nodes are also referred to as *host nodes*. T4 subarea nodes provide the SNA functions that route and control the flow of data in a subarea network. Because these functions are provided by communication controllers, T4 nodes are also referred to as *communication controller nodes*.

Peripheral Nodes. Type 2.0 (T2.0) and type 2.1 (T2.1) nodes can act as peripheral nodes attached to either T4 or T5 subarea nodes. Peripheral nodes are typically devices such as distributed processors, cluster controllers, or workstations. A T2.1 node differs from a T2.0 node by the T2.1 node's ability to support peer-oriented protocols as well as the hierarchical protocols of a simple T2.0 node. A T2.0 node requires the mediation of a T5 node in order to communicate with another node. Subarea nodes to which peripheral nodes are attached perform a *boundary function* and act as subarea *boundary nodes*.

Although Figure 18-U.22 represents nodes as particular classes of hardware, there is no architectural association between node type, or node role, and the kind of hardware that implements it. To avoid associating node types and roles with hardware implementations, network architecture diagrams use symbols to represent node types and roles. Figure 18-U.23 uses symbols to illustrate a subarea network containing the four node types acting as subarea and peripheral nodes. The network contains two type 5 subarea nodes, three type 4 subarea nodes, and seven peripheral nodes.

18-8.4.2. Peer-Oriented Roles

The APPN extensions allow greater distribution of network control by enhancing the dynamic capabilities of a node. Nodes with these extensions are referred to as *APPN nodes*, and a network of APPN nodes makes up an *APPN network*. A low-entry networking (LEN) node can also attach to an APPN network. *LEN* offers the lowest-level peer-to-peer networking capability. LEN allows simple point-to-point connections between adjacent nodes. By adjacent, we mean that the communicating nodes must be connected to one another by a single physical data link or through an SNA subarea network.

An APPN node can dynamically find the location of a partner node, place the location information in directories, compute potential routes to the partner, and select the best route from among those computed. These dynamic capabilities relieve network personnel from having to predefine those locations, directory entries, and routes. APPN nodes can include processors of varying sizes, such as the Application System/400, the Enterprise System/9370 (ES/9370) running under Distributed Processing Program Executive/370 (DPPX/370), the Personal System/2 (PS/2) running under Operating System/2 (OS/2), and VTAM running under Multiple Virtual Storage/Enterprise Systems Architecture (MVS/ESA).

Peer-oriented protocols enable nodes to communicate without mediation by a T5 node, giving them increased connection flexibility. APPN defines two possible roles for a node in an APPN network, that of an end node and that of a network node. (Network nodes have additional options that can further distinguish them.)

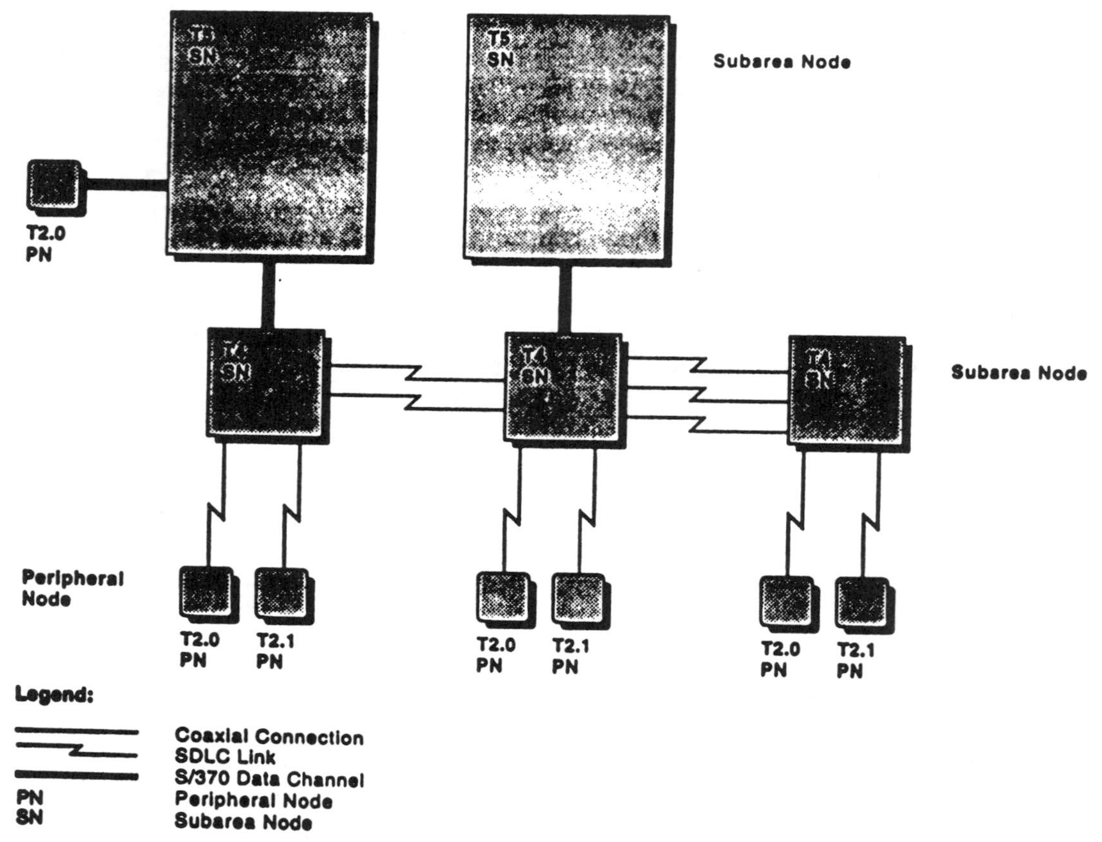

FIGURE 18-U.23. A subarea network.

T2.1 nodes can act as either APPN or LEN nodes. T5 nodes can also act as APPN or LEN nodes, but have additional capability to interconnect subarea and APPN networks by interchanging protocols between them. In this capacity they are called *interchange nodes*, discussed later. Together with its subordinate T4 nodes, a T5 node can also form for a *composite LEN node* or a *composite network node*. As composite nodes they appear as single LEN or network nodes to other LEN or APPN nodes to which they are interconnected.

If two network nodes do not support the border node option (discussed below) and are located in two separate net-ID subnetworks, CP–CP sessions cannot be established between them. For the two network nodes to communicate, a LEN connection may be established between the two. This allows the two net-ID subnetworks to communicate, but does not support any APPN function. If NN1 subnet A is to establish a session with NN2 in subnet B, all LUs in subnet A must be predefined to NN2 in subnet B. The network nodes in both subnets are APPN nodes, but because they communicate across net-ID subnetwork boundaries, they are connected to each other via LEN links.

End nodes are located in the periphery of an APPN network. An end node obtains full access to the APPN network through one of the network nodes to which it is directly attached—its *network node server*. There are two kinds of end nodes: APPN end nodes and LEN end nodes. An *APPN end node* supports APPN protocols through explicit interactions with a network node server. Such protocols support dynamic searching for resources and provide resource information for the calculation of routes by network nodes. A *LEN end node* is a LEN node attached to a network node. Although LEN nodes lack the APPN extensions, they are able to be supported in APPN networks using the services provided them by network nodes. In an APPN network, when a LEN node is connected to another LEN node, or to an APPN end node, it is referred to simply as a *LEN node*. When connected to an APPN network node, however, it is referred to as a *LEN end node*.

Network nodes, together with the links interconnecting them, form the *intermediate routing network* of an APPN network. Network nodes connect end nodes to the network and provide resource location and route selection services for them. Routes used to interconnect network users are selected based on network topology information that can change dynamically.

Figure 18-U.24 illustrates one possible APPN network configuration and contains LEN end nodes as well as APPN nodes.

Network node server is a network node that provides resource location and route selection to the LUs it serves. These LUs can be in the network node itself, or in the client end nodes. A network node server uses CP–CP sessions to provide network information for session setup in order to support LUs on served APPN end nodes. In addition, LEN end nodes can also take advantage of the services of the network node server. A LEN end node, unlike an APPN end node, must be predefined by the network operator as a client end node for which the network node acts as server. Any network node can be a network node server for end nodes

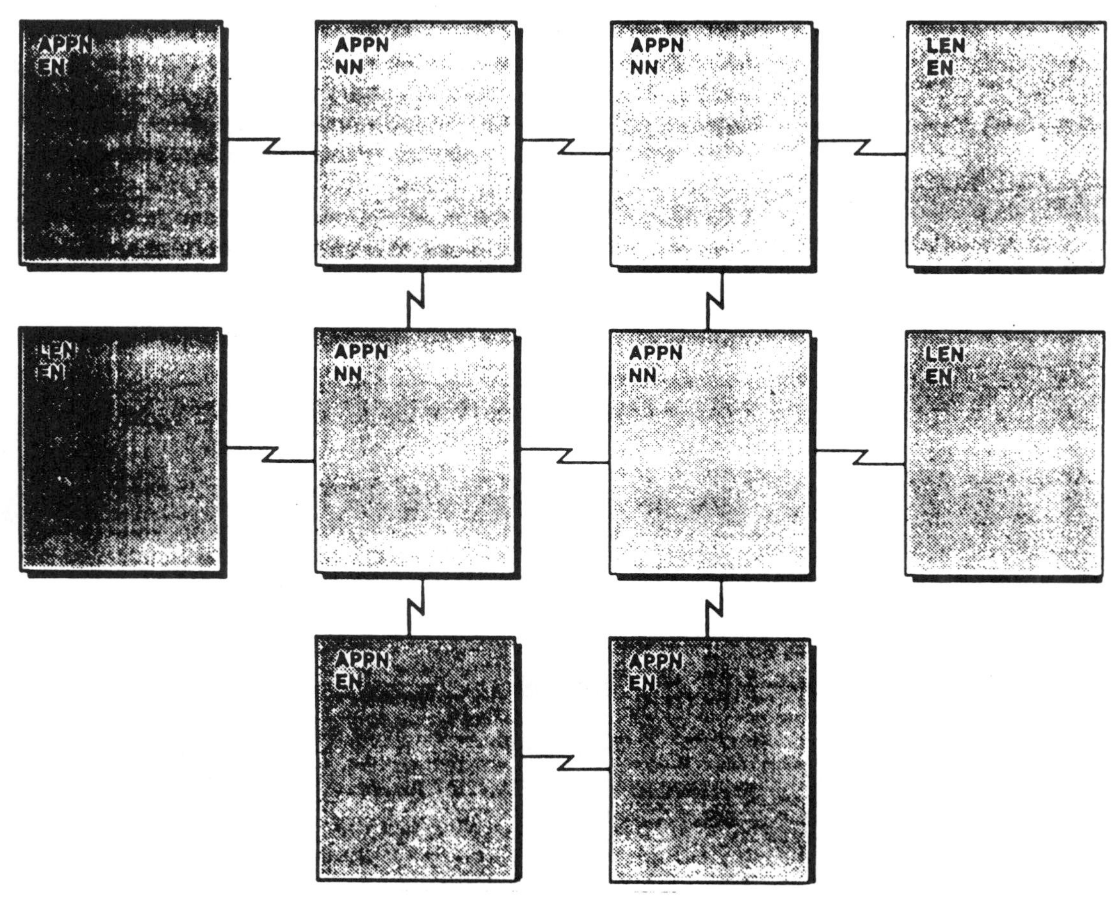

FIGURE 18-U.24. An APPN network.

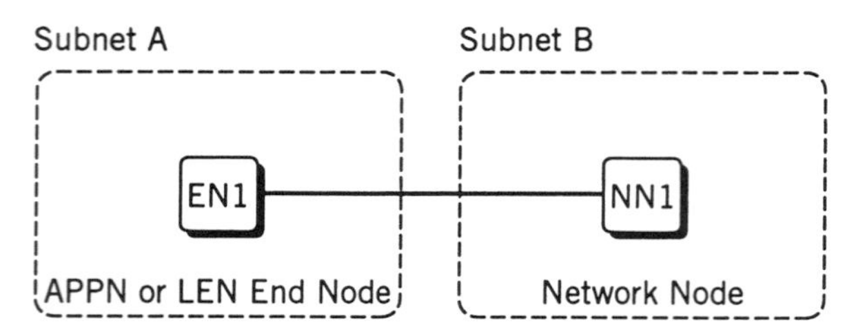

FIGURE 18-U.25. An end node attached nonnatively to its network node server.

that are attached to it. The served end nodes are defined as being in that network node server's domain.

Figure 18-U.25 illustrates an end node connecting to a network node server with a different net ID from its own. The capability allows the end node to dial into different APPN networks. This kind of attachment is called a *nonnative attachment*.

Central directory server (CDS) is a network node that builds and maintains a directory of resources from the network. The purpose of a CDS is to reduce the number of network broadcast searches to a maximum of one per resource. Network nodes and APPN end nodes can register their resources with a CDS, which acts as a focal point for resource location information.

A *subnetwork* consists of a group of interconnected nodes, within a larger composite network, that have some common attribute or characteristic, such as the same network ID, or that share a common topology database or that implement a common protocol. A *net-ID subnetwork* refers to the set of nodes having the same network ID. A *topology subnetwork* consists of all the network nodes exchanging and maintaining the same topology information. A *high-performance routing (HPR) subnetwork* consists of all the interconnected APPN nodes implementing the HPR function.

Within an APPN network it is often desirable to partition the network topology database to reduce its size at each network node and lower the overall topology data interchange traffic. Topology database replication and interchange are confined to a topology subnetwork, with each subnetwork acting independently. The presence of *border nodes* enables the topology subnetworks to be tied together into a larger composite network with the same freedom of LU–LU sessions as if the topology partitioning did not exist. Border nodes are network nodes having additional function and are of two types: *peripheral* and *extended*.

Peripheral border nodes permit APPN networks with different net IDs to interconnect, allowing session setup across net-ID subnetwork boundaries, but in addition allow the partitioning of the topology database among network nodes with the same net ID; this is known as *clustering* and the network partitions are also known as *clusters*. Each cluster is a distinct topology subnetwork.

18-8.5. Nodes with Both Hierarchical and Peer-Oriented Function

Interchange node is a VTAM product feature merging networks and subarea networks. An interchange node receives network search requests from APPN nodes and transfers them into subarea network searches, without exposing the subarea assets to the APPN part of the network. At the same time the interchange node can receive search requests from a subarea network and transfer them into APPN requests without exposing the APPN aspects to the subarea network. The

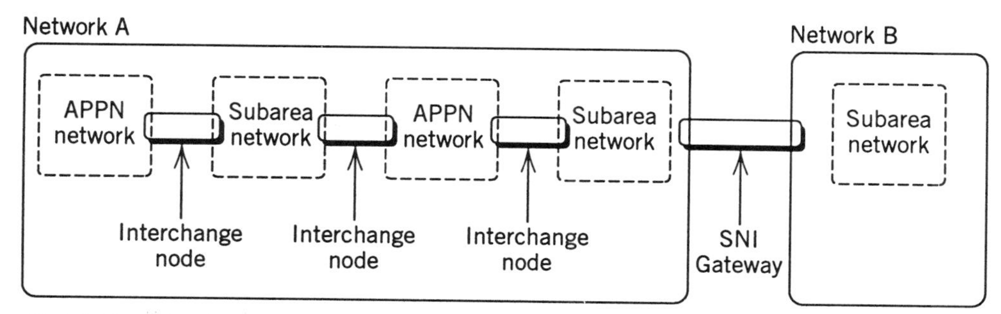

FIGURE 18-U.26. Interchange nodes interconnecting APPN and subarea networks.

interchange node maintains its subarea appearance to other subarea nodes and maintains its APPN appearance to other APPN nodes. Figure 18-U.26 illustrates an intermediate stage of migrating a subarea network to an APPN network. An interchange node permits migration of an existing subarea network to APPN on a node-by-node, link-by-link basis without a need for network-wide coordination. Also, an interchange node permits session establishment across any combination of APPN and subarea networks, including SNA network interconnection (SNI) gateways.

18-8.6. Links and Transmission Groups

Adjacent nodes in a network are connected to one another by one or more links. A *link* includes both the link stations within the two nodes it connects and the link connection between the nodes.

A *link station* is the hardware or software within a node that enables the node to attach to, and provide control over, a link connection. It exchanges information and controls signals with its partner link station in the adjacent node. Link stations use data-link control protocols to transmit data over a link connection. A *link connection* is the physical medium over which data are transmitted. Examples of transmission media include wire pairs, microwave radio, fiber-optic cables, and satellite circuits. Multiple links between the same two nodes are referred to as *parallel links*.

A *transmission group* (TG) may consist of one or more links between two nodes. A TG comprising two or more parallel links is called a *multilink TG*. Multilink TGs may be defined between two type 4 subarea nodes using SDLC links. In APPN multilink TGs are not defined, so the terms *TG* and *link* are used generally interchangeably in an APPN context. In both subarea networks and APPN networks, multiple (or parallel) TGs may connect two adjacent nodes. Data traffic is distributed dynamically over the links of a multilink TG.

Figure 18-U.27 shows parallel TGs connecting adjacent T4 nodes via TG2 and TG3, as well as a multilink TG connecting adjacent T4 nodes via TG1. The data traffic flowing between a pair of adjacent T4 nodes is distributed among parallel TGs. If one of the TGs fails, the session is broken; that is, session traffic is not automatically rerouted over the other TG. On the other hand, if at least one link in a multilink TG is still operational, session traffic is not disrupted over the TG in the case of link failure except that throughput may suffer. Similarly, a link may be restored to operation in a TG without disrupting existing sessions.

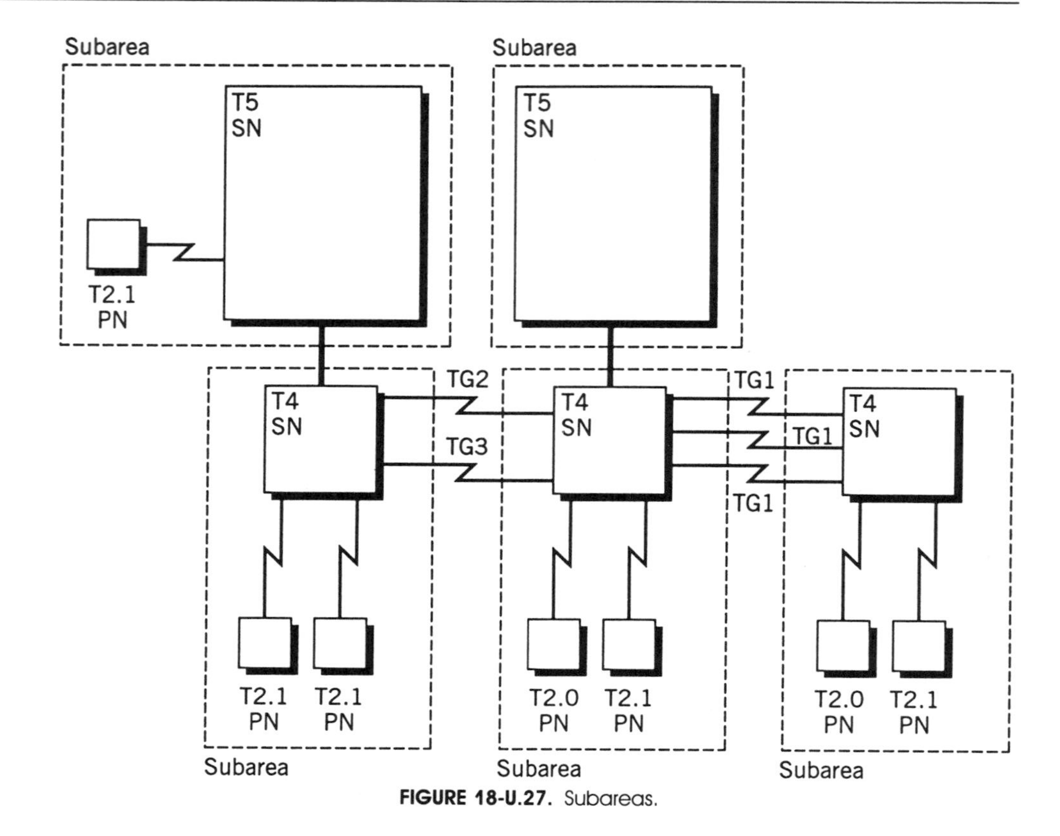

FIGURE 18-U.27. Subareas.

18-8.7. SNA Network Configurations

SNA defines the following network configurations:

- A hierarchical network consisting of subarea nodes and peripheral nodes
- A peer-oriented network consisting of APPN and LEN nodes
- A mixed network that combines one or more hierarchical subnets with one or more peer-oriented subnets

18-8.7.1. Hierarchical Network Configurations

The organization of a hierarchical network structure is determined by the way control of network services is maintained. Host nodes containing system services control points (SSCPs) are responsible for overall control of communication in the hierarchical network.

A hierarchical network might include LEN or APPN nodes that are attached as peripheral nodes. These nodes can communicate with each other through a subarea network if the boundary nodes to which they are attached support the basic SSCP-independent LU–LU protocols needed for such peer interactions.

Subareas. A *subarea* consists of one subarea node and the peripheral nodes that are attached to that subarea node. The concept of subarea applies only to subarea networks and composite networks. The network configuration in Figure 18-U.27 contains five subarea nodes and seven peripheral nodes. Because each subarea

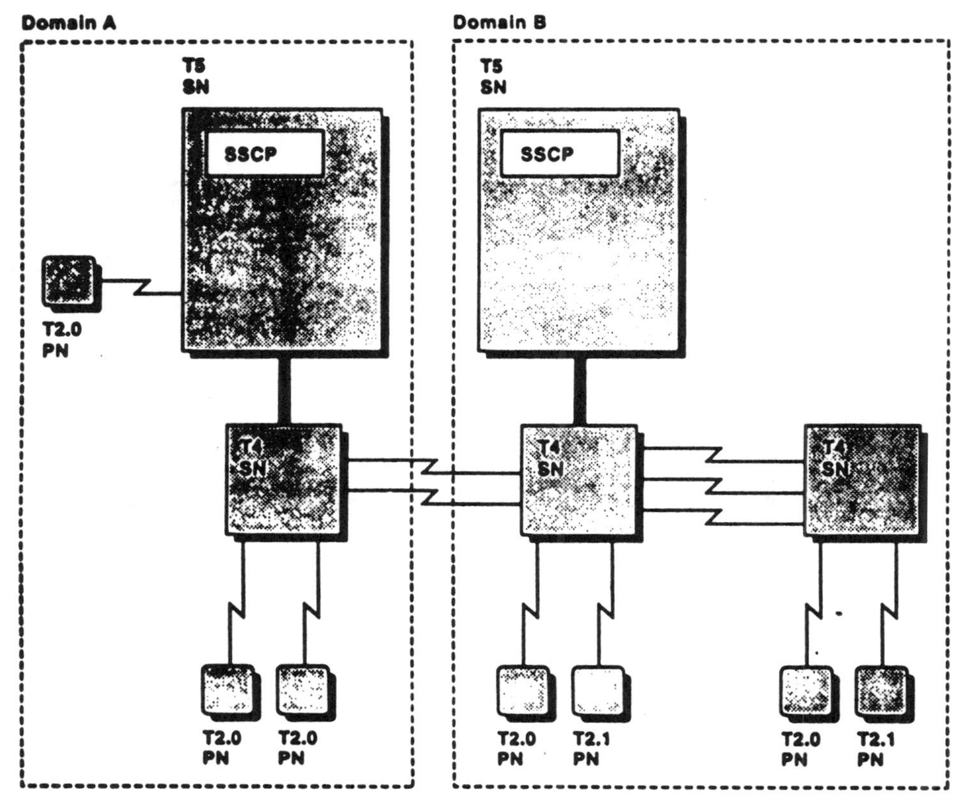

FIGURE 18-U.28. Domains in a subarea network.

node and its attached peripheral nodes constitute a subarea, this configuration contains five subareas.

Domains in a Subarea Network. A *domain* is an area of control. The concept of a domain within a subarea network differs from that within an APPN network. Within a subarea network, a domain is that portion of the network managed by the control point in a T5 subarea node. The control point in a T5 subarea node is called a *system services control point* (SSCP).

When a subarea network has only one T5 node, that node must manage all of the network resources. A subarea network that contains one T5 node is a *single-domain subarea network*. When there are multiple T5 nodes in the network, each T5 node may control a portion of the network resources. A subarea network that contains more than one T5 node is a *multiple-domain subarea network*. In a multiple-domain subarea network, the control of some resources can be shared between SSCPs. Some resources can be shared serially and some concurrently. Figure 18-U.28 illustrates two domains, A and B, joined by direct-attached T4 nodes to form a multiple-domain subarea network.

18-8.7.2. Peer-Oriented Network Configurations

An APPN network constitutes a peer-oriented SNA network. All APPN nodes are considered to be peers and do not rely on other nodes to control communication in the network the way a subarea node controls communication between peripheral nodes. There is, however, a measure of hierarchical control because a network

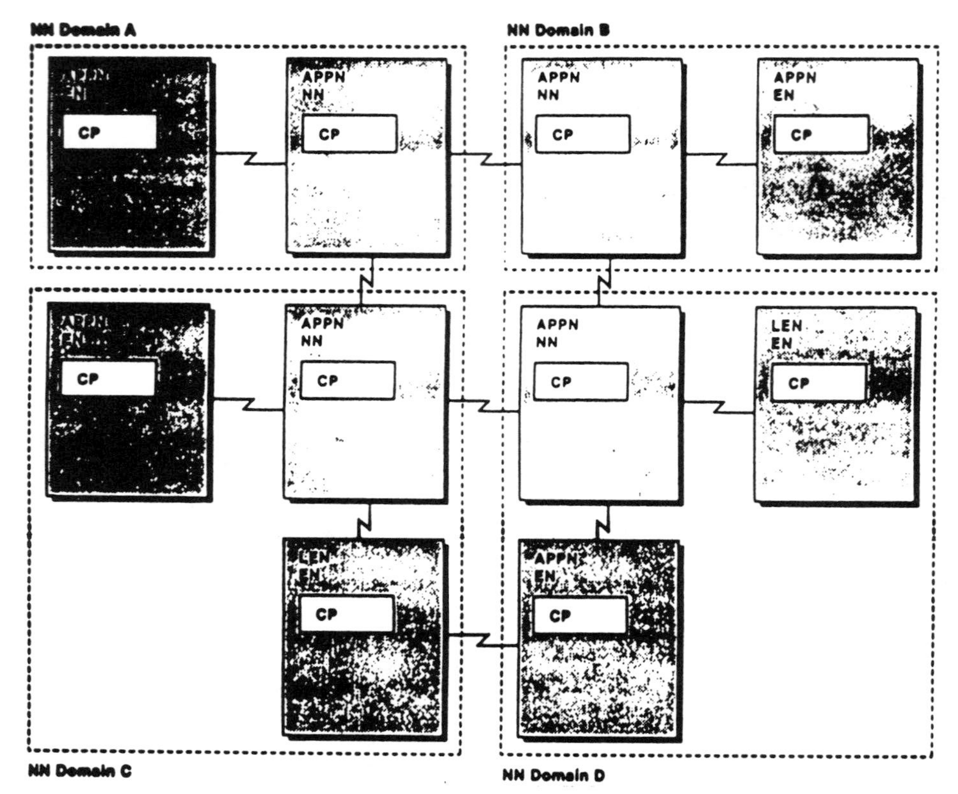

FIGURE 18-U.29. Network node domains in an APPN network.

node server provides certain network services to its attached end nodes. The difference is that, in APPN networks, hierarchical control is not determined by product or processor type because it is in subarea networks, where only large host processors contain SSCPs and node types generally reflect product types.

Domains in an APPN Network. The domain of a node in an APPN network is that portion of the network served by the control point in the node. The control point in an APPN node is called simply a *control point* (CP). An end node (EN) control point's domain consists solely of its local resources. It is included within the domain of its network node server. A network node control point's domain includes the resources in the network node and in any *client* end nodes (nodes for which the network node is acting as the network node server) attached directly to the network node.

APPN networks are by definition multiple-domain networks. Figure 18-U.29 shows an APPN network containing four network node domains: A, B, C, and D. The domains of the end nodes are included within the domains of their respective node servers.

18-8.8. The Transport Network and Network Accessible Units

Node and link components exhibit SNA layering through the functions they perform. The lower three architectural layers comprise the *transport network*, and certain components within the upper four layers are referred to as *network accessible units*.

18-8.8.1. The Transport Network

Node and link components within the lower three architectural layers (i.e., the physical control, data-link control, and path control layers) are collectively referred to as the *transport network*. The physical control layer includes both data communication equipment within nodes and the physical link connections between them. Node components within the data-link control layer activate and deactivate links on command from their control points, and they manage link-level data flow. Node components within the path control layer perform routing and congestion control. Together, the distributed components of these three lower layers transport data through the network on behalf of network accessible units.

18-8.8.2. Network Accessible Units

Certain components within the upper four architectural layers of a node use transport network services to establish temporary, logical connections with one another called *sessions*. Sessions can be established between two components residing in different nodes, or between components within the same node. Components that can establish sessions are referred to as *network accessible units* (NAUs). (*Note*: In earlier literature NAU was called a "network addressable unit.") NAUs in session with one another are referred to as *session partners*.

A set of physical connections consisting of the links and intermediate nodes between session partners constitutes a *route* for the session. During session initiation procedures, a route is selected for the session. One a session is initiated, the session traffic flows over the same route for the duration of the session.

There are three fundamental kinds of network accessible units: physical units, logical units, and control points. *Physical units* (PUs) perform local node functions such as activating and deactivating links to adjacent nodes. PUs exist only in nodes within subarea networks. (*Note*: In APPN networks these functions are performed by control points.) To perform its functions, a PU must exchange *control data* with its controlling system services control point (SSCP) over an *SSCP–PU session* initiated by the SSCP.

Logical units (LUs) provide network access for end users by helping the end users send and receive data over the network. Nodes in both subarea and APPN networks contain LUs. LUs send and receive control data and *end-user data* over the *LU–LU sessions* established between them.

In a subarea network an LU residing in a peripheral node is classified as either SSCP-dependent or SSCP-independent, depending on the protocols it uses for LU–LU session initiation. An *SSCP-dependent LU*, or simply *dependent LU*, sends a session-initiation request to its controlling SSCP over an *SSCP–LU session*. The LU is dependent on the SSCP to mediate the session initiation with the partner LU and requires an SSCP–LU session for that mediation. The session is activated when the LU receives a session-activation request from the partner LU. Dependent LUs reside in both T2.0 and T2.1 nodes.

An *SSCP-independent LU*, or simply *independent LU*, sends a session-activation request directly to the partner LU. It does not interact with an SSCP to mediate an LU–LU session and does not require an SSCP–LU session. An independent LU may reside in a T2.1 node, but not in a T2.0 node. Independent LU protocols are also referred to as *peer-session protocols*.

In APPN networks, independent LUs send session-activation requests directly to their partners, and all nodes support peer-session protocols. Two LEN nodes directly attached to one another also support peer-session protocols.

Control points (CPs) provide network control functions that include managing the resources in their domains and monitoring and reporting on the status of those resources. The functions of node CPs in subarea networks differ greatly from those in APPN networks. In subarea networks, SSCPs control the PUs and dependent LUs in their domains by exchanging control data with them over SSCP–PU and SSCP–LU sessions, respectively. They may also initiate *SSCP–SSCP sessions* for the control of cross-domain sessions. In APPN networks, a CP does not initiate sessions with LUs in its domain. It does, however, initiate sessions with adjacent CPs, called *CP–CP sessions*, in order to exchange control data for its routing and directory services.

18-8.8.3. Interconnecting Session Stages

Messages transmitted on a session carry addresses in their headers. The addresses enable *intermediate routing nodes* (the nodes situated along a session route between two partner nodes) to correlate a message with its appropriate session and to route the message accordingly.

As session data traverse a route, not only the endpoints (called *half-sessions*) but also certain components along the route called *session connectors* provide flow control and address translation functions for the session. The parts of a session that are delimited by half-sessions and/or session connectors are called *session stages*. The three points in an SNA network where session connection occurs are at the boundary function, the gateway function, and the intermediate session routing function:

1. The *boundary function* resides in a T4 or T5 node acting as a boundary node. It translates between the network addresses used by a subarea node and the local addresses used by a peripheral node.
2. The *gateway function* resides in a T4 node acting as a gateway node. It translates between the network addresses used by one network and the network addresses used by another.
3. The *intermediate session routing function* resides in an APPN network node. It translates between the session address used by one session stage and the session address used by another.

18-8.9. SNA Logical Units

All node types can contain logical units (LUs). (However, T4 nodes typically do not contain LUs except for protocol conversion for non-SNA terminals.) The LU supports sessions with control points in type 5 nodes and with LUs in other nodes. A type 6.2 LU using SSCP-independent protocols, however, does not engage in sessions with an SSCP.

End users access SNA networks through logical units. A logical unit manages the exchange of data between end users, acting as an intermediary between the end user and the network. A one-to-one relationship is not required between end users and LUs. The number of end users that can access a network through the same LU is an implementation design option.

Before end users can communicate with one another, their respective LUs must be connected in a session. In some cases multiple, concurrent sessions between the same two LUs are possible. When such sessions are activated, they are called *parallel sessions*.

SNA defines different kinds of LUs as *LU types*. LU types identify sets of SNA functions that support end-user communication. LU–LU sessions can exist only between LUs of the same LU type. For example, an LU type 2 can only communicate with another LU type 2; it cannot communicate with an LU type 3.

The LU types that SNA currently defines, the kind of configuration or application that each type represents, and the hardware or software products that typically use each type of logical unit are listed below.

LU Type 1. LU type 1 is for application programs and single- or multiple-device data processing workstations communicating in an interactive, batch data transfer or distributed data processing environment. The data streams used in LU type 1 conform to the SNA character stream or document content architecture (DCA). An example of the use of LU type 1 is an application program running under IMS/VS and communicating with an IBM 8100 information system at which the workstation operator is correcting a database that the application program maintains.

LU Type 2. LU type 2 is for application programs and display workstations communicating in an interactive environment using the SNA 3270 data stream. Type 2 LUs also use the SNA 3270 data stream for file transfer. An example of the use of LU type 2 is an application program running under IMS/VS and communicating with an IBM 3179 display station at which the 3179 operator is creating and sending data to the application program.

LU Type 3. LU type 3 is for application programs and printers using the SNA 3270 data stream. An example of the use of the LU type 3 is an application program running under CICS/VS and sending data to an IBM 3262 printer attached to an IBM 3174 establishment controller.

LU Type 4. LU type 4 is for the following:

1. Application programs and single- or multiple-device data processing or word processing workstations communicating in interactive, batch data transfer or distributed data processing environments. An example of this use of LU type 4 is an application program running under CICS/VS and communicating with an IBM 6670 information distributor.
2. Peripheral nodes that communicate with each other. An example of this use of LU type 4 is two 6670s communicating with each other.

The data streams used in LU type 4 are the SNA character string (SCS) for data processing environments and office information interchange (OII) level 2 for word processing environments.

LU Type 6.1. LU type 6.1 is for application subsystems communicating in a distributed data processing environment. An example of the use of LU type 6.1 is an application program running under CICS/VS and communicating with an application program running under IMS/VS.

LU Type 6.2. LU type 6.2 is for transaction programs communicating in a distributed data processing environment. The LU type 6.2 supports multiple concurrent sessions. The LU 6.2 data stream is either an SNA general data stream (GDS), which is a structured-field data stream, or a user-defined data stream. LU 6.2 can be used for communication between two type 5 nodes, a type 5 node and a type 2.1 node, or two type 2.1 nodes. Examples of the use of LU type 6.2 are:

1. An application program running under CICS/VS communicating with another application program running under CICS/VS.
2. An application program in an AS/400 communicating with a Personal System/2 (PS/2).

Notes on Acronyms

CICS: Customer Information Control System. A general-purpose System/370 mainframe-based transaction management system.

IMS: Information Management System. An IBM 370/390 host-based database/data communications subsystem that runs in the MVS environment.

MVS: Multiple Virtual Storage. An IBM primary operating system for the 370/390 IBM mainframes.

VS: Virtual Storage

18-8.10. Some SNA Data Formats

SNA defines the following message-unit formats that NAUs, path control elements, and data-link control elements use:

- Network accessible units use basic information units (BIUs)
- Path control elements use path information units (PIUs)
- Data-link control elements use basic link units (BLUs)

18-8.10.1. Basic Information Unit

NAUs use *basic information units* (BIUs) to exchange requests and responses with other NAUs. Figure 18-U.30 shows the format of a BIU.

Basic information units that carry requests contain a request header and a request unit. Basic information units that carry responses consist of (1) both a response header and a response unit or (2) only a response header.

Request Header. Each request that an NAU sends begins with a *request header* (RH). A request header is a 3-byte field that identifies the type of data in the associated request unit. The request header also provides information about the format of the data and specifies protocols for the session. Only NAUs use request header information.

Response Header. Each response that an NAU sends includes a *response header* (RH). Like a request header, a response header is a 3-byte field that identifies the type of data in the associated response unit. A bit called *request/response indicator* (RRI) distinguishes a response header from a request header.

Response Unit. A *response unit* (RU) contains information about the request. Positive responses to command requests generally contain a 1- to 3-byte response unit that identifies the command request. Positive responses to data requests contain response headers, but no response unit. Negative response units are 4–7 bytes long and are always returned with a negative response. Response units are identified as response RUs.

Request Header or Response Header	Request Unit or Response Unit

FIGURE 18-U.30. Basic information unit (BIU) format.

The receiving NAU returns a negative response to the request sender if:

- The sender violates an SNA protocol.
- The receiver does not understand the transmission.
- An unusual condition, such as a path outage, occurs.

The receiving NAU returns a 4- to 7-byte negative response unit to the request sender. The first 4 bytes of the response unit contain sense data explaining why the request is unacceptable. The receiving NAU sends up to three additional bytes that identify the rejected request.

18-8.10.2. Path Information Unit

The message-unit format used by path control elements is a *path information unit* (PIU). Path control elements form a PIU by adding a transmission header to a basic information unit. Figure 18-U.31 shows the format of a PIU. Path control uses the *transmission header* (TH) to route message units through the network. The transmission header contains information for the transport network.

SNA defines different header formats and identifies the different formats by a *format identification* (FID) type. Transmission headers vary in length according to their FID type. Path control uses different FID types to route data between different types of nodes.

FID 0. Path control uses this format to route data between adjacent subarea nodes for non-SNA devices. Few networks still use the FID 0; now a bit is set in the FID 4 transmission header to indicate whether the device is an SNA device or a non-SNA device.

FID 1. Path control uses this format to route data between adjacent subarea nodes if one or both of the subarea nodes do not support explicit and virtual route protocols.

FID 2. Path control uses this format to route data between a subarea boundary node and an adjacent peripheral node, or between adjacent APPN or LEN nodes.

FID 4. Path control uses this format to route data between subarea nodes if the subarea nodes support explicit and virtual route protocols.

FID 5. Path control uses this format to route data over an RTP (rapid-transport protocol) connection. The FID 5 is very similar to the FID 2.

18-8.10.3. Basic Link Unit

Data-link control uses a message-unit format called a *basic link unit* (BLU) to transmit data across a link. Data-link control forms a BLU by adding a *link header* (LH) and a *link trailer* (LT) to a PIU. Link headers and link trailers contain link control information that manages the transmission of a message unit across a link.

Transmission Header	Request Header or Response Header	Request Unit or Response Unit

FIGURE 18-U.31. Path information unit (PIU) format.

Link Header	Transmission Header	Request Header or Response Header	Request Unit or Response Unit	Link Trailer

FIGURE 18-U.32. Basic link unit (BLU) format.

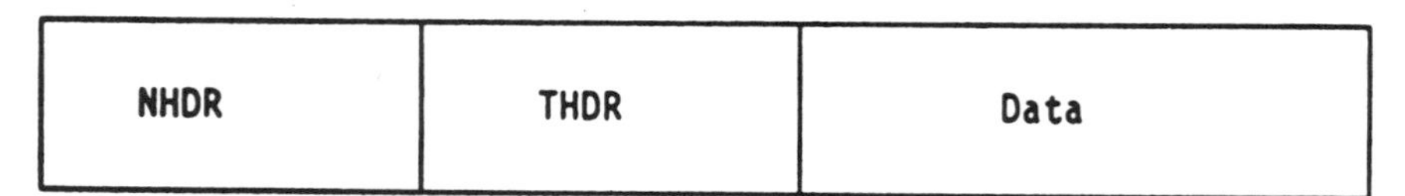

FIGURE 18-U.33. Network layer packet (NLP) format.

Only data-link control elements use link header and link trailer information. (See Chapter 6 for a brief discussion of SDLC.) Figure 18-U.32 shows the format of a BLU.

18-8.10.4. Network Layer Packet (NLP)

A *network layer packet* (NLP) is used to transport data through an HPR network. It is composed of the network layer header (NHDR), transport header (THDR), and the data. The length of a packet must not exceed the maximum packet size of any link over which the packet will flow. The maximum packet size of links is obtained during the route setup protocol.

All of the fields in the NLP are of variable lengths. Figure 18.U.33 illustrates the format of an NLP.

The following paragraphs describe how the message formats shown in Figure 18-U.34 are used.

Node D

1. End-user D gives end-user data to LU *d*.
2. LU *d* creates a request unit and affixes a request header to it; this forms a basic information unit (BIU). (Multiple BIUs may be needed to accommodate a large message.)
3. LU *d* gives the BIU to a path control element.
4. Path control affixes a transmission header to the BIU; this forms a path information unit (PIU).
5. Path control gives the PIU to a data-link control element.
6. Data-link control affixes both a link header and a link trailer to the PIU; this forms a basic link unit (BLU).
7. Data-link control transmits the BLU over link *d* to node C.

Node C

1. The data-link control element that receives the BLU removes the link header and link trailer and then gives the PIU to a peripheral path control element.
2. This path control element routes the RH and RU (BIU) to the boundary function (BF). The BF then passes the BIU to a subarea path control

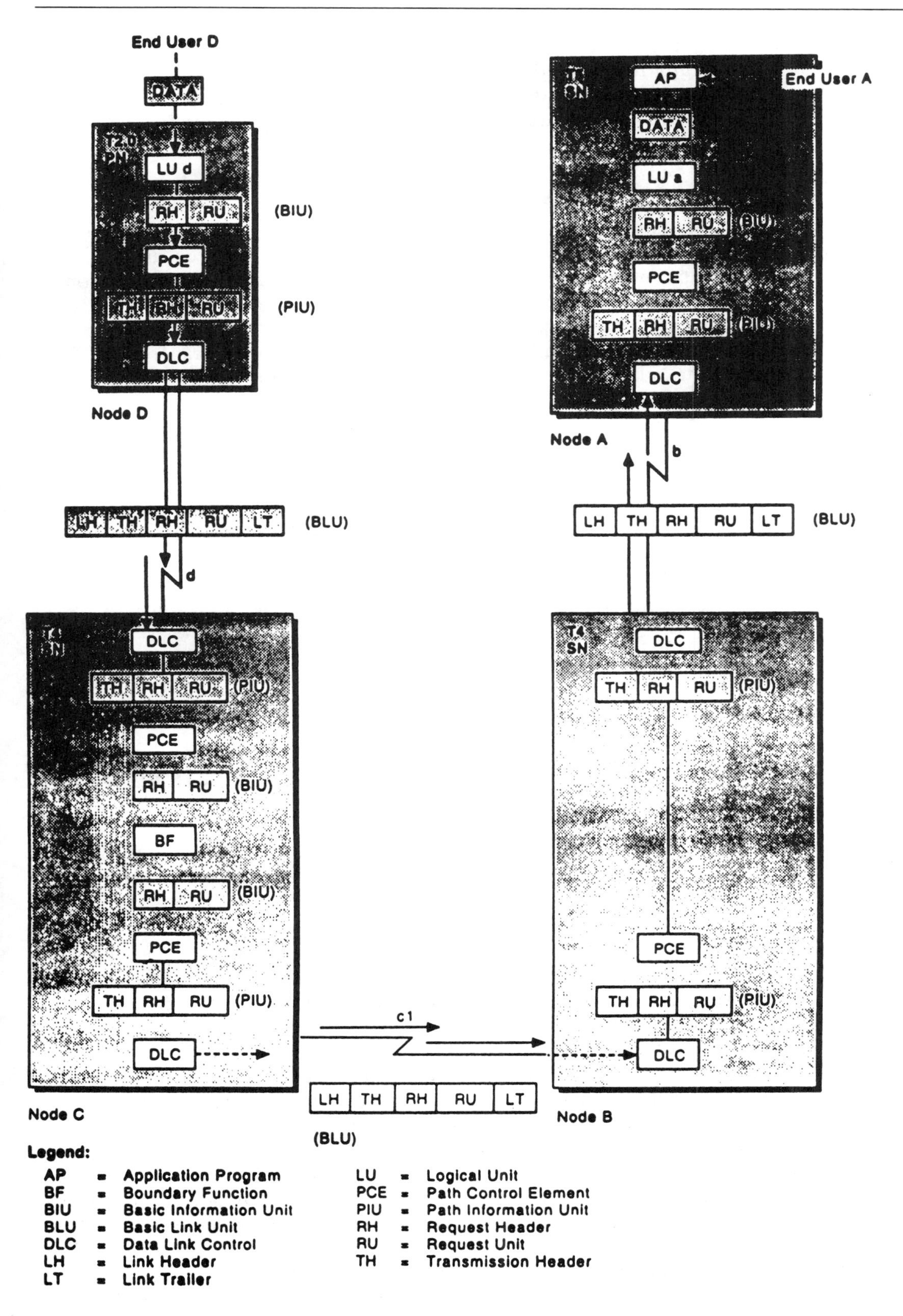

FIGURE 18-U.34. Use of SNA data formats.

element, which determines where to route the message unit text. Path control then gives the PIU to another data-link control element.
3. Data-link control affixes both a link header and a link trailer to the PIU.
4. Data-link control transmits the BLU over link *c1* to node B.

Node B

1. The data-link control element that receives the BLU removes the link header and link trailer and then gives the PIU to path control.
2. Path control uses the information in the transmission header to determine where to route the message unit next. Path control then gives the PIU to another data-link control element.
3. Data-link control affixes both a link header and a link trailer to the PIU.
4. Data-link control transmits the BLU over link *b* to node A.

Node A

1. The data-link control element that receives the BLU removes the link header and link trailer and then gives the PIU to path control.
2. Path control uses information in the transmission header to determine that node A is the destination subarea node. Path control removes the transmission header and then delivers the BIU to the destination NAU.
3. LU *a* removes the request header and gives the RU to end-user A.

The material and figures in this section have been extracted from *Systems Network Architecture Technical Overview*, 5th ed., IBM Corp., Research Triangle Park, NC, 1994.

18-9. INTERNATIONAL NUMBERING PLAN FOR PUBLIC DATA NETWORKS

18-9.1. Design Considerations

The international data number is to determine only the specific DTE/DCE interface and, in particular, to identify a country and a network, if several networks exist in the same country.

Where a number of public data networks are to be established in a country, it should not be mandatory to integrate the numbering plans of the various networks.

The number of digits comprising the code used to identify a country and a specific public data network in that country should be the same for all countries.

A national data number assigned to a DTE/DCE interface should be unique within a particular national network. This national number should form part of the international data number which also should be unique on a worldwide basis.

The numbering plan should make provision for the internetworking of data terminals on public data networks with data terminals on public telephone and telex networks and on integrated services digital networks (ISDNs).

The numbering plan should not preclude the possibility of a single national network providing an integrated telecommunications system for services of all kinds.

Where multiple RPOA (recognized private operating agency) facilities exist providing service to or within the same country, provision for the selection of a specific RPOA facility should be allowed for in the *facility request* part of the *selection* signals.

18-9.2. Characteristics and Application of the Numbering Plan

18-9.2.1. Number System

2.1.1. The 10-digit numeric character set 0–9 should be used for numbers (or addresses) assigned to DTE/DCE interfaces on public data networks. This principle should supply to both national and international data numbers.

2.1.2. Use of the above number system will make it possible for data terminals on public data networks to interwork with data terminals on public telephone and telex networks and on integrated services digital networks (ISDNs).

18-9.2.2. Data Network Identification Codes and Data Country Codes

2.2.1. A data network identification code (DNIC) could be assigned as follows:

2.2.1.1. To each public data network (PDN) within a country.

2.2.1.2. To nonzoned service, such as the public mobile satellite system (see paragraph 2.2.10).

2.2.1.3. To a public-switched telephone network (PSTN) or to an ISDN for the purpose of making calls from DTEs connected to a PDN to DTEs connected to that PSTN or ISDN.

Note: In order to facilitate the interworking of telex networks with data networks, some countries have allocated DNIC to telex networks.

2.2.1.4. To a group of PDNs within a country, when permitted by national regulations.

2.2.1.5. To a group of private data networks connected to PDNs within a country, where permitted by national regulations.

Note: For administrative purposes, including charging, a group of networks which have been assigned a single DNIC will, in the international context, be considered as a single entity.

2.2.2. In the system of data network identification codes, the first digit of such codes should be in accordance with Table 18-U.7.

Note 1: The allocation of codes for nonzoned services, other than the mobile satellite systems, is for further study.

Note 2: Digits 8, 9, and 0 are used as escape codes, not being part of the DNIC. They are defined in Section 2.6.

2.2.3. All DNICs should consist of four digits. The first three digits should always identify a country and could be regarded as a data country code (DCC). The fourth, or network, digit should identify a specific data network in the country.

TABLE 18-U.7

First Digit of Data Network Identification Code

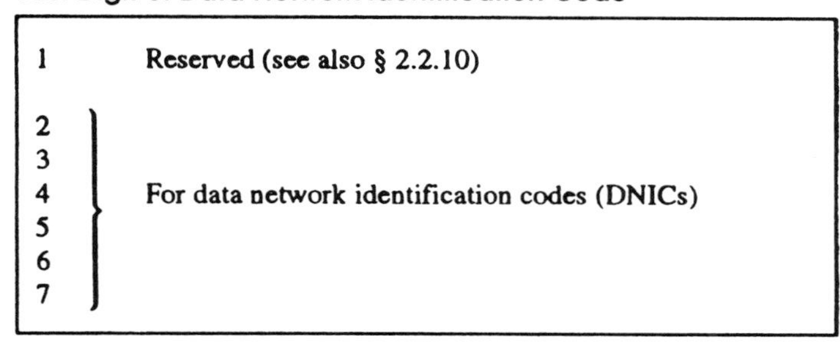

1	Reserved (see also § 2.2.10)
2 3 4 5 6 7	For data network identification codes (DNICs)

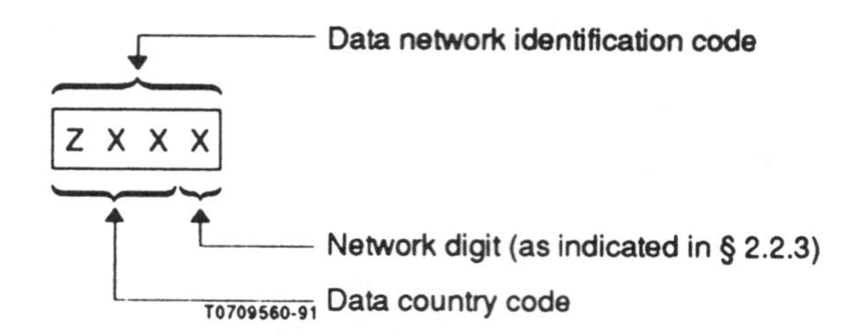

X denotes any digit from 0 through 9
Z denotes any digit from 2 through 7 as indicated in § 2.2.2

FIGURE 18-U.35. Format for data network identification codes (DNICs). (From Figure 1/X.121, ITU-T Rec. X.121, Geneva, 9/92, page 3.)

2.2.4. Each country should be assigned at least one 3-digit DCC. The DCC in conjunction with the fourth digit can identify up to 10 public data networks. The format for DNICs should be as indicated in Figure 18-U.35.

2.2.5. The system of DNICs indicated in paragraphs 2.2.2 and 2.2.4 will provide for 600 DCCs and a theoretical maximum of 6000 DNICs.

2.2.6. In the case where a country requires more than 10 DNICs, additional DCCs could be assigned to the country (see paragraph 2.2.8).

2.2.7. A list of DCCs to be used in the development of DNICs is given in Annex D to this Recommendation. This list was prepared in accordance with the requirement that the first digit of a DNIC, which is also the first digit of the embedded DCC, should be restricted to the digits 2–7 inclusive (see paragraph 2.2.2). As first digits of DCCs, the digits 2–7 are arranged to represent world zones.

2.2.8. The assignment of DCCs is to be administered by the CCITT. The assignment of network digits will be made nationally and the CCITT Secretariat notified.

The Member countries of the International Telecommunication Union not mentioned in this list who wish to take part in the international data service or those Members who require an additional or several DCCs should ask the Director of the CCITT for the assignment of available 3-digit DCCs. In their request, they may indicate the available 3-digit code(s) preferred.

Assignments by the Director of the CCITT of DCCs as well as assignments by countries of the network digits will be published in the Operational Bulletin of the International Telecommunication Union.

2.2.9. Examples indicating how DNICs could be developed are given in Annex A of ITU-T Rec. X.121.

2.2.10. *International data number for stations in the public mobile satellite systems*. The DNICs allocated to public mobile satellite systems are 111S, where the digit S indicates the ocean area. The digit S has the values as shown in Annex C.

The mobile station is identified by a unique mobile earth station number (INMARSAT mobile number) common for telephony, telex, data transmission, and other services as defined in Rec. E.215/F.125. The first digit of the mobile earth station number (INMARSAT mobile number) is the digit "T" defined in Rec. E.215/F.125 and is used for discrimination between different public mobile satellite systems (such as the INMARSAT Standard A, B, and C and aeronautical systems).

The complete international data number for mobile earth stations is composed as follows:

111S + mobile earth station number + X

where X is an optional digit which, if present, designates a particular DTE associated with the mobile earth station.

Note 1: In the INMARSAT mobile satellite systems, the use of the S digit for indicating the ocean area in which the mobile earth station is located at the time of the call is considered a temporary arrangement. It is recognized that such an arrangement should be avoided in the future, if possible, since it requires the calling user to know the exact area of a destination mobile earth station at the time of the call, and such an area may change from time to time for the mobile earth station.

Note 2: Digit "X" requires further studies regarding aeronautical and land mobile earth stations.

18-9.2.3. International Data Number

2.3.1. A data terminal on a public data network when called from another country should be addressed by the international data number assigned to its DTE/DCE interface. The international data number should consist of the DNIC of the called public data network, followed by the network terminal number (NTN) of the called DTE/DCE interface, or, for example, where an integrated numbering scheme exists within a country, the DCC followed by the national number (NN) of the called DTE/DCE interface—that is,

$$\text{International data number} = \text{DNIC} + \text{NTN, or, DCC} + \text{NN}$$

2.3.2. The NTN should consist of the full address that is used when calling the data terminal from within its serving public data network. The NN should consist of the full address used when calling the data terminal from another terminal within the national integrated numbering scheme. These numbers should consist of all the digits necessary to uniquely identify the corresponding DTE/DCE interface within the serving network and should not include any prefix (or access code) that might be employed for such calling.

Note 1: NTNs or NNs may be assigned by a PDN to DTEs connected to other public networks, when interworking capabilities are provided with that PDN.

Note 2: An example of the development of NTNs where a DNIC is assigned to a group of public or private data networks connected tp PDNs within a country is shown in Annex B of ITU-T Rec. X.121.

18-9.2.4. Number of Digits

2.4.1. International data numbers could be of different lengths but should consist of at least 5 digits but not more than 14 digits.

With the DNIC fixed at 4 digits and the DCC fixed at 3 digits, it would therefore be possible to have a network terminal number (NTN) of 10 digits maximum or to have an NN of 11 digits maximum.

Note 1: The limit of 14 digits specified above applies exclusively to the international data number information. Adequate register capacity should be made available at data switching exchanges to accommodate the above digits as well as any additional digits that might be introduced for signaling or other purposes.

Note 2: After time "T" (see Rec. E.165) the maximum number of digits of the international ISDN number will be 15. The need of extending the maximum capacity of the X.121 data number is for further study.

18-9.2.5. Prefixes

2.5.1. A prefix is an indicator consisting of one or more digits, allowing the selection of different types of address formats. Prefixes are not part of the international X.121 format and are not signaled over internetwork or international boundaries.

2.5.2. To distinguish between different address formats within a public data network (e.g., national data number and international data number formats), a prefix would generally be required. Any such prefix does not form a part of the data number. Pending further study, the use and composition of such a prefix is a national matter. However, the possible need to accommodate such a prefix with regard to digit register capacity should be noted. It is also a national matter to decide on evaluation of prefixes, escape code, and parts of the international data number of incoming path of entry for routing or other purposes.

Note: In the case of X.25 access, a prefix indicating international data number format can only be one digit.

18-9.2.6. Escape Codes

An escape code is an indicator consisting of one digit. It indicates that the following digits are a number from a different numbering plan.

An escape code when required has to be carried forward through the originating network and can be carried across internetwork and international boundaries.

Digits used for escape codes are the digits 8, 9, and 0. The allocation and their purpose are shown in Table 18-U.8. The escape codes are not part of the international data number but are part of the "international X.121 format" (see Figure 18-U.36).

18-9.2.7. Number Analysis: International Calls Between Public Networks

2.7.1. In the case of international calls between public data networks, provision should be made in originating countries to interpret the first three digits of the international data number. These digits constitute the DCC component of the

TABLE 18-U.8
Allocation of Escape Codes

8	Indicates that the digits which follow are from the F.69 numbering plan
9	Indicates that the digits which follow are from the E.164 numbering plan (Notes 2, 3, and 4)
0	Indicates that the digits which follow are from the E.164 numbering plan (Notes 1, 3, 4)

Note 1: In this case, 0 is to indicate that a digital interface between the PDN and the destination network (ISDN or integrated ISDN/PSTN) is requested.

Note 2: In this case, 9 is to indicate that an analog interface on the destination network (PSTN or integrated ISDN/PSTN) is requested.

Note 3: In the case of calls from a packet-switched public data network (PSPDN) to an integrated ISDN/PSTN which does not require a distinction between digital and analog interfaces, only a single escape code (e.g., 9 or 0) may be required. However, all PSPDNs interworking with ISDNs, PSTNs, and integrated ISDN/PSTNs should also support both 9 and 0 escape codes when acting as an originating, transit or destination network.

Note 4: The E.163 numbering plan is fully incorporated into Rec. E.164.

Note 5: Escape codes may be replaced by signaling means after time "T" (for the definition of time "T," see Rec. E.165).

Source: Table 2/X.121, ITU-T Rec. X.121, Geneva, 9/92, page 6.

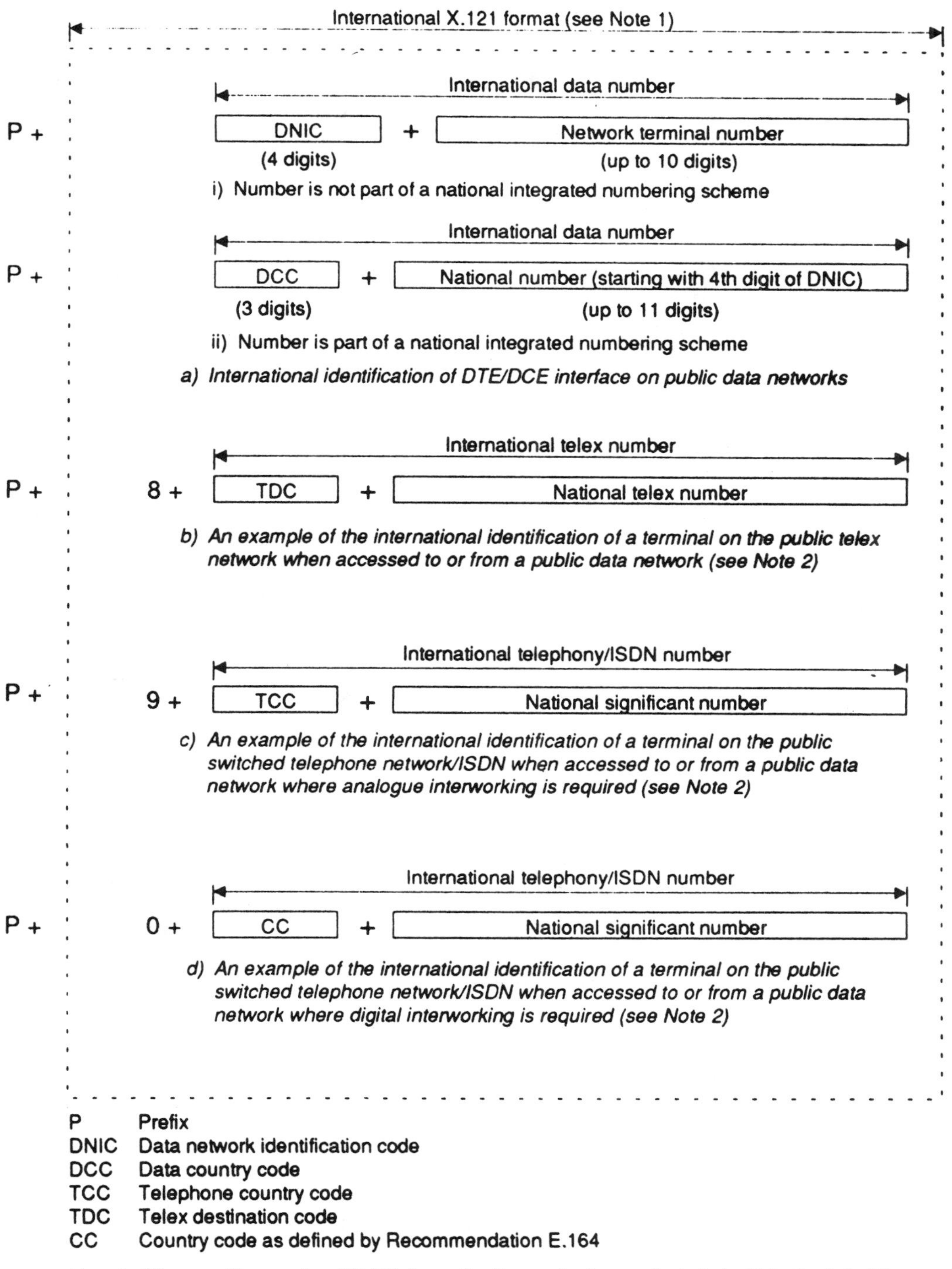

P Prefix
DNIC Data network identification code
DCC Data country code
TCC Telephone country code
TDC Telex destination code
CC Country code as defined by Recommendation E.164

Note 1 – The term "International X.121 format" refers to the formats included within the dotted lines and excludes prefixes.

Note 2 – This illustrates the case where the data terminal on the public telephone or telex networks or on the ISDN is identified by the telephony/ISDN or telex number. Other cases are possible. The various interworking scenarios are described in separate Recommendations. It should also be noted that in the case of calls from a PSPDN to an integrated ISDN/PSTN which does not require a distinction between digital and analogue interfaces, only a single escape code (e.g. 9 or 0) may be required. However all PSPDNs interworking with ISDNs, PSTNs and integrated ISDN/PSTNs should also support both 9 and 0 escape codes when acting as an originating, transit or destination network.

FIGURE 18-U.36. International X.121 format. (From Figure 2/X.121, ITU-T Rec. X.121, Geneva, 9/92, page 7.)

DNIC and identify the terminal country. This information is required in the originating country for routing purposes.

2.7.2. In originating countries, it might also be necessary to interpret the fourth, or network, digit of a DNIC and, if the originating network requires it, the first digit after the DNIC. Such interpretation would provide the identity of a specific network in a country where several public data networks are in service. This information might be required for billing purposes or for the selection of specific routes to called networks. An example of the requirement for interpretation of the fifth digit is the use of this digit in the mobile satellite systems for selection of a particular mobile system (digit "T," see paragraph 2.2.10).

Note 1: With regard to number analysis and routing in the case of interworking with PSTN and ISDN, see Recs. X.110 and X.122.

Note 2: With regard to RPOA selection, see Section 18-9.1.7.

2.7.3. Countries receiving international calls for public data networks should receive the complete international data number. However, where a country of destination indicates that it does not wish to receive the DCC component of the DNIC, arrangements should be made to suppress the DCC.

2.7.4. For destination countries with more than 10 public data networks, interpretation of the first three digits of the DNIC (i.e., the DCC) would identify the group of networks within which the called network is included. Interpretation of the fourth, or network, digit of the DNIC would identify the called network in that group. Interpretation of the first three digits would also make it possible to verify that an incoming call has in fact reached the correct country.

2.7.5. In the case of destination countries where there are fewer than 10 public data networks, the first three digits of the DNIC could provide the verification indicated in paragraph 2.7.4. Interpretation of the fourth, or network, digit of the DNIC would identify the specific network being called.

2.7.6. In transit countries the complete international data number must be received. Interpretation of the first three digits would identify the called country. Interpretation of the fourth or network digit would identify a specific data network in the called country. Interpretation of the fourth digit might be required for billing purposes or for route selection beyond the transit country. It might also be necessary in the transit network to analyze the fifth digit to allow selection of a particular public mobile system (e.g., digit "T"; see paragraph 2.2.10).

2.7.7. Where a data call is to be routed beyond a transit country through a second transit country, the complete international data number should always be sent to the second transit country. Where the data call is to be routed by a transit country to the country of destination, the arrangements indicated in paragraph 2.7.3 should apply.

18-9.2.8. Numbering Plan Interworking

Details on numbering plan interworking are outlined in Rec. X.122/E.166 (see also Recs. E.165, X.301, and I.330).

Transit cases are considered in these Recommendations. For routing aspects see also Rec. X.110.

18-9.2.9. Directories and Letterheads

2.9.1. Directories for public data networks should include information on the procedures to be followed for making international data calls. A diagram, such as that of Figure 18-U.36, could assist the customer in these procedures.

2.9.2. With regard to the prefix shown in Figure 18-U.36, it should be noted that the same prefix (designated P) could be used for all four types of calls. The choice of prefix is, however, a national matter.

2.9.3. With regard to RPOA selection (see Section 18-9.1.7), it should be noted that an RPOA facility request designator would be used either in international data calls or within certain countries. Provision of this facility as well as the designation of the RPOA facility selection designator is a national matter in the originating country.

2.9.4. With regard to the publication of international data numbers on letterheads or other written material, it is recommended that the NTN or NN should be easily distinguished within the international number—that is, that there be a space between the 4-digit DNIC and the NTN or between the 3-digit DCC and the NN, where the fourth digit of the DNIC is included in the NN.

18-9.3. Development of Data Network Identification Codes

Example 1. In this example, it is assumed (for illustrative purposes only) that the Netherlands has established its first public data network. To develop the DNIC for this network, it would be necessary for the Netherlands to assign to it a network digit to follow the listed DCC 204 (see Annex D). Assuming that the Netherlands selected the digit 0 as the network digit, the DNIC for this initial network would be 2040.

Example 2. In this example, it is assumed (for illustrative purposes only) that five public data networks have been established in Canada. To develop the DNIC for these networks, it would be necessary for Canada to assign to each of these networks a network digit to follow the listed DCC 302 (see Annex D). Assuming that Canada assigned the network digits 0–4 to the five networks, the resulting DNICs would be 3020, 3021, 3022, 3023, and 3024.

Example 3. In this example, it is assumed (for illustrative purposes only) that eight public data networks have been established in the United States of America. It is also assumed that network digits 0–7 would be assigned by the United States of America to follow the listed DCC 310 (see Annex D). The DNICs thus formed for these eight networks would be 3100, 3101, 3102, 3103, 3104, 3105, 3106, and 3107.

If, some time later, four additional public data networks were to be introduced in the United States of America, two of the four new networks could be assigned network digits 8 and 9 in association with DCC 310, to produce the DNICs 3108 and 3109.

For the remaining two public data networks, the United States of America would have to ask the CCITT for an additional DCC. A request for a code next in sequence (i.e., 311) could be made if this code appeared to be spare. If code 311 could be made available, it would be assigned to the United States of America. If it was not available, a spare code in the "300" series of DCCs would be assigned. Assuming DCC 311 was available and issued to the United States of America, the two remaining public data networks could be assigned network digits 0 and 1 in association with DCC 311, to produce the DNICs 3110 and 3111.

The DNICs for the 12 public data networks would then be 3100, 3101, 3102, 3103, 3104, 3105, 3106, 3107, 3108, 3109, 3110, and 3111.

Example 4. In this example, it is assumed (for illustrative purposes only) that a public data network is to be established in each of two Caribbean islands that are part of the group of islands known as the French Antilles. The islands concerned are Guadeloupe and Martinique.

To develop the DNICs for these public data networks, it is assumed that the French Administration would assign network digit 0 to the network in Guadeloupe and network digit 1 to the network in Martinique and associate these network digits with the listed DCC 340 for the French Antilles (see Annex D). The DNICs thus formed would be 3400 for Guadeloupe and 3401 for Martinique.

This example indicates that the system of DNICs is appropriate for application to groups of islands or regions of a country because DCC could provide for up to 10 public data networks dispersed over several islands or regions. At the same time such island or regional networks would be distinguishable from each other.

18-9.4. List of DNICs for Nonzoned Systems: Public Mobile Satellite Systems

Code	Area	System
1110	Spare	
1111	Atlantic Ocean	INMARSAT mobile satellite data transmission system
1112	Pacific Ocean	INMARSAT mobile satellite data transmission system
1113	Indian Ocean	INMARSAT mobile satellite data transmission system
1114	Atlantic Ocean—West	INMARSAT mobile satellite data transmission system
1115	Spare	
1116	Spare	
1117	Spare	
1118	Spare	
1119	Spare	

18-9.5. List of Data Country or Geographical Area Codes

Zone 2

Code	Country or Geographical Area
202	Greece
204	Netherlands (Kingdom of the)
205	Netherlands (Kingdom of the)
206	Belgium
208	France
209	France
212	Monaco
214	Spain
216	Hungaria (Republic of)
220	Yugoslavia (Federal Republic of)
222	Italy
225	Vatican City State
226	Romania (Socialist Republic of)
228	Switzerland (Confederation of)
230	Czech and Slovak Federal Republic
232	Austria
234	United Kingdom of Great Britain and Northern Ireland
235	United Kingdom of Great Britain and Northern Ireland
236	United Kingdom of Great Britain and Northern Ireland
237	United Kingdom of Great Britain and Northern Ireland
238	Denmark

Zone 2 (*Continued*)

Code	Country or Geographical Area
240	Sweden
242	Norway
244	Finland
246	Lithuania (Republic of)
247	Latvia (Republic of)
248	Estonia (Republic of)
250	Russian Federation
251	Russian Federation
255	Ukraine
257	Belarus (Republic of)
259	Moldova Republic of)
260	Poland (Republic of)
262	Germany (Federal Republic of)
263	Germany (Federal Republic of)
264	Germany (Federal Republic of)
265	Germany (Federal Republic of)
266	Gibraltar
268	Portugal
270	Luxembourg
272	Ireland
274	Iceland
276	Albania (Republic of)
278	Malta (Republic of)
280	Cyprus (Republic of)
282	Georgia (Republic of)
283	Armenia (Republic of)
284	Bulgaria (Republic of)
286	Turkey
288	Faroe Islands
290	Greenland
292	San Marino (Republic of)

Zone 2, Spare Codes: 48

Zone 3

Code	Country or Geographical Area
302	Canada
303	Canada
308	St. Pierre and Miquelon
310	United States of America
311	United States of America
312	United States of America
313	United States of America
314	United States of America
315	United States of America
316	United States of America
330	Puerto Rico
332	Virgin Islands (USA)
334	Mexico
338	Jamaica
340	French Antilles
342	Barbados
344	Antigua and Barbuda
346	Cayman Islands
348	British Virgin Islands
350	Bermuda
352	Grenada
354	Montserrat
356	St. Kitts

Zone 3 (*Continued*)

Code	Country or Geographical Area
358	St. Lucia
360	St. Vincent and the Grenadines
362	Netherlands Antilles
364	Bahamas (Commonwealth of the)
366	Dominica
368	Cuba
370	Dominican Republic
372	Haiti (Republic of)
374	Trinidad and Tobago
376	Turks and Calcos Islands

Zone 3, Spare Codes: 67

Zone 4

Code	Country or Geographical Area
400	Azerbaijani Republic
401	Kazakhstan (Republic of)
404	India (Republic of)
410	Pakistan (Islamic Republic of)
412	Afghanistan (Islamic State of)
413	Sri Lanka (Democratic Socialist Republic of)
414	Myanmar (Union of)
415	Lebanon
416	Jordan (Hashemite Kingdom of)
417	Syrian Arab Republic
418	Iraq (Republic of)
419	Kuwait (State of)
420	Saudi Arabia (Kingdom of)
421	Yemen (Republic of)
422	Oman (Sultanate of)
423	Yemen (Republic of)
424	United Arab Emirates
425	Israel (State of)
426	Bahrain (State of)
427	Qatar (State of)
428	Mongolia
429	Nepal
430	United Arab Emirates (Abu Dhabi)
431	United Arab Emirates (Dubai)
432	Iran (Islamic Republic of)
434	Uzbekistan (Republic of)
436	Tajikistan (Republic of)
437	Kyrgyztan (Republic of)
438	Turkmenistan (Republic of)
440	Japan
441	Japan
442	Japan
443	Japan
450	Korea (Republic of)
452	Viet Nam (Socialist Republic of)
453	Hong Kong
454	Hong Kong
455	Macao
456	Cambodia
457	Lao People's Democratic Republic
460	China (People's Republic of)
466	Taiwan, China
467	Democratic People's Republic of Korea
470	Bangladesh (People's Republic of)

Zone 4—Continued

Code	Country or Geographical Area
472	Maldives (Republic of)
480	Korea (Republic of)
481	Korea (Republic of)

Zone 4, Spare Codes: 54

Zone 5

Code	Country or Geographical Area
502	Malaysia
505	Australia
510	Indonesia (Republic of)
515	Philippines (Republic of the)
520	Thailand
525	Singapore (Republic of)
528	Brunei Darussalam
530	New Zealand
534	Commonwealth of the Northern Marianas (USA)
535	Guam
536	Nauru (Republic of)
537	Papua New Guinea
539	Tonga (Kingdom of)
540	Solomon Islands
541	Vanuatu (Republic of)
542	Fiji
543	Wallis and Futuna Islands
544	American Samoa
545	Kiribati (Republic of)
546	New Caledonia and Dependencies
547	French Polynesia
548	Cook Islands
549	Western Samoa
550	Micronesia (Federated States of)

Zone 5, Spare Codes: 76

Zone 6

Code	Country or Geographical Area
602	Egypt (Arab Republic of)
603	Algeria (People's Democratic Republic of)
604	Morocco (Kingdom of)
605	Tunisia
606	Libya (Socialist People's Libyan Arab Jamahiriya)
607	Gambia (Republic of the)
608	Senegal (Republic of)
609	Mauritania (Islamic Republic of)
610	Mali (Republic of)
611	Guinea (Republic of)
612	Côte d'Ivoire (Republic of)
613	Burkina Faso
614	Niger (Republic of the)
615	Togolese Republic
616	Benin (Republic of)
617	Mauritius (Republic of)
618	Liberia (Republic of)
619	Sierra Leone
620	Ghana
621	Nigeria (Federal Republic of)
622	Chad (Republic of)

Zone 6—Continued

Code	Country or Geographical Area
623	Central African Republic
624	Cameroon (Republic of)
625	Cape Verde (Republic of)
626	Sao Tome and Principe (Democratic Republic of)
627	Equatorial Guinea (Republic of)
628	Gabonese Republic
629	Congo (Republic of the)
630	Zaire (Republic of)
631	Angola (People's Republic of)
632	Guinea-Bissau (Republic of)
633	Seychelles
634	Sudan (Republic of the)
635	Rwandese (Republic of)
636	Ethiopia
637	Somali Democratic Republic
638	Djibouti (Republic of)
639	Kenya (Republic of)
640	Tanzania (United Republic of)
641	Uganda (Republic of)
642	Burundi (Republic of)
643	Mozambique (Republic of)
645	Zambia (Republic of)
646	Madagascar (Democratic Republic of)
647	Reunion (French Department of)
648	Zimbabwe (Republic of)
649	Namibia (Republic of)
650	Malawi
651	Lesotho (Kingdom of)
652	Botswana (Republic of)
653	Swaziland (Kingdom of)
654	Comoros (Islamic Federal Republic of the)
655	South Africa (Republic of)

Zone 6, Spare Codes: 47

Zone 7

Code	Country or Geographical Area
702	Belize
704	Guatemala (Republic of)
706	El Salvador (Republic of)
708	Honduras (Republic of)
710	Nicaragua
712	Costa Rica
714	Panama (Republic of)
716	Peru
722	Argentine Republic
724	Brazil (Federative Republic of)
730	Chile
732	Colombia (Republic of)
734	Venezuela (Republic of)
736	Bolivia (Republic of)
738	Guyana

Code	**Zone 7** (*Continued*) Country or Geographical Area
740	Ecuador
742	Guiana (French Department of)
744	Paraguay (Republic of)
746	Suriname (Republic of)
748	Uruguay (Eastern Republic of)
	Zone 7, Spare Codes: 80

Section 18-9 has been extracted from ITU-T Rec. X.121, Geneva, 9/92 [Ref. 16].

REFERENCES FOR CHAPTER 18 UPDATE 1995

1. Ulyess Black, *TCP/IP and Related Protocols*, McGraw-Hill, New York, 1992.
2. "Packet-Switched Signalling System between Public Networks Providing Data Transmission Services," *CCITT Rec. X.75, Fascicle VIII.3*, IXth Plenary Assembly, Melbourne, 1988.
3. "Internet Protocol," *RFC* 791, DDN Network Information Center, SRI International, Menlo Park, CA, September 1981.
4. "Simple Network Management Protocol," *RFC* 1098, DDN Network Information Center, SRI International, Menlo Park, CA, April 1989.
5. "Assigned Numbers," *RFC* 1060, DDN Network Information Center, SRI International, Menlo Park, CA, March 1990.
6. *Internet Protocol Transition Workbook*, SRI International, Menlo Park, CA, March 1982.
7. "An Ethernet Address Resolution Protocol," *RFC* 826, DDN Network Information Center, SRI International, Menlo Park, CA, November 1982.
8. "A Reverse Address Resolution Protocol," *RFC* 903, DDN Network Information Center, SRI International, Menlo Park, CA, June 1984.
9. "Internet Control Message Protocol," *RFC* 792, DDN Network Information Center, SRI International, Menlo Park, CA, September 1981.
10. "Transmission Control Protocol," *RFC* 793, DDN Network Information Center, SRI International, Menlo Park, CA, September 1981.
11. *Military Standard Transmission Control Protocol*, MIL-STD-1778, U.S. DoD, Washington, DC, August 1983.
12. "ISO Transport Protocol Specification," *ISO DP* 8073, *RFC* 905, Bolt, Beranek and Newman (BBN), Cambridge, MA, April 1984.
13. *The New IEEE Standard Dictionary of Electrical and Electronics Terms*, 5th ed., IEEE Press, New York, 1993.
14. *Systems Network Architecture Technical Overview*, IBM, Research Triangle Park, NC, January 1994.
15. Edwin R. Coover, "Systems Network Architecture—SNA Networks," IEEE Computer Soc. Press, Los Alamitos, CA, 1992.
16. "International Numbering Plan for Public Data Networks," CCITT Rec. X.121, ITU, Geneva, September 1992.

20

Broadband ISDN (B-ISDN)

This new section updates Section 20.3, which begins on page 1783 and ends on page 1784 of the Manual.

20-3.1. Introduction to ATM

ATM is an outgrowth of the several data transmission format systems such as frame relay, DQDB, and SMDS, although some may argue this point. Whereas these predecessor formats ostensibly were to satisfy the needs of the data world,* ATM (according to some) provides an optimum format or protocol family for data, voice, and image communications, where cells of each can be intermixed as shown in Figure 20-U.1. It would really seem to be more of a compromise. Typically, these ATM cells can be transported on SONET, SDH, E1/T1, and other popular digital formats. Cells can also be transported contiguously without an underlying digital network format.

Philosophically, voice and data are worlds apart regarding time sensitivity. Voice cannot wait for long processing and ARQ delays. Most types of data can. Thus ATM must distinguish the type of service such as constant bit rate (CBR) and variable bit rate (VBR) services. Voice service is typical of CBR service.

Signaling is another area of major philosophical difference. In data communications, "signaling" is carried out within the header of a data frame (or packet). As a minimum the signaling will have the destination address, and quite often the source address as well. And this signaling information will be repeated over and over again on a long data file that is heavily segmented. On a voice circuit, a connectivity is set up and the destination address, and possibly the source address, are sent just once during call setup. There is also some form of circuit supervision to keep the circuit operational throughout the duration of a telephone call. ATM is a compromise, stealing a little from each of these separate worlds.

Like voice telephony, ATM is fundamentally a connection-oriented telecommunication system. Here we mean that a connection must be established between two stations before data can be transferred between them. An ATM connection specifies the transmission path, allowing ATM cells to self-route through an ATM network. Being connection-oriented also allows ATM to specify a guaranteed quality of service (QoS) for each connection.

By contrast, most LAN protocols are connectionless. This means that LAN nodes simply transmit traffic when they need to, without first establishing a specific connection or route with the destination node.

In that ATM uses a connection-oriented protocol, bandwidth is allocated only when the originating end user requests a connection. This allows ATM to

*DQDB can also transport voice in its PA (prearbitrated) segments.

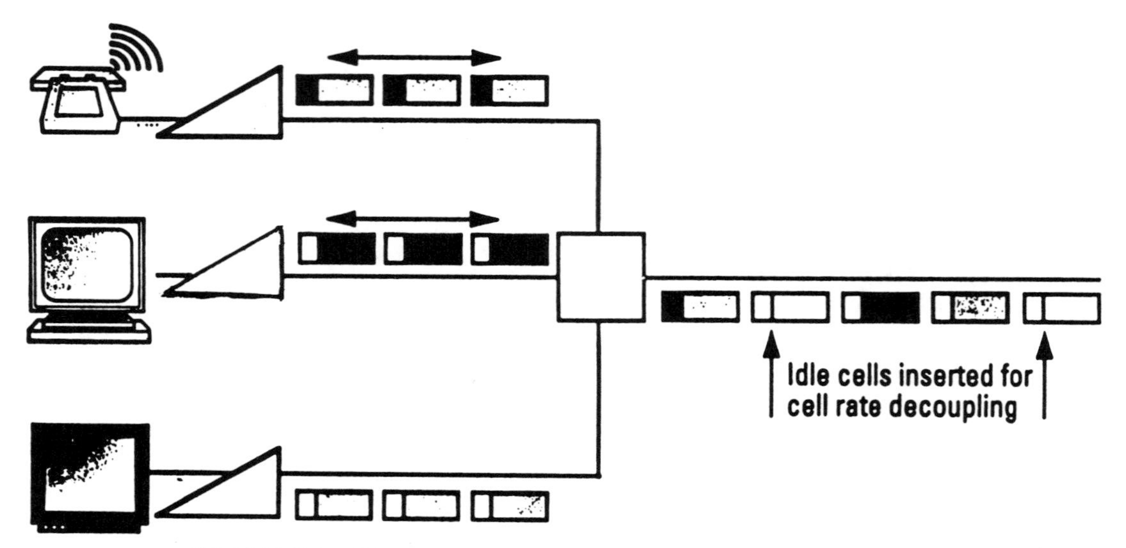

FIGURE 20-U.1. ATM links simultaneously carry a mix of voice, data, and image information.

efficiently support a network's aggregate demand by allocating bandwidth on demand based on immediate user need. Indeed it is this concept which lies in the heart of the word *asynchronous*. Here the meaning is that ATM cells are entered into a system at varying rates per unit of time up to some maximum which is determined by the maximum capacity of the underlying bearer system.

20-3.2. User – Network Interface (UNI) Configuration and Architecture

ATM is the underlying technology of broadband ISDN (B-ISDN). At times in this section, we will use the terms ATM and B-ISDN interchangeably. Figures 20-U.2 and 20-U.3 interrelate the two. Figure 20-U.2 relates the B-ISDN access reference configuration with ATM user–network interface (UNI). Figure 20-U.3 is the traditional ITU-T Rec. I.121 [Ref. 1] B-ISDN protocol reference model showing the extra layer necessary for switched service (SVC).

ATM Forum [Ref. 2] provides the following definitions applicable to Figure 20-U.3:

U-PLANE. The user plane provides for the transfer of user application information. It contains physical layer, ATM layer, and multiple ATM adaptation layers required for different service users such as CBR and VBR service.

C-PLANE. The control plane protocols deal with call establishment and call release and other connection control functions necessary for providing switched services. The C-plane structure shares the physical and ATM layers with the U-plane as shown in Figure 20-U.3. It also includes ATM adaptation layer (AAL) procedures and higher-layer signaling protocols.

M-PLANE. The management plane provides management functions and the capability to exchange information between the U-plane and the C-plane. The M-plane contains two sections: layer management and plane management. The layer management performs layer-specific management functions, while the plane management performs management and coordination functions related to the complete system.

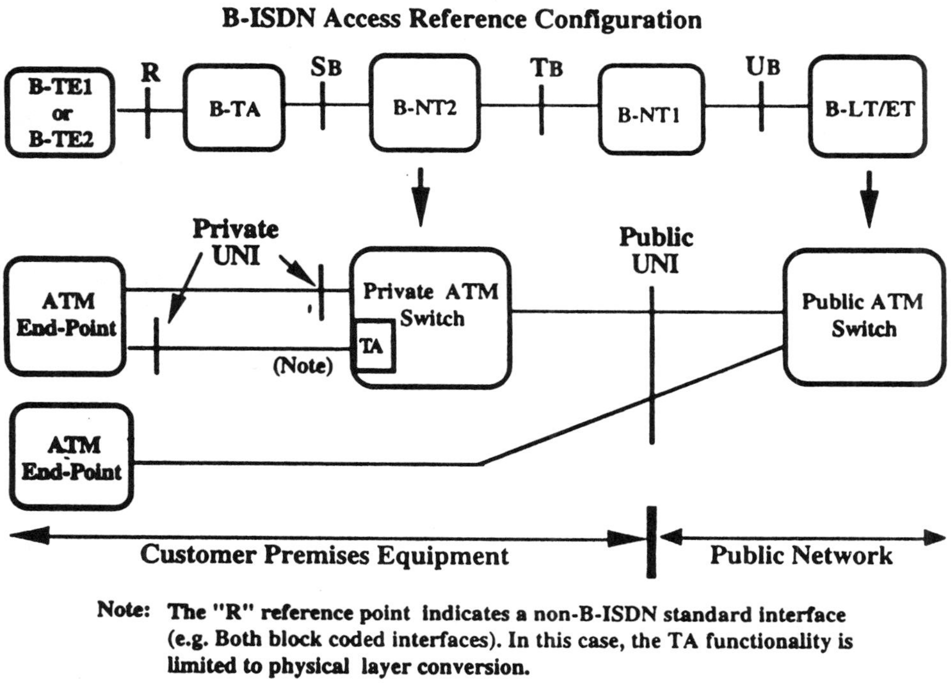

B-ISDN Access Reference Configuration
B-TE1
or
B-TE2
R
B-TA
SB
B-NT2
TB
B-NT1
UB
B-LT/ET
Private
UNI
Public
UNI
ATM
End-Point
Private ATM
Switch
TA
(Note)
Public ATM
Switch
ATM
End-Point
Customer Premises Equipment
Public Network
Note: The "R" reference point indicates a non-B-ISDN standard interface
(e.g. Both block coded interfaces). In this case, the TA functionality is
limited to physical layer conversion.

FIGURE 20-U.2. User–network interface configuration. (From ATM Forum [Ref. 2].) Courtesy of Prentice-Hall, Inc.

Plane Management
Layer Management
Control Plane
User Plane
Higher
Layers
Higher
Layers
ATM Adaptation Layer
ATM Layer
Physical Layer
Protocol
layers at
the UNIs
Protocol layers involved
for initial deployment (PVC)
(except for ILMI support)
Extra layers necessary
for switched service (SVC)

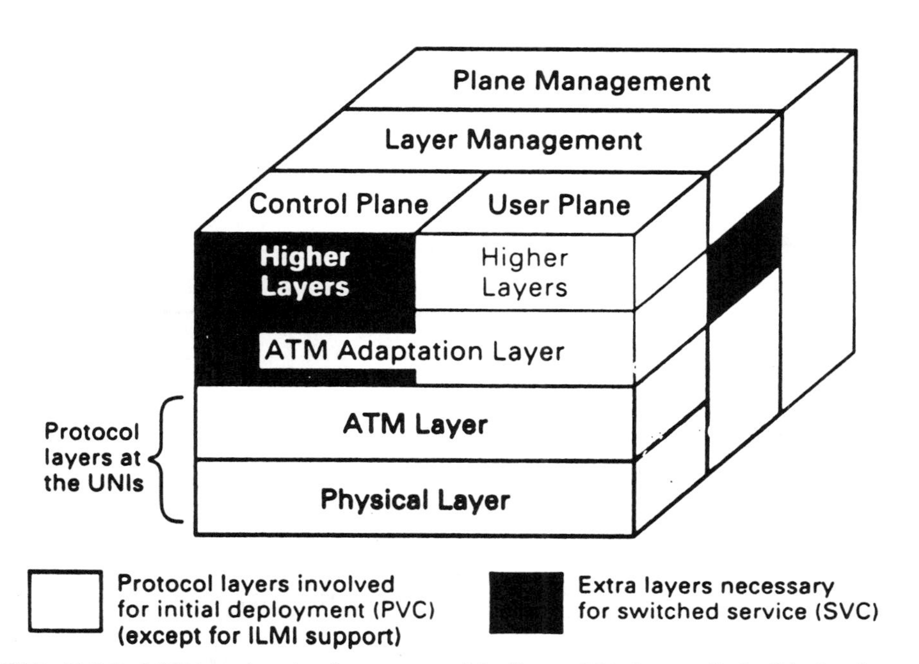

FIGURE 20-U.3. B-ISDN protocol reference model. (From ATM Forum [Ref. 2].) Courtesy of Prentice-Hall, Inc.

We return to Figure 20-U.3 and B-ISDN/ATM layering and layer descriptions in Section 20-3.5.

20-3.3. The ATM Cell: Key to Operation

20-3.3.1. ATM Cell Structure

The ATM cell consists of 53 octets, five of which make up the header and 48 of which are in the payload or "info" portion of the cell. This basic structure is shown in Figure 20-U.4.

Figure 20-U.5 shows the detailed structure of the cell header at the UNI and at the NNI.

Now let's examine each of the bit fields making up the ATM header.

Generic Flow Control (GFC). The GFC field contains 4 bits. When the GFC function is not used, the value of this field is 0000. This field is not used with the network–network interface (NNI) header. It has local significance only and can be used to provide standardized local flow control functions on the customer site. In

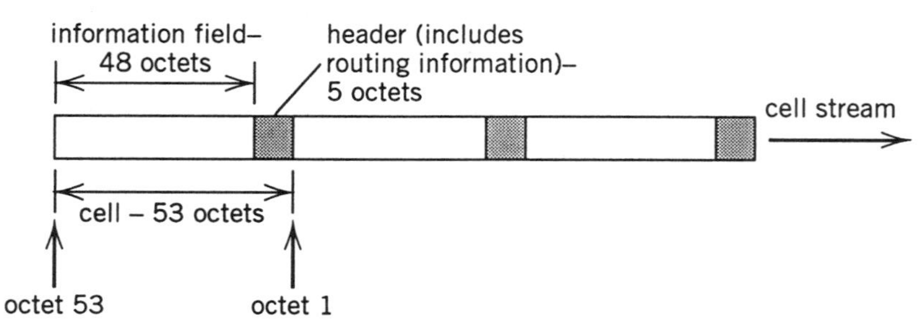

FIGURE 20-U.4. Basic ATM cell structure.

Cell Header at User-to-Network Interface (UNI)

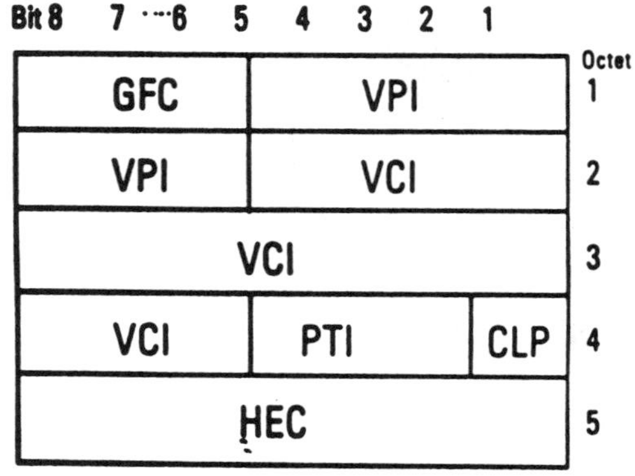

GFC Generic Flow Control
VPI Virtual Path Identifier
VCI Virtual Channel Identifier
PTI Payload Type
CLP Cell Loss Priority
HEC Header Error Control

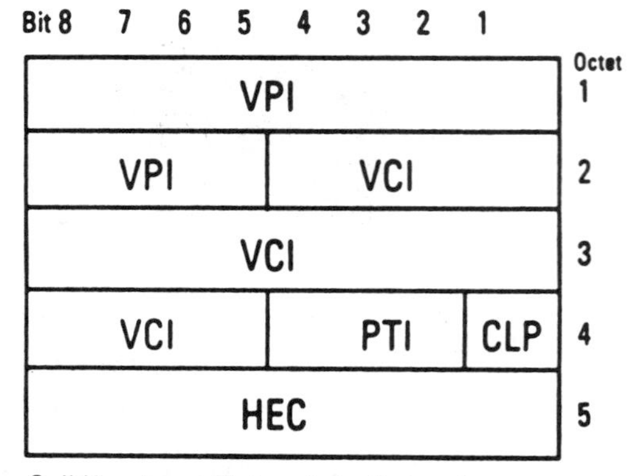

Cell Header at Network-to-Network Interface (NNI)

FIGURE 20-U.5. Header structure at UNI; at NNI.

fact the value encoded in the GFC is not carried end-to-end and will be overwritten by ATM switches.

Two modes of operation have been defined for operation of the GFC field. These are *uncontrolled access* and *controlled access*. The "uncontrolled access" mode of operation is used in early ATM environment. This mode has no impact on the traffic which a host generates. Each host transmits the GFC field set to all zeros (000). In order to avoid unwanted interactions between this mode and the "controlled access" mode where hosts are expected to modify their transmissions according to the activity of the GFC field, it is required that all customer premise equipment (CPE) and public network equipment monitor the GFC field to ensure that attached equipment is operating in "uncontrolled mode." A count of the number of nonzero GFC fields should be measured for nonoverlapping intervals of 30,000 ± 10,000 cell times. If 10 or more nonzero values are received within this interval, an error is indicated to layer management [Ref. 2].

Routing Field (VPI / VCI). Twenty-four bits are available for routing a cell. There are 8 bits for virtual path identifier (VPI) and 16 bits for virtual channel identifier (VCI). Preassigned combinations of VPI and VCI values are given in Table 20-U.1.

TABLE 20-U.1
Combinations of Preassigned VPI, VCI, and CLP Values at the UNI

Use	VPI	VCI	PT	CLP
Meta-signaling (refer to Rec. I.311)	XXXXXXXX (Note 1)	00000000 00000001 (Note 5)	0A0	C
General broadcast signaling (refer to Rec. I.311)	XXXXXXXX (Note 1)	00000000 00000010 (Note 5)	0AA	C
Point-to-point signaling (refer to Rec. I.311)	XXXXXXXX (Note 1)	00000000 00000101 (Note 5)	0AA	C
Segment OAM F4 flow cell (refer to Rec. I.610)	YYYYYYYY (Note 2)	00000000 00000011 (Note 4)	0A0	A
End-to-end OAM F4 flow cell (refer to Rec. I.610)	YYYYYYYY (Note 2)	00000000 00000100 (Note 4)	0A0	A
Segment OAM F5 flow cell (refer to Rec. I.610)	YYYYYYYY (Note 2)	ZZZZZZZZ ZZZZZZZZ (Note 3)	100	A
End-to-end OAM F5 flow cell (refer to Rec. I.610)	YYYYYYYY (Note 2)	ZZZZZZZZ ZZZZZZZZ (Note 3)	101	A
Resource management cell (refer to Rec. I.371)	YYYYYYYY (Note 2)	ZZZZZZZZ ZZZZZZZZ (Note 3)	110	A
Unassigned cell	00000000	00000000 00000000	BBB	0

The GFC field is available for use with all of these combinations.

A Indicates that the bit may be 0 or 1 and is available for use by the appropriate ATM layer function.

B Indicates the bit is a "don't care" bit.

C Indicates the originating signaling entity shall set the CLP bit to 0. The value may be changed by the network.

Note 1: XXXXXXXX: Any VPI value. For VPI value equal to 0, the specific VCI value specified is reserved for user signaling with the local exchange. For VPI values other than 0, the specified VCI value is reserved for signaling with other signaling entities (e.g., other users or remote networks).

Note 2: YYYYYYYY: Any VPI value.

Note 3: ZZZZZZZZ ZZZZZZZZ: Any VCI value other than 0.

Note 4: Transparency is not guaranteed for the OAM F4 flows in a user-to-user VP.

Note 5: The VCI values are preassigned in every VPC at the UNI. The usage of these values depends on the actual signaling configurations. (See Rec. I.311.)

Source: Table 2/I.361, ITU-T Rec. I.361, 3/93, page 3.

Other preassigned values of VPI and VCI are for further study according to the ITU-T organization. The VCI value of zero is not available for user virtual channel identification. The bits within the VPI and VCI fields used for routing are allocated using the following rules:

- The allocated bits of the VPI field are contiguous.
- The allocated bits of the VPI field are the least significant bits of the VPI field, beginning at bit 5 of octet 2.
- The allocated bits of the VCI field are contiguous.
- The allocated bits of the VCI field are the least significant bits of the VCI field, beginning at bit 5 of octet 4.

Payload Type (PT) Field. Three bits are available for PT identification. Table 20-U.2 gives the payload type identifier (PTI) coding. The main purpose of the PTI is to discriminate between user cells (i.e., cells carrying user information) and nonuser cells. The first four code groups (000–011) are used to indicate user cells. Within these four, 2 and 3 (010 and 011) are used to indicate congestion has been experienced. The fifth and sixth code groups (100 and 101) are used for VCC-level management functions.

Any congested network element, upon receiving a user data cell, may modify the PTI as follows. Cells received with PTI = 000 or PTI = 010 are transmitted with PTI = 010. Cells received with PTI = 001 or PTI = 011 are transmitted with PTI = 011. Noncongested network elements should not change the PTI.

Cell Loss Priority (CLP) Field. Depending on network conditions, cells where the CLP is set (i.e., CLP value is 1) are subject to discard prior to cells where the CLP is not set (i.e., CLP value is 0). The concept here is identical with that of frame relay and the DE (discard eligibility) bit. ATM switches may tag CLP = 0 cells detected by the usage parameter control (UPC) to be in violation of the traffic contract by changing the CLP bit from 0 to 1.

Header Error Control (HEC) Field. The HEC is an 8-bit field and it covers the entire cell header. The code used for this function is capable of either single-bit error correction or multiple-bit error detection. Briefly, the transmitting side

TABLE 20-U.2
PTI Coding

	PTI Coding	Interpretation
Bits	432	
	000	User data cell, congestion not experienced. ATM-user-to-ATM-user indication = 0
	001	User data cell, congestion not experienced. ATM-user-to-ATM-user indication = 1
	010	User data cell, congestion experienced. ATM-user-to-ATM-user indication = 0
	011	User data cell, congestion experienced. ATM-user-to-ATM-user indication = 1
	100	OAM F5 segment associated cell
	101	OAM F5 end-to-end associated cell
	110	Resource management cell
	111	Reserved for future functions

Source: ITU-T Rec. I.361, page 4.

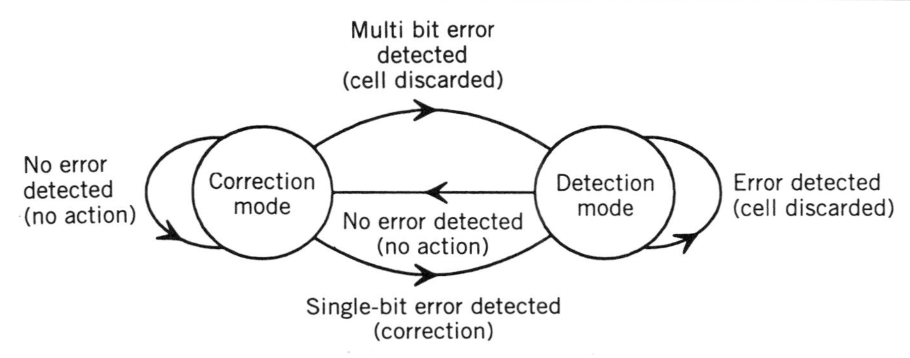

FIGURE 20-U.6. HEC: receiver modes of operation.

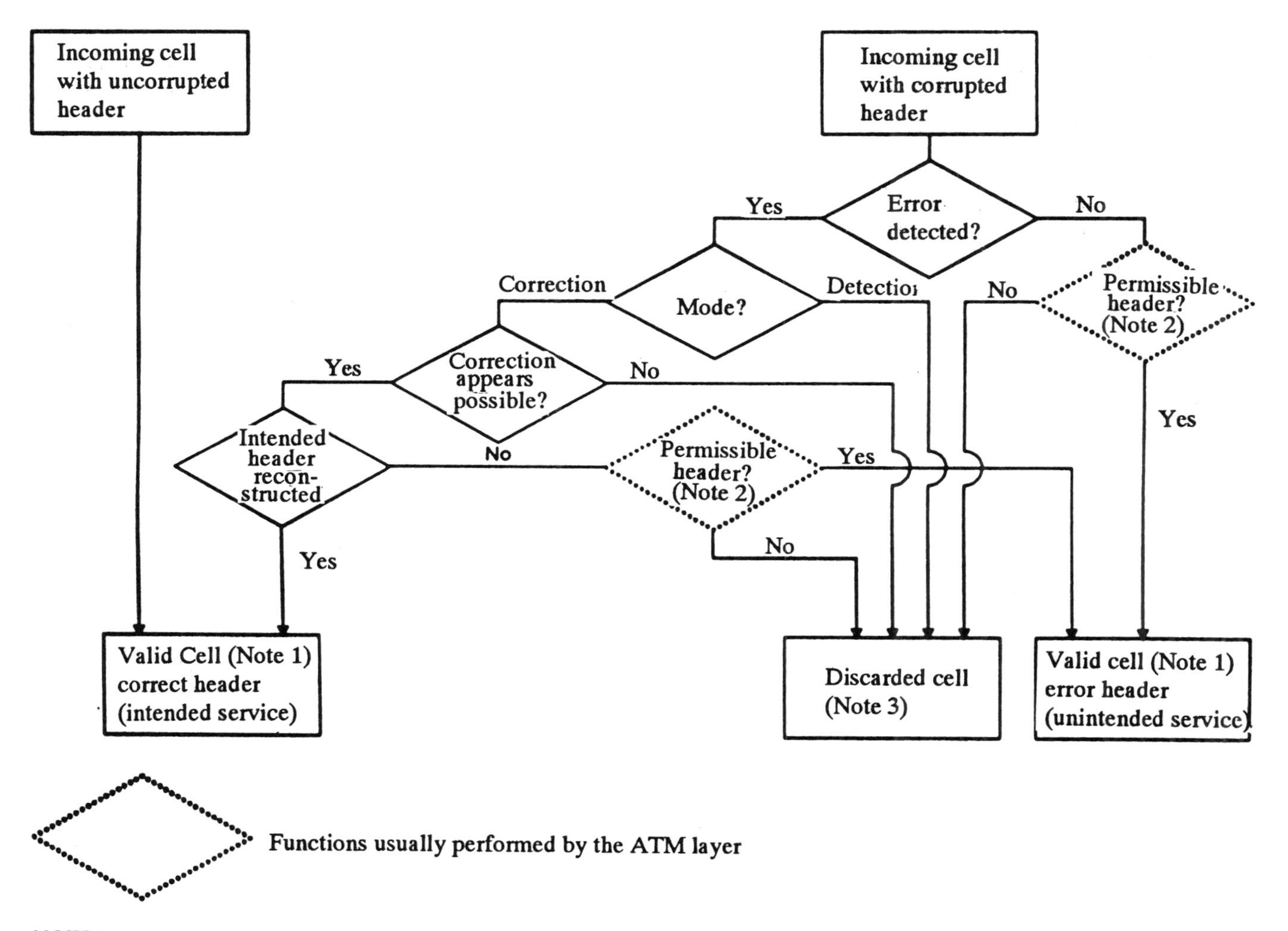

NOTES

1 **Definition of "valid cell"**: A cell where the header is declared by the header error control process to be free of errors (Recommendation I.113).

2 **An example of an impermissible header** is a header whose VPI/VCI is neither allocated to a conneciton nor pre-assigned to a particular function (idle cell, OAM cell, etc.). In many instances, the ATM layer will decide if the cell header is permissible.

3 **A cell is descarded if its header is declared** to be invalid; or if the header is declared to be valid and the resulting header is impermissible.

FIGURE 20-U.7. Consequences of errors in an ATM cell header. (From Figure 12/I.432, ITU-T Rec. I.432, Geneva 3/93, page 18 [Ref. 3].)

TABLE 20-U.3
Header Pattern for Idle Cell Identification

	Octet 1	Octet 2	Octet 3	Octet 4	Octet 5
Header pattern	00000000	00000000	00000000	00000001	HEC = Valid code 01010010

Source: Table 4/I.432, page 19, ITU-T Rec. I.432, Geneva 3/93.

computes the HEC field value. The receiver has two modes of operation, as shown in Figure 20-U.6. In the default mode there is the capability of single-bit error correction. Each cell header is examined and, if an error is detected, one of two actions takes place. The action taken depends on the state of the receiver. In the *correction mode*, only single-bit errors can be corrected and the receiver switches to the *detection mode*. In the "detection mode," all cells with detected header errors are discarded. When a header is examined and found not to be in error, the receiver switches to the "correction mode." The term *no action* in Figure 20-U.6 means no correction is performed and no cell is discarded.

Figure 20-U.7 is a flow chart showing the consequence of errors in the ATM cell header. The error protection function provided by the HEC provides for both recovery from single-bit errors and a low probability of delivery of cells with errored headers under bursty error conditions. ITU-T Rec. I.432 [Ref. 3] states that error characteristics of fiber-optic transmission systems appear to be a mix of single-bit errors and relatively large burst errors. Thus, for some transmission systems the error correction capability might not be invoked.

20-3.3.2. Idle Cells

Idle cells cause no action at a receiving node except for cell delineation including HEC verification. They are inserted and discarded for cell rate decoupling. Idle cells are identified by the standardized pattern for the cell header, as shown in Table 20-U.3.

The content of the information field is 01101010 repeated 48 times for an idle cell.

20-3.4. Cell Delineation and Scrambling

20-3.4.1. Delineation and Scrambling Objectives

Cell delineation allows identification of the cell boundaries. The cell HEC field achieves cell delineation. Keep in mind that the ATM signal must be self-supporting in that it has to be transparently transported on every network interface without any constraints from the transmission systems used. Scrambling is used to improve security and robustness of the HEC cell delineation mechanism discussed below. In addition, it helps the randomizing of data in the information field for possible improvement in transmission performance.

Any scrambler specification must not alter the ATM header structure, header error control, and cell delineation algorithm.

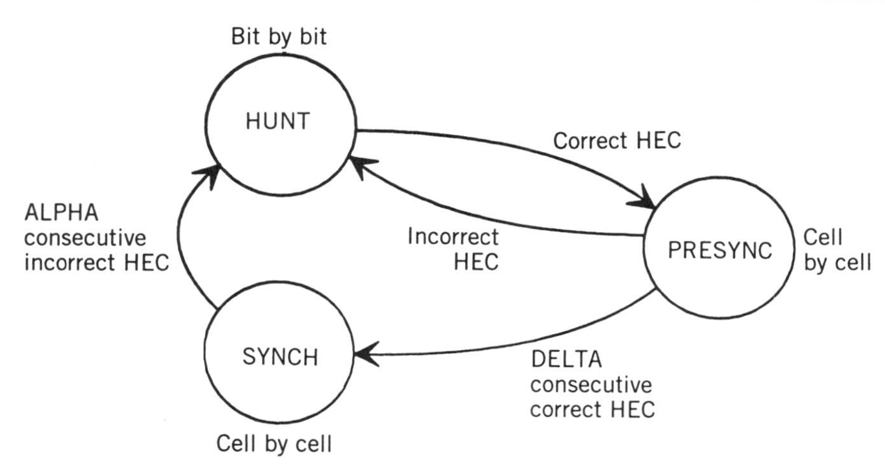

FIGURE 20-U.8. Cell delineation state diagram. (From Figure 13/I.432, ITU-T Rec. I.432, page 20 [Ref. 3].)

20-3.4.2. Cell Delineation Algorithm

Cell delineation is performed by using the correlation between the header bits to be protected (32 bits or 4 octets) and the HEC octet which are the relevant control bits (8 bits) introduced in the header using a shortened cyclic code with the generating polynomial $X^8 + X^2 + X + 1$.

Figure 20-U.8 shows the state diagram of the HEC cell delineation method. A discussion of the figure is given below.

1. In the HUNT state, the delineation process is performed by checking bit by bit for the correct HEC (i.e., syndrome equals zero) for the assumed header field. For the cell-based* physical layer, prior to scrambler synchronization, only the last six bits of the HEC are used for cell delineation checking. For the SDH-based interface,* all 8 bits are used for acquiring cell delineation. Once such an agreement is found, it is assumed that one header has been found, and the method enters the PRESYNC state. When octet boundaries are available within the receiving physical layer prior to cell delineation as with the SDH-based interface, the cell delineation process may be performed octet by octet.
2. In the PRESYNC state, the delineation process is performed by checking cell by cell for the correct HEC. The process repeats until the correct HEC has been confirmed *Delta* times consecutively. If an incorrect HEC is found, the process returns to the HUNT state.
3. In the SYNCH state, the cell delineation will be assumed to be lost if an incorrect HEC is obtained *Alpha* times consecutively.

The parameters *Alpha* and *Delta* are chosen to make the cell delineation process as robust and secure as possible while satisfying QoS (quality of service) requirements. Robustness depends on *Alpha* when it is against false misalignments due to bit errors, and robustness depends on *Delta* when it is against false delineation in the resynchronization process.

*Only cell-based and SDH-based interfaces are covered by current ITU-T Recommendations. Besides these, we will cover cells riding on other transport means at the end of this chapter.

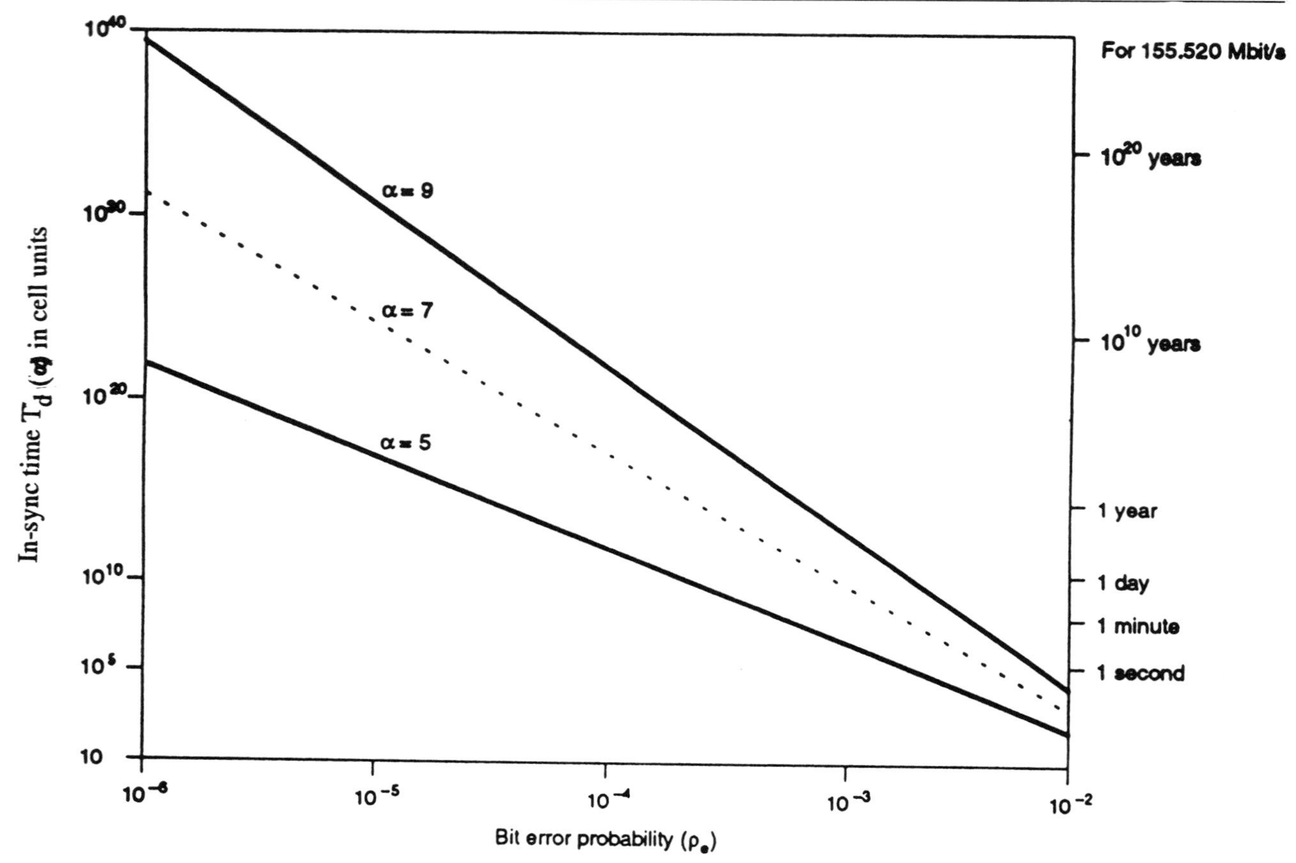

FIGURE 20-U.9. In-sync time versus bit error probability. (From Figure B.1/I.432, ITU-T Rec. I.432, page 33 [Ref. 3].)

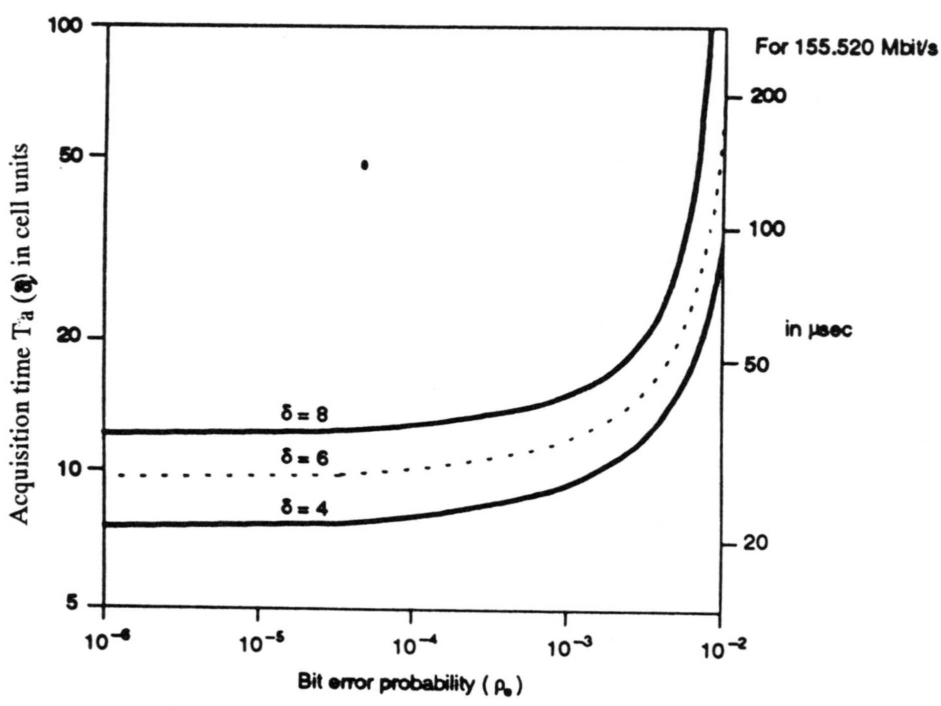

FIGURE 20-U.10. Acquisition time versus bit error probability. (From Figure B.2/I.432, ITU-T Rec. I.432, page 34 [Ref. 3].)

For the SDH-based physical layer, values of *Alpha* = 7 and *Delta* = 6 are suggested by the ITU-T organization (Rec. I.432 [Ref. 3]); and for cell-based physical layer, values of *Alpha* = 7 and *Delta* = 8. Figures 20-U.9 and 20-U.10 give performance information of the cell delineation algorithm in the presence of random bit errors for various values of *Alpha* and *Delta*.

20-3.5. ATM Layering and B-ISDN

The B-ISDN reference model is given in Figure 20-U.3, and its several planes are described.

20-3.5.1. Functions of Individual ATM / B-ISDN Layers

Figure 20-U.11 illustrates B-ISDN/ATM layering and sublayering of the protocol reference model. It identifies the functions of the physical layer, the ATM layer and the AAL, and related sublayers.

Physical Layer. The physical layer consists of two sublayers. The physical medium (PM) sublayer includes only physical medium dependent functions. The transmission convergence (TC) sublayer performs all functions required to transform a flow of cells into a flow of data units (i.e., bits) which can be transmitted and received over a physical medium. The service data unit (SDU) crossing the boundary between the ATM layer and the physical layer is a flow of valid cells. The ATM layer is unique (i.e., is independent of the underlying physical layer). The data flow inserted in the transmission system payload is PM-independent and self-supported.

<table>
<tr><td rowspan="6">Layer Management</td><td>Higher layer functions</td><td colspan="2">Higher layers</td></tr>
<tr><td>Convergence</td><td>CS</td><td rowspan="2">AAL</td></tr>
<tr><td>Segmentation and reassembly</td><td>SAR</td></tr>
<tr><td>Generic flow control
Cell header generation/extraction
Cell VPI/VCI translation
Cell multiplex and demultiplex</td><td colspan="2">ATM</td></tr>
<tr><td>Cell rate decoupling
HEC header sequence generation/verification
Cell delineation
Transmission frame adaptation
Transmission frame generation/recovery</td><td>TC</td><td rowspan="2">Physical Layer</td></tr>
<tr><td>Bit timing
Physical medium</td><td>PM</td></tr>
</table>

CS Convergence sublayer
PM Physical medium
SAR Segmentation and reassembly sublayer
TC Transmission convergence

FIGURE 20-U.11. B-ISDN / ATM functional layering.

The physical layer merges the ATM cell flow with the appropriate information for cell delineation (according to the cell delineation mechanism described above) and carries the operations and maintenance (OAM) information relating to this cell flow.

The PM sublayer provides bit transmission capability including bit transfer and bit alignment as well as line coding and electrical–optical transformation. Of course, the principal function is the generation and reception of waveforms suitable for the medium, along with insertion and extraction of bit timing information and line coding where required. The primitives identified at the border between the PM and TC sublayers are a continuous flow of logical bits or symbols with this associated timing information.

Transmission Convergence Sublayer Functions. Among the important functions of this sublayer is the generation and recovery of transmission frame. Another function is transmission frame adaptation which includes the actions necessary to structure the cell flow according to the payload structure of the transmission frame (transmit direction) and to extract this cell flow out of the transmission frame (receive direction). The transmission frame may be a cell equivalent (i.e., no external envelope is added to the cell flow), and SDH/SONET envelope, an E-1/T-1 envelope, and so on. In the transmit direction, the HEC sequence is calculated and inserted in the header. In the receive direction, we include cell header verification. Here cell headers are checked for errors and, if possible, header errors are corrected. Cells are discarded where it is determined that headers are errored and are not correctable.

Another transmission convergence function is cell rate decoupling. This involves the insertion and removal of idle cells in order to adapt the rate of valid ATM cells to the payload capacity of the transmission system. In other words, cells must be generated to exactly fill the payload of SDH/SONET (for example), whether the cells are idle or busy.

Section 20-3.11 gives several examples of transporting cells using the convergence sublayer.

The ATM Layer. Table 20-U.4 shows the ATM layer functions supported at the UNI (U-plane). The ATM layer is completely independent of the physical medium. One important function of this layer is *encapsulation*. This includes cell header generation and extraction. In the transmit direction, the cell header generation function receives a cell information field from a higher layer and generates an appropriate ATM cell header except for the HEC sequence. This function can also

TABLE 20-U.4
ATM Layer Functions Supported at the UNI

Functions	Parameters
Multiplexing among different ATM connections	VPI/VCI
Cell rate decoupling (unassigned cells)	Preassigned header field values
Cell discrimination based on predefined header field values	Preassigned header field values
Payload type discrimination	PT field
Loss priority indication and selective cell discarding	CLP field Network congestion state
Traffic shaping	Traffic descriptor

include the translation from a service access point (SAP) identifier to a virtual path (VP) and virtual circuit (VC) identifier.

In the receive direction, the cell header extraction function removes the ATM cell header and passes the cell information field to a higher layer. As in the transmit direction, this function can also include a translation of a VP and VC identifier into an SAP identifier.

In the case of the NNI, the GFC is applied at the ATM layer. The flow control information is carried in assigned and unassigned cells. Cells carrying this information are generated in the ATM layer.

In a switch the ATM layer determines where the incoming cells should be forwarded to, resets the corresponding connection identifiers for the next link, and forwards the cell. The ATM layer also handles traffic management functions and buffers incoming and outgoing cells. It indicates to the next higher layer (the AAL) whether or not there is congestion during transmission. The ATM layer monitors both transmission rates and conformance to the service contract—called *traffic shaping* and *traffic policing*.

Cell Rate Decoupling (ATM Forum Interpretation). The cell rate decoupling function at the sending entity adds unassigned cells to the assigned cell stream (i.e., cells with valid payloads) to be transmitted, transforming a noncontinuous stream of assigned cells into a continuous stream of assigned and unassigned cells. At the receiving entity the opposite operation is performed for both unassigned and invalid cells. The rate at which the unassigned cells are inserted/extracted depends on the bit rate (and rate variation) of assigned cells, and/or the physical layer transmission rate of the unassigned and invalid cells are recognized by specific header patterns which are shown in Table 20-U.5.

Physical layers that have synchronous cell timeslots generally require cell rate decoupling (typically SONET/SDH, DS3, etc.), whereas physical layers that have

TABLE 20-U.5
Predefined Header Field Values

Use	Value (Notes 1–4)			
	Octet 1	Octet 2	Octet 3	Octet 4
Unassigned cell indication	00000000	00000000	00000000	0000xxx0
Meta-signaling (default) (Notes 5, 7)	00000000	00000000	00000000	00010a0c
Meta-signaling (Notes 6, 7)	0000yyyy	yyyy0000	00000000	00010a0c
General broadcast signaling (default) (Note 5)	00000000	00000000	00000000	00100aac
General broadcast signaling (Note 6)	0000yyyy	yyyy0000	00000000	00100aac
Point-to-point signaling (default) (Note 5)	00000000	00000000	00000000	01010aac
Point-to-point signaling (Note 6)	0000yyyy	yyyy0000	00000000	01010aac
Invalid pattern	xxxx0000	00000000	00000000	0000xxx1
Segment OAM F4 flow cell (Note 7)	0000aaaa	aaaa0000	00000000	00110a0a
End-to-end OAM F4 flow cell (Note 7)	0000aaaa	aaaa0000	00000000	01000a0a

Note 1: "a" indicates that the bit is available for use by the appropriate ATM layer function.
Note 2: "x" indicates "don't care" bits.
Note 3: "y" indicates any VPI value other than 00000000.
Note 4: "c" indicates that the originating signaling entity shall set the CLP bit to 0. The network may change the value of the CLP bit.
Note 5: Reserved for user signaling with the local exchange.
Note 6: Reserved for signaling with other signaling entities (e.g., other users or remote networks).
Note 7: The transmitting ATM entity shall set bit 2 of octet 4 to zero. The receiving ATM entity shall ignore bit 2 of octet 4.
Source: ATM Forum, Figure 3-7, page 57, Ref. 2. Courtesy Prentice-Hall, Inc.

TABLE 20-U.6
ATM Layer Management Functions at the UNI

Functions	Parameters
Fault management	
Alarm surveillance (VP)	OAM cells
Connectivity verification (VP, VC)	OAM cells
Invalid VPI/VCI detection	VPI/VCI

Source: ATM Forum, Figure 3-10, page 61 [Ref. 2]. Courtesy of Prentice-Hall, Inc.

asynchronous cell timeslots do not require this function because no continuous flow of cells needs to be provided.

Cell Discrimination Based on Predefined Header Field Values. The predefined header field values defined at the UNI are given in Table 20-U.5.

Additional Notes to Table 20-U.5. Meta-signaling cells are used by the meta-signaling protocol for establishing and releasing signaling virtual channel connections. For virtual channels allocated permanently (PVC), meta-signaling is not used.

General broadcast signaling cells are used by the ATM network to broadcast signaling information independent of service profiles.

The virtual path connection (VPC) operation flow (F4 flow) is carried via specially designated OAM cells. F4 flow OAM cells have the same VPI value as the user-data cells transported by the VPC but are identified by two unique preassigned virtual channels within this VPC. At the UNI, the virtual channel identified by a VCI value = 3 is used for VP level management functions between ATM nodes on both sides of the UNI (i.e., single VP link segment), while the virtual channel identified by a VCI value = 4 can be used for VP level end-to-end (User ↔ User) management functions.

ATM Layer Management (M-Plane). Management functions at the UNI require some level of cooperation between customer premise equipment and network equipment. To minimize the coupling required between equipment on both sides of the UNI, the functional requirements have been reduced to a minimal set. The ATM layer management functions supported at the UNI are grouped into the categories listed in Table 20-U.6.

Fault management contains alarm surveillance and connectivity verification functions. OAM cells are used for exchanging related operation information.

ATM Layer Management Information Flows. Figure 20-U.12 shows the OAM flows defined for the exchange of operations information between nodes (including customer premise equipment). At the ATM layer, the F4–F5 flows are carried via OAM cells. The OAM cell flow used for end-to-end management functions may be carried transparently through the private ATM switch and made available to the user.

The F4 flow is used for segment* or end-to-end (VP termination) management at the VP level using VCI values 3 and 4. The F5 flow is used for segment or end-to-end (VC termination) management at the VC level using PT codes 4 and 5 (100 and 101, Table 20-U.2).

*In this case the segment is defined as the link between the ATM nodes on either side of the UNI.

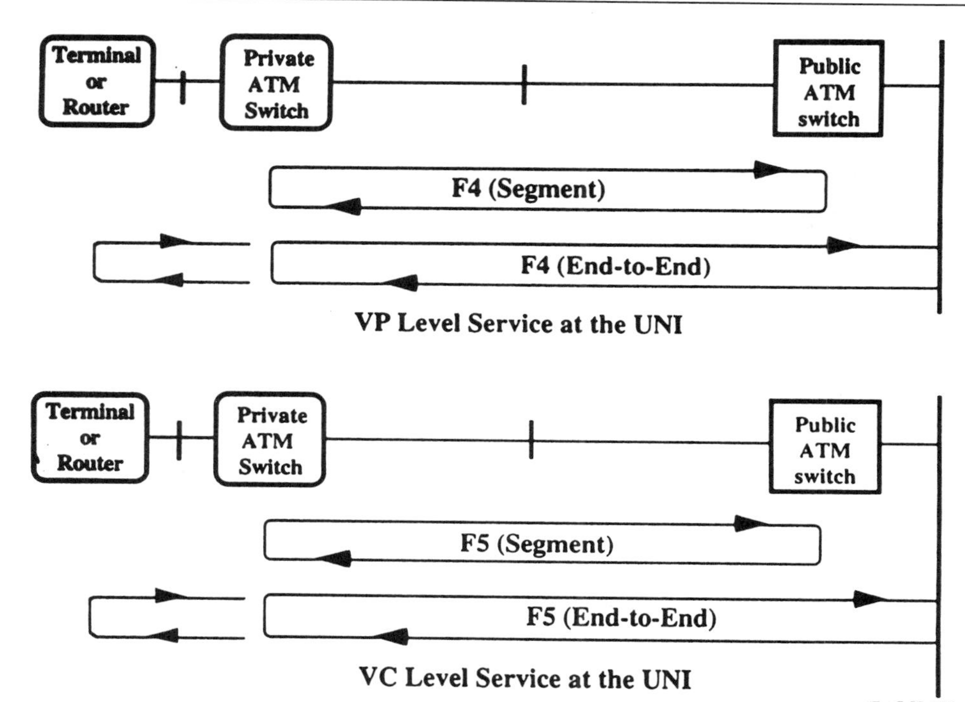

FIGURE 20-U.12. ATM layer OAM flows at the UNI. (From ATM Forum, Figure 3-11, page 62 [Ref. 2]. Courtesy of Prentice-Hall, Inc.)

In case of only virtual path (VP) visibility (i.e., VPC service at the public network interface), the OAM operation information exchange is limited to the F4 flow. Under this scenario, any of the VC level OAM functions and information exchange (F5 flow) are user-specific and ignored by the network. It is, however, possible to have VP level service at the public UNI while maintaining VC visibility at the private UNI. In this case, the private ATM switch would terminate the F4 flow but could carry transparently the user end-to-end F5 flow. For VC visibility (VCC service), the OAM operation information exchange specified at the public UNI could be limited to the F5 flow or could invoke both F4 and F5 flows.

The ATM Adaptation Layer (AAL). The basic purpose of the AAL is to isolate the higher layers from the specific characteristics of the ATM layer by mapping the higher-layer protocol data units (PDUs) into the information field of the ATM cell and vice versa.

Sublayering of the AAL. To support services above the AAL, some independent functions are required of the AAL. These functions are organized in two logical sublayers: the convergence sublayer (CS) and the segmentation and reassembly sublayer (SAR). The prime functions of these sublayers are as follows:

- The segmentation of higher-layer information into a size suitable for the information field of an ATM cell.
- Reassembly of the contents of ATM cell information fields into higher-layer information.
- CS: Here the prime function is to provide the AAL service at the AAL-SAP (SAP = service access point). This sublayer is service-dependent.

Service Parameters	Class A	Class B	Class C	Class D
Timing Compensation	Required		Not Required	
Bit Rate	Constant	Variable		
Connection Mode	Connection-oriented			Connectionless
Example	Circuit Emulation	Variable Bit Rate Video	CO Data Transfer	CL Data Transfer
AAL Type	Type 1	Type 2	Type 3 Type 5	Type 4

FIGURE 20-U.13. Service classification for AAL. CO = connection-oriented; CL = connectionless. (Courtesy of Hewlett-Packard Company [Ref. 4].)

Service Classification for the AAL. Service classification is based on the following parameters:

- Timing relation between source and destination (this refers to urgency of traffic): required or not required.
- Bit rate: constant or variable.
- Connection mode: connection-oriented or connectionless.

When we combine these parameters, four service classes emerge as shown in Figure 20-U.13.

Examples of services in the classes shown in Figure 20-U.13 are as follows:

- Class A: Constant bit rate such as uncompressed voice or video.
- Class B: Variable bit rate video and audio, connection-oriented synchronous traffic.
- Class C: Connection-oriented data transfer, variable bit rate, asynchronous traffic.
- Class D: Connectionless data transfer, asynchronous traffic such as SMDS.

AAL Categories or Types. There are five different AAL types or categories. The simplest of these is AAL-0. It just transmits cells down a pipe. That pipe is commonly a fiber-optic link. Ideally we would like the bit rate to be some multiple of 53 × 8, or 424, bits. For example, 424 Mbps would handle 1 million cells per second.

AAL-1. AAL-1 is used to provide transport for synchronous bit streams. Its primary application is to adapt ATM cell transmission to typically E1/T1 and SDH/SONET circuits. Typically, AAL-1 is for voice communications [plain old telephone service (POTS)]. AAL-1 checks that mis-sequencing of information does not occur by verifying a 3-bit sequence counter and allows for regeneration of the original clock timing of the data received at the far end of the link. The SAR-PDU

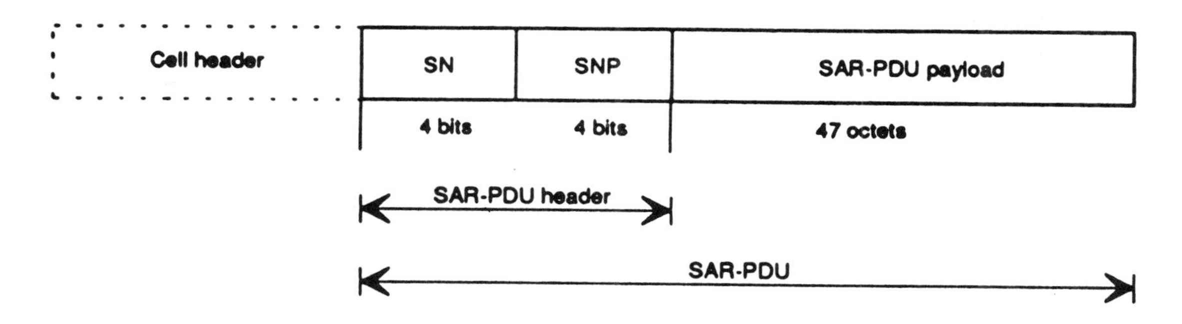

SN Sequence number (4 bits); to detect lost or misinserted cells. A specific value of the sequence number may indicate a special purpose, e.g. the existence of convergence sublayer functions. The exact counting scheme is for further study.

SNP Sequence number protection (4 bits). The SNP field may provide error detection and correction capabilities. The polynomial to be used is for further study.

FIGURE 20-U.14. SAR-PDU format for AAL-1. (Figure 1/I.363, ITU-T Rec. I.363, page 3 [Ref. 5].)

format of AAL-1 is shown in Figure 20-U.14. The 4-bit sequence number (SN) is broken down into a 1-bit CSI (convergence sublayer indicator) and sequence count. The sequence number protection (SNP) contains a 3-bit CRC and a parity bit. Clock recovery is via a synchronous residual time stamp (SRTS) and common network clock by means of a 4-bit residual time stamp extracted from CSI of cells with odd sequence numbers. The residual time stamp is transmitted over eight cells. Alarm indication in this adaptation layer is via a check of the one's density. When the one's density of the received cell stream becomes significantly different than the density used for the particular PCM line coding scheme in use, it is determined that the system has lost signal, and alarm notifications are given.

AAL-2. AAL-2 handles the variable bit rate (VBR) scenario such as MPEG* video. Functions in AAL-2 include:

(a) Segmentation and reassembly (SAR) of user information
(b) Handing of cell delay variation
(c) Handling of lost and misinserted cells
(d) Source clock frequency recovery at the receiver
(e) Monitoring and handling of AAL-PCI bit errors (PCI = protocol control information)
(f) Monitoring of user information field for bit errors and possible corrective action

AAL-2 is still in ITU-T definitive stages. However, an example of an SAR-PDU format for AAL-2 is given in Figure 20-U.15. The following indications from the AAL-2 user plane are passed to the management plane:

- Errors in the transmission of user information
- Loss of timing/synchronization
- Lost or misinserted cells
- Cells with errored AAL-PCI

*MPEG is a set of video compression schemes. MPEG stands for Motion Picture Experts Group.

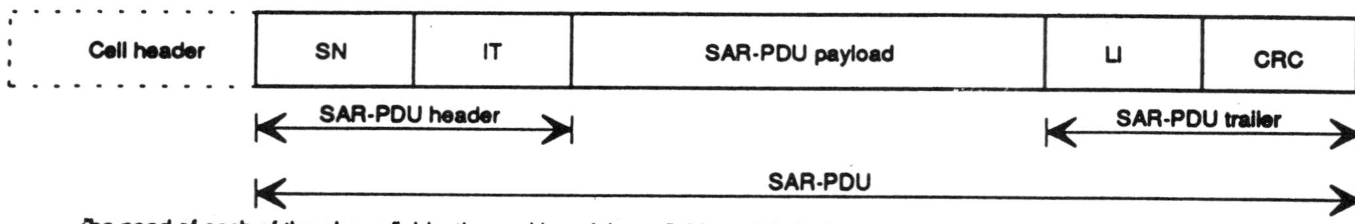

The need of each of the above fields, the position of those fields and their size are for further study.

SN Sequence number, to detect lost or misinserted cells. A specific value of the sequence number may indicate a specified purpose

IT Information type, used to indicate beginning of message (BOM), continuation of message (COM), end of message (EOM), timing information and also component of the video or audio signal

LI Length indicator, check to indicate that the number of octets of the CS-PDU are included in the SAR-PDU payload field

CRC Cyclic redundancy check code, to correct up to two correlated bit errors

FIGURE 20-U.15. Example of an SAR-PDU format for AAL-2. (From Figure 2/I.363, ITU-T Rec. I.363 1/93, page 5 [Ref. 5].)

The AAL-2 convergence sublayer (CS) performs the following functions:

(a) Clock recovery is carried out for variable bit rate audio and video services by means of the insertion of a time stamp or real-time synchronization word in the CS-PDU.
(b) Sequence number processing is performed to detect the loss or misinsertion of ATM-SDUs. The handling of lost and misinserted ATM-SDUs is also performed in this sublayer.
(c) For audio and video services, forward error correction may also be performed.

AAL-3/4. Initially, in ITU-T Rec. I.363 1991, there were two separate AALs, one for connection-oriented variable bit rate data services (AAL-3) and one for connectionless service (AAL-4). As the specifications evolved, the same procedures turned out to be necessary for both of these services, and the specifications were merged to become the AAL-3/4 standard. AAL-3/4 is used for ATM shipping of SMDS, CBDS (connectionless broadband data services, an ETSI initiative), and frame relay.

AAL-3/4 has been designed to take variable-length frames/packets and segment them into cells. The segmentation is done in a way that protects the transmitted data from corruption if cells are lost or mis-sequenced.

Variable-length packets (up to 64 kbytes) from SMDS/CLNAP (CLNAP stands for connectionless network access protocol) or frame relay frames are padded to an integral word length and encapsulated with a header and a trailer to form what is called the *convergence sublayer PDU* (CS-PDU) and then is segmented into cells. The passing is done to make sure that fields align themselves to 32-bit boundaries, allowing the efficient implementation of the operations in hardware at lower layers. The added header and trailer contain a tag to match the end and the packet length so that the receiving end may allocate the buffer for this packet upon reception of the first fragment. In practice, this buffer allocation size (BAsize) is mostly used as a length check for integrity verification, because the most efficient algorithm is used to allocate fixed-size maximum length buffers. This AAL CS-PDU is shown in Figure 20-U.16.

Figure 20-U.17 shows the SAR-PDU format for AAL-3/4 from ITU-T Rec. I.363, and Figure 20-U.18 shows this same format when used for transmitting SMDS frames.

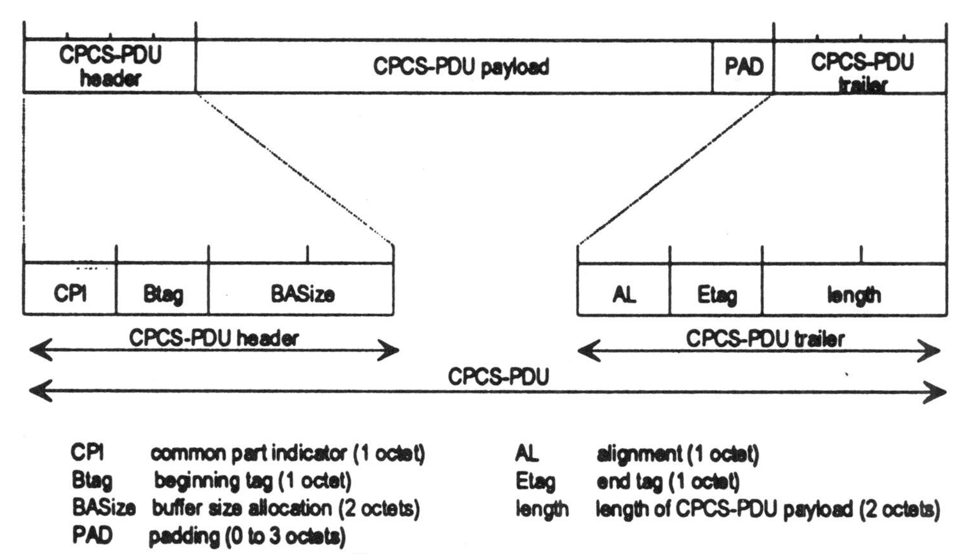

FIGURE 20-U.16. Convergence sublayer protocol data unit (AAL-3/4). From Ref. 4, couresy of Hewlett-Packard Company.

Turning to Figure 20-U.18, the segmented portions are 44 octets long (except possibly for the last segment). These portions are then encapsulated with another header (2 octets) and trailer (2 octets) to become a segmentation and reassembly PDU (SAR-PDU) which is inserted into cell payloads. The header at this level with the segment type field identifies what kind of a cell it is (i.e., BOM, beginning of message; COM, continuation of message; or EOM, end of message) so that the individual CS-PDUs can be delineated. The header also includes a sequence number for protection against misordered delivery. It also includes the MID (message identification in SMDS, multiplexing identifier for ATM). The SAR-PDU trailer contains a length indicator to identify how much of the payload is filled. It also has a CRC-10 error check to protect against cell corruption. A complete message contains a BOM cell, zero or more COM cells, and an EOM cell. If the entire message can fit into one cell, it is called a *single-segment message* (SSM), where the CS-PDU is less than 44 octets long.

AAL-3/4 has several measures to ensure the integrity of the data which have been segmented and transmitted as cells. The contents of the cell are protected by the CRC-10; sequence numbers protect against misordering. Still another measure

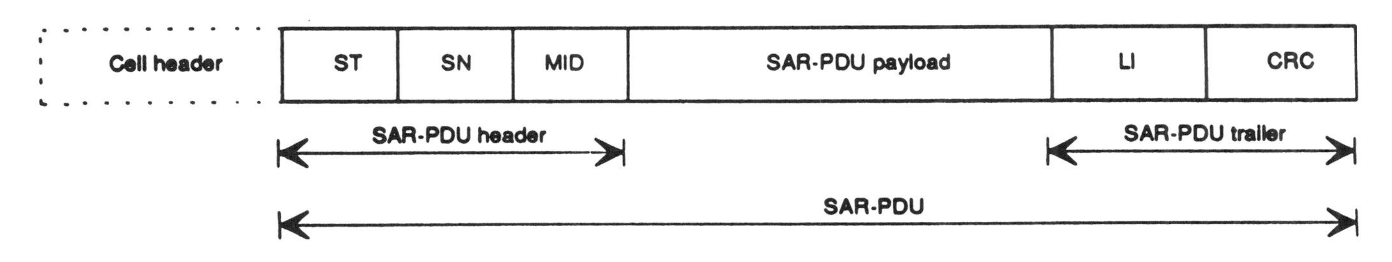

ST Segment type (2 bits)
SN Sequence number (4 bits)
MID Multiplexing identification (10 bits)
LI Length indicator (6 bits)
CRC Cyclic redundancy check code (10 bits)

FIGURE 20-U.17. SAR-PDU format for AAL-3/ 4. (From Figure 6/I.363, ITU-T Rec. I.363, page 13 [Ref. 5].)

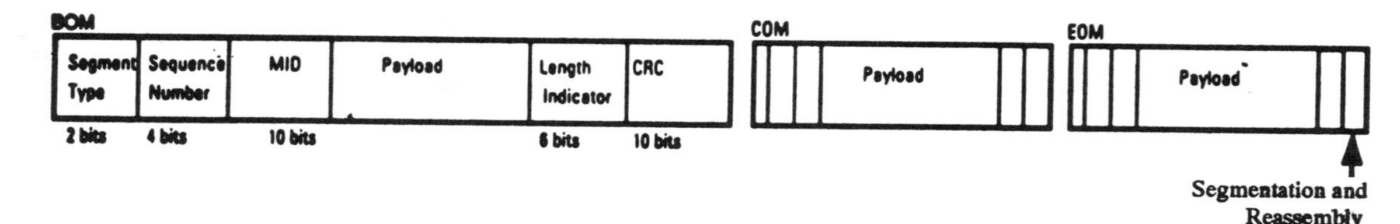

FIGURE 20-U.18. AAL-3/4 SAR-PDU as applied to the transmission of SMDS frames. (Courtesy of Hewlett-Packard Company [Ref. 4].)

to ensure against corrupted PDUs being delivered is EOM/BOM protection. If the EOM of one CPCS-PDU and the BOM of the next are dropped for some reason, the resulting cell stream could be interpreted as a valid PDU. To protect against these kinds of errors, the BEtag numeric values in the CPCS-PDU headers and trailers are compared to ensure that they match. (CPCS = common part convergence sublayer). Two modes of service are defined for AAL-3/4:

1. *Message Mode Service.* This service provides for the transport of one or more fixed-size AAL service data units in one or more convergence sublayer protocol data units (CS-PDUs).
2. *Streaming Mode Service.* Here the AAL SDU is passed across the AAL interface in one or more AAL interface data units (IDUs). The transfer of these AAL-IDUs across the AAL interface may occur separated in time, and this service provides the transport of variable-length AAL-SDUs. The streaming mode service includes an abort service by which the discarding of an AAL-SDU partially transferred across the AAL interface can be requested. In other words, in the streaming mode a single packet is passed to the AAL layer and transmitted in multiple CPCS-PDUs, as and when pieces of the packet are received. Streaming mode may be used in intermediate switches or ATM-to-SMDS routers so they can begin re-transmitting a packet being received before the entire packet has arrived. This reduces the latency experienced by the entire packet.

AAL-5. This type of AAL was designed specifically to carry data traffic typically found in today's LANs. AAL-5 evolved after AAL-3/4, which was found to be too complex and inefficient for LAN traffic. Thus AAL-5 got the name "SEAL" as an acronym for simple and efficient AAL layer. Only a small amount of overhead is added to the CPCS-PDU, and no extra overhead is added when the AAL-5 segments them into SAR-PDUs. There is no AAL-level cell multiplexing. In AAL-5 all cells belonging to an AAL-5 CPCS-PDU are sent sequentially.

As shown in Figure 20-U.19, the CPCS-PDU has only a payload and a trailer. The trailer contains padding, a length field, and a CRC-32 field for error detection.

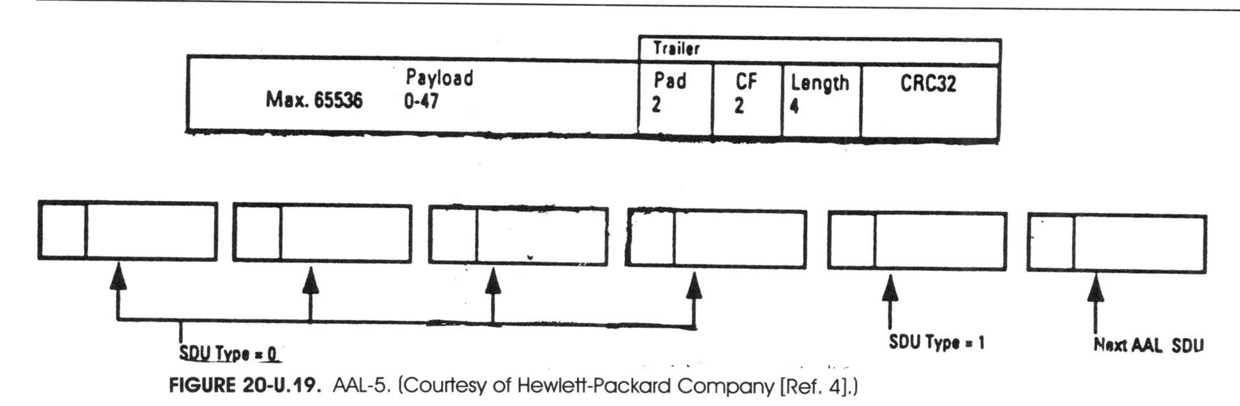

FIGURE 20-U.19. AAL-5. (Courtesy of Hewlett-Packard Company [Ref. 4].)

The CPCS-PDUs are padded to become integral multiples of 48, ensuring that there will never be a need to send partially filled cells after segmentation. A bit in the PTI field in the cell headers is used to indicate when the last cell of a PDU is transmitted, so that one PDU can be distinguished from the one that follows.

20-3.6. Services: Connection-Oriented and Connectionless

The issues such as routing decisions and architectures have a major impact on connection-oriented services, where B-ISDN/ATM end nodes have to maintain or get access to lookup tables which translate destination addresses into circuit paths. These circuit path lookup tables which differ at every node must be maintained in a quasi-real-time fashion. This will have to be done by some kind of routing protocol.

One way to resolve this problem is to make it an internal network problem and use a connectionless service as described in ITU-T Rec. I.364 [Ref. 6]. We must keep in mind that ATM is basically a connection-oriented service. Here we are going to adapt it to provide a connectionless service.

20-3.6.1. Functional Architecture

The provision of connectionless data service in the B-ISDN is carried out by means of ATM switches and connectionless service functions (CLSF). ATM switches support the transport of connectionless data units in the B-ISDN between specific functional groups where the CLSF handles the connectionless protocol and provides for the adaptation of the connectionless data units into ATM cells to be transferred in a connection-oriented environment. As shown in Figure 20-U.20, CLSF functional groups may be located outside the B-ISDN, in a private connectionless network or in a specialized service provider, or inside the B-ISDN.

The ATM switching is performed by the ATM nodes (ATM switch/cross-connect) which are a functional part of the ATM transport network. The CLSF functional group terminates the B-ISDN connectionless protocol and includes functions for the adaptation of the connectionless protocol to the intrinsically connection-oriented ATM layer protocol. These latter functions are performed by the ATM adaptation layer Type 3/4 (AAL-3/4), while the CLSF group terminations are carried out by the services layer above the AAL called the connectionless network access protocol (CLNAP).

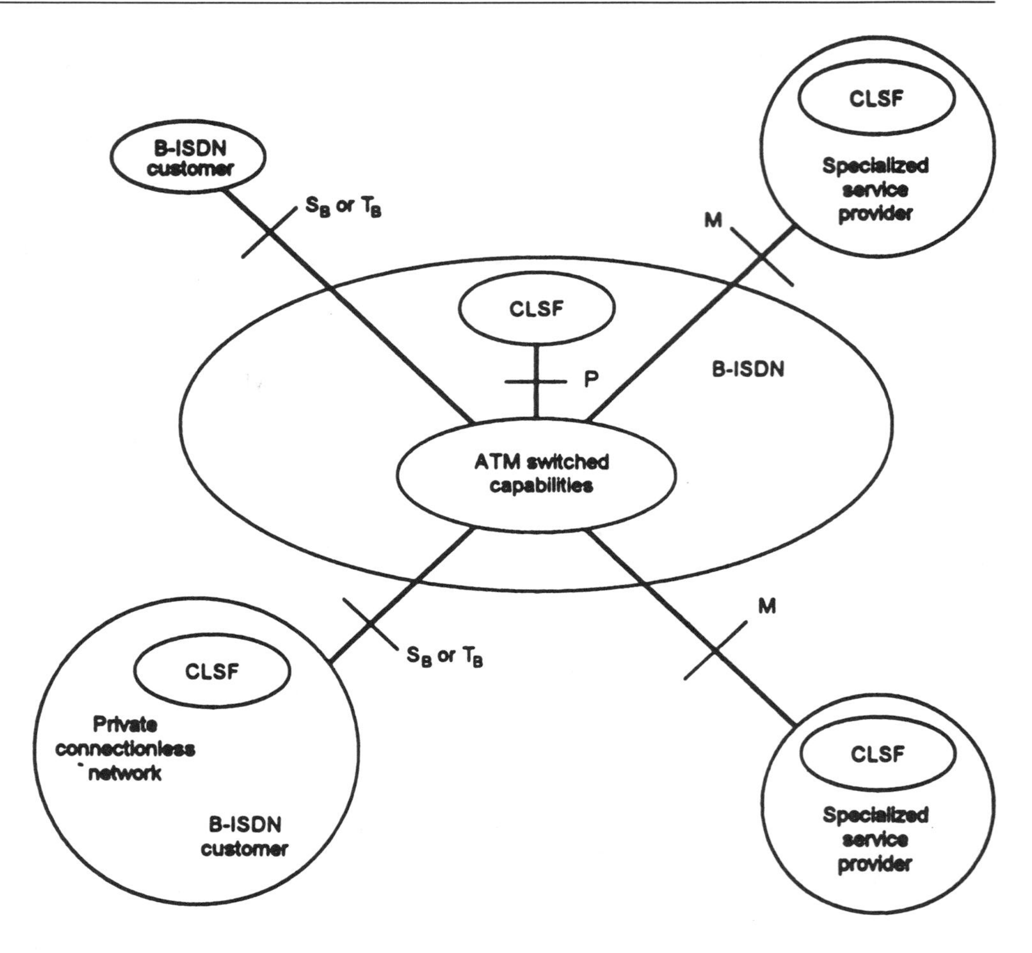

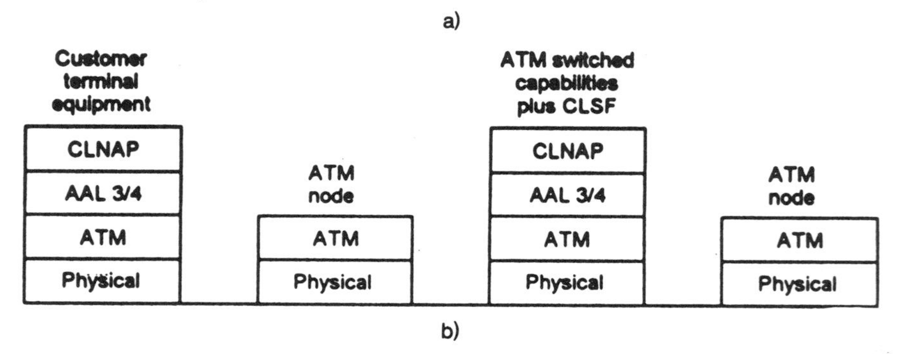

FIGURE 20-U.20. a) Reference configuration for the provision of the CL (connectionless) data service in the B-ISDN. (From Figure 1/I.364, ITU-T Rec. 5.364, Geneva, Mar 1993, page 2 [Ref. 6].) b) General protocol structure for provision of CL data service in B-ISDN.

The CL protocol includes functions such as routing, addressing, and quality of service (QoS) selection. In order to perform the routing of CL data units, the CLSF has to interact with the control/management planes of the underlying ATM network.

The general protocol structure for the provision of connectionless (CL) data service is shown in Figure 20-U.20. Figure 20-U.21 shows the protocol architecture

CLNAP user layer
CLNAP
Type 3/4 AAL
ATM
Physical

FIGURE 20-U.21. Protocol architecture for supporting connectionless service.

for supporting connectionless layer service. The CLNAP (connectionless network access protocol) layer uses the Type 3/4 AAL unassured service and includes the necessary functionality to provide the connectionless layer service.

The connectionless service layer provides for transparent transfer of variable-size data units from source to one or more destinations in a manner such that lost or corrupted data units are not retransmitted. This transfer is performed using a connectionless technique, including embedding destination and source addresses into each data unit.

20-3.6.2. CLNAP Protocol Data Unit (PDU) Structure and Encoding

Figure 20-U.22 shows the detailed structure of the CLNAP-PDU which contains the following fields:

Destination address and source address. These 8-octet fields each contain a 4-bit *address type* subfield, followed by the 60-bit *address* subfield. The "address-type" subfield indicates whether the "address" subfield contains a publicly administered 60-bit individual address or a publicly administered 60-bit group address. The "address" subfield indicates to which CLNAP-entity(ies) the CLNAP-PDU is destined; and in the case of the source address, it indicates the CLNAP-entity that sourced the CLNAP-PDU. The encoding of the "address-type" and "address" is shown in Figures 20-U.23 and 20-U.24. The address is structured in accordance with ITU-T Rec. E.164 [Ref. 7].

Higher-layer protocol identifier (HLPI). This 6-bit field is used to identify the CLNAP user layer entity which the CLNAP-SDU is to be passed to at the destination node. It is transparently carried end-to-end by the network.

PAD length. This 2-bit field gives the length of the PAD field (0–3 octets). The number of PAD octets is such that the total length of the user information field and the PAD field together is an integral multiple of four octets (32 bits).

CRC indication bit (CIB). This 1-bit field indicates the presence (if CIB = 1) or absence (if CIB = 0) of a 32-bit CRC field.

Header extension length (HEL). This 3-bit field can take on any value from 0 to 5 and indicates the number of 32-bit words in the header extension field.

Reserved. This 16-bit field is reserved for future use. Its default value is 0.

Header extension. This variable-length field can range from 0 to 20 octets. Its length is indicated by the value of the header extension length field (see above). In the case where the header extension length (HEL) is not equal to zero, all unused octets in the header extension are set to zero. The information carried in the header extension is structured into information entities. An information entity

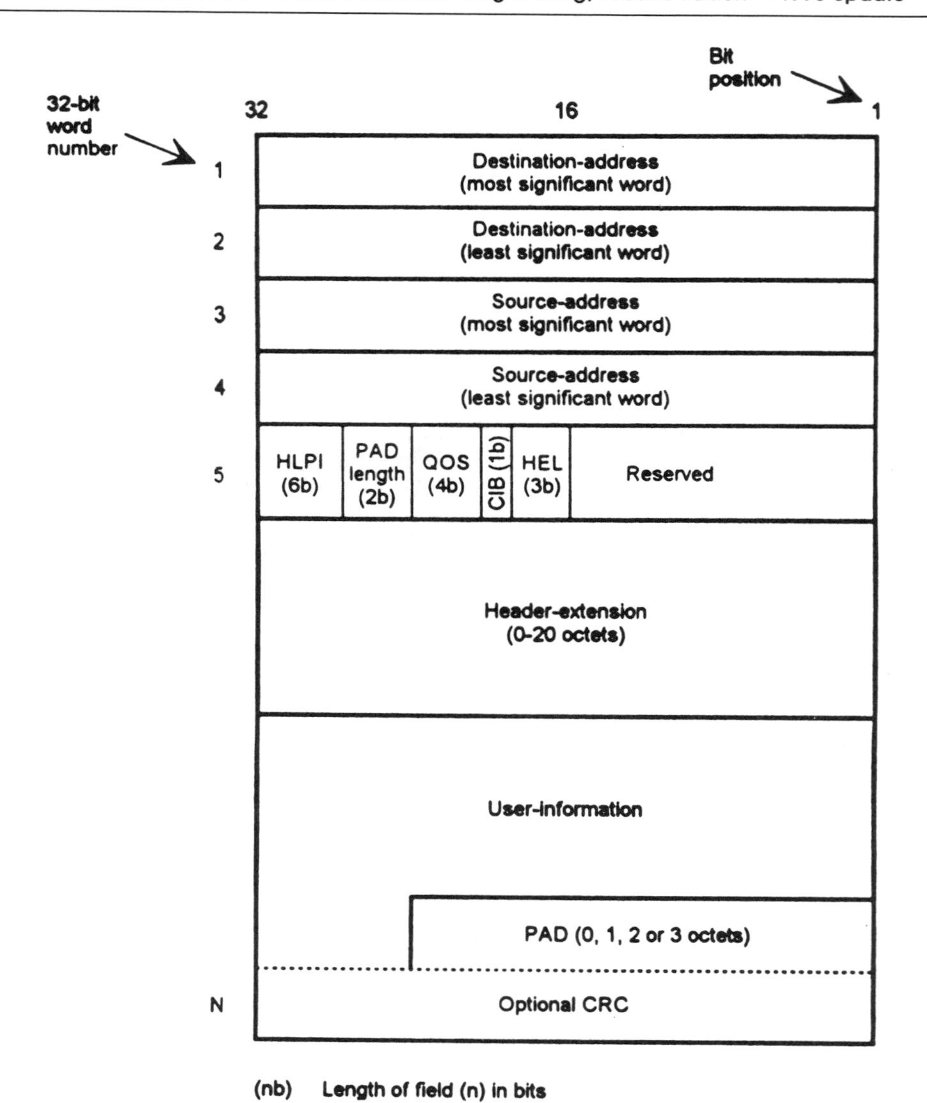

FIGURE 20-U.22. Structure of the CLNAP-PDU. (From Figure 5/I.364, ITU-T Rec. I.364, Geneva, Mar. 1993, page 7 [Ref. 6].)

Address type	Meaning
1100	60-bit publicly administered individual address
1110	60-bit publicly administered group address

FIGURE 20-U.23. Destination address field.

Address type	Meaning
1100	60-bit publicly administered individual address

FIGURE 20-U.24. Source address field.

(element) consists of (in this order) element length, element type, and element payload:

ELEMENT LENGTH. This is a 1-octet field and contains the combined lengths of the element length, element type, and element payload in octets.

ELEMENT TYPE. This is also a 1-octet field and contains a binary coded value which indicates the type of information found in the element payload field.

ELEMENT PAYLOAD. This is a variable-length field and contains the information indicated by the element type field.

User information. This field is variable length up to 9188 octets and is used to carry the CLNAP-SDU.

PAD. This field is 0, 1, 2, or 3 octets in length and is coded as all zeros. Within each CLNAP-PDU the length of this field is selected such that the length of the resulting CLNAP-PDU is aligned on a 32-bit boundary.

CRC. This optional 32-bit field may be present or absent as indicated by the CIB field. The field contains the result of a standard CRC-32 calculation performed over the CLNAP-PDU with the "reserved" field always treated as if it were coded as all zeros.

20-3.7. Some Aspects of a B-ISDN / ATM Network

20-3.7.1. ATM Routing and Switching

An ATM transmission path supports virtual paths (VPs), and inside virtual paths are virtual channels (VCs) as shown in Figure 20-U.25.

As we discussed in Section 20-3.3.1, each ATM cell contains a label in its header to explicitly identify the VC to which the cell belongs. This label consists of two parts: a virtual channel identifier (VCI) and a virtual path identifier (VPI).

Virtual Channel Level. Virtual channel (VC) is a generic term used to describe a unidirectional communication capability for the transport of ATM cells. A VCI identifies a particular VC link for a given virtual path connection (VPC). A specific value of VCI is assigned each time a VC is switched in the network. A VC link is a unidirectional capability for the transport of ATM cells between two consecutive ATM entities where the VCI value is translated. A VC link is originated or terminated by the assignment or removal of the VCI value.

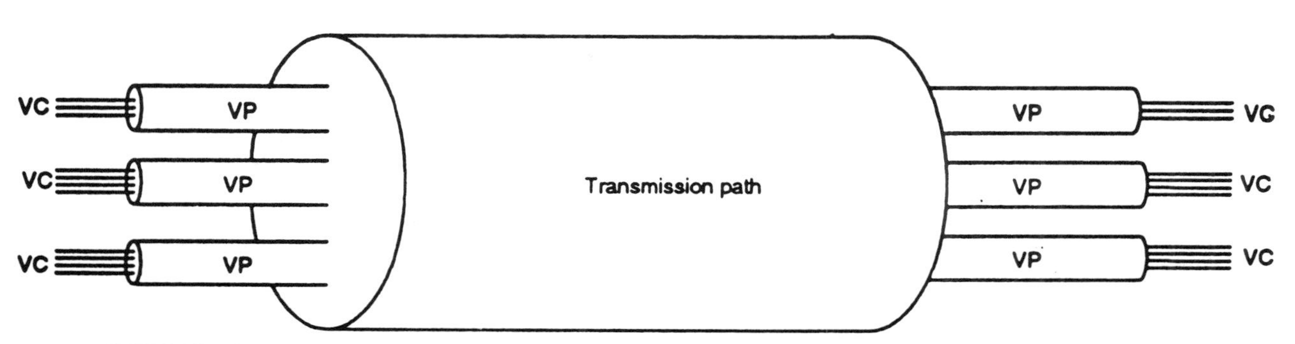

FIGURE 20-U.25. Relationship between the VC, the VP, and the transmission path.

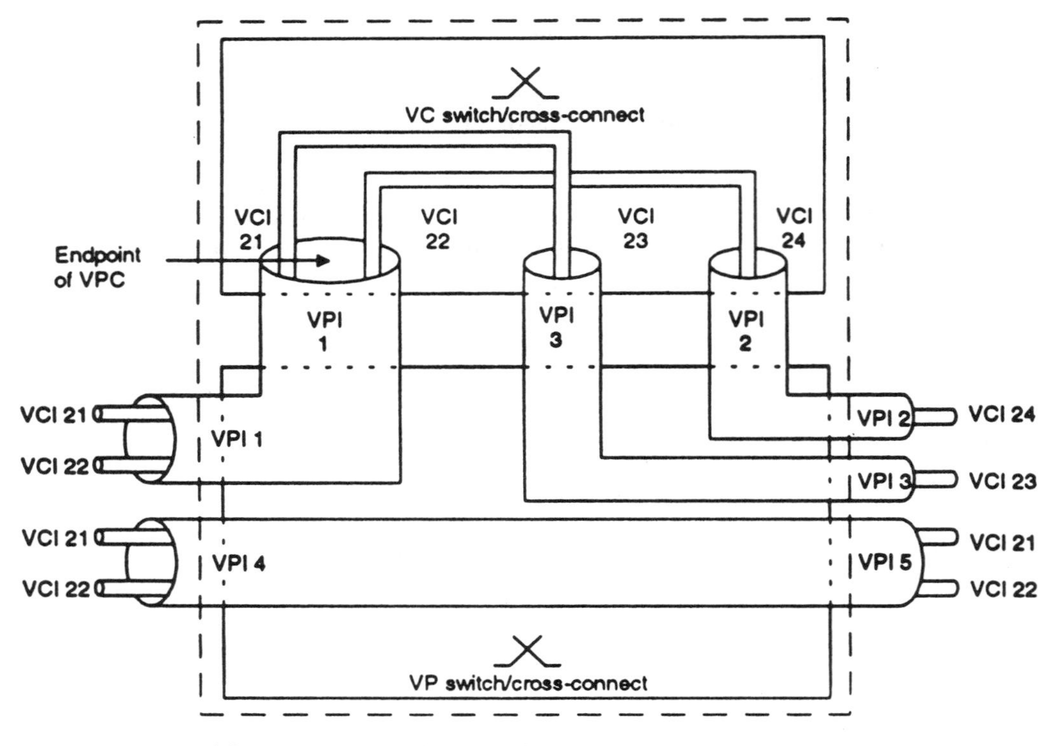

a) Representation of VC and VP switching

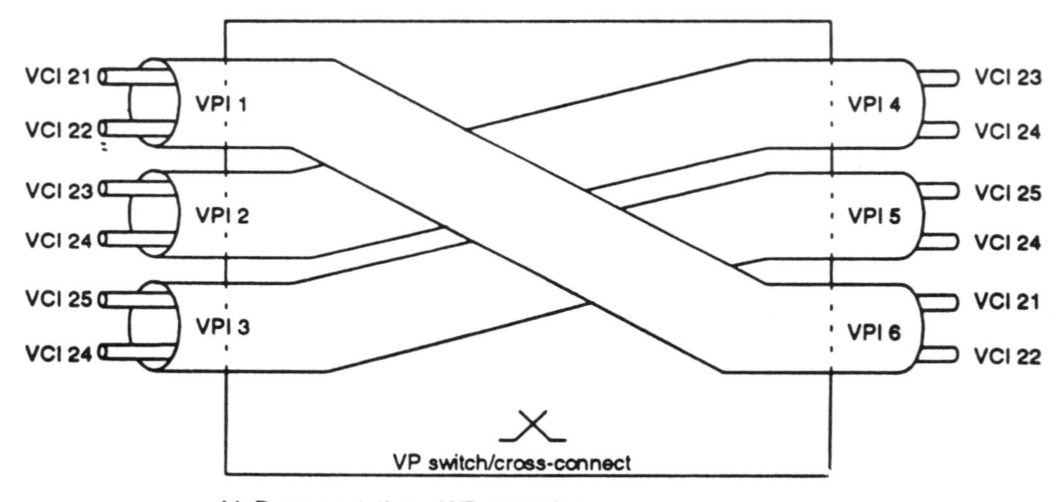

b) Representation of VP switching

FIGURE 20-U.26. Representation of the VP and VC switching hierarchy. (From Figure 4/I.311, ITU-T Rec. I.311, 3/93, page 5 [Ref. 8].)

Routing functions of VC are done at a VC switch/cross-connect.* The routing involves translation of the VCI values of the incoming VC links into the VCI values of the outgoing VC links.

*A VC cross-connect is a network element which connects VC links. It terminates VPCs and translates VCI values and is directed by management plane functions, not by control plane functions.

Virtual channel links are concatenated to form a virtual channel connection (VCC). A VCC extends between two VCC endpoints or, in the case of point-to-multipoint arrangements, more than two VCC endpoints. A VCC endpoint is the point where the cell information field is exchanged between the ATM layer and the user of the ATM layer service.

At the VC level, VCCs are provided for the purpose of user–user, user–network, or network–network information transfer. Cell sequence integrity is preserved by the ATM layer for cells belonging to the same VCC.

Virtual Path Level. Virtual path (VP) is a generic term for a bundle of virtual channel links; all the links in a bundle have the same endpoints.

A VPI identifies a group of VC links, at a given reference point, that share the same VPC. A specific value of VPI is assigned each time a VP is switched in the network. A VP link is a unidirectional capability for the transport of ATM cells between two consecutive ATM entities where the VPI value is translated. A VP link is originated or terminated by the assignment or removal of the VPI value.

Routing functions for VPs are performed at a VP switch/cross-connect. This routing involves translation of the VPI values of the incoming VP links into the VPI values of the outgoing VP links. VP links are concatenated to form a VPC. A VPC extends between two VPC endpoints or, in the case of point-to-multipoint arrangements, more than two VPC endpoints. A VPC endpoint is the point where the VCIs are originated, translated, or terminated. At the VP level, VPCs are provided for the purpose of user–user, user–network, and network–network information transfer.

When VPCs are switched, the VPC supporting the incoming VC links is terminated first and a new outgoing VPC is then created. Cell sequence integrity is preserved by the ATM layer for cells belonging to the same VPC. Thus cell sequence integrity is preserved for each VC link within a VPC.

Figure 20-U.26 is a representation of the VP and VC switching hierarchy where the physical layer is the lowest layer composed of, from bottom up, a regenerator section level, digital section level, and transmission path level. The ATM layer resides just above the physical layer and is composed of the VP level and just above that is the VC level.

20-3.8. Signaling Requirements

20-3.8.1. Setup and release of VCCs

The setup and release of VCCs at the user–network interface (UNI) can be performed in various ways:

- Without using signaling procedures. Circuits are set up at subscription with permanent or semipermanent connections.
- By meta-signaling procedures where a special VCC is used to establish or release a VCC used for signaling. Meta-signaling is a simple protocol used to establish and remove signaling channels. All information interchanges in meta-signaling are carried out via single-cell messages.
- User-to-network signaling procedures such as a signaling VCC to establish or release a VCC used for end-to-end connectivity.
- User-to-user signaling procedures such as a signaling VCC to establish or release a VCC within a pre-established VPC between two UNIs.

20-3.8.2. Signaling Virtual Channels

Requirements for Signaling Virtual Channels. For a point-to-point signaling configuration, the requirements for signaling virtual channels are as follows:

1. One virtual channel connection in each direction is allocated to each signaling entity. The same VPI/VCI value is used in both directions. A standardized VCI value is used for point-to-point signaling virtual channel (SVC).
2. In general, a signaling entity can control, by means of associated point-to-point SVCs, user-VCs belonging to any of the VPs terminated in the same network element.
3. As a network option, the user-VCs controlled by a signaling entity can be constrained such that each controlled user-VC is in either upstream or downstream VPs containing the point-to-point SVCs of the signaling entity.

For point-to-multipoint signaling configurations, the requirements for signaling virtual channels are as follows:

1. *Point-to-Point Signaling Virtual Channels.* For point-to-point signaling, one virtual channel connection in each direction is allocated to each signaling entity. The same VPI/VCI value is used in both directions.
2. *General Broadcast Signaling Virtual Channels.* The general broadcast signaling virtual channel (GBSVC) may be used for call offering in all cases. In cases where the "point" does not implement service profiles or where "the multipoints" do not support service profile identification, the GBSVC is used for call offering. The specific VCI value for general broadcast signaling is reserved per VP at the UNI. Only when meta-signaling is used in a VP is the GBSVC activated in the VP.
3. *Selective Broadcast Signaling Virtual Channels.* Instead of the GBSVC, a virtual channel connection for selective broadcast signaling (SBS) can be used for call offering, in cases where a specific service profile is used. No other uses for SBSVCs are foreseen.

20-3.8.3. Meta-Signaling

Meta-Signaling Requirements. A meta-signaling channel manages signaling virtual channels only within its own VP pair. In VPI = 0, the meta-signaling virtual channel is always present and has a standardized VCI value.

Meta-signaling VC is activated at VP establishment. The signaling virtual channel (SVC) is assigned and removed when necessary. A specific VCI value for meta-signaling is reserved per VP at the UNI. For a VP with point-to-multipoint signaling configuration, meta-signaling is required and the meta-signal VC within this VP is activated.

The user negotiates the SVC bandwidth parameter value. The meta-signaling virtual channel (MVSC) bandwidth has a default value. The bandwidth can be changed by mutual agreement between a network operator and user.

Meta-Signaling Functions at the User Access. In order to establish, check, and release the point-to-point and selective broadcast signaling virtual channel connections, meta-signaling procedures are provided. For each direction, meta-signaling is carried out in a permanent virtual channel connection having a standardized

VCI value. The channel is called the meta-signaling virtual channel. The meta-signaling protocol is terminated in the ATM layer management entity.

The meta-signaling function is required to:

- Manage the allocation of capacity to signaling channels
- Establish, release, and check the status of signaling channels
- Provide a means to associate a signaling endpoint with a service profile if service profiles are supported
- Provide a means to distinguish between simultaneous requests

Meta-signaling should be able to be supported on any VP; however, meta-signaling can only control signaling VCs within its VP.

20-3.9. Quality of Service (QoS)

20-3.9.1. ATM Service Quality Review

A basic performance measure for any digital data communication system is bit error rate (BER). Well-designed fiber-optic links will predominate now and into the foreseeable future. We may expect BERs from such links on the order of 1×10^{-10} and with end-to-end performance better than 1×10^{-9}.* Thus other performance issues may dominate the scene. These may be called ATM unique QoS items, namely:

- Cell transfer delay
- Cell delay variation
- Cell loss ratio
- Mean cell transfer delay
- Cell error ratio
- Severely errored cell block ratio
- Cell misinsertion rate

Definitions

Cell Event

1. A "cell exit event" occurs when the first bit of an ATM cell has completed transmission out of an end-user device to a private ATM network element across the "private UNI" measurement point, or out of a private ATM network element to a public ATM network element across the "public UNI" measurement point, or out of an end-user device to a public ATM network across the "public UNI" measurement point.
2. A "cell entry event" occurs when the last bit of an ATM cell has completed transmission into an end-user device from a private ATM network element across the "private UNI" measurement point, or into a private ATM network from a public ATM network element across the "public UNI" measurement point, or into an end-user device from a public ATM network element across the "public UNI" measurement point.

*Opinion of the reference publication (Ref. 9), not necessarily of the author.

ATM Cell Transfer Outcome. The following are possible cell transfer outcomes between measurement points for transmitted cells (ITU-T definitions):

1. *Successful Cell Transfer Outcome*. The cell is received corresponding to the transmitted cell with a specified time T_{max}. The binary content of the received cell conforms exactly to the corresponding cell payload, and the cell is received with a valid header field after header error control procedures are completed.
2. *Errored Cell Outcome*. The cell is received corresponding to the transmitted cell within a specified time T_{max}. The binary content of the received cell payload differs from that of the corresponding transmitted cell, or the cell is received with an invalid header field after the header error control procedures are completed.
3. *Lost Cell Outcome*. No cell is received corresponding to the transmitted cell within a specified time T_{max} (Examples: "never showed up" or "late.")
4. *Misinserted Cell Outcome*. A received cell for which there is no corresponding transmitted cell.
5. *Severely Errored Cell Block Outcome*. When M or more lost cell outcomes, misinserted cell outcomes, or errored cell outcomes are observed in a receiver cell block of N cells transmitted consecutively on a given connection.

20-3.9.2. Cell Transfer Delay

Cell transfer delay is defined as the elapsed time between a cell exit event at the measurement point 1 (e.g., at the source UNI) and the corresponding cell entry event at measurement point 2 (e.g., the destination UNI) for a particular connection. The cell transfer delay between two measurement points is the sum of the total inter-ATM node transmission delay and the total ATM node processing delay between MP_1 and MP_2.

In addition to the normal delay that one would expect for a cell to traverse a network, extra delay is added in the ATM network at each ATM switch. One cause of this delay is asynchronous digital multiplexing. Where this method is employed, two cells directed toward the same output of an ATM switch or cross-connect can result in contention.

One or more cells are held in a buffer until the contention is resolved. Thus the second cell suffers additional delay. Delay of a cell depends on the traffic intensity within a switch which influences the probability of contention.

The asynchronous path of each ATM cell also contributes to cell delay. Cells can be delayed one or many cell periods, depending on traffic intensity, switch sizing, and the transmission path taken through the network.

20-3.9.3. Cell Delay Variation

ATM traffic by definition is asynchronous, magnifying transmission delay. Delay is also inconsistent across the network. It can be a function of time (i.e., a moment in time), network design/switch design (such as buffer size), and traffic characteristics at that moment of time. The result is cell delay variation (CDV).

CDV can have several deleterious effects. The dispersion effect, or spreading out, of cell interarrival times can impact signaling functions or the reassembly of cell user data. Another effect is called *clumping*. This occurs when the interarrival times between transmitted cells shorten. One can imagine how this could affect the instantaneous network capacity and how it can impact other services using the network.

There are two performance parameters associated with CDV: 1-point CDV and 2-point CDV.

The 1-point CDV describes variability in the pattern of cell arrival events observed at a single measurement point with reference to the negotiated peak rate 1/T as defined in ITU-T Rec. I.371 [Ref. 9].

The 2-point CDV describes variability in the pattern of cell arrival events as observed at the output of a connection portion (MP_2) with reference to the pattern of the corresponding events observed at the input to the connection portion (MP_1).

20-3.9.4. Cell Loss Ratio

Cell loss may not be uncommon in an ATM network. There are two basic causes of cell loss: error in cell header or network congestion.

Cells with header errors are automatically discarded. This prevents (a) misrouting of errored cells and (b) the possibility of privacy and security breaches.

Switch buffer overflow can also cause cell loss. It is in these buffers that cells are held in prioritized queues. If there is congestion, cells in a queue may be discarded selectively in accordance with their level of priority. Here enters the cell loss priority (CLP) bit discussed in Section 20-3.1. Cells with this bit set to 1 are discarded in preference to other, more critical cells. In this way, buffer fill can be reduced to prevent overflow.

Cell loss ratio is defined for an ATM connection as

$$\frac{\text{Lost cells}}{\text{Total transmitted cells}}$$

Lost and transmitted cells counted in severely errored cell blocks should be excluded from the cell population in computing cell loss ratio.

20-3.9.5. Mean Cell Transfer Delay

Mean cell transfer delay is defined as the arithmetic average of a specified number of cell transfer delays for one or more connections.

20-3.9.6. Cell Error Ratio

Cell error ratio is defined as follows for an ATM connection:

$$\frac{\text{Errored cells}}{\text{Successfully transferred cells} + \text{Errored cells}}$$

Successfully transferred cells and error cells contained in cell blocks counted as severely errored cell blocks should be excluded from the population used in calculating cell error ratio.

20-3.9.7. Severely Errored Cell Block Ratio

The severely errored cell block ratio for an ATM connection is defined as

$$\frac{\text{Severely errored cell blocks}}{\text{Total transmitted cell blocks}}$$

A cell block is a sequence of N cells transmitted consecutively on a given connection. A severely errored cell block outcome occurs when more than M

errored cells, lost cells, or misinserted cell outcomes are observed in a received cell block.

For practical measurement purposes, a cell block will normally correspond to the number of user information cells transmitted between successive OAM cells. The size of a cell block is to be specified.

20-3.9.8. Cell Misinsertion Rate

The cell misinsertion rate for an ATM connection is defined as

$$\frac{\text{Misinserted cells}}{\text{Time interval}}$$

Severely errored cell blocks should be excluded from the population when calculating the cell misinsertion rate. Cell misinsertion on a particular connection is most often caused by an undetected error in the header of a cell being transmitted on a different connection. This performance parameter is defined as a rate (rather than a ratio) because the mechanism producing misinserted cells is independent of the number of transmitted cells received on the corresponding connection.

20-3.10. Traffic Control and Congestion Control

20-3.10.1. Generic Functions

The following functions form a framework for managing and controlling traffic and congestion in ATM networks and are to be used in appropriate combinations:

1. *Network Resource Management (NRM).* Provision is used to allocate network resources in order to separate traffic flows in accordance with service characteristics.
2. *Connection Admission Control (CAC).* CAC is defined as a set of actions taken by the network during the call setup phase or during the call renegotiation phase in order to establish whether a VC or VP connection request can be accepted or rejected, or whether a request for re-allocation can be accommodated. Routing is part of connection admission control actions.
3. *Feedback Controls.* These are a set of actions taken by the network and by users to regulate the traffic submitted on ATM connections according to the state of network elements.
4. *Usage/Network Parameter Control (UPC/NPC).* This is a set of actions taken by the network to monitor and control traffic, in terms of traffic offered and validity of the ATM connection, at the user access and network access, respectively. Their main purpose is to protect network resources from malicious as well as unintentional misbehavior which can affect the QoS of other already established connections by detecting violations of negotiated parameters and taking appropriate actions.
5. *Priority Control.* The user may generate different priority traffic flows by using the CLP. A congested network element may selectively discard cells with low priority, if necessary, to protect as far as possible the network performance for cells with higher priority.

Figure 20-U.27 is a reference configuration for traffic and congestion control.

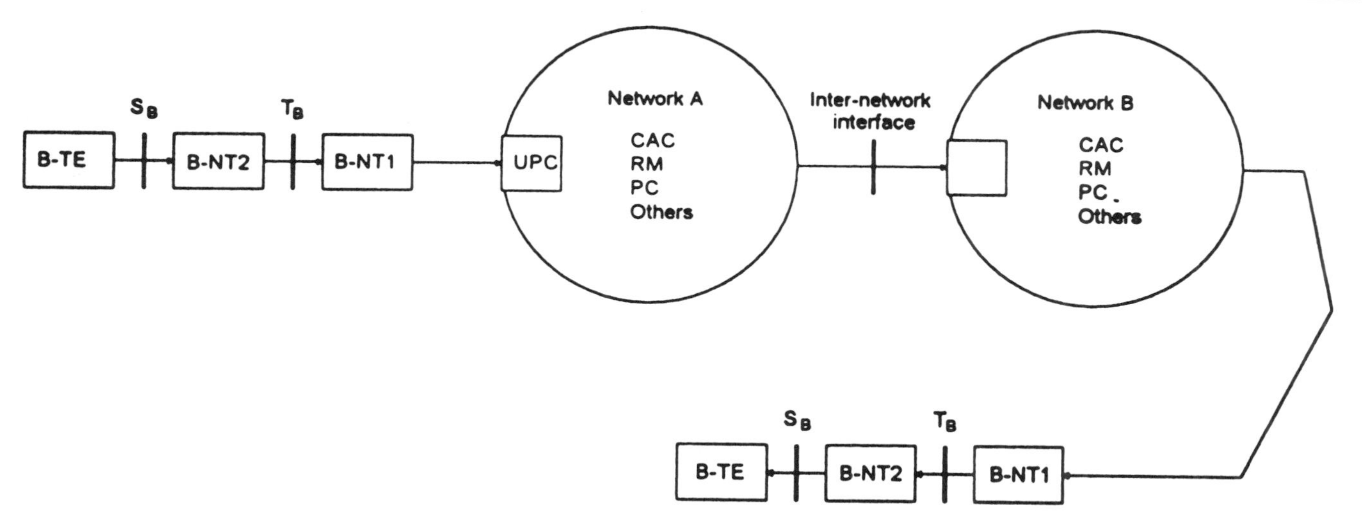

UPC Usage parameter control
CAC Connection admission control
PC Priority control
NPC Network parameter control
RM Resource management
Others For further study

Notes

1 NPC may apply as well at some intra-network NNIs.

2 The arrows are indicating the direction of the cell flow.

FIGURE 20-U.27. Reference configuration for traffic control and congestion control. (From Figure 1/I.371, ITU-T Rec. I.371, 3/93, page 3 [Ref. 9].)

20-3.10.2. Events, Actions, Time Scales, and Response Times

Figure 20-U.28 shows the time scales over which various traffic control and congestion control functions can operate. The response time defines how quickly the controls react. For example, call discarding can react on the order of the insertion time of a cell. Similarly, feedback controls can react on the time scale of round-trip propagation times. Because traffic control and resource management functions are needed at different time scales, no single function is likely to be sufficient.

20-3.10.3. Quality of Service, Network Performance, and Cell Loss Priority

QoS at the ATM layer is defined by a set of parameters such as cell delay, cell delay variation sensitivity, cell loss ratio, and so forth.

A user requests a specific ATM layer QoS from the QoS classes which a network provides. This is part of the *traffic contract* at connection establishment. It is a commitment for the network to meet the requested QoS as long as the user complies with the traffic contract. If the user violates the traffic contract, the network need not respect the agreed-upon QoS.

A user may request at most two QoS classes for a single ATM connection, which differ with respect to the cell loss ratio objectives. The cell loss priority (CLP) bit in the ATM header allows for two cell loss ratio objectives for a given ATM connection.

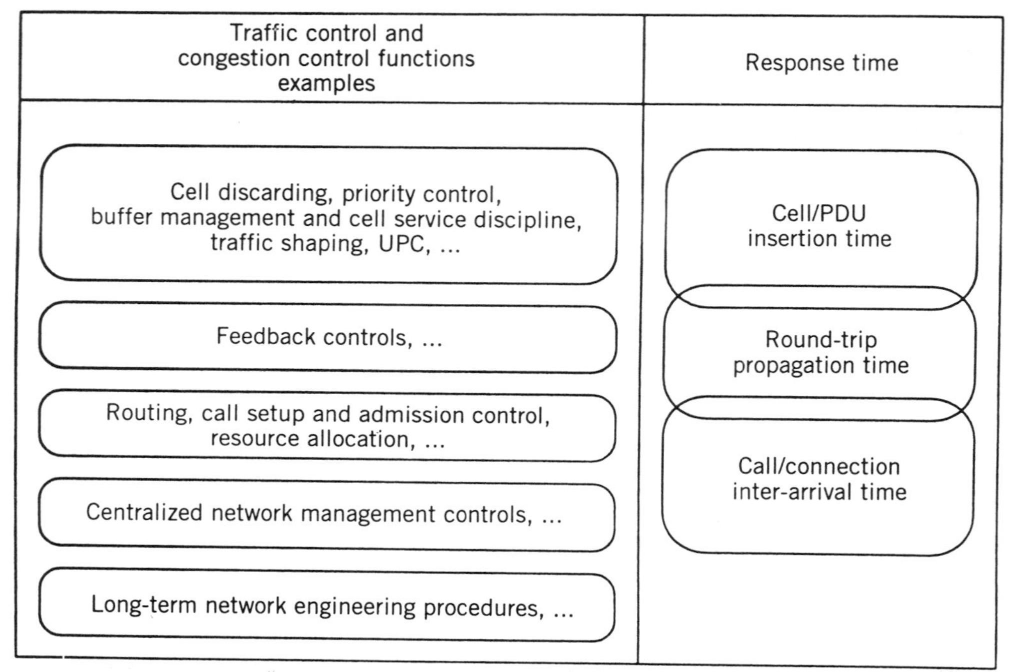

FIGURE 20-U.28. Control response times. (From Figure 2/I.371, ITU-T Rec. I.371, 3/93, page 4 [Ref. 9].)

Network performance objectives at the ATM SAP (service access point) are intended to capture the network ability to meet the requested ATM layer QoS. It is the role of upper layers, including the AAL, to translate this ATM layer QoS to any specific application-requested QoS.

20-3.10.4. Traffic Descriptors and Parameters

Traffic parameters describe traffic characteristics of an ATM connection. Traffic parameters are grouped into source traffic descriptors for exchanging information between the user and the network. Connection admission control procedures use source traffic descriptors to allocate resources and derive parameters for the operation of the UPC/NPC.

We now define the terms *traffic parameters* and *traffic descriptors*:

Traffic Parameters. A traffic parameter is a specification of a particular traffic aspect. It may be qualitative or quantitative. Traffic parameters, for example, may describe peak cell rate, average cell rate, burstiness, peak duration, and source type (such as telephone, videophone). Some of these traffic parameters are interdependent, such as burstiness with average and peak cell rate.

Traffic Descriptors. The ATM traffic descriptor is the generic list of traffic parameters which can be used to capture the intrinsic traffic characteristics of an ATM connection. A *source traffic descriptor* is the set of traffic parameters belonging to the ATM traffic descriptor used during the connection setup to capture the intrinsic traffic characteristics of the connection requested by the source.

Connection traffic descriptor specifies the traffic characteristics of the ATM connection at the public or private UNI. The connection traffic descriptor is the set of traffic parameters in the source traffic descriptor, cell delay variation (CDV) tolerance, and the conformance definition that is used to unambiguously specify the conforming cells of an ATM connection. Connection admission control (CAC) procedures will use the connection traffic descriptor to allocate resources and to derive parameter values for the operation of the UPC. The connection traffic descriptor contains the necessary information for conformance testing of cells of the ATM connection at the UNI.

Any traffic parameter and the CDV tolerance in a connection traffic descriptor should fulfill the following requirements:

- It should be understandable by the user or terminal equipment, and conformance testing should be possible as stated in the traffic contract.
- It should be useful in resource allocation schemes meeting network performance requirements as described in the traffic contract.
- It should be enforceable by the UPC.

20-3.10.5. User–Network Traffic Contract

Operable Conditions. CAC and UPC/NPC procedures require the knowledge of certain parameters to operate efficiently. For example, they should take into account the source traffic descriptor, the requested QoS, and the CDV tolerance (defined below) in order to decide whether the requested connection can be accepted.

The source traffic descriptor, the requested QoS for any given ATM connections, and the maximum CDV tolerance allocated to the CEQ (customer equipment) define the traffic contract at the T_B reference point (see Figure 20-U.27). Source traffic descriptors and QoS are declared by the user at connection setup by means of signaling or subscription. Whether the maximum allowable CDV tolerance is also negotiated on a subscription or on a per-connection basis is for further study by the ITU-T organization.

The CAC and UPC/NPC procedures are operator specific. Once the connection has been accepted, the value of the CAC and UPC/NPC parameters are set by the network on the basis of the network operator's policy.

ITU-T Rec. I.371 [Ref. 9] notes that all ATM connections handled by network connection-related functions (CRF) have to be declared and enforced by the UPC/NPC. ATM layer QoS can only be assured for compliant ATM connections. As an example, individual VCCs inside user end-to-end VPC are neither declared nor enforced at the UPC, and hence no ATM layer QoS can be assured for them.

Source Traffic Descriptor, Quality of Service, and Cell Loss Priority. If a user requests two levels of priority for an ATM connection, as indicated by the CLP bit value, the intrinsic traffic characteristics of both cell flow components have to be characterized in the source traffic descriptor. This is by means of a set of traffic parameters associated with the CLP = 0 component and a set of traffic parameters associated with the CLP = 0 + 1 component.

Impact of Cell Delay Variation on UPC/NPC and Resource Allocation. ATM layer functions such as cell multiplexing may alter the traffic characteristics of ATM connections by introducing cell delay variation, as shown in Figure 20-U.29. When cells from two or more ATM connections are multiplexed, cells of a given ATM connection may be delayed while cells of another ATM connection are being inserted at the output of the multiplexer. Similarly, some cells may be delayed while physical layer overhead or OAM cells are inserted. Therefore, some randomness affects the time interval between reception of ATM cell data requests at the

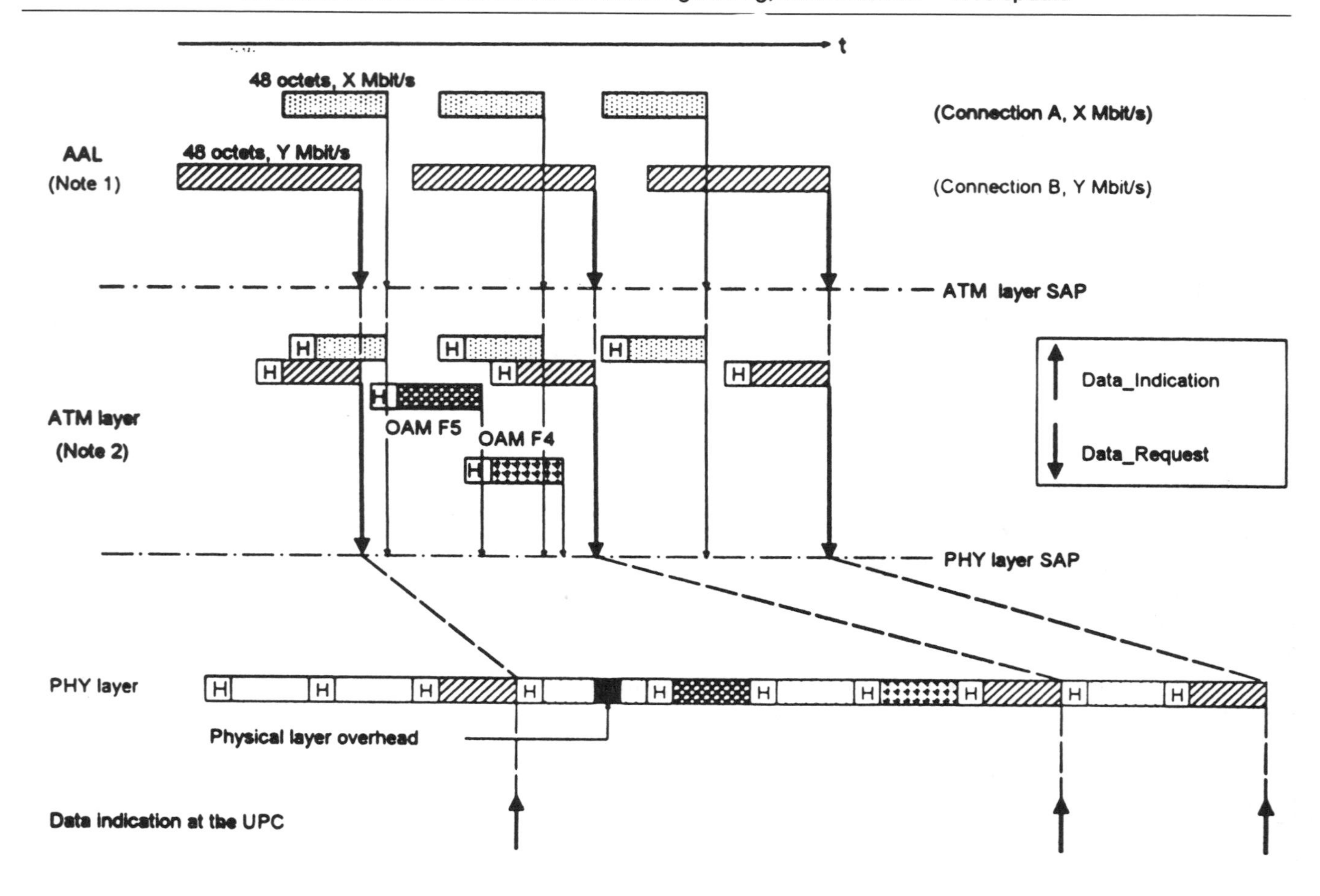

NOTES

1 ATM SDUs are accumulated at the upper layer service bit rate. Besides, CDV may also originate in AAL multiplexing.

2 GFC delay and delay variation is part of the delay and delay variation introduced by the ATM layer.

3 CDV may also be introduced by the network because of random queuing delays which are experienced by each cell in concentrators, switches and cross-connects.

FIGURE 20-U.29. Origins of cell delay variation. (From Figure 3/I.371, ITU-T Rec. I.371, page 7 [Ref. 9].)

endpoint of an ATM connection to the time that an ATM cell data indication is received at the UPC/NPC. Besides, AAL multiplexing may cause CDV.

The UPC/NPC mechanism should not discard or tag cells in an ATM connection if the source conforms to the source traffic descriptor negotiated at connection establishment. However, if the CDV is not bounded at a point where the UPC/NPC function is performed, it is not possible to design a suitable UPC/NPC mechanism and to allocate resources properly. Therefore, it is required that a maximum allowable value of CDV be standardized edge-to-edge (e.g., between the ATM connection endpoint and T_B, between T_B and an internetwork interface, and between internetwork interfaces; see Figure 12.29).

UPC/NPC should accommodate the effect of the maximum CDV allowed on ATM connections within the limit resulting from the accumulated CDV allocated to upstream subnetworks including customer equipment (CEQ). Traffic shaping partially compensates for the effects of CDV on the peak cell rate of the ATM connection. Examples of traffic-shaping mechanisms are respacing cells of individual ATM connections according to their peak cell rate or suitable queue service schemes.

Values of the cell delay variation are network performance issues. The definition of a source traffic descriptor and the standardization of a maximum allowable CDV may not be sufficient for a network to allocate resources properly. When allocating resources, the network should take into account the worst-case traffic passing through the UPC/NPC in order to avoid impairments to other ATM connections. The worst-case traffic depends on the specific implementation of the UPC/NPC. The tradeoffs between UPC/NPC complexity, worst-case traffic, and optimization of network resources are made at the discretion of network operators. The quantity of available network resources and the network performance to be provided for meeting QoS requirements can influence these tradeoffs.

Cell Conformance and Connection Compliance. Conformance applies to the cells as they pass the UNI and are in principle tested according to some combination of generic cell rate algorithms (GCRAs). The first cell of the connection initializes the algorithm, and from then on each cell is either conforming or not conforming. Because in all likelihood even with the best intentions a cell or two may be nonconforming, it is inappropriate for the network operator to only commit to the QoS objectives for connections, all of whose cells are conforming. Thus the term *compliant*, which is not precisely defined, is used for connection in which some of the cells may be nonconforming.

The precise definition of a compliant connection is left to the network operator. For any definition of a compliant connection, a connection for which all cells are conforming is identified as compliant.

Based on action of the UPC function, the network may decide whether a connection is compliant or not. The commitment by the network operator is to support the QoS for all connections that are compliant.

For compliant connections at the public UNI, the agreed QoS is supported for at least the number of cells equal to the conforming cells according to the conformance definition. For noncompliant connections, the network need not respect the agreed-upon QoS class. The conformance definition that defines conformity at the public UNI of the cells of the ATM connection uses a GCRA configuration in multiple instances to apply to (a) particular combinations of the CLP = 0 and CLP = 1 + 0 cell streams with regard to the peak cell rate and (b) particular combinations of CLP = 0, CLP = 1, and CLP = 0 + 1 cell streams with regard to the sustainable cell rate and burst tolerance. For example, the conformance definition may use the GCRA twice, once for peak cell rate of the aggregate (CLP = 0 + 1) cell stream and once for the sustainable cell rate of the CLP = 0 cell stream. The network operator may offer a limited set of alternative conformance definitions (all based on GCRA) from which the user may choose for a given ATM connection.

Generic Cell Rate Algorithm (GCRA). The GCRA is a virtual scheduling algorithm or a continuous-state leaky bucket algorithm as defined by the flow chart in Figure 20-U.30. The GCRA is used to operationally define the relationship between peak cell rate (PCR) and the CDV tolerance and the relationship between sustained cell rate (SCR) and the burst tolerance. In addition, for the cell flow of an ATM connection, the GCRA is used to specify the conformance at the public or private UNI to declared values of the above two tolerances, as well as declared values of the traffic parameters "PCR" and "SCR and burst tolerance."

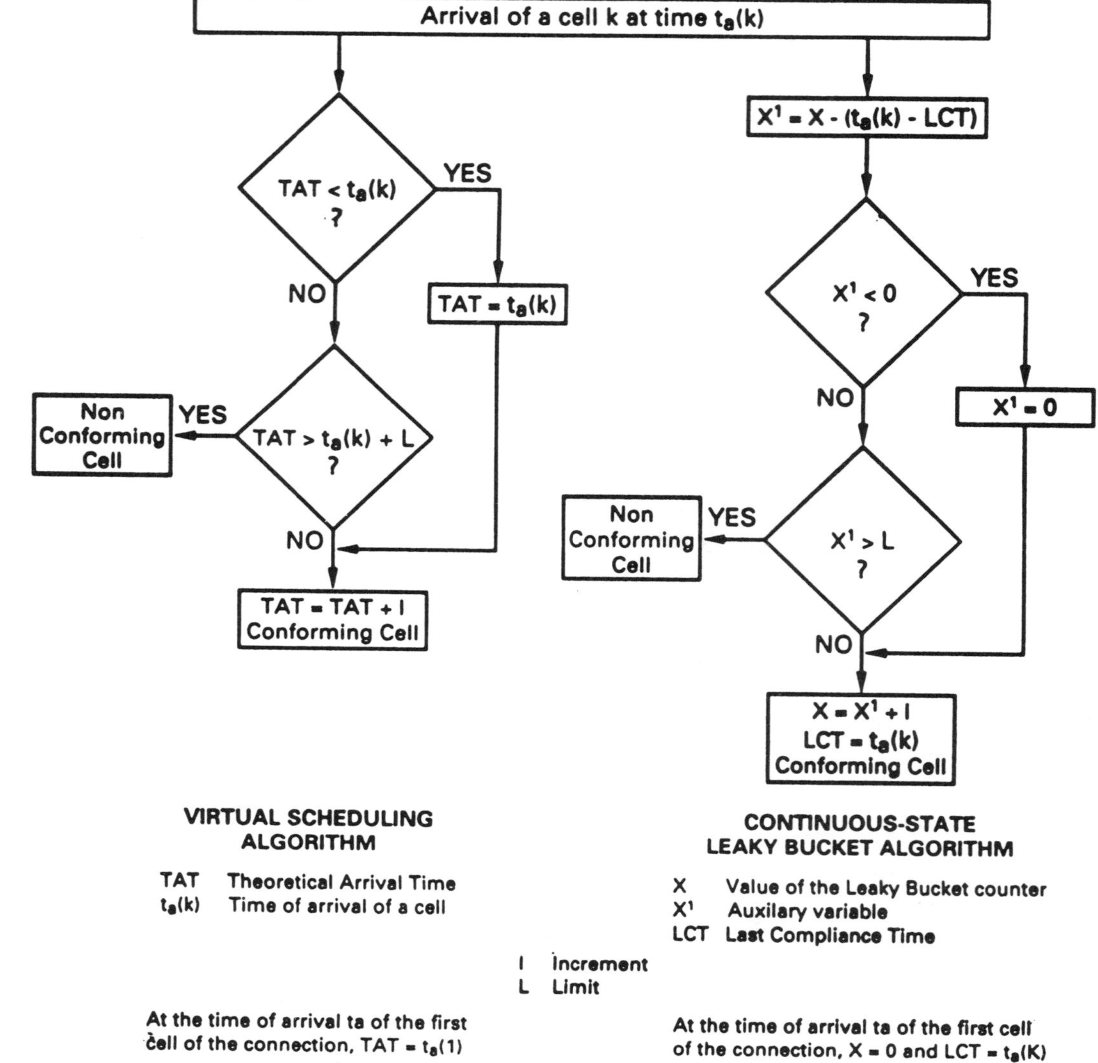

FIGURE 20-U.30. Equivalent versions of the generic cell rate algorithm (GCRA). (From ATM Forum, Figure 3-15, page 79 [Ref. 2], courtesy of Prentice-Hall, Inc.)

For each cell arrival, the GCRA determines whether the cell is conforming with the traffic contract of the connection, and thus the GCRA is used to provide the formal definition of traffic conformance to the traffic contract. Although traffic conformance is defined in terms of the GCRA, the network provider is not obligated to use this algorithm (or this algorithm with the same parameter values) for the UPC. Rather, the network provider may use any UPC as long as the operation of the UPC does not violate the QoS objectives of a compliant connection.

The GCRA depends only on two parameters: the increment I and the limit L. These parameters have been denoted by T and τ, respectively, above and in ITU-T Rec. I.371 [Ref. 9] but have been given more generic labels (by the ATM Forum) herein because the GCRA is used in multiple instances. We will now use the ATM Forum notation "GCRA (I, L)," which means "the GCRA with the value of the increment parameter set equal to I and the value of the limit parameter set equal to L."

The GCRA is formally defined in Figure 20-U.30, which ATM Forum took as a generic version of Figure 1 in Annex 1 of I.371. The two algorithms in Figure 20-U.30 are equivalent in the sense that for any sequence of cell arrival times $[t_a(k), k \geq 1]$, the two algorithms determine the same cells to be conforming and thus the same cells to be nonconforming. The two algorithms are easily compared if one notices that at each arrival epoch, $t_a(k)$, and after the algorithms have been executed, TAT = X + LCT. See Figure 20-U.30.

The virtual scheduling algorithm updates a theoretical arrival time (TAT), which is the "nominal" arrival time of the cell assuming equally spaced cells when the source is active. If the actual arrival time of a cell is not "too" early relative to the TAT (in particular, if the actual arrival time is after TAT-L), then the cell is conforming; otherwise the cell is nonconforming.

The continuous-state leaky bucket algorithm can be viewed as a finite-capacity bucket whose real-valued content drains out at a continuous rate of 1 unit of content per time-unit and whose content is increased by the increment I for each conforming cell. Equivalently, it can be viewed as the work load in a finite-capacity queue or as a real-valued counter. If at a cell arrival the content of the bucket is less than or equal to the limit value, L, then the cell is conforming; otherwise the cell is nonconforming. The capacity of the bucket (i.e., the upper bound on the counter) is $L + I$.

Traffic Contract Parameter Specification. Peak cell rate for CLP = 0 + 1 is a mandatory traffic parameter to be explicitly or implicitly declared in any source traffic descriptor. In addition to the peak cell rate of an ATM connection, it is mandatory for the user to declare either explicitly or implicitly the cell delay variation tolerance τ within the relevant traffic contract.

Peak Cell Rate (PCR). The following definition applies to ATM connections supporting both CBR and VBR services [Ref. 9]:

> *The peak cell rate in the source traffic descriptor specifies an upper bound on the traffic that can be submitted on an ATM connection. Enforcement of this bound by the UPC/NPC allows the network operator to allocate sufficient resources to ensure that the performance objectives (e.g., for cell loss ratio) can be achieved.*

For switched ATM connections, the PCR for CLP = 0 + 1 and the QoS class must be explicitly specified for each direction in the connection establishment SETUP message.

The CDV tolerance must be either explicitly specified at subscription time or implicitly specified.

The sustainable cell rate (SCR) and burst tolerance is an optional traffic parameter set in the source traffic descriptor. If either SCR or burst tolerance is specified, then the other must be specified within the relevant traffic contract.

20-3.11. Transporting ATM Cells

20-3.11.1. In the DS3 Frame

One of the most popular higher-speed digital transmission systems in North America is DS3 operating at a nominal transmission rate of 45 Mbps. It is also

PLCP Framing		PO	POH	PLCP Payload	
A1	A2	P11	Z6	First ATM Cell	
A1	A2	P10	Z5	ATM Cell	
A1	A2	P9	Z4	ATM Cell	
A1	A2	P8	Z3	ATM Cell	
A1	A2	P7	Z2	ATM Cell	
A1	A2	P6	Z1	ATM Cell	
A1	A2	P5	X	ATM Cell	
A1	A2	P4	B1	ATM Cell	
A1	A2	P3	G1	ATM Cell	
A1	A2	P2	X	ATM Cell	
A1	A2	P1	X	ATM Cell	
A1	A2	P0	C1	Twelfth ATM Cell	Trailer
1 Octet	1 Octet	1 Octet	1 Octet	53 Octets	13 or 14 Nibbles

Object of BIP-8 Calculation (POH + PLCP Payload)

POI	Path Overhead Indicator
POH	Path Overhead
BIP-8	Bit Interleaved Parity - 8
X	Unassigned - Receiver required to ignore
A1, A2	Frame Alignment

Bellcore TA-NWT-001112

Bellcore TR-TSV-000773

FIGURE 20-U.31. Format of DS3 PLCP frame. (Courtesy of Hewlett-Packard Company, [Ref. 4].)

being widely implemented for transport of SMDS. The system used to map ATM cells into the DS3 format is the same as used for SMDS.

DS3 uses the physical layer convergence protocol (PLCP) to map ATM cells into its bit stream. A DS3 PLCP frame is shown in Figure 20-U.31.

There are 12 cells in a frame. Each cell is preceded by a 2-octet framing pattern (A1, A2), to enable the receiver to synchronize the cells. After the framing pattern there is an indicator consisting of one of 12 fixed-bit patterns used to identify the cell location within the frame (POI). This is followed by an octet of overhead information used for path management. The entire frame is then padded with either 13 or 14 nibbles (a nibble = 4 bits) of trailer to bring the transmission rate up to the exact DS3 bit rate. The DS3 frame has 125-μsec duration.

DS3 has to contend with network slips (added/dropped frames to accommodate synchronization alignment). Thus PLCP is padded with a variable number of stuff (justification) bits to accommodate possible timing slips. The C1 overhead octet indicates the length of padding. The bit interleaved parity (BIP) checks the payload and overhead functions for errors and performance degradation. This performance information is transmitted in the overhead.

20-3.11.2. DS1 Mapping

One approach to mapping ATM cells into a DS1 frame is to use a similar procedure as used on DS3 with PLCP. In this case, only 10 cells are bundled into a frame, and two of the Z overheads are removed. The padding in the frame is set at 6 octets. The entire frame takes 3 msec to transmit and spans many DS1 ESF (extended superframe) frames. This mapping is shown in Figure 20-U.32. Note the reference to L2_PDU, taken directly from SMDS sources (Bellcore). One must also consider the arithmetic. Each DS1 timeslot is 8 bits long, 1 octet. There are then 24 octets in a DS1 frame. This, of course, can lead to the second method of

1	1	1	1	← 53 Octets →	
A1	A2	P9	Z4	L2_PDU	
A1	A2	P8	Z3	L2_PDU	
A1	A2	P7	Z2	L2_PDU	
A1	A2	P6	Z1	L2_PDU	
A1	A2	P5	F1	L2_PDU	
A1	A2	P4	B1	L2_PDU	
A1	A2	P3	G1	L2_PDU	
A1	A2	P2	M2	L2_PDU	
A1	A2	P1	M1	L2_PDU	Trailer = 6 Octets
A1	A2	P0	C1	L2_PDU	

OH Byte	Function
A1, A2	Framing Bytes
P9-P0	Path Overhead Identifier Bytes
PLCP Path Overhead Bytes	
Z4-Z1	Growth Bytes
F1	PLCP Path User Channel
B1	BIP-8
G1	PLCP Status
M2-M1	SMDS Control Information
C1	Cycle/Stuff Counter Byte

3 msec

FIGURE 20-U.32. DS1 mapping with PLCP. (Courtesy of Hewlett-Packard Company [Ref. 4].)

carrying ATM cells in DS1, by directly mapping in ATM cells octet for octet (timeslot). This is done in groups of 53 octets (1 cell) and would, by necessity, cross DS1 frame boundaries to accommodate an ATM cell.

20-3.11.3. E1 Mapping

E1 PCM has a 2048-Mbps transmission rate. An E1 frame has 256 bits representing 32 channels or timeslots, 30 of which carry traffic. Timeslots (TS) 0 and 16 are reserved. TS0 is used for synchronization, and TS16 is used for signaling. The E1 frame is shown in Figure 20-U.33. From bits 9 to 128 and from bits 137 to 256 may be used for ATM cell mapping.

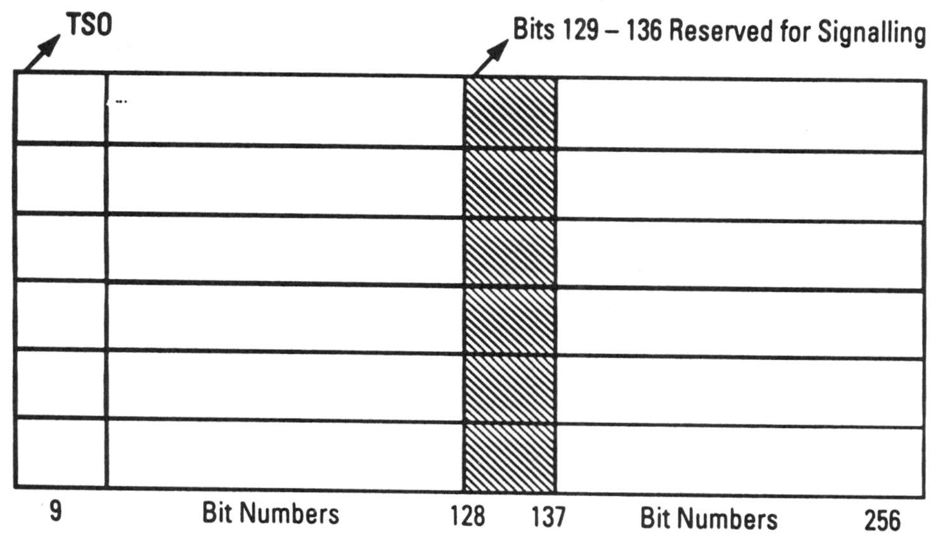

FIGURE 20-U.33. Mapping ATM cells directly into E1. (Courtesy of Hewlett-Packard Company [Ref. 4].)

ATM cells can also be directly mapped into special E3 and E4 frames. The first has 530 octets available for cells (i.e., 10 cells), and the second has 2160 octets (not evenly divisible).

20-3.11.4. Mapping ATM Cells into SDH

At STM-1 (155.520 Mbps). SDH is described in Chapter 11. Figure 20-U.34 shows the mapping procedure. The ATM cell stream is first mapped into the C-4 and then mapped into the VC-4 container along with the VC-4 path overhead. The ATM cell boundaries are aligned with the STM-1 octet boundaries. Because the C-4 capacity (2340 octets) is not an integer multiple of the cell length (53 octets), a cell may cross a C-4 boundary.

The AU-4 pointer (octets H1 and H2 in the SOH) is used for finding the first octet of the VC-4.

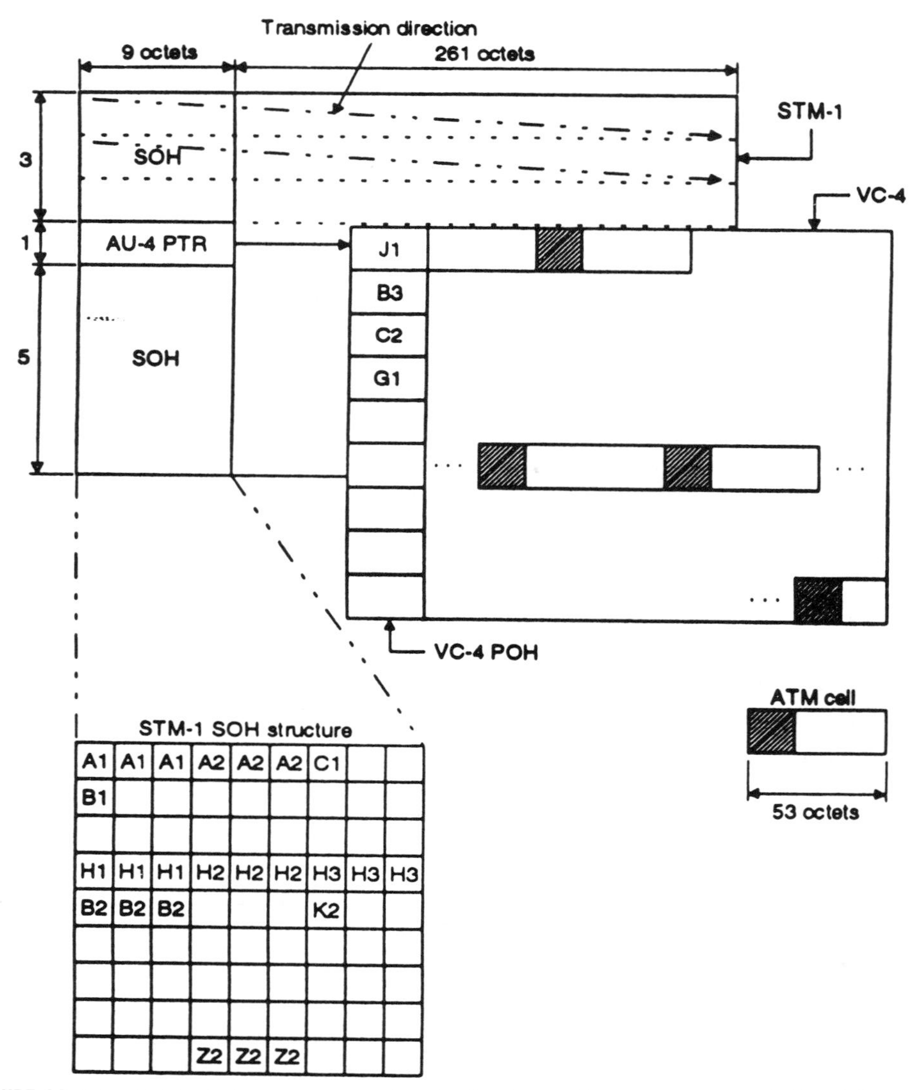

FIGURE 20-U.34. 155.520-Mbps frame structure for SDH-based UNI. (From Figure 8/I.432, ITU-T Rec. I.432, 3/93, page 13 [Ref. 3].)

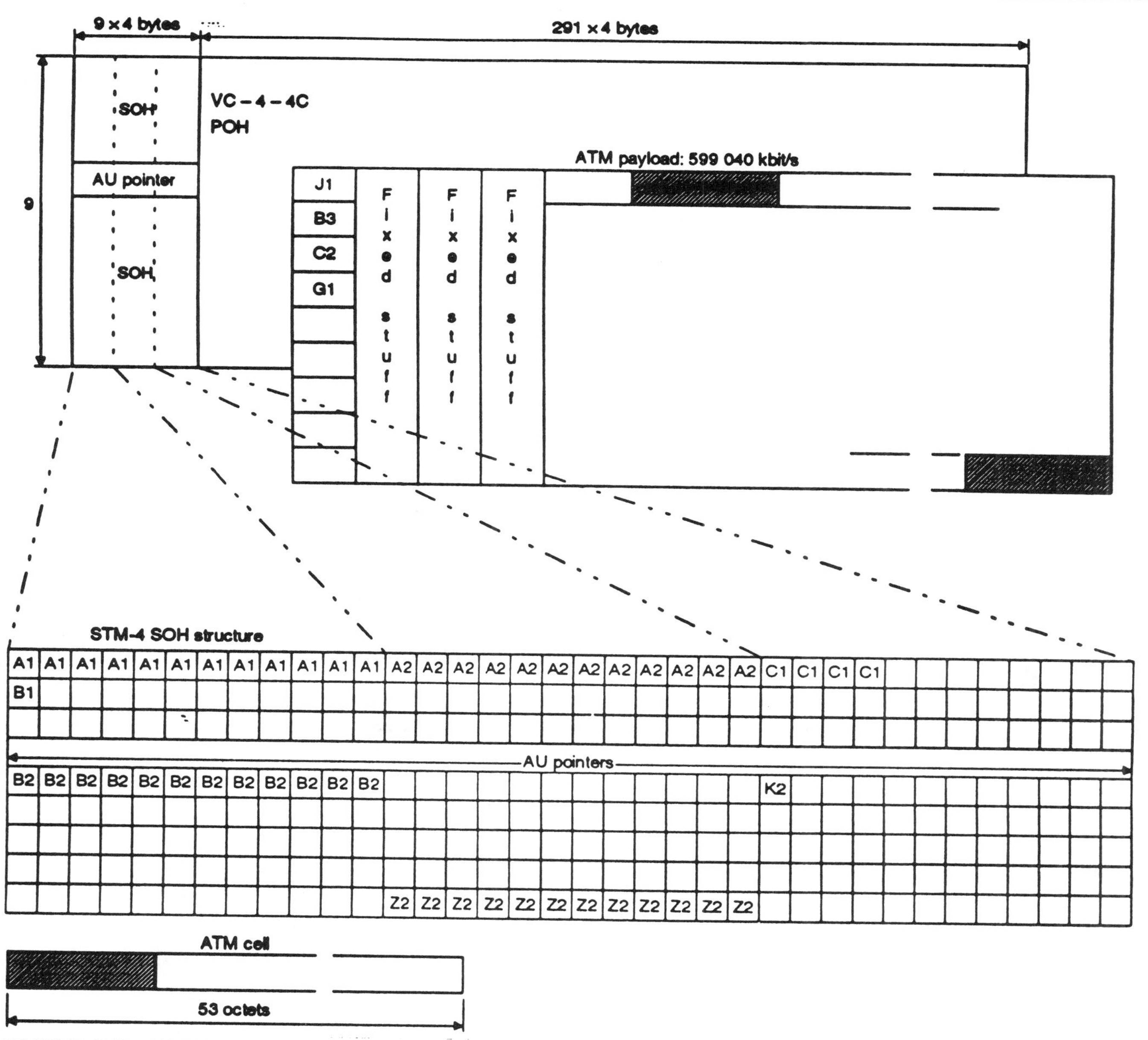

FIGURE 20-U.35. 622.080 Mbps frame structure for SDH-based UNI. (From Figure 10/I.432, ITU-T Rec. I.432, 3/93, page 15 [Ref. 3].)

At STM-4 (622.080 Mbps). As shown in Figure 20-U.35, the ATM cell stream is first mapped into C-4-4c and then packed into the VC-4-4c container along with the VC-4-4c path overhead. The ATM cell boundaries are aligned with STM-4 octet boundaries. Because the C-4-4c capacity (9360 octets) is not an integer multiple of the cell length (53 octets), a cell may cross a C-4-4c boundary.

The AU-pointers are used for finding the first octet of the VC-4-4c.

20-3.11.5. Mapping ATM Cells into SONET

ATM cells are mapped directly into the SONET payload (49.54 Mbps). As with SDH, the payload in octets is not an integer multiple of the cell length, and thus a

- via HEC Sync

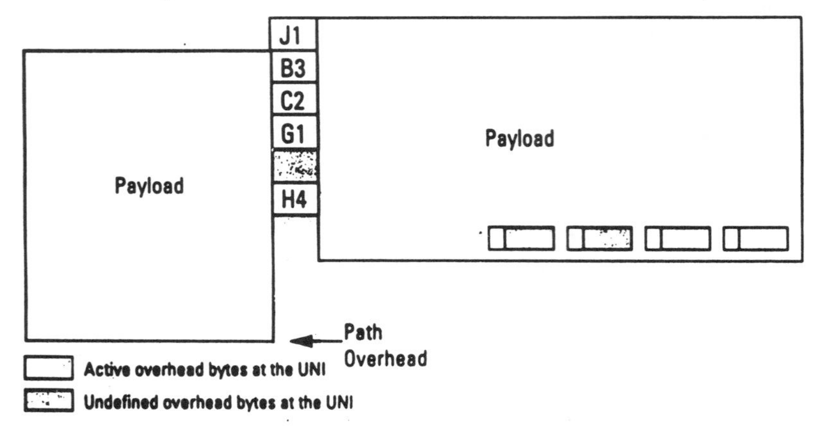

FIGURE 20-U.36. Mapping ATM cells into a SONET STS-1 frame. (Courtesy of Hewlett-Packard Company [Ref. 4].)

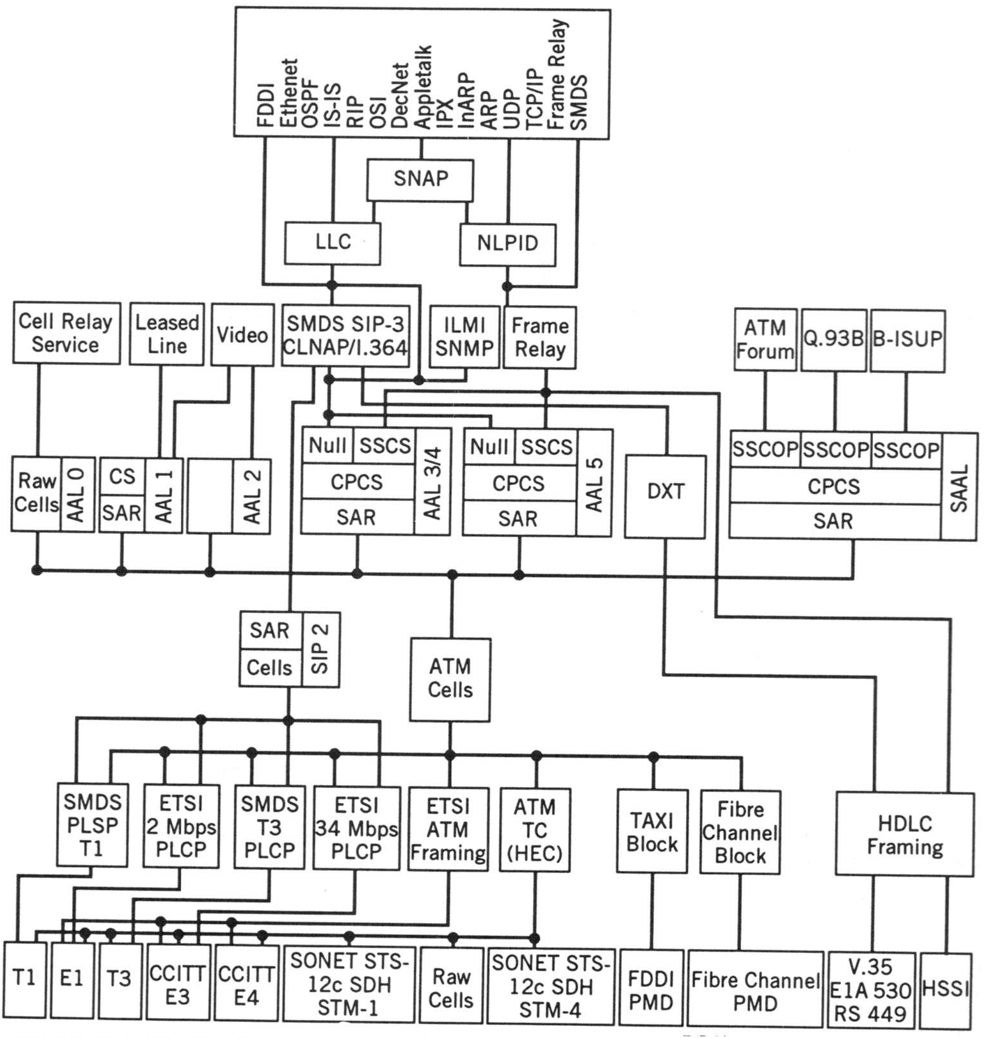

FIGURE 20-U.37. Tying together the evolving broadband network scene. ETSI, European Telecommunications Standardization Institute. (Courtesy of Hewlett-Packard Company [Ref. 4].)

cell may cross an STS frame boundary. This mapping concept is shown in Figure 20-U.36. The H4 pointer can indicate where the cells begin inside an STS frame. Another approach is to identify cell headers and, thus, the first cell in the frame.

20-3.12. Tying It All Together

Figure 20-U.37 brings together the entire evolving broadband network and technologies groupings.

Insert on page 1788 of the Manual, between Reference list and Supplementary Reference.

REFERENCES TO SECTION 20-3 (UPDATED SECTION)

1. *Broadband Aspects of ISDN*, CCITT Rec. I.121, CCITT, Geneva, 1991.
2. *ATM User–Network Interface Specification*, Version 3.0, The ATM Forum, PTR Prentice-Hall, Englewood Cliffs, NJ, 1993.
3. *B-ISDN User–Network Interface—Physical Layer Specification*, ITU-T Rec. I.432, ITU Telecommunication Standardization Sector, Geneva, March 1993.
4. *Broadband Testing Technologies*, an H-P seminar, Hewlett-Packard, Burlington, MA, October 1993.
5. *B-ISDN ATM Adaptation Layer (AAL) Specification*, CCITT Rec. I.363, CCITT, Geneva, 1991.
6. *Support of Broadband Connectionless Data Service on B-ISDN*, ITU-T Rec. I.364, ITU Telecommunication Standardization Sector, Geneva, March 1993.
7. *Numbering Plan for the ISDN Era*, CCITT Rec. E.164, CCITT, Geneva, 1991.
8. *B-ISDN General Network Aspects*, ITU-T Rec. I.311, ITU Telecommunication Standardization Sector, Geneva, March 1993.
9. *Traffic Control and Congestion Control in B-ISDN*, ITU-T Rec. I.371, ITU Telecommunication Standardization Sector, Geneva, March 1993.

22

Television Transmission

This new section follows Section 22-9.2, which ends on page 1903 of the Manual.

22-9.3. Conference Television: Video codec at $p \times 64$ kbps

1. Scope

CCITT Rec. H.261, as described in this section, details information on video coding and decoding methods for the moving picture component of audiovisual services at transmission rates of $p \times 64$ kbps, where p can range from 1 to 30.*

2. Brief Specification

A block diagram of the codec is given in Figure 22-U.1.

2.1. Video Input and Output. To permit a single Recommendation to cover use in and between regions using 625- and 525-line television standards, the source coder operates on pictures based on a common intermediate format (CIF). The standards of the input and output television signals (which may, for example, be composite or component, analog or digital) and the methods of performing any necessary conversion to and from the source coding format are not subject to Recommendation.

2.2. Digital Output and Input. The video coder provides a self-contained digital bit stream which may be combined with other multifacility signals (for example, as defined in Rec. H.221). The video decoder performs the reverse process.

2.3. Sampling Frequency. Pictures are sampled at an integer multiple of the video line rate. This sampling clock and the digital network clock are asynchronous.

2.4. Source Coding Algorithm. A hybrid of interpicture prediction to utilize temporal redundancy and transform coding of the remaining signal to reduce spatial redundancy is adopted. The decoder has motion compensation capability, allowing optional incorporation of this technique in the coder.

2.5. Bit Rate. This Recommendation is primarily intended for use at video bit rates between approximately 40 kbit/sec and 2 Mbit/sec.

2.6. Symmetry of Transmission. The codec may be used for bidirectional or unidirectional visual communication.

*We will call this *conference television*.

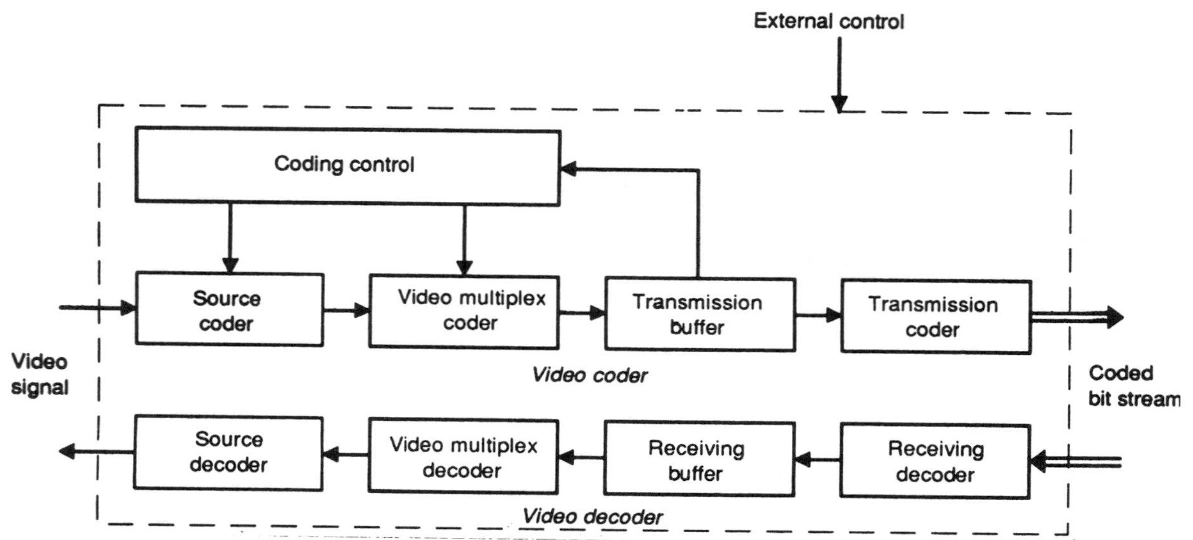

FIGURE 22-U.1. Block diagram of the video codec. (From Figure 1/H.261, CCITT Rec. H.261, Geneva, 1990, page 2.)

2.7. Error Handling. The transmitted bit stream contains a Bose–Chaudhuri–Hocquenham (BCH) (511,493) forward error correction code. Use of this by the decoder is optional.

2.8. Multipoint Operation. Features necessary to support switched multipoint operation are included.

3. Source Coder

3.1. Source Format. The source coder operates on noninterlaced pictures occurring 30,000/1001 (approximately 29.97) times per second. The tolerance on picture frequency is ± 50 ppm.

Pictures are coded as luminance and two color difference components (Y, C_B, and C_R). These components and the codes representing their sampled values are as defined in CCIR Rec. 601.

Black = 16
White = 235
Zero color difference = 128
Peak color difference = 16 and 240
These values are nominal ones and the coding algorithm functions with input values of 1 to 254.
Two picture-scanning formats are specified.

In the first format (CIF), the luminance sampling structure is 352 pels per line and 288 lines per picture in an orthogonal arrangement. Sampling of each of the two color difference components is at 176 pels per line and 144 lines per picture, orthogonal. Color difference samples are sited such that their block boundaries coincide with luminance block boundaries as shown in Figure 22-U.2. The picture area covered by these numbers of pels and lines has an aspect ratio of 4 : 3 and corresponds to the active portion of the local standard video input.

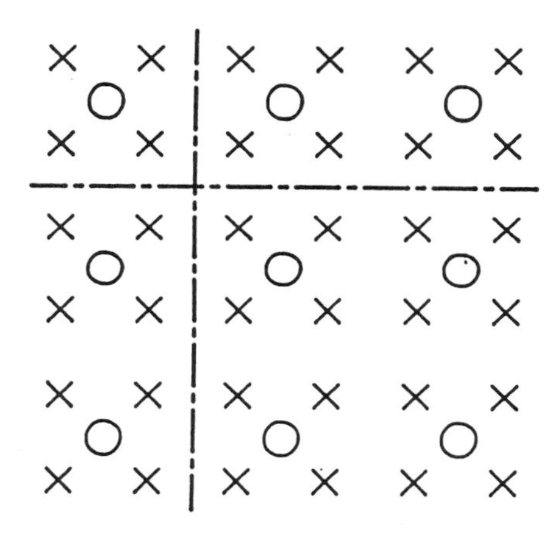

FIGURE 22-U.2. Positioning of luminance and chrominance samples. (From Figure 2/H.261, CCITT Rec. H.261, Geneva, 1990, page 4.)

Note: The number of pels per line is compatible with sampling the active portions of the luminance and color difference signals from 525- or 625-line sources at 6.75 and 3.375 MHz, respectively. These frequencies have a simple relationship to those in CCIR Rec. 601.

The second format, quarter-CIF (QCIF), has half the number of pels and half the number of lines stated above. All codecs must be able to operate using QCIF. Some codecs can also operate with CIF.

Means shall be provided to restrict the maximum picture rate of encoders by having at least 0, 1, 2, or 3 nontransmitted pictures between transmitted ones. Selection of this minimum number and CIF or QCIF shall be by external means (for example via Rec. H.221).

3.2. Video Source Coding. The source coder is shown in a generalized form in Figure 22-U.3. The main elements are prediction, block transformation, and quantization.

The prediction error (INTER mode) or the input picture (INTRA mode) is subdivided into 8 pel by 8 line blocks which are segmented as transmitted or nontransmitted. Furthermore, four luminance blocks and two spatially corresponding color difference blocks are combined to form a macroblock as shown in Figure 22-U.10.

The criteria for choice of mode and transmitting a block are not subject to recommendation and may be varied dynamically as part of the coding control strategy. Transmitted blocks are transformed, and resulting coefficients are quantized and variable-length-coded.

3.2.1. Prediction. The prediction is interpicture and may be augmented by motion compensation (see Subsection 3.2.2) and a spatial filter (see Subsection 3.2.3).

3.2.2. Motion Compensation. Motion compensation (MC) is optional in the encoder. The decoder will accept one vector per macroblock. Both horizontal and vertical components of these motion vectors have integer values not exceeding ± 15. The vector is used for all four luminance blocks in the macroblock. The

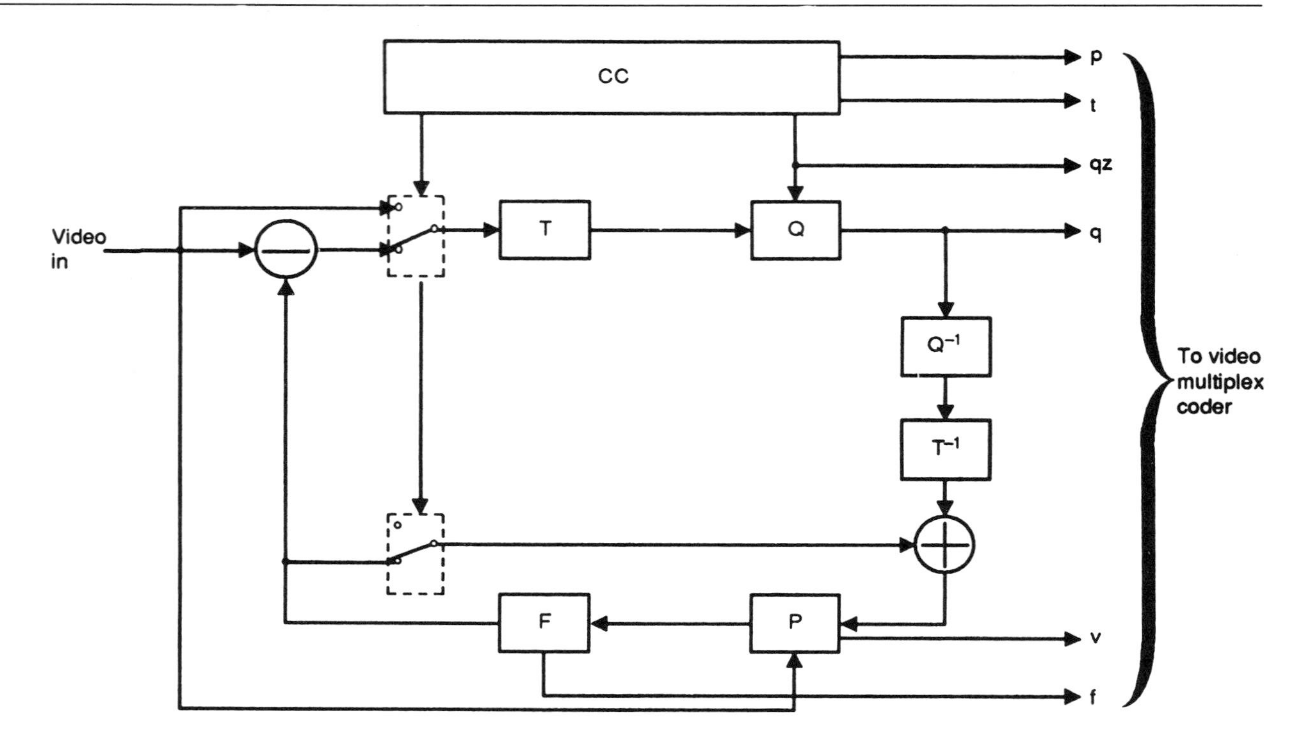

T Transform
Q Quantizer
P Picture memory with motion compensated variable delay
F Loop filter
CC Coding control
p Flag for INTRA/INTER
t Flag for transmitted or not.
qz Quantizer indication
q Quantizing index for transform coefficients
v Motion vector
f Switching on/off of the loop filter

FIGURE 22-U.3. Source coder. (From Figure 3/H.261, CCITT Rec. H.261, Geneva, 1990, page 5.)

motion vector for both color difference blocks is derived by halving the component values of the macroblock vector and truncating the magnitude parts toward zero to yield integer components.

A positive value of the horizontal or vertical component of the motion vector signifies that the prediction is formed from pels in the previous picture which are spatially to the right or below the pels being predicted.

Motion vectors are restricted such that all pels referenced by them are within the coded picture area.

3.2.3. Loop Filter. The prediction process may be modified by a two-dimensional spatial filter (FIL) which operates on pels within a predicted 8-by-8 block.

The filter is separable into one-dimensional horizontal and vertical functions. Both are nonrecursive with coefficients of $1/4, 1/2, 1/4$ except at block edges where one of the taps would fall outside the block. In such cases the 1-D filter is changed to have coefficients of $0, 1, 0$. Full arithmetic precision is retained with rounding to 8-bit integer values at the 2-D filter output. Values whose fractional part is one-half are rounded up.

The filter is switched on/off for all six blocks in a macroblock according to the macroblock type (MTYPE; see Subsection 4.2.3).

3.2.4. Transformer. Transmitted blocks are first processed by a separable two-dimensional discrete cosine transform of size 8 by 8. The output from the inverse transform ranges from -256 to $+255$ after clipping to be represented with 9 bits. The transfer function of the inverse transform is given by

$$f(x,y) = 1/4 \sum_{u=0}^{7} \sum_{v=0}^{7} C(u)C(v)F(u,v)\cos\left[\pi(2x+1)u/16\right]\cos\left[\pi(2y+1)v/16\right]$$

with $u, v, x, y = 0, 1, 2, \ldots, 7$
where

x, y = spatial coordinates in the pel domain

u, v = coordinates in the transform domain

$C(u) = 1/\sqrt{2}$ for $u = 0$, otherwise 1

$C(v) = 1/\sqrt{2}$ for $v = 0$, otherwise 1

Note: Within the block being transformed, $x = 0$ and $y = 0$ refer to the pel nearest the left and top edges of the picture, respectively.

The arithmetic procedures for computing the transforms are not defined, but the inverse one should meet the error tolerance specified in Annex A to CCITT Rec. H.261.

3.2.5. Quantization. The number of quantizers is 1 for the INTRA dc coefficient and 31 for all other coefficients. Within a macroblock the same quantizer is used for all coefficients except the INTRA dc one. The decision levels are not defined. The INTRA dc coefficient is nominally the transform value linearly quantized with a step size of 8 and no dead zone. Each of the other 31 quantizers is also nominally linear but with a central dead zone around zero and with a step size of an even value in the range 2 to 62.

The reconstruction levels are as defined in Subsection 4.2.4.

Note: For the smaller quantization step sizes, the full dynamic range of the transform coefficients cannot be represented.

3.2.6. Clipping of Reconstructed Picture. To prevent quantization distortion of transform coefficient amplitudes causing arithmetic overflow in the encoder and decoder loops, clipping functions are inserted. The clipping function is applied to the reconstructed picture which is formed by summing the prediction and the prediction error as modified by the coding process. This clipper operates on resulting pel values less than 0 or greater than 255, changing them to 0 and 255, respectively.

3.3. Coding Control. Several parameters may be varied to control the rate of generation of coded video data. These include processing prior to the source coder, the quantizer, block significance criterion, and temporal subsampling. The proportions of such measures in the overall control strategy are not subject to recommendation.

When invoked, temporal subsampling is performed by discarding complete pictures.

3.4. Forced Updating. This function is achieved by torcing the use of the INTRA mode of the coding algorithm. The update pattern is not defined. For control of accumulation of inverse transform mismatch error, a macroblock should be forcibly updated at least once per every 132 times it is transmitted.

4. Video Multiplex Coder

4.1. Data Structure. Unless specified otherwise, the most significant bit is transmitted first. This is bit 1 and is the leftmost bit in the code tables in this Recommendation. Unless specified otherwise, all unused or spare bits are set to "1." Spare bits must not be used until their functions are specified by the CCITT.

4.2. Video Multiplex Arrangement. The video multiplex is arranged in a hierarchical structure with four layers. From top to bottom the layers are:

- Picture
- Group of blocks (GOB)
- Macroblock (MB)
- Block

A syntax diagram of the video multiplex coder is shown in Figure 22-U.4. Abbreviations are defined in a later section.

4.2.1. Picture Layer. Data for each picture consist of a picture header followed by data for GOBs. The structure is shown in Figure 22-U.5.

4.2.1.1. Picture Start Code (PCS) (20 bits). A word of 20 bits. Its value is 0000 0000 0000 0001 0000.

4.2.1.2. Temporal Reference (TR) (5 bits). A 5-bit number which can have 32 possible values. It is formed by incrementing its value in the previously transmitted picture header by one plus the number of nontransmitted pictures (at 29.97 Hz) since that last transmitted one. The arithmetic is performed with only the five LSBs.

4.2.1.3. Type Information (PTYPE) (6 bits). Information about the complete picture:

Bit 1	Split screen indicator; "0" off, "1" on.
Bit 2	Document camera indicator: "0" off, "1" on.
Bit 3	Freeze picture release: "0" off, "1" on.
Bit 4	Source format: "0" QCIF, "1" CIF.
Bits 5 to 6	Spare.

4.2.1.4. Extra Insertion Information (PEI) (1 bit). A bit which when set to "1" signals the presence of the following optional data field.

4.2.1.5. Spare Information (PSPARE) (0/8/16... bits). If PEI is set to "1," then 9 bits follow consisting of 8 bits of data (PSPARE) and then another PEI bit to indicate if a further 9 bits follow, and so on. Encoders must not insert PSPARE until specified by the CCITT. Decoders must be designed to discard PSPARE if PEI is set to 1. This will allow the CCITT to specify future backward compatible additions in PSPARE.

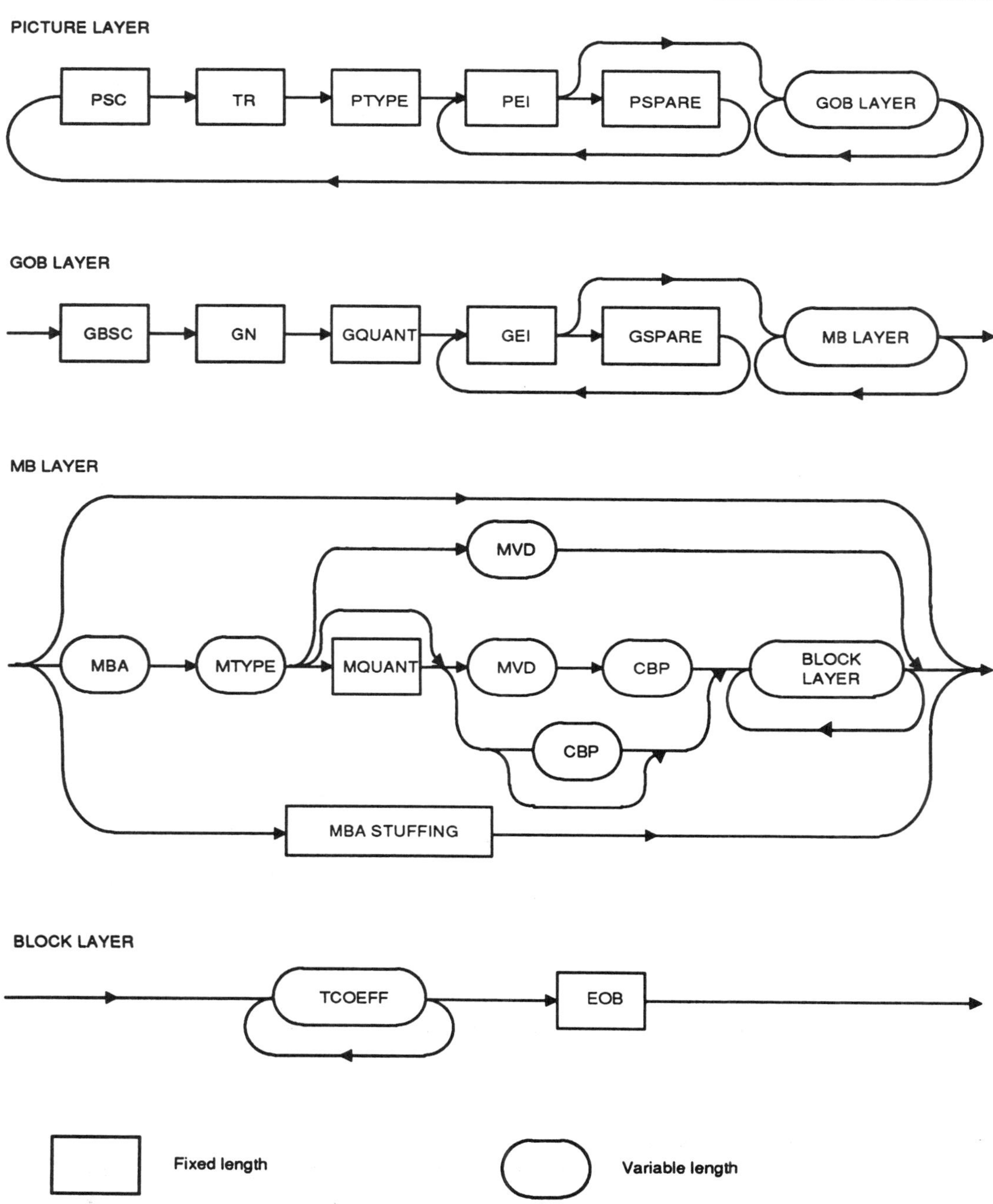

FIGURE 22-U.4. Syntax diagram for the video multiplex coder. (From Figure 4/H.261, CCITT Rec. H.261, Geneva, 1990, page 8.)

FIGURE 22-U.5. Structure of picture layer. (From Figure 5/H.261, CCITT Rec. H.261, Geneva, 1990, page 9.)

1	2
3	4
5	6
7	8
9	10
11	12

CIF

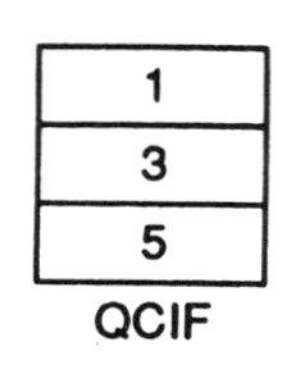

FIGURE 22-U.6. Arrangement of GOBs in a picture. (From Figure 6/H.261, CCITT Rec. H.261, Geneva, 1990, page 10.)

4.2.2. Group of Blocks (GOB) Layer. Each picture is divided into groups of blocks (GOBs). A GOB comprises one-twelfth of the CIF or one-third of the QCIF picture areas (see Figure 22-U.6). A GOB relates to 176 pels by 48 lines of Y and the spatially corresponding 88 pels by 24 lines of each C_B and C_R.

Data for each group of blocks consist of a GOB header followed by data for macroblocks. The structure is shown in Figure 22-U.7. Each GOB header is transmitted once between picture start codes in the CIF or QCIF sequence numbered in Figure 22-U.6, even if no macroblock is present in that GOB.

4.2.2.1. Group of Blocks Start Code (GBSC) (16 bits). A word of 16 bits, 0000 0000 0000 0001.

4.2.2.2. Group Number (GN) (4 bits). A GN consists of four bits indicating the position of the group of blocks. The bits are the binary representation of the number in Figure 22-U.6. Group numbers 13, 14, and 15 are reserved for future use. Group number 0 is used in the PSC.

4.2.2.3. Quantizer Information (GQUANT) (5 bits). A GQUANT is a fixed-length codeword of 5 bits which indicates the quantizer to be used in the group of blocks until overridden by any subsequent MQUANT. The codewords are the natural binary representations of the values of QUANT (Section 4.2.4) which, being half the step sizes, range from 1 to 31.

4.2.2.4. Extra Insertion Information (GEI) (1 bit). A GEI is a bit which, when set to "1," signals the presence of the following optional data field.

4.2.2.5. Spare Information (GSPARE) (0/8/16... bits). If GEI is set to "1," then 9 bits follow consisting of 8 bits of data (GSPARE) and then another GEI bit to indicate if a further 9 bits follow, and so on. Encoders must not insert GSPARE until specified by the CCITT. Decoders must be designed to discard GSPARE if GEI is set to 1. This will allow the CCITT to specify future "backward" compatible additions in GSPARE.

Note: Emulation of start codes may occur if the future specification of GSPARE has no restrictions on the final GSPARE data bits.

FIGURE 22-U.7. Structure of group of blocks layer. (From Figure 7/H.261, CCITT Rec. H.261, Geneva, 1990, page 10.)

1	2	3	4	5	6	7	8	9	10	11
12	13	14	15	16	17	18	19	20	21	22
23	24	25	26	27	28	29	30	31	32	33

FIGURE 22-U.8. Arrangement of macroblocks in a GOB. (From Figure 8/H.261, CCITT Rec. H.261, Geneva, 1990, page 11.)

4.2.3. Macroblock Layer. Each GOB is divided into 33 macroblocks as shown in Figure 22-U.8. A macroblock relates to 16 pels by 16 lines of Y and relates to the spatially corresponding 8 pels by 8 lines of each C_B and C_R.

Data for a macroblock consist of an MB header followed by data blocks (see Figure 22-U.9). MQUANT, MVD, and CBP are present when indicated by MTYPE.

4.2.3.1. Macroblock Address (MBA) (Variable Length). An MBA is a variable-length codeword indicating the position of a macroblock within a group of blocks. The transmission order is shown in Figure 22-U.8. For the first transmitted macroblock in a GOB, MBA is the absolute address in Figure 22-U.8. For subsequent macroblocks, MBA is the difference between the absolute addresses of the macroblock and the last transmitted macroblock. The code table for MBA is given in Table 22-U.1.

An extra codeword is available in the table for bit stuffing immediately after a GOB header or a coded macroblock (MBA stuffing). This codeword should be discarded by decoders.

The VLC for start code is also shown in Table 22-U.1.

MBA is always included in transmitted macroblocks. Macroblocks are not transmitted when they contain no information for that part of the picture.

4.2.3.2. Type Information (MTYPE) (Variable Length). Variable-length codewords which gives information about the macroblock and which data elements are present. Macroblock types, including elements and VLC words, are listed in Table 22-U.2.

4.2.3.3. Quantizer (MQUANT) (5 bits). An MQUANT is a codeword of 5 bits signifying the quantizer to be used for this and any following blocks in the group of blocks until overridden by any subsequent MQUANT. MQUANT is present only if so indicated by MTYPE. Codewords for MQUANT are the same as for GQUANT.

4.2.3.4. Motion Vector Data (MVD) (Variable Length). Motion vector data are included for all MC macroblocks. MVD is obtained from the macroblock vector by subtracting the vector of the preceding macroblock. For this calculation the vector

MBA	MTYPE	MQUANT	MVD	CBP	Block data

FIGURE 22-U.9. Structure of macroblock layer. (From Figure 9/H.261, CCITT Rec. H.261, Geneva, 1990, page 11.)

TABLE 22-U.1
VLC Table for Macroblock Addressing

MBA	Code	MBA	Code
1	1	17	0000 0101 10
2	011	18	0000 0101 01
3	010	19	0000 0101 00
4	0011	20	0000 0100 11
5	0010	21	0000 0100 10
6	0001 1	22	0000 0100 011
7	0001 0	23	0000 0100 010
8	0000 111	24	0000 0100 001
9	0000 110	25	0000 0100 000
10	0000 1011	26	0000 0011 111
11	0000 1010	27	0000 0011 110
12	0000 1001	28	0000 0011 101
13	0000 1000	29	0000 0011 100
14	0000 0111	30	0000 0011 011
15	0000 0110	31	0000 0011 010
16	0000 0101 11	32	0000 0011 001
		33	0000 0011 000
		MBA stuffing	0000 0001 111
		Start code	0000 0000 0000 0001

Source: Table 1/H.261, CCITT Rec. H.261, Geneva, 1990, page 12.

of the preceding macroblock is regarded as zero in the following three situations:

1. Evaluating MVD for macroblocks 1, 12, and 23;
2. Evaluating MVD for macroblocks in which MBA does not represent a difference of 1;
3. MTYPE of the previous macroblock was not MC.

MVD consists of a variable-length codeword for the horizontal component followed by a variable-length codeword for the vertical component. Variable-length codes are given in Table 22-U.3.

TABLE 22-U.2
VLC Table for MTYPE

Prediction	MQUANT	MVD	CBP	TCOEFF	VLC
Intra				x	0001
Intra	x			x	0000 001
Inter			x	x	1
Inter	x		x	x	0000 1
Inter + MC		x			0000 0000 1
Inter + MC		x	x	x	0000 0001
Inter + MC	x	x	x	x	0000 0000 01
Inter + MC + FIL		x			001
Inter + MC + FIL		x	x	x	01
Inter + MC + FIL	x	x	x	x	0000 01

Note 1: "x" means that the item is present in the macroblock.
Note 2: It is possible to apply the filter in a non-motion-compensated macroblock by declaring it as MC + FIL but with a zero vector.
Source: Table 2/H.261, CCITT Rec. H.261, Geneva, 1990, page 13.

TABLE 22-U.3
VLC Table for MVD

MVD	Code
−16 & 16	0000 0011 001
−15 & 17	0000 0011 011
−14 & 18	0000 0011 101
−13 & 19	0000 0011 111
−12 & 20	0000 0100 001
−11 & 21	0000 0100 011
−10 & 22	0000 0100 11
−9 & 23	0000 0101 01
−8 & 24	0000 0101 11
−7 & 25	0000 0111
−6 & 26	0000 1001
−5 & 27	0000 1011
−4 & 28	0000 111
−3 & 29	0001 1
−2 & 30	0011
−1	011
0	1
1	010
2 & − 30	0010
3 & − 29	0001 0
4 & − 28	0000 110
5 & − 27	0000 1010
6 & − 26	0000 1000
7 & − 25	0000 0110
8 & − 24	0000 0101 10
9 & − 23	0000 0101 00
10 & − 22	0000 0100 10
11 & − 21	0000 0100 010
12 & − 20	0000 0100 000
13 & − 19	0000 0011 110
14 & − 18	0000 0011 100
15 & − 17	0000 0011 010

Source: Table 3/H.261, CCITT Rec. H.261, Geneva, 1990, page 14.

TABLE 22-U.4
VLC Table for CBP

CBP	Code	CBP	Code
60	111	35	001 1100
4	1101	13	0001 1011
8	1100	49	0001 1010
16	1011	21	0001 1001
32	1010	41	0001 1000
12	1001 1	14	0001 0111
48	1001 0	50	0001 0110
20	1000 1	22	0001 0101
40	1000 0	42	0001 0100
28	0111 1	15	0001 0011
44	0111 0	51	0001 0010
52	0110 1	23	0001 0001
56	0110 0	43	0001 0000
1	0101 1	25	0000 1111
61	0101 0	37	0000 1110
2	0100 1	26	0000 1101
62	0100 0	38	0000 1100
24	0011 11	29	0000 1011
36	0011 10	45	0000 1010
3	0011 01	53	0000 1001
63	0011 00	57	0000 1000
5	0010 111	30	0000 0111
9	0010 110	46	0000 0110
17	0010 101	54	0000 0101
33	0010 100	58	0000 0100
6	0010 011	31	0000 0011 1
10	0010 010	47	0000 0011 0
18	0010 001	55	0000 0010 1
34	0010 000	59	0000 0010 0
7	0001 1111	27	0000 0001 1
11	0001 1110	39	0000 0001 0
19	0001 1101		

Source: Table 4/H.261, CCITT Rec. H.261, Geneva, 1990, page 15.

Advantage is taken of the fact that the range of motion vector values is constrained. Each VLC word represents a pair of difference values. Only one of the pair will yield a macroblock vector falling within the permitted range.

4.2.3.5. Coded Block Pattern (CBP) (Variable Length). CBP is present if indicated by MTYPE. The codeword gives a pattern number signifying those blocks in the macroblock for which at least one transform coefficient is transmitted. The pattern

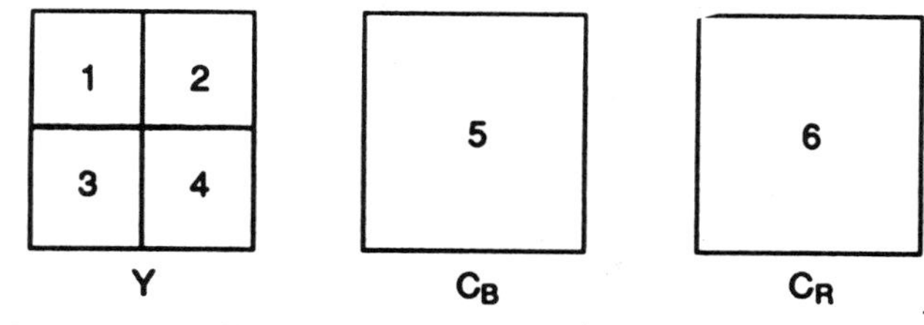

FIGURE 22-U.10. Arrangement of blocks in a macroblock. (From Figure 10/H.261, CCITT Rec. H.261, Geneva, 1990, page 16.)

FIGURE 22-U.11. Structure of block layer. (From Figure 11/H.261, CCITT Rec. H.261, Geneva, 1990, page 16.)

TCOEFF	EOB

number is given by

$$32 \cdot P_1 + 16 \cdot P_2 + 8 \cdot P_3 + 4 \cdot P_4 + 2 \cdot P_5 + P_6$$

where $P_n = 1$ if any coefficient is present for block n, else 0. Block numbering is given in Figure 22-U.10.

The codewords for CBP are given in Table 22-U.4.

4.2.4. Block Layer. A macroblock comprises four luminance blocks and one of each of the two color difference blocks (see Figure 22-U.10).

Data for a block consist of codewords for transform coefficients followed by an end of block (see Figure 22-U.11). The order of block transmission is as in Figure 22-U.10.

4.2.4.1. Transform Coefficients (TCOEFF). Transform coefficient data are always present for all six blocks in a macroblock when MTYPE indicates INTRA. In other cases, MTYPE and CBP will indicate those blocks which have coefficient data transmitted for them. The quantized transform coefficients are sequentially transmitted according to the sequence given in Figure 22-U.12.

The most commonly occurring combinations of successive zeros (RUN) and the following value (LEVEL) are encoded with variable-length codes. Other combinations of (RUN, LEVEL) are encoded with a 20-bit word consisting of 6 bits ESCAPE, 6 bits RUN, and 8 bits LEVEL. For the variable-length encoding there are two code tables, one being used for the first transmitted LEVEL in INTER, INTER + MC and INTER + MC + FIL blocks, the second for all other LEVELs except the first one in INTRA blocks which is fixed-length coded with 8 bits. Codes are given in Table 22-U.5.

The most commonly occurring combinations of zero-run and the following value are encoded with variable-length codes as listed in the table below. End of block (EOB) is in this set. Because CBP indicates those blocks with no coefficient data, EOB cannot occur as the first coefficient. Hence EOB can be removed from the VLC table for the first coefficient.

1	2	6	7	15	16	28	29
3	5	8	14	17	27	30	43
4	9	13	18	26	31	42	44
10	12	19	25	32	41	45	54
11	20	24	33	40	46	53	55
21	23	34	39	47	52	56	61
22	35	38	48	51	57	60	62
36	37	49	50	58	59	63	64

→ Increasing cycles per picture width

↓ Increasing cycles per picture height

FIGURE 22-U.12. Transmission order for transform coefficients. (From Figure 12/H.261, CCITT Rec. H.261, Geneva, 1990, page 16.)

TABLE 22-U.5
VLC Table for TCOEFF

Run	Level	Code	Run	Level	Code
EOB		10	4	1	0011 0s
0	1	1s[a] If first coefficient in block	4	2	0000 0011 11s
0	1	11s Not first coefficient in block	4	3	0000 0001 0010 s
0	2	0100 s	5	1	0001 11s
0	3	0010 1s	5	2	0000 0010 01s
0	4	0000 110s	5	3	0000 0000 1001 0s
0	5	0010 0110 s	6	1	0001 01s
0	6	0010 0001 s	6	2	0000 0001 1110 s
0	7	0000 0010 10s	7	1	0001 00s
0	8	0000 0001 1101 s	7	2	0000 0001 0101 s
0	9	0000 0001 1000 s	8	1	0000 111s
0	10	0000 0001 0011 s	8	2	0000 0001 0001 s
0	11	0000 0001 0000 s	9	1	0000 101s
0	12	0000 0000 1101 0s	9	2	0000 0000 1000 1s
0	13	0000 0000 1100 1s	10	1	0010 0111 s
0	14	0000 0000 1100 0s	10	2	0000 0000 1000 0s
0	15	0000 0000 1011 1s	11	1	0010 0011 s
1	1	011s	12	1	0010 0010 s
1	2	0001 10s	13	1	0010 0000 s
1	3	0010 0101 s	14	1	0000 0011 10s
1	4	0000 0011 00s	15	1	0000 0011 01s
1	5	0000 0001 1011 s	16	1	0000 0010 00s
1	6	0000 0000 1011 0s	17	1	0000 0001 1111 s
1	7	0000 0000 1010 1s	18	1	0000 0001 1010 s
2	1	0101 s	19	1	0000 0001 1001 s
2	2	0000 100s	20	1	0000 0001 0111 s
2	3	0000 0010 11s	21	1	0000 0001 0110 s
2	4	0000 0001 0100 s	22	1	0000 0000 1111 1s
2	5	0000 0000 1010 0s	23	1	0000 0000 1111 0s
3	1	0011 1s	24	1	0000 0000 1110 1s
3	2	0010 0100 s	25	1	0000 0000 1110 0s
3	3	0000 0001 1100 s	26	1	0000 0000 1101 1s
3	4	0000 0000 1001 1s	Escape		0000 01

[a]Never used in INTRA macroblocks.
Source: Table 5/H.261, CCITT Rec. H.261, Geneva, 1990, page 17.

The last bit "s" denotes the sign of the level: "0" for positive and "1" for negative.

The remaining combinations of (RUN, LEVEL) are encoded with a 20-bit word consisting of 6 bits ESCAPE, 6 bits RUN, and 8 bits LEVEL. Use of this 20-bit word form encoding the combinations listed in the VLC table is not prohibited.

Run is a 6-bit fixed-length code

Run	Code
0	0000 00
1	0000 01
2	0000 10
.	.
.	.
63	1111 11

Level is an 8-bit fixed-length code

Level	Code
−128	FORBIDDEN
−127	1000 0001
.	.
−2	1111 1110
−1	1111 1111
0	FORBIDDEN
1	0000 0001
2	0000 0010
.	.
127	0111 1111

For all coefficients other than the INTRA dc one, the reconstruction levels (REC) are in the range −2048 to 2047 and are given by clipping the results of the following formulas:

$$\left.\begin{aligned} REC &= QUANT \cdot (2 \cdot level + 1);\ level > 0 \\ REC &= QUANT \cdot (2 \cdot level - 1);\ level < 0 \end{aligned}\right\} \quad QUANT = \text{“odd”}$$

$$\left.\begin{aligned} REC &= QUANT \cdot (2 \cdot level + 1) - 1;\ level > 0 \\ REC &= QUANT \cdot (2 \cdot level + 1) + 1;\ level < 0 \end{aligned}\right\} \quad QUANT = \text{“even”}$$

$$REC = 0;\ level = 0$$

Note: QUANT ranges from 1 to 31 and is transmitted by either GQUANT or MQUANT (see Table 22-U.6).

For INTRA blocks the first coefficient is nominally the transform dc value linearly quantized with a step size of 8 and no dead zone. The resulting values are represented with 8 bits. A nominally black block will give 0001 0000 and a nominally white one 1110 1011. The code 0000 0000 is not used. The code 1000 0000 is not used, the reconstruction level of 1024 being coded as 1111 1111 (see Table 22-U.7).

Coefficients after the last nonzero one are not transmitted. EOB (end-of-block code) is always the last item in blocks for which coefficients are transmitted.

TABLE 22-U.6
Reconstruction Levels (REC)

	QUANT												
Level	1	2	3	4	·	8	9	·	17	18	·	30	31
−127	−255	−509	−765	−1019	·	−2039	−2048	·	−2048	−2048	·	−2048	−2048
−126	−253	−505	−759	−1011	·	−2023	−2048	·	−2048	−2048	·	−2048	−2048
·	·	·	·	·	·	·	·	·	·	·	·	·	·
−2	−5	−9	−15	−19	·	−39	−45	·	−85	−89	·	−149	−155
−1	−3	−5	−9	−11	·	−23	−27	·	−51	−53	·	−89	−93
0	0	0	0	0	·	0	0	·	0	0	·	0	0
1	3	5	9	11	·	23	27	·	51	53	·	89	93
2	5	9	15	19	·	39	45	·	85	89	·	149	155
3	7	13	21	27	·	55	63	·	119	125	·	209	217
4	9	17	27	35	·	71	81	·	153	161	·	269	279
5	11	21	33	43	·	87	99	·	187	197	·	329	341
·	·	·	·	·	·	·	·	·	·	·	·	·	·
56	113	225	339	451	·	903	1017	·	1921	2033	·	2047	2047
57	115	229	345	459	·	919	1035	·	1955	2047	·	2047	2047
58	117	233	351	467	·	935	1053	·	1989	2047	·	2047	2047
59	119	237	357	475	·	951	1071	·	2023	2047	·	2047	2047
60	121	241	363	483	·	967	1089	·	2047	2047	·	2047	2047
·	·	·	·	·	·	·	·	·	·	·	·	·	·
125	251	501	753	1003	·	2007	2047	·	2047	2047	·	2047	2047
126	253	505	759	1011	·	2023	2047	·	2047	2047	·	2047	2047
127	255	509	765	1019	·	2039	2047	·	2047	2047	·	2047	2047

Note: Reconstruction levels are symmetrical with respect to the sign of level except for 2047/ − 2048.
Source: From unnumbered table on page 20, CCITT Rec. H.261, ITU Geneva, 1990.

TABLE 22-U.7
Reconstruction Levels for INTRA Mode dc Coefficient

FLC	Reconstruction Level into Inverse Transform
0000 0001 (1)	8
0000 0010 (2)	16
0000 0011 (3)	24
.	.
.	.
0111 1111 (127)	1016
1111 1111 (255)	1024
1000 0001 (129)	1032
.	.
.	.
1111 1101 (253)	2024
1111 1110 (254)	2032

Note: The decoded value corresponding to FLC "n" is 8n, except FLC 255 gives 1024.
Source: Table 6/H.261, CCITT Rec. H.261, Geneva, 1990, page 21.

4.3. Multipoint Considerations. The following facilities are provided to support switched multipoint operation.

4.3.1. Freeze Picture Request. This causes the decoder to freeze its displayed picture until a freeze picture release signal is received or a timeout period of at least six seconds has expired. The transmission of this signal is via external means (for example, by Rec. H.221).

4.3.2. Fast Update Request. This causes the encoder to encode its next picture in INTRA mode with coding parameters such as to avoid buffer overflow. The transmission method for this signal is via external means (for example, by Rec. H.221).

4.3.3. Freeze Picture Release. This is a signal from an encoder which has responded to a fast update request and allows a decoder to exit from its freeze picture mode and display decoded pictures in the normal manner. This signal is transmitted by bit 3 of PTYPE (see Subsection 4.2.1) in the picture header of the first picture coded in response to the fast update request.

5. Transmission Coder

5.1. Bit Rate. The transmission clock is provided externally (for example from an I.420 interface).

5.2. Video Data Buffering. The encoder must control its output bitstream to comply with the requirements of the hypothetical reference decoder defined in Annex B of CCITT Rec. H.261.

When operating with CIF, the number of bits created by coding any single picture must not exceed $256 \cdot K$ bits. $K = 1024$.

When operating with QCIF, the number of bits created by coding any single picture must not exceed $64 \cdot K$ bits.

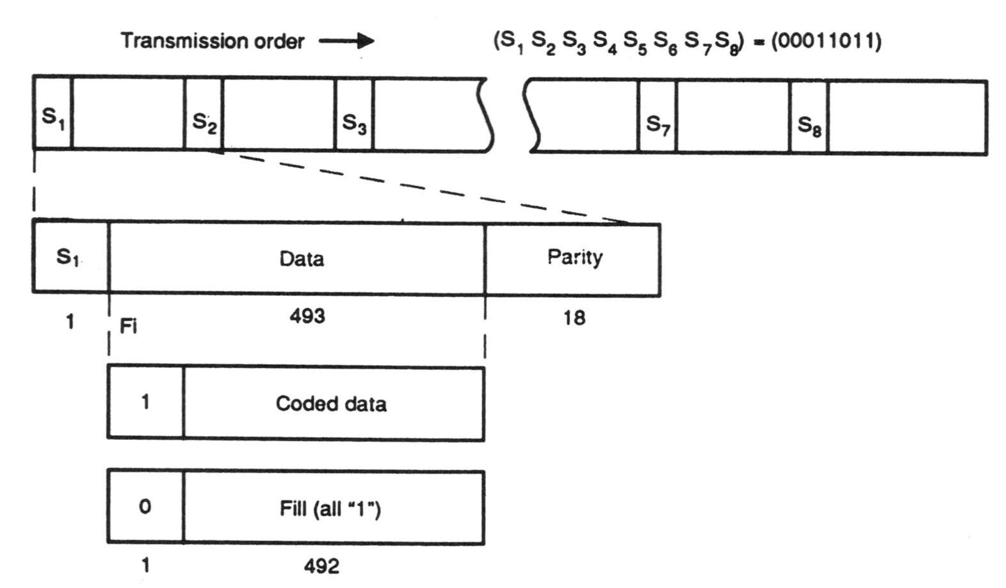

FIGURE 22-U.13. Error correcting frame. (From Figure 13/H.261, CCITT Rec. H.261, Geneva, 1990, page 22.)

In both the above cases the bit count includes the picture start code and all other data related to that picture including PSPARE, GSPARE, and MBA stuffing. The bit count does not include error correction framing bits, fill indicator (Fi), fill bits, or error correction parity information described in Section 5.4.

Video data must be provided on every valid clock cycle. This can be ensured by use of either the fill bit indicator (Fi) and subsequent fill all 1s bits in the error corrector block framing (see Figure 22-U.13) or MBA stuffing (see Subsection 4.2.3), or both.

5.3. Video Coding Delay. This item is included in this Recommendation because the video encoder and video decoder delays need to be known to allow audio compensation delays to be fixed when H.261 is used to form part of a conversational service. This will allow lip synchronization to be maintained. Annex C recommends a method by which the delay figures are established. Other delay measurement methods may be used, but they must be designed in a way to produce results similar to those of the method given in Annex C.

5.4. Forward Error Correction for Coded Video Signal

5.4.1. Error Correcting Code. The transmitted bit stream contains a BCH (511,493) forward error correction code. Use of this by the decoder is optional.

5.4.2. Generator Polynomial

$$g(x) = (x^9 + x^4 + 1)(x^9 + x^6 + x^4 + x^3 + 1)$$

Example: For the input data of "01111 . . . 11" (493 bits) the resulting correction parity bits are "011011010100011011" (18 bits).

5.4.3. Error Correction Framing. To allow the video data and error correction parity information to be identified by a decoder, an error correction framing pattern is included. This consists of a multiframe of eight frames, each frame comprising 1 bit framing, 1 bit fill indicator (Fi), 492 bits of coded data (or fill all 1s), and 18 bits parity. The frame alignment pattern is

$$(S_1S_2S_3S_4S_5S_6S_7S_8) = (00011011).$$

See Figure 22-U.13 for the frame arrangement. The parity is calculated against 493 bits including fill indicator (Fi).

The fill indicator (Fi) can be set to zero by an encoder. In this case only 492 consecutive fill bits (fill all 1s) plus parity are sent and no coded data are transmitted. This may be used to meet the requirement in Section 5.2 to provide video data on every valid clock cycle.

5.4.4. Relock Time for Error Corrector Framing. Three consecutive error correction framing sequences (24 bits) should be received before frame lock is deemed to have been achieved. The decoder should be designed such that frame lock will be re-established within 34,000 bits after an error corrector framing phase change.

Note: This assumes that the video data do not contain three correctly phased emulations of the error correction framing sequence during the relocking period.

6. codec Delay Measurement Method

The video encoder and video decoder delays will vary depending on implementation. The delay will also depend on the picture format (QCIF, CIF) and data rate in use. This subsection specifies the method by which the delay figures are established for a particular design. To allow correct audio delay compensation, the overall video delay needs to be established from a user perception point of view under typical viewing conditions.

Figure 22-U.14 shows the measuring points to be discussed. A video sequence lasting more than 100 sec is connected to the video coder input (point A) in Figure 22-U.14. The video sequence should have the following characteristics:

- It should contain a typical moving scene consistent with the intended purpose of the video codec.
- It should produce a minimum coded picture rate of 7.5 Hz at the bit rate in use.
- It should contain a visible identification mark at intervals throughout the length of the sequence. The visible identification should change every 97

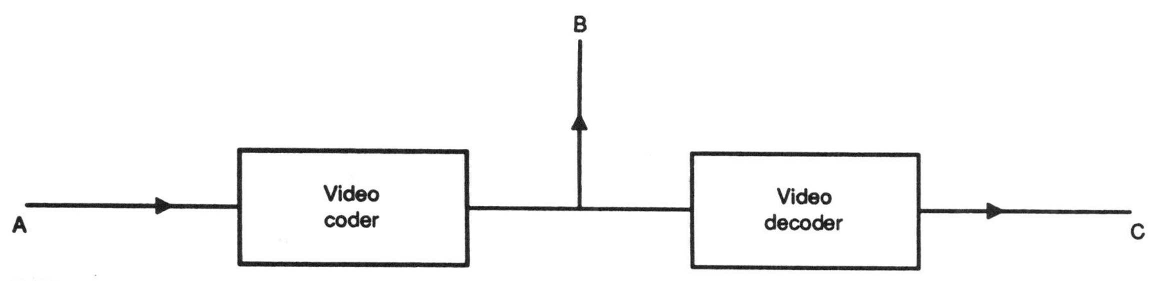

FIGURE 22-U.14. Measuring points. (From Figure C-1/H.261, CCITT Rec. H.261, Annex C, Geneva, 1990, page 26.)

video input frames and be located within the picture area represented by the first GOB in the picture. For example, the first block in the picture could change from black to white at intervals of 97 video frame periods. The identification mark should be chosen so that it can be detected at point B and does not significantly contribute to the overall coding performance.

The codec and video sequence should be arranged so that the bit stream contains less than 10% stuffing (MBA stuffing + error correction fill bits).

The encoder delay is obtained by measuring the time from when the visible identification changes at point A to the time that the change is detected at point B. Similarly, the decoder delay is obtained by taking measurements at points B and C.

Several measurements should be made during the sequence length and the average period obtained. Several tests should be made to ensure that a consistent average figure can be obtained for both encoder and decoder delay times.

Average results should be obtained for each combination of picture format and bit rate within the capability of the particular codec design.

Note: Due to pre- and post-temporal processing, it may be necessary to take a mid-level for establishing the transition of the identification mark at points B and C.

Section 22-9.3 was extracted from CCITT Rec. H.261, Geneva, 1990.

25

Standard Time and Frequency

This updates Section 25-6.2, whose text begins and ends on page 1997 of the Manual and whose tables and figures appear on pages 1998–2006.

25-6.2. Standard Frequencies and Time Signals

Table 25-7 gives characteristics of standard-frequency and time-signal emissions in the allocated bands as of November 1991. Figure 25-8 gives information on the daily emission schedule; Figure 25-9 gives information on the hourly modulation schedule.

Table 25-8 provides characteristics of standard-frequency and time-signal emissions in additional bands valid as of November 1991.

The material in Section 25-6.2 was extracted from ITU-R Rec. 768, Standard Frequencies and Time Signals, Geneva, 1992.

Note to Annex 1, ITU-R Rec. 768. The characteristics of stations appearing in the following tables are valid as of November 1991. For information concerning changes which may have occurred since that date, reference may be made to the Annual Report of the time section of the Bureau International des Poids et Mesures (BIPM) or directly to the respective authority of each service listed in the tables and figures that follow.

This new section follows Section 25.9.8.1, which ends on page 2044 of the Manual.

25-10. A NOTE ON STANDARD TIME AND DATE CONVENTIONS

1. Introduction

Chapter 25 deals with standard time and frequency and is based on numerous standards such as CCIR (ITU-R) recommendations. This section describes how these standards relate to each other and discusses the need for broadcasting to disseminate time and frequency.

Time and date information is used to label the actual or nominal point of origin of material (a document or a television or radio program) or the actual or anticipated point of receipt. The difference represents the propagation delay. There is also a requirement for indicating local clock-time and date in their own right, or for use together with such labels for assisting decisions or controlling processes associated with broadcasting.

TABLE 25-7

Characteristics of Standard-Frequency and Time-Signal Emissions in the Allocated Bands, Valid as of November 1991

Station			Type of Antenna(s)	Carrier Power (kW)	Number of Simultaneous Transmissions	Period of Operation		Standard Frequencies used		Duration of Emission		Uncertainty of Frequency and Time Intervals (Parts in 10^{12}) (Note 1)	Method of DUT1 Indication
Call Sign	Approximate Location	Latitude and Longitude				Days/ Week	Hours/ Day	Carrier (MHz)	Modulation (Hz)	Time Signal (min)	Audio Modulation (min)		
ATA	New Delhi, India	28°34′N 77°19′E	Horizontal folded dipole	8 (PEP)	3	7	24 (Note 2)	5, 10, 15	1, 1000	Continuous	4/15	±10	
BPM (Note 3)	Pucheng, China	35°00′N 109°31′E	Omnidirectional	10–20	2	7	24 (Note 4)	2.5, 5, 10, 15	1, 1000	20/30 (UTC) 4/30 (UT1)	Nil	±10	Direct emission of UT1 time signal
HLA	Taejon, Taedok Science Town, Republic of Korea	36°23′N 127°22′E	Vertical (conical monopole)	2	1	5 (Note 5)	7 (Note 6)	5	1	Continuous	Continuous	±10	CCIR code by double pulse
IAM (Note 7)	Roma, Italy	41°47′N 12°27′E	Vertical λ/4	1	1	6	2	5	1	Continuous	Nil	±10	CCIR code by double pulse
IBF (Note 7)	Torino, Italy	45°02′N 07°46′E	Vertical λ/4	5	1	7	$2\frac{3}{4}$	5	1	Continuous	Nil	±10	CCIR code by double pulse
JJY (Note 7)	Sanwa, Sashima, Ibaraki, Japan	36°11′N 139°51′E	(Note 8)	2	5	7	24 (Note 9)	2.5, 5, 8, 10, 15	1 (Note 10) 1000 (Note 11)	Continuous	30/60	±10	CCIR code by lengthening
LOL (Note 7)	Buenos Aires, Argentina	34°37′S 58°21′W	Horizontal 3-wire folded dipole	2	3	7	5	5, 10, 15	1, 440, 1000	Continuous	3/5	±20	CCIR code by lengthening
OMA (Note 7)	Prague, Czech and Slovak Federal Republic	50°07′N 14°35′E	T	1	1	7	24	2.5	1, 1000 (Note 12)	15/30	4/15	±1000	
RCH (Note 7)	Tashkent	41°19′N 69°15′E	Horizontal dipole	1	2	7	21	2.5, 5, 10	1, 10	40/60	Nil	±50	CCIR code by double pulse, additional information dUT1 (Note 13)
RID (Note 7)	Irkutsk	52°26′N 104°02′E	Horizontal dipole	1 1 1	3	7	24	5.004 10.004 15.004	1, 10	40/60	Nil	±50	CCIR code by double pulse, additional information dUT1 (Note 13)
RTA (Note 7)	Novosibirsk	55°04′N 82°58′E	Horizontal dipole	5	1	7	$20\frac{1}{2}$	10, 15	1, 10	40/60	Nil	±50	CCIR code by double pulse, additional information dUT1 (Note 13)

RWM (Note 7)	Moscow	55°48'N 38°18'E	Horizontal dipole	5 5 8	3	7	24	4.996 9.996 14.996	1, 10	40/60	Nil	±50	CCIR code by double pulse, additional information dUT1 (Note 13)
VNG	Llandilo, New South Wales, Australia	33°43'S 150°48'E	Omnidirectional	10	1	7	24	5	1, 1000 (Note 14)	Continuous	Nil	±100	CCIR code by 45 cycles of 900 Hz immediately following the normal second markers
WWV (Note 7)	Fort Collins, Colorado, USA	40°41'N 105°02'W	Vertical λ/2 dipoles	2.5–10	5	7	24	2.5, 5, 10, 15, 20 (Note 15)	1, 440, 500, 600	Continuous (Note 16)	Continuous (Note 17)	±10	CCIR code by double pulse, additional information on UT1 corrections
WWVH (Note 7)	Kekaha, Kauai, Hawaii, USA	21°59'N 159°46'W	Vertical λ/2 dipole arrays	2.5–10	4	7	24	2.5, 5, 10, 15 (Note 15)	1, 440, 500, 600	Continuous (Note 16)	Continuous (Note 17)	±10	CCIR code by double pulse, additional information on UT1 corrections

The daily transmission schedule and hourly modulation schedule are given, where appropriate, in the form of Figs. 25-8 and 25-9 supplemented by the following notes:

Note 1: This value applies at the transmitter; to realize the quoted uncertainty at the point of reception it could be necessary to observe the received phase time frequency over a sufficiently long period in order to eliminate noise and random effects.

Note 2: 5 MHz: 1800–0900 hr UTC; 10 MHz: 24 hr; 15 MHz: 0900–1800 hr UTC.

Note 3: Call sign in Morse and language.

Note 4: 2.5 MHz: 0730–0100 hr UTC; 15 MHz: 0100–0900 hr UTC; 5 MHz and 10 MHz: continuous.

Note 5: Monday to Friday (except national holidays in Korea).

Note 6: 0100 to 0800 hr UTC. Pulses of 9 cycles of 1800-Hz modulation. 59th and 29th second pulses omitted. Hour identified by 0.80-sec long 1500-Hz tone. Beginning of each minute, identified by a 0.8-sec long 1800-Hz tone, voice announcement of hours and minutes each minute following 52nd second pulse. BCD time code given on 100-Hz subcarrier.

Note 7: These stations have indicated that they follow the UTC system as specified in Rec. 460. Since 1 January 1972 the frequency offset has been eliminated and the time signals remain within about 0.8 sec of UT1 by means of occasional 1-sec steps as directed by the International Earth Rotation Service.

Note 8: Vertical $\lambda/4$ for 2.5 MHz, horizontal $\lambda/2$ dipole for 5 and 8 MHz, and vertical $\lambda/2$ dipoles for 10 and 15 MHz.

Note 9: Interrupted from 35 to 39 minutes of each hour.

Note 10: Pulse consists of 8 cycles of 1600-Hz tone. First pulse of each minute preceded by 655 msec of 600-Hz tone.

Note 11: 1000-Hz tone modulation between the minutes of 0–5, 10–15, 20–25, 30–35, 40–45, 50–55 except 40 msec before and after each second's pulse.

Note 12: In the period from 1800 to 0600 hr UTC, audio-frequency modulation is replaced by time signals.

Note 13: The additional information about the value of the difference UT1 − UTC is transmitted by code dUT1. It provides more precisely the difference UTI − UTC in multiples of 0.02 sec. The total value of the correction is DUT1 + dUT1. Possible values of dUT1 are transmitted by marking of p second pulses between the 21st and 24th seconds of the minute, so that dUT1 = $+0.02$ sec $\times p$. Negative values of dUT1 are transmitted by marking of q second pulses between the 31st and 34th second of the minute, so that dUT1 = -0.02 sec $\times q$.

Note 14: Pulses of 50 cycles of 1000-Hz tone, shortened to 5 cycles from the 55th to the 58th second; the 59th pulse is omitted; the minute marker is 500 cycles. At the 5th, 10th, 15th, etc., minutes, pulses from the 50th to the 58th second are shortened to 5 cycles. Voice identification on 5000 kHz between the 20th and 50th seconds in the 15th, 30th, 45th, and 60th minutes. A BCD time incorporating time of day and day number of the year is transmitted between the 20th and 46th second with a binary "0" represented by 100 cycles and a binary "1" by 200 cycles of 1000-Hz tone. The minute information for the next minute is given from the 21st to the 28th second, hour information from the 29th to the 35th second, and day of the year from the 36th to the 46th second; parity bits are included at the end of each code sequence.

Note 15: As of 1 February 1977, transmissions on 25 MHz from WWV and 20 MHz from WWVH were discontinued, but may be resumed at a later date.

Note 16: In addition to other timing signals and time announcements, a modified IRIG-H time code is produced at a 1-pps rate and radiated continuously on a 100-Hz subcarrier on all frequencies. A complete code frame is 1 min. The 100-Hz subcarrier is synchronous with the code pulses, so that 10-msec resolution is obtained. The code contains DUT1 values; UTC time expressed in year, day of year, hour, and minute; and status indicators relating to impending leap seconds and Daylight Saving Time.

Note 17: Except for voice announcement periods and the 5-min semi-silent period each hour.

TABLE 25-8

Characteristics of Standard-Frequency and Time-Signal Emissions in Additional Bands, Valid as of November 1991

Station			Type of Antenna(s)	Carrier Power (kW)	Number of Simultaneous Transmissions	Period of Operation		Standard Frequencies used		Duration of Emission		Uncertainty of Frequency and Time Intervals (Parts in 10^{12}) (Note 1)	Method of DUT1 Indication
Call Sign	Approximate Location	Latitude and Longitude				Days/ Week	Hours/ Day	Carrier (kHz)	Modulation (Hz)	Time Signal (min)	Audio Modulation (min)		
	Allouis, France	47°10′N 02°12′E	Omni-directional	1000 to 2000	1	7	24	162	1 (Note 2)	Continuous	A3E broadcast continuously	±2	No DUT1 transmission
CHU (Note 3)	Ottawa, Canada	45°18′N 75°45′W	Omni-directional	3, 10, 3	3	7	24	3330, 7335, 14,670	1 (Note 4)	Continuous	Nil	±5	CCIR code by split pulses
	Donebach, F.R. of Germany	49°34′N 09°11′E	Omni-directional	250	1	7	24	153	Nil	Nil	A3E broadcast continuously	±2	
DCF77 (Note 3)	Mainflingen, F.R. of Germany	50°01′N 09°00′E	Omni-directional	20 (Note 5)	1	7	24	77.5	1	Continuous (Note 6)	Continuous (Note 7	±0.5	No DUT1 transmission
	Droitwich, United Kingdom	52°16′N 02°09′W	T	400	1	7	22	198 (Note 8)	Nil	Nil	A3E broadcast continuously	±20	
	Westerglen, United Kingdom	55°58′N 03°50′W	T	50	1	7	22	198 (Note 8)	Nil	Nil	A3E broadcast continuously	±20	
	Burghead, United Kingdom	57°42′N 03°28′W	T	50	1	7	22	198 (Note 8)	Nil	Nil	A3E broadcast continuously	±20	
GBR (Notes 3, 9, 10)	Rugby, United Kingdom	52°22′N 01°11′W	Omni-directional	750 60 (Note 5)	1	7	22 (Note 11)	16.0	Nil	Nil	Nil	±10	
HBG (Note 12)	Prangins, Switzerland	46°24′N 06°15′E	Omni-directional	20	1	7	24	75	1 (Note 13)	Continuous	Nil	±1	No DUT1 transmission
JJF2 (Note 3) JG2AS	Sanwa, Sashima, Ibaraki, Japan	36°11′N 139°51′E	Omni-directional	10	1	7	24 (Note 14)	40	1 (Note 15)	Continuous (Note 16)	Nil	±10	
MSF	Rugby, United Kingdom	52°22′N 01°11′W	Omni-directional	25 (Note 5)	1	7	24 (Note 17)	60	1 (Note 18)	Continuous	Nil	±2	CCIR code by double pulse
	Milano, Italy	45°20′N 09°12′E	Omni-directional	600	1	7	24	900	Nil	Nil	A3E broadcast continuously	±2	
NAA (Notes 3, 10, 19)	Cutler, Maine, USA	44°39′N 67°17′W	Omni-directional	1000 (Note 5)	1	7	24 (Note 20)	24.0 (Note 21)	Nil	Nil	Nil	±10	
NAU (Notes 3, 10, 19)	Aguada, Puerto Rico	18°23′N 67°11′W	Omni-directional	100 (Note 22)	1	7	24	28.5	Nil	Nil	Nil	±10	
NTD (Notes 3, 10, 19)	Yosami, Japan	34°58′N 137°01′E	Omni-directional	50 (Note 5)	1	7	24 (Note 23)	17.4	Nil	Nil	Nil	±10	
NLK (Notes 3, 10, 19)	Jim Creek, Washington, USA	48°12′N 121°55′W	Omni-directional	125 (Note 5)	1	7	24 (Note 24)	24.8	Nil	Nil	Nil	±10	
NPM (Notes 3, 10, 19)	Lualualei, Hawaii, USA	21°25′N 158°09′W	Omni-directional	600 (Note 5)	1	7	24 (Note 25)	23.4	Nil	Nil	Nil	±10	
NSS (Notes 3, 10, 19)	Annapolis, Maryland, USA	38°59′N 76°27′W	Omni-directional	400 (Note 5)	1	7	24 (Note 26)	21.4	Nil	Nil	Nil	±10	
NWC (Notes 3, 10, 19)	Exmouth, Australia	21°49′S 114°10′E	Omni-directional	1000 (Note 5)	1	7	24 (Note 27)	22.3	Nil	Nil	Nil	±10	

OMA	Podebrady, Czech and Slovak Federal Republic	50°08′N 15°08′E	T	5	1	7	24	50	1 (Note 28)	23 hours per day (Note 29)	Nil	±1000	No DUT1 transmission
RBU (Note 3)	Moskva	55°48′N 38°18′E	Omni-directional	10	1	7	24	$66\frac{2}{3}$	10, 100, 312.5	Continuous DXXXW (Note 30)	Continuous (Note 31)	±5	CCIR code by double pulse (Note 32)
RTZ (Note 3)	Irkutsk	52°26′N 104°02′E	Omni-directional	10	1	7	23	50	1, 10	6/60	Nil	±5	CCIR code by double pulse (Note 32)
RW-166	Irkutsk	52°18′N 104°18′E	Omni-directional	40	1	7	23	198		Nil	A3E broadcast continuously	±5	
RW-76	Novosibirsk	55°04′N 82°58′E	Omni-directional	150	1	7	22	270		Nil	A3E broadcast continuously	±5	
SAJ	Stockholm, Sweden	59°15′N 18°06′E	Omni-directional	0.02 (e.r.p.)	1	3 (Note 33)	2 (Note 34)	150,000	Nil	10 (Note 35)		±2	
UNW3	Molodechno	54°26′N 26°48′E	Omni-directional	—	1	7	2	25.5, 25.1, 25.0, 23.0, 20.5	1, 10, 40 (Note 36)	40 min twice per day (Note 37)	Nil	±10	
UPD8	Arkhangelsk	67°24′N 41°32′E	Omni-directional	—	1	7	2	25.5, 25.1, 25.0, 23.0, 20.5	1, 10, 40 (Note 36)	40 min twice per day (Note 38)	Nil	±10	
UQC3	Khabarovsk	48°30′N 134°51′E	Omni-directional	300	1	7	2	25.0, 25.1, 25.5, 23.0, 20.5	1, 10, 40 (Note 36)	40 min 3 time per day (Note 39)	Nil	±10	
USB2	Beshkeck	43°04′N 73°39′E	Omni-directional	—	1	7	3	25.5, 25.1, 25.0, 23.0, 20.5	1, 10, 40 (Note 36)	40 min 3 times per day (Note 40)	Nil	±10	
UTR3	Gorky	56°11′N 43°58′E	Omni-directional	300	1	7	2	25.0, 25.1, 25.5, 23.0, 20.5	1, 10, 40 (Note 36)	40 min 3 times per day (Note 41)	Nil	±10	
VNG	Llandilo, New South Wales, Australia	33°43′S 150°48′E	Omni-directional	10 3 5	2–3	7	24 (Note 42)	8,538 12,984 16,000	1, 1000 (Note 43)	Continuous	Nil	±100	CCIR code by 45 cycles of 900 Hz immediately following the normal second markers
WWVB (Note 3)	Fort Collins, Colorado, USA	40°40′N 105°03′W	Top-loaded vertical	13 (Note 5)	1	7	24	60	1 (Note 44)	Continuous	Nil	±10	No CCIR code
	Motala, Sweden	58°26′N 14°59′E	Omni-directional	300	1	7	17	189	Nil	21 sec once per day (Note 46)	A3E broadcast continuously	±50 (Note 8)	CCIR code by decreased audio modulation frequency
EBC	San Fernando, Cadiz, Spain	36°28′N 06°12′W	Omni-directional	1	1	7	1	12,008 6,840	(Note 47)	10	(Note 48)	±100	CCIR code by double pulse

Note 1: This value applies at the transmitter; to realize the quoted uncertainty at the point of reception, it could be necessary to observe the received phase time frequency over a sufficiently long period in order to eliminate noise and random effects.

Note 2: Phase modulation of the carrier by +1 and −1 radian in 0.1 sec every second except the 59th second of each minute. This modulation is doubled to indicate binary 1. The numbers of the minute, hour, day of the month, day of the week, month, and year are transmitted each minute from the 21st to the 58th second, in accordance with the French legal time scale. In addition, a binary 1 at the 17th second indicates that the local time is 2 hr ahead of UTC (summer time), a binary 1 at the 18th second indicates when the local time is 1 hr ahead of UTC (winter time), a binary 1 at the 14th second indicates the current day is a public holiday (Christmas, 14 July, etc.), and a binary 1 at the 13th second indicates that the current day is the eve of a public holiday.

Note 3: These stations have indicated that they follow one of the systems referred to in Rec. 460.

Note 4: Pulses of 300 cycles of 1000-Hz tone: the first pulse in each minute is prolonged.

Note 5: Figures give the estimated *radiated* power.

Note 6: At the beginning of each second (except the 59th second), the carrier amplitude is reduced to 25% for a duration of 0.1 or 0.2 sec corresponding to "binary 0" or "binary 1," respectively. The number of the minute, hour, day of the month, day of the week, month, and year are transmitted in BCD code from the 21st to the 58th second. The time signals are generated by the Physikalisch-Technische Bundesanstalt (PTB) and are in accordance with the legal time of the Federal Republic of Germany which is UTC (PTB) + 1 hr (Central European Time CET) or UTC (PTB) + 2 hr (Central European Summer Time CEST). In addition, CET and CEST are indicated by a binary 1 at the 18th or 17th second, respectively. To achieve a more accurate time transfer and a better use of the frequency spectrum available, an additional pseudo-random phase shift keying of the carrier is superimposed on the AM second markers.

Note 7: Call sign is given by modulation of the carrier with 250-Hz tone three times every hour at the minutes 19, 39, and 59, without interruption of the time-signal sequence.

Note 8: No coherence between carrier frequency and time signals.

Note 9: FSK is used, alternatively with CW; both carriers are frequency controlled.

Note 10: MSK (minimum shift keying) in use: a phase-stable carrier can be recovered after suitable multiplication and mixing in the receiver. It will be recalled that the use of minimum shift keying means that no discrete component exists at the respective carrier frequencies which are given in the table. The MSK signal can be expressed as

$$S(t) = \cos[2\pi f_c t + a_n(\pi t/2T) + \varphi_n]$$

where $a_n = i(-1)$ for mark (space) and $\varphi_n = 0, \pi$ (modulo 2π). If the transmission is to be useful as a frequency reference it is necessary to recover a phase coherent carrier free from the $\pi/2$ increments introduced by the modulation. There are two approaches. The MSK signal is considered as a continuous-phase frequency shift keying (CPFSK) with a modulation index of 0.5. Squaring the signal followed by bandpass filtering at center frequency $2f_c$ produces a CPFSK signal with spectral components at $2f_c + 2f_b$ and $2f_c - 2f_b$, corresponding to mark and space, respectively. The components can be extracted by means of two phase-locked loops (PLL) and the reference carrier recovered by multiplication, division, and filtering. The other approach treats the MSK signal as a form of phase-shift keying (PSK), MSK being obtained by transformations from binary PSK (BPSK) or quadrature PSK (QPSK). The carrier recovery techniques available for PSK such as Costas-loop can thus be applied to MSK; such a demodulator has been realized in a single-chip form.

Note 11: Maintenance period from 1000 to 1400 hr UTC each Tuesday.

Note 12: Coordinated time signals.

Note 13: Interruption of the carrier during 100 msec at the beginning of each second; double pulse each minute; triple pulse each hour; quadruple pulse every 12 hr.

Note 14: JJF2: telegraph; JG2AS: in the absence of telegraph signals.

Note 15: There are two types of formats: One is the transmission of the carrier frequency for 500-msec duration at the beginning of each second—except for the 59th second, which is for 200-msec duration. The second format is generated in a slow time code (1 bit/sec) which consists of a transmitted carrier frequency for 500-msec and 800-msec duration, corresponding to "binary 1" and "binary 0," respectively. The duration of the "position mark" at each 9th second and that of the frame reference marker is 200 msec. The number of the minute, hour, day of the year, and the time offset to DUT1 are transmitted in BCD code from the 1st through the 43rd second.

Note 16: In absence of telegraphic traffic.

Note 17: The transmission is interrupted during the maintenance period from 1000 to 1400 hr UTC (on the first Tuesday of each month).

Note 18: Carrier interrupted for 100 msec at each second and 500 msec at each minute; fast time code, 100 bit/sec, BCD NRZ emitted during min-interruption giving month, day of month, hour, and minute. Slow time code, 1 bit/sec, BCD PWM emitted from seconds 17 to 51 giving year, month, day of month, day of week, hour, and minute together with 8-bit identifier from 52 seconds to 59. CCIR DUT1 code by double pulse.

Note 19: This station is primarily for communication purposes; while these data are subject to change, the changes are announced in advance to interested users by the U.S. Naval Observatory, Washington, D.C., USA.

Note 20: From 1200 to 2000 hr UTC each Sunday while NSS is off the air (until 15 July).

Note 21: As of 23 January 1984, until further notice.

Note 22: Became operational on 14 August 1984, 74 kW.

Note 23: 2300 to 0900 hr UTC just first Thursday-Friday, 2300 to 0700 hr UTC all other Thursday-Fridays. Half power 2200 to 0200 hr UTC each Monday and Friday.

Note 24: Except from 1600 to 2400 hr UTC each Thursday. During Daylight Saving Time 1500 to 2300 hr UTC each Thursday.

Note 25: 2.5 MHz: 0000–1000 hr UTC; 5 MHz: 0900–0100 hr UTC; 10 MHz: continuous; 15 MHz: 0100–0900 hr UTC.

Note 26: Off the air until 2100 hr UTC on 15 July, except for 14 hr each Sunday to cover the period when NAA is off the air.

Note 27: From 0000 to 0800 hr, usually each Monday.

Note 28: A1A telegraphy signals.

Note 29: From 1000 to 1100 hr UTC, transmission without keying except for call-sign OMA at the beginning of each quarter-hour.

Note 30: The standard frequencies and time signals are DXXXW-type emissions and are made up of carrier sine-wave oscillations with the frequency of $66\frac{2}{3}$ kHz, which are interrupted for 5 msec every 100 msec; 10 msec after an interruption the carrier oscillations are narrow-band phase-modulated for 80 msec by sine-wave signals with subcarriers of 100 or 312.5 Hz and a modulation index of 0.698. Amplitude-modulated signals with a repetition frequency of 10 Hz are used to transmit time markers. Signals with a subcarrier of 312.5 Hz are used to indicate second and minute markers, and also "1s" in the binary code for the transmission of time-scale information; signals with a frequency of 100 Hz are used to indicate "0s" in the binary code.

Note 31: N0N signals may be transmitted in individual cases.

Note 32: The additional information about the value of the difference UT1 − UTC is transmitted by code dUT1. It provides more precisely the difference UT1 − UTC down to multiples of 0.02 sec. The total value of the correction is DUT1 + dUT1. Possible values of dUT1 are transmitted by marking of p second pulses between the 21st and 24th seconds of the minute, so that dUT1 = $+0.02$ sec $\times\ p$. Negative values of dUT1 are transmitted by marking of q second pulses between the 31st and 34th second of the minute, so that dUT1 = -0.02 sec $\times\ q$.

Note 33: Each Monday, Wednesday, and Friday.

Note 34: From 0930 to 1130 hr UTC. When summer time, add 1 hr to the times given.

Note 35: Second pulses of 8 cycles of 1-kHz modulation during 5 min beginning at 1100 hr UTC and 1125 hr UTC. When summer time, add 1 hr to the instants given.

Note 36: Two types of signal are transmitted during a duty period:
(a) A1A signals with carrier frequency 25 kHz, duration 0.0125; 0.025; 0.1; 1, and 10 sec with repetition periods of 0.025, 0.1, 1, 10, and 60 sec, respectively.
(b) N0N signals with carrier frequencies 25.0, 25.1, 25.5, 23.0, 20.5 kHz. The phases of these signals are matched with time markers of the transmitted scale.

Note 37: From 0706 to 0747 hr and 1306 to 1347 hr UTC in winter.
From 0606 to 0647 hr and 1206 to 1247 hr UTC in summer.

Note 38: From 2106 to 2147 hr and 1106 to 1147 hr UTC in winter.
From 0206 to 0247 hr and 0806 to 0847 hr UTC in summer.

Note 39: From 0206 to 0247 hr, 0806 to 0847 hr and 1406 to 1447 hr UTC in winter.
From 0106 to 0147 hr, 0706 to 0747 hr and 1306 to 1347 hr UTC in summer.

Note 40: From 0406 to 0447 hr, 1006 to 1047 hr and 1606 to 1647 hr UTC in winter.
From 0306 to 0347 hr, 0906 to 0947 hr and 1506 to 1547 hr UTC in summer.

Note 41: From 0506 to 0547 hr and 1096 to 1947 hr UTC in winter.
From 0406 to 0447 hr and 1806 to 1847 hr UTC in summer.

Note 42: 8638 kHz and 12,984 kHz continuous; 16,000 kHz from 2200 to 1000 hr UTC.

Note 43: Pulses of 50 cycles of 1000-Hz tone, shortened to 5 cycles from the 55th to the 58th second; the 59th pulse is omitted; the minute marker is 500 cycles. At the 5th, 10th, 15th, etc., minutes, pulses from the 50th to the 58th second are shortened to 5 cycles. Voice identification on 5000 kHz and 16,000 kHz between the 20th and 50th seconds in the 15th, 30th, 45th, and 60th minutes. Morse identification "VNG" on 8638 kHz and 1,984 kHz in the 15th, 30th, 45th and 60th minutes. A BCD time incorporating time of day and day number of the year is transmitted between the 20th and 46th second with a binary "0" represented by 100 cycles and a binary "1" by 200 cycles of 1000-Hz tone. The minute information for the next minute is given from the 21st to the 28th second, hour information from the 29th to the 35th second, and day of the year from the 36th to the 46th second; parity bits are included at the end of each code sequence.

Note 44: Time code used which reduces carrier by 10 dB at the beginning of each second. The code contains information on the year, day of year, hour, minute, UT1 value, and status indicators for impending leap seconds and Daylight Saving Time.

Note 45: A1A time signals of 0.1-sec duration (minute marker of 0.5-sec duration) followed by code pulses from 0.25 to 0.3 sec for information about DUT1, dUT1 and time of the day (minute, hour) in UTC.

Note 46: A3E time signals of 0.1-sec duration between 11 hr 58 min 55 sec and 11 hr 59 min 16 sec UTC. The minute marker is of 0.5-sec duration. When summer time, add 1 hr to the instants given.

Note 47: Seconds pulses of a duration of 0.1 sec, modulated at 1000 Hz.
Minutes pulses of a duration of 0.5 sec, modulated at 1250 Hz.

Note 48: Minutes 00 to 10, 12,008 kHz, A2A.
15 to 25, 12,008 kHz, J3E.
30 to 40, 6840 kHz, A2A.
45 to 55, 6840 kHz, J3E.

During the minute immediately preceding each of the periods indicated, transmission of call sign in slow Morse twice.

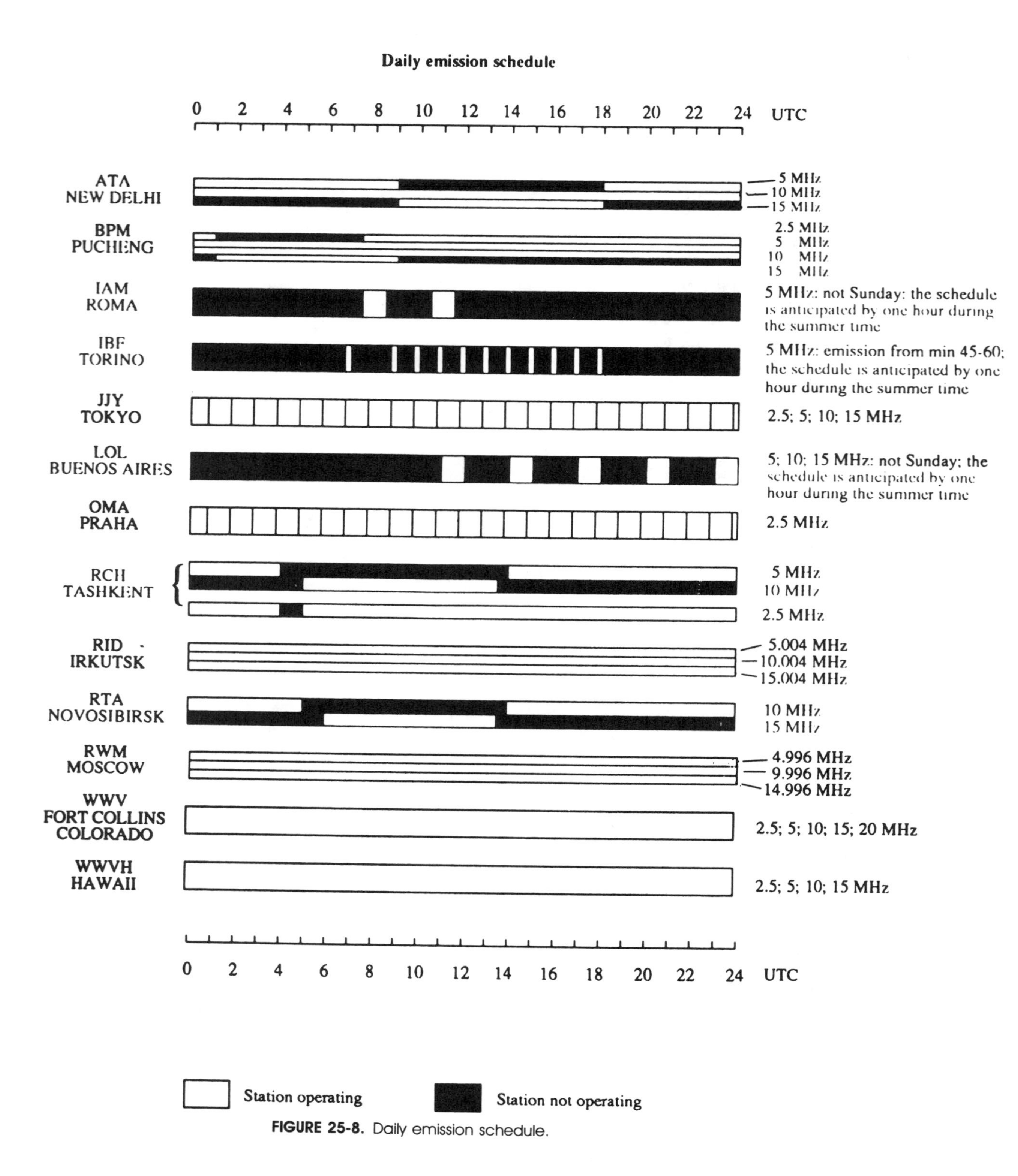

FIGURE 25-8. Daily emission schedule.

Hourly modulation schedule

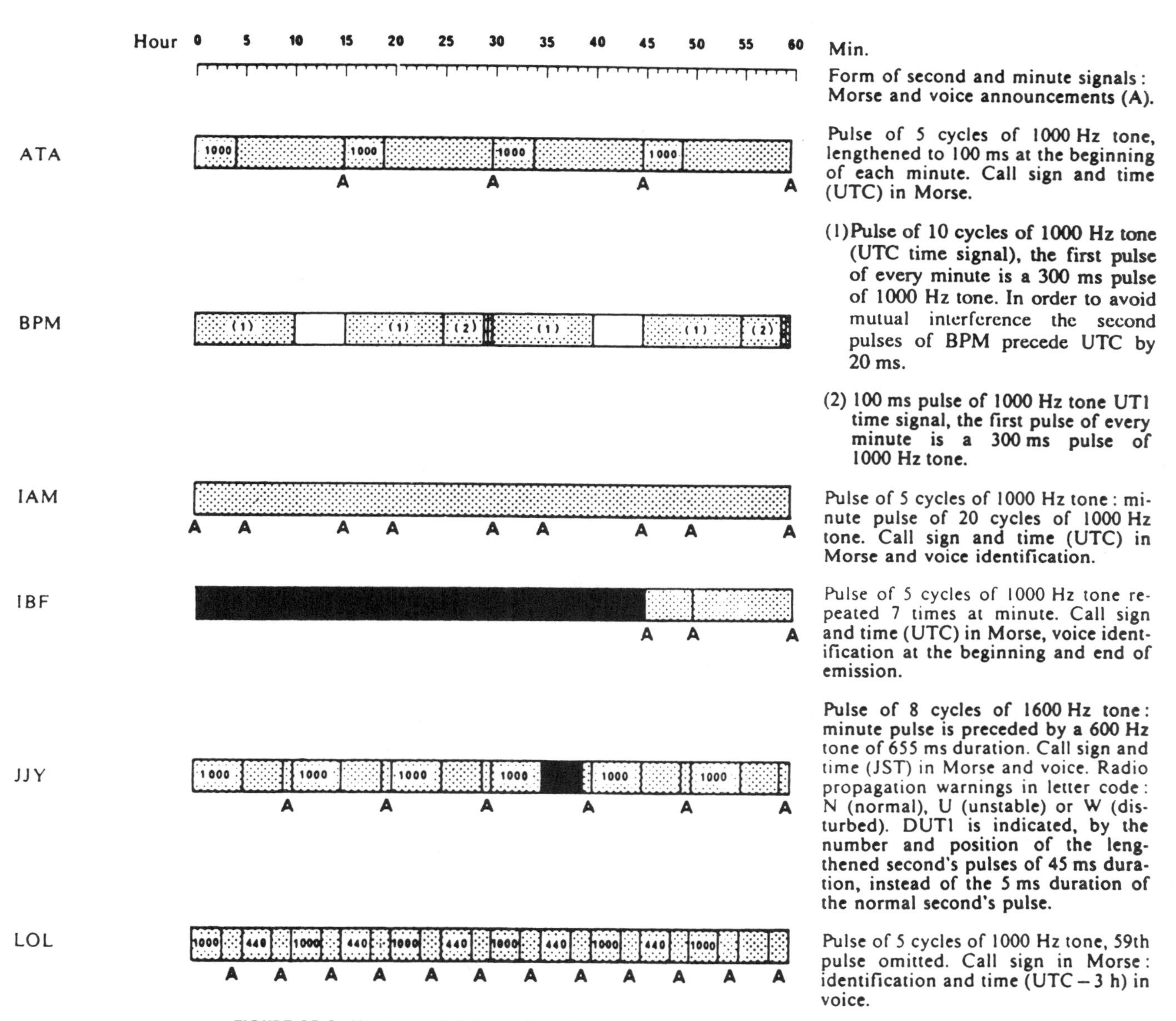

FIGURE 25-9. Hourly modulation schedule.

The broadcasting and telecommunications environment provides the possibility of sending signals worldwide within 1 sec of time. It is therefore necessary to accommodate the variations in local time (and date) in any method of coding intended to be consistent worldwide. There are also known discontinuities in local time (the duplicated hour at the end of "summer time," and the "leap second") to be taken into account.

2. Standard Time

The standard unit of time is the second, obtained by defining the frequency of the cesium atomic transition as 9,192,631,770 Hz. For the purpose of creating a

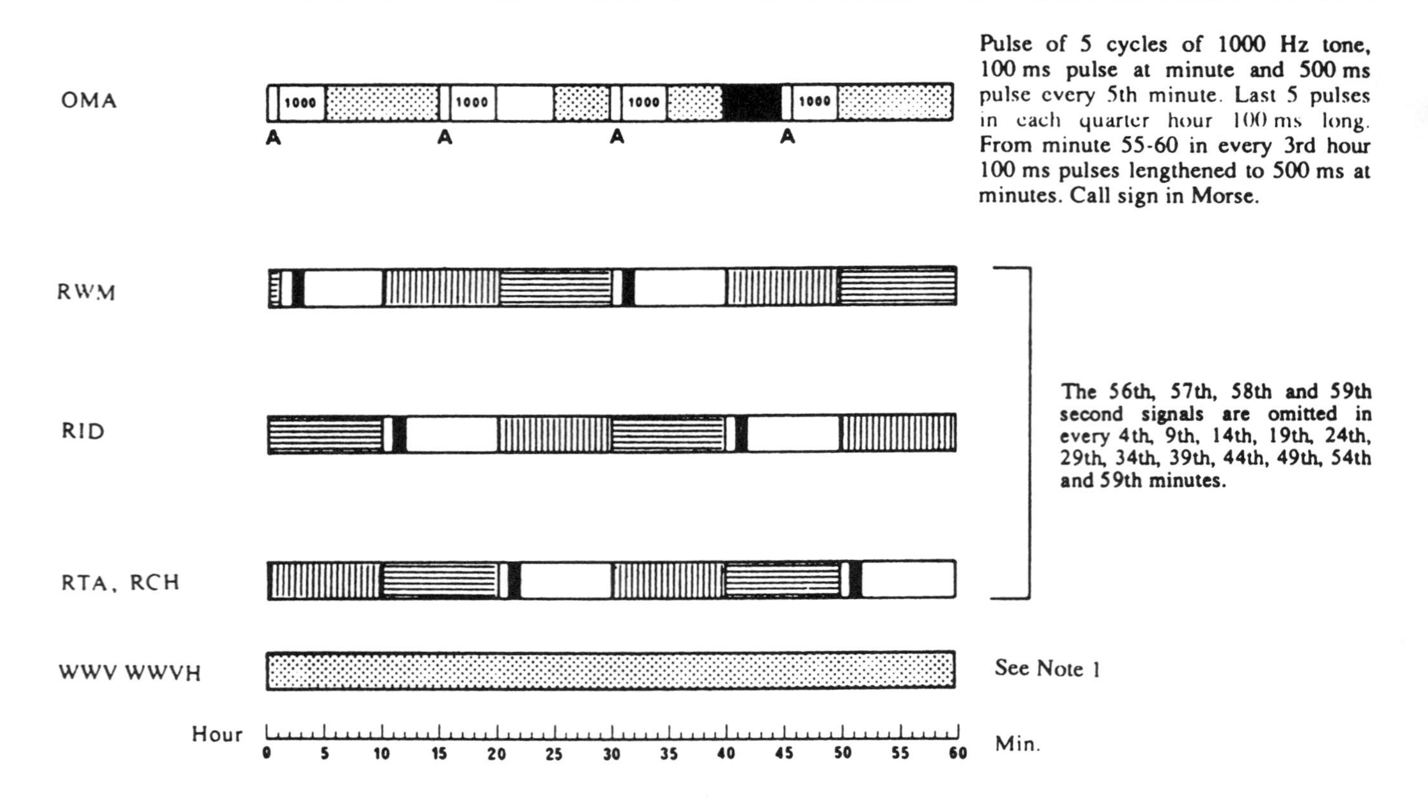

Note 1 – Pulse of 5 cycles of 1 000 Hz (WWV) or 6 cycles of 1 200 Hz (WWVH) tone, lengthened to 0.8 s at the beginning of each minute. An 0.8 s pulse of 1 500 Hz begins each hour at both stations. 29th and 59th pulses each minute are omitted. Voice time announcements preceding each minute. 45-second audio tones alternating between 500 and 600 Hz each minute, except when special announcements or station identification messages are given in voice. One 45-second segment of 440 Hz is included each hour at one minute (WWVH) or two minutes (WWV) past the hour. A modified IRIG-H time code, giving the year, day of year, hour, minute, DUT1 value, and information on impending leap seconds and Daylight Saving Time, is broadcast continuously on a 100 Hz sub-carrier. DUT1 information is provided by the number and position of doubled second pulses each minute. All modulations interrupted for 40 ms around each second's pulse.

FIGURE 25-9. Hourly modulation schedule (continued).

regular time scale, these seconds are counted to give days, hours, and minutes since 1 January 1958. This is known as International Atomic Time (TAI). This time scale, based on a physical property, drifts out of step with a time scale, such as Universal Time (UT) or Greenwich mean time (GMT), obtained from astronomical observation. Since the origin of TAI was set in agreement with UT at the beginning of 1958, TAI has advanced by about 21 sec with respect to UT. In order to provide a time scale with seconds coincident to those of TAI, but within a close tolerance (± 0.8 sec) of UT, a version of TAI offset by a whole number of seconds is maintained by the Bureau International des Poids et Mesures (BIPM). This is known as Coordinated Universal Time (UTC). The tolerance is maintained by occasionally adding (or, in principle, deleting) a single second to make a 61-sec (or 59-sec) minute. The preferred occasions are at the end or middle of the year, with at least 8 weeks' notice. For example, one of these "leap seconds" occurred at 0000 hr UTC on 1 July 1982 when the UTC seconds marker sequence was

30 June 1982	23 hr 59 min 59 sec
	23 hr 59 min 60 sec
1 July 1982	00 hr 00 min 00 sec

All of the standard time signals used by broadcasters worldwide are derived from the UTC time scale, and the times are often, wrongly, referred to in terms of Greenwich mean time (GMT) and an offset. For reasons given above, the UTC time signal known in the United Kingdom as "the Greenwich time signal" will sometimes differ from true GMT by more than half a second. This confusion of name is of little practical consequence in everyday life, but it is significant to astronomers, navigators, and lawyers.

Recommendation 460 recommends "that all ... time signal emissions conform as closely as possible to Coordinated Universal Time (UTC) ... from 1 January 1975."

2.1. Time Offsets

In practice, all countries refer their national time or times to UTC with an offset. There are 38 different offsets currently in use. Except for Nepal (+5 hr 40 min), all the offsets are multiples of half an hour and range from −11 hr (Samoa) to +14 hr (Anadyr, Russian Federation, in summer time). Many countries advance their time by 1 hr during local summer (dependent on hemisphere); exceptionally, Cook Islands advance by half an hour. There are various dates and times for summer time changes. Within some countries (Australia, Canada) there are some time zones differing by half an hour. There are some states (Queensland, Australia; Arizona and Indiana, USA) which, unlike their neighbors, do not adopt summer time.

It would probably be sufficient to provide a method of signaling a local time offset with a six-bit code giving half-hour steps in a range −12 to +15 hr. In some applications, the local offset of the program source or of the transmitter would be signaled; in other cases, the local offset applicable to the receiver site would be required.

3. Date

The change of date varies, of course, with local time, so a common broadcast standard for date would be referred to UTC and it would be corrected, if necessary, by the operation of the local offset.

There are several calendars in use worldwide, but a simple common reference, the Modified Julian Date (MJD), has been defined for this purpose. This is a five-digit decimal number increasing by one at midnight UTC. The origin of the count is 17 November 1858, because at midday on that date the Julian Day (used by astronomers to give continuity from 4713 B.C.) reached the figure 2,400,000. A more convenient reference is 31 January 1982, when the MJD was 45,000. It is a simple matter to calculate time intervals, even over many days, by use of MJD and UTC (provided the occasional leap second is known or can be neglected).

Recommendation 457 recommends that for modern timekeeping and dating requirements, a decimal day count should be used wherever necessary; the calendar day should be counted from 0000 hr TAI, UTC, or UT, and it should be specified by a number with five significant figures.

Although not defined in any standards, it is convenient to use the idea of a local day number which is advanced or delayed by the local time offset and which changes at local midnight.

3.1. Week Number

For many commercial purposes and for planning broadcast program schedules, it is convenient to work in terms of day of the week, week number, and year.

There is an international standard (ISO 2015) for the numbering of weeks. This can be summarized by saying that weeks begin on Mondays, and that week 1 of a year contains the first Thursday of January. The week number can be associated with a day of the week (conventionally, Monday = 1 to Sunday = 7) and a year to specify a particular date. Note that occasional years (about 5 in 28) have 53 weeks, and that the "week-year" of a date in the inclusive range 29 December to 3 January may differ from the "calendar" year. The relation between week number and MJD is given in Section 4.

Although the ISO week numbering system is in general use worldwide, other week number systems remain within certain organizations. In some cases the week number of Monday accords with ISO, but the week is taken to run from Saturday to Friday, for example. In other cases, even the years containing 53 weeks are different.

3.2. Calendar Date

The various calendar systems in use are well known and, in most cases, well defined. In these cases, it is possible to generate a formula for conversion between calendar systems, the convenient intermediate standard being the MJD. The information for conversion between MJD and the Gregorian calendar is given in Section 4.

Certain calendar systems depend on a suitably qualified person witnessing an event (e.g., the first sighting of a crescent moon, or the sighting of a particular type of fish off a Pacific island), and these can only be related to the MJD after the event.

4. Conversion Between Time and Date Conventions

The types of conversion which may be required are summarized in Figure 25-U.1.

The conversion between MJD + UTC and the local day number + local time is simply a matter of adding or subtracting the local offset. This process may, of course, involve a "carry" or "borrow" from the UTC affecting the MJD. The other five conversion routes shown in the diagram are as follows:

Symbols Used

MJD:	Modified Julian date
Y:	Years since 1900 (e.g., for 2003, Y = 103)
M:	Month from January = 1 to December = 12
D:	Day of month from 1 to 31
WY:	"Week number" year from 1900
WN:	Week number according to ISO 2015
WD:	Day of the week from Monday = 1 to Sunday = 7
K, L, M′, W, Y′:	Intermediate variables
INT:	Integer part, ignoring remainder
MOD 7:	Remainder (0–6) after dividing integer by 7
*:	Multiplication

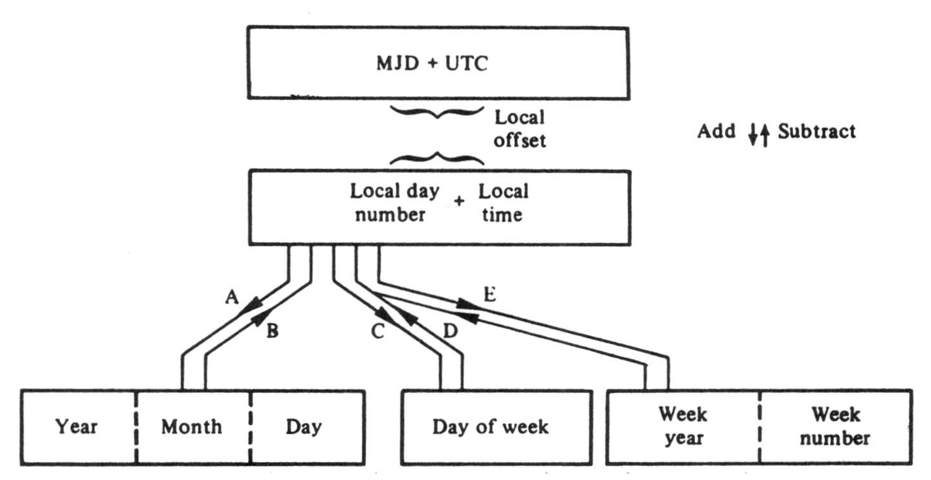

FIGURE 25-U.1. Conversion between time and date conventions.

A: To find Y, M, D from MJD:

$$Y' = INT((MJD - 15{,}078.2)/365.25)$$

$$M' = INT((MJD - 14{,}956.1 - INT(Y'*365.25))/30.6001)$$

$$D = MJD - 14{,}956 - INT(Y'*365.25) - INT(M'*30.6001)$$

If $M' = 14$ or $M' = 15$, then $K = 1$; else $K = 0$

$$Y = Y' + K$$

$$M = M' - 1 - K*12$$

B: To find MJD from Y, M, D:

If $M = 1$ or $M = 2$, then $L = 1$; else $L = 0$

$$MJD = 14{,}956 + D + INT((Y - L)*365.25) + INT((M + 1 + L*12)*30.6001)$$

C: To find WD from MJD:

$$WD = ((MJD + 2) \text{MOD}\ 7) + 1$$

D: To find MJD from WY, WN, WD:

$$MJD = 15{,}012 + WD + 7*(WN + INT((WY*1{,}461/28) + 0.41))$$

E: To find WY, WN from MJD:

$$W = INT((MJD/7) - 2{,}144.64)$$

$$WY = INT((W*28/1{,}461) - 0.0079)$$

$$WN = W - INT((WY*1{,}461/28) + 0.41)$$

Example:

MJD = 45,218	W = 4,315
Y = (19)82	WY = (19)82
M = 9 (September)	WN = 36
D = 6	WD = 1 (Monday)

Note: These formulae are applicable from 1 March 1990 to 28 February 2100 inclusive.

Section 25-10 was taken from ITU-R Rec. 808, RBT Series, Geneva, 1992.

Index

An *italic* term denotes extensive coverage of a subject. **Boldface** denotes a definition.

AAL-0, 278
AAL-1, 278–279
AAL-2, 279–280
AAL-3/4, 280–282
AAL-5, 282–283
AAL (ATM adaptation layer), 277–283
AAL categories or types, 278–279
AAL-PCI bit errors, 279
AAL-SAP, 277
AAL-3/4 SAR-PDU, 282
AAL-SDU, 282
Absolute group delay, 43, 44
ACK control bit, TCP, 216
Acknowledge number, 216
Acquisition time *vs.* bit error probability, 272
Active open, TCP, 213
Adaptation of intermediate rates to 64 kbps, 185
Adaptation based on CCITT Rec. V.110, 177–197
Adaptation for asynchronous rates up to 19,200 bps, 187–188
Adaptation of 48-kbps user rate to 64 kbps, 185
Adaptation of 56-kbps user rate to 64 kbps, 186
Adaptation of V-series data signaling rates to intermediate rate, 177–185
Adapting standard data rates to the 64-kbps digital channel, 176–197
Additional noise contributed, PCM, 44
Addresses determine IP routing, 205
Address formats, internet protocol, 208
Address resolution protocol (ARP), 203, 209
Adjustment of actual relative levels, PCM, 36–37
Advanced peer-to-peer networking (APPN), 232
Advanced peer-to-peer networking (APPN), SNA, 230
Advanced Research Projects Agency, 201
Adverse propagation conditions, 105
Aerodynamic characteristics vary with wind angle, 134
Aggregate interfering power, 120
A-law, μ-law, 52
Allocated bits of the VCI field, 268
Allocation of escape codes, 252
Allowable beam twist and sway for cross-polarization limited systems, 141–147
Allowable bit error ratios at output of hypothetical digital path, 104–107
Allowable twist and sway, cross-polarization limited systems, 146
Alpha and delta parameters, ATM, 271, 273
Alternative conformance definitions, 299
Alternative frame structure for the adaptation of 56-kbps user rate to 64 kbps, 186
Ambient electron density, 89
Ambipolar diffusion, 124, 126, 127
Analog network propagation time, 3
Analog-to-analog channels, group delay, 42
Analog-to-analog channels, PCM, 40
Analog-to-digital channels, PCM, 40, 41
Analog-to-digital channels, 2-wire, 53
Annual average flux, meteors, 122
Annual worst-month time percentage, 86
Antenna beamwidth nomogram, 143
Antenna boresight axis, 133
Antenna considerations, meteor burst, 129–130
Antenna diameter variation as a function of attenuation, 118
Antenna gain formula, 116
Antenna noise, 108
Antenna noise characteristics, 112
Antenna noise temperature, 116, 117
Antenna noise temperature as a function of elevation angle, 113
Antenna radiation standards A and B, FCC, 147
Antennas, passive repeaters and towers, 133–147
Applications, EIA-530, 161
Applications for an unbalanced digital interface, 169
APPN (advanced peer-to-peer networking), SNA, 230
APPN end node, 234
APPN network(s), 233, 239, 240, 241, 242
APPN nodes, 233, 237
ARP (address resolution protocol), 203, 209
Assigned internet protocol numbers, 207
Asynchronous, ATM, 264
Asynchronous accommodation utilizes floating mode, 17
Asynchronous character-mode interfacing, V.32 bis, 156
Asynchronous digital multiplexing, 292
Asynchronous mapping of 1544 kbps tributary, 29
Asynchronous mapping of 44,736 kbps tributary into VC-3, 21
Asynchronous mapping of 139,264 kbps tributary into VC-4, 20
Asynchronous mapping of 1544 kbps into VC-11, 26–27
Asynchronous mapping of 2048 kbps into VC-12, 24–25
Asynchronous mapping of 6312 kbps into VC-2, 23–24
Asynchronous mapping of 34,368 kbps into VC-3, 21, 22
Asynchronous-to-synchronous conversion, V.110, 188
Asynchronous-to-synchronous conversion method, 199–200
Asynchronous transfer mode (ATM), 263–307
Asynchronous user rates, V.110, 187
Async-to-sync converter into a synchronous DCE, 198
ATM adaptation layer (AAL), 277–283
ATM (asynchronous transfer mode), 263–307
ATM/B-ISDN layering, 273
ATM cell: key to operation, 266–270
ATM cell crossing VC-4-4c boundary, 305
ATM cell flow, 274
ATM cell mapping, into E1, 303
ATM cells in DS3 frame, 301–302

ATM cells mapping into SDH, 304–305
ATM cell transfer outcome, 292
ATM end-point, 265
ATM Forum, 264, 301
ATM forum notation, 300
ATM header, consequences of error, 269
ATM into E3 and E4, 304
ATM layer, 265
ATM layer, 274–277
ATM layer functions supported at the UNI, 274
ATM layer information flows, 276–277
ATM layering and B-ISDN, 273–283
ATM layer management (M-plane), 276
ATM layer OAM flows at the UNI, 277
ATM layer QoS, 296, 297
ATM links simultaneously carry a mix of voice, data and image information, 264
ATM node processing delay, 292
ATM nodes, 283
Atmospheric attenuation factor, 116
Atmospheric constituents, 108
ATM routing and switching, 287–289
ATM SAP, 296
ATM-to-SMDS routers, 282
ATM unique QoS items, 291
Attenuation:
  backbone UTP cable, 66
  elementary cable section, 73
  horizontal UTP cable, 59–60
  optical fiber for buildings, 67
Attenuation and group-delay distortion in trunks and exchanges, 9–10
Attenuation coefficient:
  factory length specifications, 76
  mono-mode fiber, 72
Attenuation distortion, 2/4-wire exchanges, 9
Attenuation distortion and group delay distortion (PCM), 6–7
Attenuation due to vegetation, 85
Attenuation/frequency distortion, PCM, 39, 40, 41
Attenuation prediction process model, satellite path, 80
Attenuation at UHF through timber-framed buildings, 86
AU-3 frame, 14
AU-*n* frame, 15
AU-*n* pointer, 11
AU-*n* pointer value, 12
AU-*n*/TU-3 pointer coding, 14
AU-4 pointer, ATM, 304
AU-3 pointer adjustment operation, negative justification, 18
AU-4 pointer adjustment operation, negative justification, 17
Average channel capacity, meteor burst, 130
Average trail height, 128
  meteor burst, 127

Backbone cabling distances, 58
Backbone UTP cable, characteristic impedance, backbone cable, 64–67
Balanced interchange circuit, EIA-530, 165
Balanced mode attenuation, 150-ohm STP-A cable, 63
Balanced/unbalanced conversion, 168
Balance impedances, PCM, 54
Balance return loss, 1
Balance test network, 53
Band 17.7 to 19.7 GHz in North America, 102
Basic ATM cell structure, 266
Basic information unit (BIU), 244–245
Basic information unit (BIU) format, 244
Basic link unit, SNA, 245–246
Basic link unit (BLU), SNA, 245–246
Basic transmission loss, meteor burst, 126–127
BAsize, ATM, 282
BAsize (buffer size allocation), 281
BCH coding, conference TV, 310
Beamwidth nomogram, 143
Bending radii, mono-mode fiber, 71
Bending radius, 150-ohm STP-A cable, 62
BER allowable at output of hypothetical reference path, 104–107
BERs on ATM links, 291
BIP (bit interleaved parity), 302
B-ISDN access reference configuration, 265
B-ISDN/ATM functional layering, 273
B-ISDN/ATM network, 287–289
B-ISDN protocol reference model, 264, 265
Bit assignment, V.110, 182, 185
Bit/byte synchronous mapping of 1544 kbps, 28, 29, 30
Bit/byte synchronous mapping of 2048 kbps, 25, 26, 27
Bit error ratio, 105
Bit error ratio objectives, 105–106
Bit rate, conference TV, 309, 323
Bit rate adaptation of synchronous data rates up to 19.2 kbps, 177–185
Bit sequences:
  EIA-561, 159
  EIA-574, 157
Bit-serialized data, EIA-530, 161
Bit synchronous mapping of 1544 kbps tributary, 29
Bit synchronous mapping of 6312 kbps tributary, 24
Block layer, conference TV, 320
Block significance criterion, conference TV, 313
Block structure of VC-4, 20
Blueprint of SNA networking, 231
BOM (beginning of message) cell, 281
Border nodes, SNA, 236
Boundary function, SNA, 232, 242
600-bps to 8-kbps intermediate rate, 182
1200-bps to 8-kbps intermediate rate, 182
2400-bps to 8-kbps intermediate rate, 182
12,000-bps to 32-kbps intermediate rate, 183
Braided shielded core sizing, 62
Breaking strength, 150-ohm STP-A cable, 62
Break signal:
  V.14, 199–200
  V.110, 188
Broadband ISDN, 263–307
Broadband network scene, tying together, 306
Btag (beginning tag), ATM, 281, 282
Burst overlapping in TDMA networks, 121
Burst tolerance, ATM, 301
Bursty error conditions, ATM, 270

Cable capacitance, 176
Cable diameter, 150-ohm STP-A cable, 62
Cable engineering, EIA-562, 176
Cable specifications for commercial building telecommunication applications, 57–67
CAC (connection admission control), 297
  ATM, 294
Calculation of long-term attenuation statistics from point rainfall rate, 79–80
Calendar date, 338
Call discarding, 295
Called networks, 254
Capacitance model per unit length, EIA-562, 175
Capacitance unbalance, 150-ohm STP-A cable, 63
  pair to ground, backbone UTP cable, 65
Capacitive cable model, EIA-562, 174
Capacitive unbalance, 56
  wire pair cable, 59
Carrier-to-interference (C/I) ratio, 97
Cassegrain antenna, 112
Category I circuits, EIA-530, 162
Category II circuits, EIA-530, 162–163
CBDS (connectionless broadband data services), 280
C-4 boundary crossing by ATM cell, 304
CBR (constant bit rate service), 263, 264
CDS (central directory server), SNA, 236
CDV, maximum allowed, 298
CDV (cell delay variation), 292–293
CDV (cell delay variation), 296
CDV tolerance, 297, 299, 301
Cell arrival events, 293
Cell block is a sequence of N cells, 293
Cell conformance and connection compliance, 299
Cell delay variation, origins of, 298
Cell delay variation (CDV), 279
Cell delay variation (CDV), 292–293
Cell delay variation impact on UPC/NPC and resource allocation, 297
Cell delineation algorithm, 271–273
Cell delineation and scrambling, 270–272
Cell delineation state diagram, 271
Cell discrimination based on predefined header field values, 276
Cell entry event, 291
Cell error ratio, 293
Cell exit event, 291
Cell header at network-to-network interface, 266
Cell loss priority (CLP), 266
Cell loss priority (CLP) field, 268
Cell loss ratio, 293

Cell loss ratio objectives, 295
Cell misinsertion rate, 294
Cell rate decoupling, 264
Cell rate decoupling, 275
Cell transfer delay, 292
Center-fed antenna, required XPD, 145
Central directory server (CDS), SNA, 236
Cesium atomic transition, 335
Channel arrangements, radio-frequency, 97
Channel-associated signaling, 25
Channel-associated signaling methods, 30
Channel spacings and center gap, 104
Characteristic impedance:
  backbone UTP cable, 65–66
  horizontal UTP cable, 60
  150-ohm STP-A cable, 63
Checksum, TCP header, 217
Checksum parameter, transport protocol, 226
Choice of frequency, meteor burst, 130–131
Chromatic dispersion coefficient, 72
Chromatic dispersion coefficient, 76
CIB (crc indication bit), 285
C/I (carrier-to-interference) ratio, 97
CICS/VS, SNA, 243
CIF, conference TV, 316, 323, 325
CIF (common intermediate format), 309, 310, 311
CIF format, conference TV, 310, 311
Circuit 106, V.110, 190–191
Circuit 109, V.110, 191
Circuit grounding (signal common), EIA-530, 163
Circular aperture deflection angle nomogram, 144
Cladding diameter:
  dispersion shifted fiber, 74
  mono-mode fiber, 69
Cladding noncircularity, dispersion shifted fiber, 74
Classes and options, transport protocol, 222–223
Classes of service:
  EIA-530, 161
  EIA-561, 159
  EIA-574, 157
CL (connectionless) protocol, 284
Clear-sky figure of merit, 116, 118
Clear-sky noise temperature, 112
Clipping of reconstructed picture, 313
CLNAP (connectionless network access protocol), 280, 283, 284, 285
CLNAP entity, 285
CLNAP-PDU structure, 286
CLNAP protocol data unit structure and encoding, 285–287
Clouds and fog attenuation, 81
CLP = 0, CLP = 1, 299
CLP = 0 + 1, 301
CLP = 0 + 1 component, 297
CLP bit, 295, 297
CLP (cell loss priority), 266
CLP (cell loss priority) field, 268
CLSF (connectionless service functions), 283
Clumping, ATM, 292
Clusters/network partitions, SNA, 236
Co-channel arrangement, 18 GHz, 99, 101
CODEC delay measurement method, 325
Coded block pattern (CBP), conference TV, 319
Code distortions, equivalent total, 5
Coding, V.32 bis, 150–152
Coding of bits 4-1 of octet 7, TPDU format, 228
Coding control, conference TV, 313–314
Coding E bits for network-independent clocking, 196
Coherence bandwidth, meteor burst, 131
Coincident pointer, 17
Color code(s):
  backbone cable, 64–65
  horizontal UTP cable, 58
Color codes for horizontal 150-ohm STP-A cable, 61
Color difference components, 310
Combinations of preassigned VPI, VCI and CLP values at the UNI, 267
COM (continuation of message), 281
Common mode attenuation, 150-ohm STP-A cable, 63
Common part indicator (CPI), 282
Communication between two transmission control layers, SNA, 230
Communication by meteor-burst propagation, 122–131
Complex nominal impedances, 39
Complex nominal impedance at 2-wire port, 1
Compliant, its meaning in ATM, 299
Composite LEN node, SNA, 234
Composite network node, SNA, 234
Concatenation indication, SDH, 14
Concatenation and separation, transport protocol, 221
Conductor resistance, EIA-562, 174
Conference television, video CODEC at pX64, 309–326
Conformance applies to cells, 299
Congestion control functions, ATM, 295
Conical horn, wind forces on, 134
Connection admission control (CAC), 294
Connection admission control procedures, ATM, 296
Connection closing, TCP, 213
Connection establishment, transport protocol, 221
Connectionless data transfer, AAL, 278
Connection-oriented and connectionless services, 283–285
Connection oriented data transfer, 278
Connection-oriented protocol, ATM, 263
Connection-oriented transfer, 219
Connection-oriented variable bit rate services, 280
Connection traffic descriptor, 296–297
Connector contact assignments:
  EIA-530, 166
  EIA-561, 160
  EIA-574, 158
Connect TA to line state, 193
Consequences of errors in an ATM header, 269
Constant bit rate, ATM, 278
Constant bit rate service (CBR), 263, 264
Continuous jitter, V.110, 197
Continuous-state leaky bucket algorithm, 299, 300, 301
Contributions to noise temperature of an earth-station receiving antenna, 108, 112–115
Contributions to one-way propagation time, 4
Control flags, TCP header, 217
Controlled access, ATM, 267
Control point, APPN-SNA, 240, 241
Control response times, ATM, 296
Convergence sublayer, 280
Convergence sublayer protocol data unit, 281
Conversion of annual statistics to worst-month statistics, 86–88
Conversion between floating and locked tributary unit modes, 34
Conversion between floating and locked TU modes, 31
Conversion between time and date conventions, 338–340
Conversion of VC-11 to VC-12 for transport of TU-12, 33
Coordinated universal time (UTC), 336
Coordination between D bits and S bits, V.110, 180
Coordination between S bits and D bits, V.110, 179
Core assembly, UTP backbone cable, 65
Core concentricity error, 70
Core shield, 150-ohm STP-A horizontal cable, 62
Core shield resistance:
  backbone cable, 64
  backbone UTP cable, 67
Core wrap:
  backbone cable, 65
  150-ohm horizontal STP-A cable, 62
Correction mode, ATM, 270
Corrugated horn, 112
Cosine transform, conference TV, 313
Cosmic effects, 112
Cosmic noise, 108
CP-CP sessions, 242
CPCS (common part convergence sublayer) PDU, 281, 282, 283
C-plane, 264
CRC, CLNAP, 287
CRC indication bit (CIB), 285
CR (connection request), transport protocol, 226
Cross polarization discrimination (XPD), 97–98, 141
Crosstalk, PCM, 48–52
Crosstalk, wire pair cable, 56
Crosstalk level, 50, 52

Crosstalk measurement between two channels, PCM, 48
Crosstalk measurements of individual primary multiplexers, 51
CR TPDU, 226, 227
CR TPDU structure, 227
CSI (convergence sublayer indicator), 279
CS-PDU (CS, convergence sublayer), 280, 281
Cumulative error probability, 105
Cumulative statistics, scintillation, 95
Cumulative time distribution, 112
Customer data service unit interface, V.110, 177
Cutoff wavelength, 69, 70
  dispersion shifted fiber, 74
Cylindrical shrouds, wind forces on, 134

Daily emission schedule, 334
Data bits, V.110, 181
Data country code (DCC), 250
Datagram and network address, 204
Datagrams, 205
Data interchange circuits, EIA-562, 174
Data link layers and TCP/IP, 201–204
Data network identification codes, format, 250
Data network identification codes (DNICs), 250, 254, 255, 256
Data networks, 201–261
Data offset, TCP header, 217
Data rate ranges of the start-stop characters at the converter input, 197–198
Data rates:
  EIA-530, 161
  V.14, 197
Data rates on 64 kbps digital channel, 176–197
Data signaling rates:
  EIA-561, 159
  EIA-574, 157
Data structure, conference TV, 314
Data/telegraph transmission, 149–200
Data transfer, transport protocol, 221
Data transfer state, V.110, 194
Date, 337–338
Day-to-day variability in incidence, meteors, 122
D bits and S bits coordination, V.110, 180
DCC (data country code), 255, 256
DCE connector, EIA-530, 166
DC resistance:
  backbone UTP cable, 65
  wire pair cable, 59
DC resistance unbalance:
  backbone UTP cable, 65
  wire pair cable, 59
DDS (digital data system), 176–177
Decoding side adjustment, 36
Default routing, IP, 208
Definition of mechanical interface, EIA-530, 164
Deflection angle, 140
Deflection angle for a circular aperture, nomogram, 144
Deflection angle for rectangular aperture, nomogram, 142
Degraded minutes, 104
Delay estimation for circuits, 3–5
Delineation and scrambling objectives, ATM, 270–271
Dependence of 4-GHz equatorial ionospheric scintillation on monthly mean sunspot cycle, 95
Destination address, IP datagram, 206
Destination address field, CLNAP, 286
Destination-address and source-address, CLNAP, 285
Destination countries, 254
Destination port, TCP, 216
Destination unreachable message, ICMP, 210
Desynchronizer protection, 19
Detailed TPDU types, 227–228
Detection mode, ATM, 270
Development of data network identification codes (DNICs), 255–256
Dicke-type radiometers, 112
Dielectric strength:
  backbone UTP cable, 67
  150-ohm STP-A cable, 64
Differentially encoded bits, 150
Differential quadrant coding for 4800 bps, 154
Differential quadrant coding with trellis coding, 150
Digital cross-connect system, 35
Digital data system (DDS), 176–177
Digital 0 dBm0, 50
Digital generator, standard, 33
Digital local exchanges, 1
Digital network transmission time, 3
Digital output and input, conference TV, 309
Digital processes (PCM), 5, 6
Digital process (PCM) *vs.* qdu, 7
Digital radio relay, 98, 99, 100
Digital satellite networks, 121
Digital termination system, 102
Digital test point, 35, 53
Digital test point T, 35
Directories and letterheads, 255
Direct routing to locally attached devices, IP, 208
Disconnect mode, V.110, 194
Discontinuities in local time, 335
Dispersion coefficient, dispersion shifted fiber, 77
Dispersion-shifted single-mode fiber optic cable, 73–78
Distance metric, IP, 209
Diversity gain, site diversity, 83
Diversity improvement factor, site diversity, 83
DM (degraded minutes), 105
DNIC (data network identification code), 250, 254, 255
DNIC development, 255–256
Domains in an APPN network, 240
Domains in a subarea network, SNA, 239
Don't fragment flag, 211
Doppler effects, meteor burst, 131
DS1 frame boundaries, ATM, 303
DS3 frame for ATM cells, 301–302
DS1 mapping, ATM cells, 302–303
DS1 mapping with PLCP, 303
DTE/DCE, EIA-561, 159–160
DTE interface, V.32 bis, 152–155
Dual polarized antenna, 141
DUT1 indication, 328
DXC (digital crossconnect), 35
Dynamic alignment of the VC-n, 11
Dynamic fatigue parameter, dispersion shifted fiber, 76
Dynamic pressure of wind, 134

Earth-station antenna characteristics above 10 GHz, 115–118
E-bit usage, V.110, 181
Echo and stability, 1, 3
Echo and stability at 2-wire ports, PCM, 53–54
Echoing area, 125
Echo-time constant, 126
Effective length of meteor trail, 123
EIA-530, 161–176
EIA/TIA-562, 169–176
Eight-position nonsynchronous interface between DTE and DCE, 159–161
Electrical characteristics, EIA-530, 161–162
Electrical characteristics, EIA-562, 170–174
Electrical characteristics of interchange circuits, V.32 bis, 156
Electrical characteristics for an unbalanced digital interface, 169–176
Electrical input impedance, receiver EIA-562, 173
Electrical/mechanical characteristics of interchange circuits, V.110, 191
Electrical specifications, wire pair cable, 56
Electromagnetic wave propagation, 79–96
Electron density fluctuations, scintillation, 89
Elementary cable sections, dispersion shifted fiber, 77–78
Elementary cable sections, mono-mode fiber, 73
Elevation angle affects antenna noise temperature, 113
Elevation dependence of antenna noise temperature, 114
E1 mapping of ATM cells, 303–304
Emulation of start codes, conference TV, 316
Encapsulation, ATM, 274
Encoder delay, conference TV, 326
Encoding side adjustment, 36
Encrypt/decrypt at transmission control layer of SNA, 230
End nodes, SNA, 234
End-to-end flow control (TA to TA), V.110, 189–190
EOM/BOM protection, 282
EOM (end of message), 281
Equatorial scintillations, 94
Error burst objectives, 107
Error bursts, 106, 107
Error correcting code, conference TV, 324
Error correcting frame, conference TV, 324, 325

Error correction fill bits, conference TV, 326
Error correction framing bits, 324
Errored cell outcome, 292
Errored seconds, 104, 105
Errored seconds (ES), 106
Error handling, conference TV, 310
Errors in an ATM header, 269
Escape codes, data numbering, 252
ES (errored second) performance, 106
Estimating the useful burst rate, 128–129
EtherType assignments, 202, 203
EUTELSAT, 116, 117
Events, actions, time scales and response times, ATM, 295
Excess attenuation due to rainfall, CCIR method for satellite paths, 79–85
Excess attenuation due to rainfall on line-of-sight paths, 79
Excess attenuation due to vegetation, 85
Exponential decay of meteor trail, 123
Exposure coefficient, 135
Extra insertion information (GEI), 316
Extra insertion information (PEI), conference TV, 314

Factory length specifications, dispersion shifted fiber, 76–77
Factory length specifications, mono-mode fiber, 71–72
Fading and fading period, scintillation, 89
Fail-safe operation:
 EIA-530, 163–164
 EIA-562, 174
Faraday rotation, 131
Far-end and near-end crosstalk measured with analog test signal, 50, 51
Fast update request, conference TV, 323
Fault condition on interchange circuits:
 V.32 bis, 156
 V.110, 192
Fault management, ATM, 376
FCC radiation standard A, B, 147
Feedback controls, ATM, 294
FEXT (far-end crosstalk), 52
F4 flow, ATM, 276
Fiber characteristics, dispersion-shifted, 74–77
Fiber characteristics, mono mode fiber, 69–72
Fiber cutoff wavelength, 74
Fiber materials, 75
Fiber optics transmission, 69–78
Fiber properties, 71
FID (format identification) type, 245
Figures of merit, 118
File access systems, BTAM, RTAM, TCAM, 228
Fill bits used, 196
Fill indicator, conference TV, 324, 325
Finite-capacity queue, 301
First step rate adaptation, 178
Fixed-length code, conference TV, 321
Fixed part, TPDU format, 225
Fixed service sharing parameters below 3 GHz, 108
Fixed service system parameters for frequency sharing above 10 GHz, 109, 110
Fixed service system parameters for frequency sharing at 30 to 60 GHz, 110
Fixed stuff bit, 21, 22, 23, 24, 25, 26
Flat-plate, wind forces on, 135
Floating and locked mode conversion, 30–31
Floating and locked tributary unit modes, conversion between, 34
Floating mode, SDH, 17
Floating tributary unit mode, 27, 28, 29, 30
Flourcarbon materials, twisted pair cable, 56
Flow control, transport protocol, 221
Flow control, V.110, 188–190
Flow control mechanism, TCP, 213
Flow of valid cells, 273
Forced updating, conference TV, 313
Format identification (FID) type, 245
Format of DS3 PLCP frame, 302
Format of variable part, transport protocol, 226
Four service classes, AAL, 278
Four-wire exchanges, attenuation and group-delay distortion, 9
Fragmentation meaning segmentation, 205
Fragmentation and reassembly of TCP segments, 212
Fragment offset, 206
Fragment zero, ICMP, 209
Frame rate of the AUG, 13
Frame structure, V.110, 178, 179
Frame synchronization, V.110, 178, 186
Frame synchronization and additional signaling capacity, V.110, 185
Frame synchronization pattern, V.110, 193
Freeze picture request, release, 323
Frequency and polarization scaling of rain attenuation statistics, 81–82
Frequency choice, meteor burst, 130–131
Frequency-dependence of scintillation, 90–91
Frequency-dependent coefficients, 81
Frequency-dependent gain, site diversity, 84
Frequency justification, SDH, 13–14
Frequency offset, SDH, 13
Frequency reuse, 98, 120
Functional architecture, ATM connectionless, 283–284
Functional description of interchange circuits, EIA-530, 167
Functional description of interchange circuits, EIA-561, 160–161
Functional descriptions of interchange circuits, EIA-574, 158
Functions of individual ATM/B-ISDN layers, 273–283
Functions of the transport layer, transport protocol, 220–221

Gain relative to gain at 1020 Hz, PCM, 45
Gain varies with input level, PCM, 47, 48
Gateway function, SNA, 232, 242
Gateway routing tables, 211
Gauge block, 150-ohm STP-A cable, 62
GBSVC (general broadcast signaling virtual channel), 290
GCRA defined, 301
GCRA (generic cell rate algorithm), 299–301
GCRA (I,L), 300
General broadcast signaling virtual channel (GBSVC), 290
General mapping scheme, V.110, 179
Generator characteristics, EIA-562, 170
Generator and receiver connections at interface, EIA-530, 162
Generator output signal waveform template, 173
Generator polynomial, conference TV, 324
Generic cell rate algorithm (GCRA), equivalent versions, 300
Generic cell rate algorithms (GCRAs), 299
Generic flow control (GFC), 266–267
Generic functions, ATM traffic control, 294–295
Geographical area codes, data numbering, 255–261
Geometric considerations, scintillation, 93–94
Geostationary network in the FSS, 120
GFC, ATM, 275
GFC delay and delay variation, 298
GFC (generic flow control), 266–267
23 GHz band channel arrangements, 103
18 GHz band RF channel arrangement, 98–103
GOB (group of blocks) layer, conference TV, 316
GOB header, 316, 317
GOB layer, conference TV, 315
Go and return channels, 101
Go-return separation, 103
Go-to-return crosstalk, 51, 52
 PCM, 50
GPA-generating polynomial, 156
GQUANT, conference TV, 322
Graceful close, 213
Graceful connection close, TCP, 215
Grazing incidence reflection, 126
Grid antenna without ice, wind force coefficients for, 138
Group of blocks layer, structure, 316
Group delay, PCM, 41–43
Group-delay distortion, 2/4-wire exchanges, 9
Group delay distortion with frequency, 42–43
Group-delay distortion, planning values at exchanges/trunks, 8–9
Group number (GN), conference TV, 316
G/T losses, 115
G/T specification, 115
Guard bands suitable for subchannel arrangements, 101
Guidelines for channelization, 23 and 25 GHz bands, 103, 104
Gust response factor(s), 136, 141

Half-loop loss, 53

Hardware and software components of SNA network, 231
H3 byte, SDH, 13
Header checksum:
ICMP, 210
IP datagram, 206
Header error control (HEC), 266
Header error control (HEC) field, 268–269
Header field values, predefined, 275
Header length, IP, 206
Header pattern for idle cell identification, 270
Header structure at UNI, NNI-ATM, 266
HEC cell delineation mechanism, 270
HEC (header error control), 268
HEC octet, 271
HEC sequence calculated, 274
Height distribution of underdense meteors, 128
HEL (header extension length), CLNAP, 285
Hierarchical network configurations, SNA, 238
Hierarchical networks, SNA, 232
Hierarchical roles, SNA, 232–233
Hierarchy of connected network resources, SNA, 229
High bit error ratio, 105
Higher-layer protocol identifier (HLPI), 285
Higher-layer signaling protocols, 264
High-latitude scintillations, 95
High performance routing (HPR), SNA, 232
High-speed data transmission services, 105
High-speed 25-position interface for DTE and DCE, 161–176
High-velocity propagation lines, 3
High-voltage withstanding VGM pairs, 57
HLPI (higher-layer protocol identifier), 285
Homogeneous pattern, frequency arrangement, 103
Hop count, IP, 209
Horizontal 62.5/125 micron optical fiber cable, 67
Horizontal 150-ohm STP-A cable, 61–64
Horizontal UTP cable, 58
Horizontal UTP cable NEXT, 60
Horn reflector antenna, wind force coefficients for, 138
Horn reflector antennas, 141
Hot spots, 129
Hourly modulation schedule, 335
HPR (high-performance routing), SNA, 232, 236
HRDP (hypothetical reference digital path), 104
HRX (hypothetical reference connection), 105, 106
HUNT state, ATM, 271
Hypothetical decoder, conference TV, 323
Hypothetical reference digital path for radio-relay, 104

I bits inverted, 15
IBM system network architecture (SNA), 228–248
ICMP fields, 211
ICMP (internet control message protocol), 209–211
ICMP message format, 210
Identification of cell boundaries, 270
Identifier, IP datagram, 206
Idle cell identification, header pattern, 270
Idle cells, 270
Idle channel noise, 43
Idle or ready state, V.110, 192
IEEE 802 frame showing LLC and TCP/IP functions, 202, 204
IHL (internet header length), 210
Impact of cell delay variation on UPC/NPC and resource allocation, 297–298
Impairments due to digital processes, transmission, 5–9
Impedance, wire pair cable, 56
Impedance unbalance about earth, PCM, 37–38
IMS/VS, SNA, 243
Incident meteor trails, 128
Incompatibility between standard data rate and standard digital channel rate, 177
Incorporation of upper-layer PDUs into data link layer showing relationship with TCP and IP, 203
Information duty cycle, 130
Information transmission capacity, optical fiber cable, 67–68
Initial trail radius, 124
Input signals below 300 Hz at analog port, PCM, 44
Input signals above 4600 Hz at analog ports, 43
Insertion/deletion of stop elements, 199
Instantaneous statistics and spectrum behavior, scintillation, 91
Insulated conductor:
backbone cable, 64
horizontal 150-ohm STP-A cable, 61
100-ohm UTP cable, 58
Insulation resistance, wire pair cable, 57
In-sync time *vs.* bit error probability, ATM, 272
Intelligible and unintelligible crosstalk, 45
Interchange cable:
capacitance model per unit length, EIA-562, 175
Interchange capacitance limits, EIA-562, 175
Interchange circuit(s):
EIA-530, 167
EIA-562, 171
Interchange circuits, V.110, 190–192
Interchange circuits by category:
EIA-561, 160
EIA-574, 158
Interchange circuits for V.32 bis, 155
Interchange equivalent circuits, generator point, 170
Interchange node(s), SNA, 234, 236, 237
Interchange nodes interconnecting APPN and subarea networks, 237
Interchange voltage, EIA-562, 171
Interchannel crosstalk, analog-to-analog channels, PCM, 50
Interconnecting cable guidelines, EIA-562, 174–175
Interconnecting session stages, SNA, 242
Interface data units (IDUs), 282
Interface mechanical characteristics, 159–160
EIA-530, 164, 166–167
Interference, meteor burst, 131
Interference caused by other networks, 120–121
Interference from signaling, PCM, 52–53
Interference in a satellite network for an HRP, 120–121
Interference noise, 121
Interference noise power, 120
Interleaved arrangement, 18 GHz, 100
Intermediate routing network, SNA, 235
Intermediate session routing function, SNA, 242
Intermediate transmission requirements, 102
International atomic time, 336
International data number, 248, 250, 251, 254
International data service, 250
International numbering plan for public data networks, 248–261
International telephone connection and qdu, 8
International telephony/ISDN number, 253
Internet control message protocol (ICMP), 209–211
Internet from one LAN to another LAN via a WAN, 202
Internet header, 211
Internet protocol address formats, 208
Interoperation of EIA-530 and CCITT Rec. V.35, 167–169
Intra and inter building distance guidelines, 58
Intra-network NNIs, 295
Introduction to ATM, 263–264
Inversion of N bits, 14
Inverted D bits, SDH, 13
Ionization trail, 124
Ionized trail, 128
Ionospheric absorption, 131
Ionospheric effects influencing radio systems involving spacecraft, 88–96
Ionospheric scintillation and rain fading, 96
Ionospheric scintillations, 92
Ionospheric wind motions, 131
IP datagram description, 205–206
IP datagram format, 205
IP fields, ICMP, 211
IP operation detailed, 205–211
IP protocol field numbering, 207
IP router, 202
IP routing, 208–211
IP routing function, 204–205
ISDN basic user-network interface, 190
ISDN D-channel signaling protocol, 194
ISO transport protocol, 219–228

Jacket, backbone multipair cable, 65

Jacket and jacket splitting cord, 150-ohm STP-A cable, 62
Jackets, wire pair (cable), 56
Justification, positive and negative, 13, 14
Justification byte, 13
Justification control bit(s), 18, 19, 21, 22, 23, 24, 25, 26
Justification decision, 28
Justification opportunity, 21, 22, 23, 24, 25, 26

64-kbps hypothetical reference digital path, 104

LD-CELP distortion, 6
Leaky-bucket algorithm, ATM, 299
Leap second(s), 335, 337
LED transmitters, optical fiber for buildings, 67, 68
LEN end node, SNA, 235, 236
Length indicator field, TPDU format, 225
LEN (low entry networking), SNA, 234
LEN nodes, SNA, 234, 235
Levels at voice-frequency ports, PCM, 36
Limit of reflector movement, 140
Limit of structure movement, 146
Line signals, V.32 bis, 149–150
Link, SNA, 237
Link connection, SNA, 237
Link header, SNA, 245
Link parameters, typical LOS microwave, 108–111
Links and transmission groups, SNA, 237
Link station, SNA, 237
Link trailer, SNA, 245
Lip synchronization maintained, 324
List of data country or geographical area codes, 256–261
List of DNICs for non-zoned systems public mobile satellite systems, 256
LLC header, TCP/IP, 202
Load capacity, PCM, 36, 37
Local area network twisted pair, 55–57
Local clock-time and date, 327
Local flow control of the DTE interface. 189
Local flow control protocol, 190
Local magnetic meridian, 91
Local time offset, 337
Locked and floating mode, 18
Locked mode, SDH, 25
Locked tributary unit mode, 27, 28, 29, 30
Logical units (LUs), SNA, 241–244
Longitudinal conversion loss parameters, PCM, 37, 38, 39
Longitudinal interference, 38
Lookup table for destination addresses/paths, 283
Loop filter, conference TV, 312
LOS microwave link examples, 108–111
LOS microwave parameters above 10 GHz, 109, 110
LOS microwave system parameters for bands 30 to 60 GHz, 111
Losses due to water and oxygen in atmosphere, 114
Loss of frame synchronization, V.110, 185, 194–195
Lost cell outcome, 292
Low entry networking (LEN), SNA, 234
Low-level activating signal, 48, 49
L2_PDU, ATM, 302, 303
LU-LU protocols, SNA, 238
LU-LU sessions, 241, 242
Luminance sampling structure, 310
LU types 1, 2, 3, 4, 6.1 and 6.2, 243
LU types, SNA, 242

Macroblock address, layer, 317
Macroblocks arrangement in GOB, conference TV, 317
Macroblock vector, conference TV, 319
Magnetic activity, 94
Main frequency pattern, 98
Majority vote, 19, 21, 23, 24, 25, 26
Majority voting, SDH, 13
Mapping ATM cells into SDH, 304–305
Mapping ATM cells into SONET, 305–306, 307
Mapping of 32-bit internet address into 48-bit IEEE 802 address, 203
Mapping 6312 kbps into VC-2, 23–24
Mapping tributaries into VCs, 17–31
Mapping tributaries into VC-4, 18–21
Mapping tributaries into VC-11, 26–31
Mapping of tributaries into VC-12, 24–26
Mapping VC-4 into STM-1, 20
Margin of converter input, 199
Masterblock, arrangement of blocks in, 319
Material properties of fiber, 71
Material properties of fiber, dispersion shifted, 75–76
Maximum cable length allowed, EIA-530, 163
Maximum long-term interference, 108
Maximum permissible levels of interference in a geostationary satellite network for an HRP, 120–121
MBA stuffing, conference TV, 326
2M + 3 bits all of start polarity, 200
155.520 Mbps frame structure for SDH-based UNI, ATM, 304
622.080 Mbps frame structure for SDH-based UNI, ATM, 305
Mean cell transfer delay, 293
Meaning of circuit conditions, EIA-530, 162
Measured antenna temperature, 112
Measurement of phase differences, V.110, 195–196
Measurement on 4-wire ports with analog test signal, PCM, 49
Measurement on 4-wire ports with digital test signal, 50
Measuring half-loop loss, 53
Measuring points, CODEC delay, 325
Mechanical interface, EIA-530, 164, 166–167
Medium capacity systems, 18 GHz, 101
Message formats, ICMP, 209–210
Message mode service, AAL-3/4, 282
Metallic systems, outside plant, 55–67
Meta-signaling, 267, 275, 276, 289
Meta-signaling, 290–291
Meta-signaling function is required to, 291
Meta-signaling functions at the user access, 290–291
Meta-signaling protocol, 291
Meta-signaling requirements, 290
Meteor-burst communication, 122–131
Meteor-burst path ray geometry, 124
Meteor flux temporal variations, 122–123
Meteor ionization trails, 128
Meteor trails, effective length and radius, 123–124
Mid-latitude scintillations, 94
MID (message identification), ATM, 281
Minimal zero-dispersion wavelength, 73
Misinserted cell outcome, 292
Misinserted cells, 294
Misrouting of errored cells, 293
Mis-sequencing of information, ATM, 278
MJD + UTC, 338
Mode field concentricity error:
  dispersion shifted fiber, 74
  mono-mode fiber, 69–70
Mode field diameter:
  dispersion shifted fiber, 74
  mono-mode fiber, 69
Mode field noncircularity, dispersion shifted fiber, 74
Modelling/scaling rules for system application, scintillation, 89–90
Model of the transport layer, 223
Modem operating at 14,400 bps, 149–157
Modified Julian date (MJD), 337
Modulation rate, V.32 bis, 149
Monitoring and recovery, frame synchronization, 185
Mono-mode optical fiber, 69–73
Month-to-month variation in sporadic meteor flux rate, 122
Morphology, scintillation, 94–95
Motion compensation, conference TV, 311
Motion vector data (MVD), conference TV, 317
Motion vectors, conference TV, 312
MPEG (motion picture experts group), 279
M-plane, 264
MQUANT, conference TV, 322
M to 2M + 3 bits, 188
MTYPE, conference TV, 320
Multiframe alignment signal, 27
Multiframe indicator byte, 31
Multiframe information, V.110, 181
Multiple-domain subarea network, SNA, 239
Multiplexing mechanism, TCP, 213
Multiplexing techniques, 11–54
Multipoint considerations, conference TV, 323
Multipoint operation, conference TV, 310
Multiservice requirements, 102
Mutual capacitance:
  backbone UTP cable, 65
  wire pair cable, 59
MVD for macroblocks, conference TV, 318

MVSC (meta-signaling virtual channel), 290

Nakagami m-coefficient and scintillation index, 92
National data number, 248
NAU (network accessible unit), 240
NAU returns a negative response, 245
NCP (network control program), SNA, 229
NDF, SDH, 17
NDF (new data flag), SDH, 14
Near-end crosstalk, wire pair cable, 60
NEC requirements, wire pair cable, 57
Negative justification, AU-4 pointer adjustment operation, 17
Negative justification byte, 14, 18
Negative justification opportunity, 12, 13, 14
Negative justification required, 15
Negotiation, transport protocol, 223
Net diversity gain, site diversity, 84
Net filter discrimination (NFD), 97
Net-ID subnetwork boundaries, SNA, 234
Network accessible unit (NAU), 240–242
Network connection choice, transport protocol, 223
Network independent clock information, V.110, 181
Network-independent clocking state diagram, 196
Network independent clocks, V.110, 195–197
Networking blueprint of SNA, 230
Network interconnection gateways, SNA, 237
Network layer header (NHDR), SNA, 246
Network layer packet (NLP), SNA, 246, 248
Network node domains in an APPN network, 240
Network nodes, SNA, 235
Network node server, SNA, 234, 235
Network resource management (NRM), ATM, 294
Network roles, SNA, 232
Network service access points (NSAPs), transport protocol, 223
Network service primitives, transport protocol, 220
New data flag (NDF), 14–15
NEXT loss:
  backbone UTP cable, 66
  150-ohm STP-A cable, 64
NEXT loss UTP cable, 61
NEXT (near-end crosstalk), 51
NFD (net filter discrimination), 97
Nine-position nonsynchronous interface between data terminal equipment and data circuit terminating equipment, 157–159
1550-nm bend loss, 75
1550-nm bend performance, dispersion shifted fiber, 75
1550-nm bend sensitivity, 71
1550 nm loss performance, mono-mode fiber, 71
NNI (network-to-network interface), 266
NN (national number), 255
Node, SNA, 232
Nodes with both hierarchical and peer-oriented function, SNA, 236–237
Node types, SNA, 232
Noise from ground, 112
Noise temperature of an earth-station, contributions to, 108, 112–115
Nominal balance impedance, 53
Nominal clear sky conditions, 115
Nominal impedance, PCM ports, 37
Nominal reference frequency, PCM test, 33
Noncircularity:
  dispersion shifted fiber, 74
  mono-mode fiber, 70
Nonnative attachment, SNA, 236
Nonsynchronous communication, EIA-574, 157
Nonuniform spectral distribution noise power, 121
Nonuse of explicit flow control with class 2, transport protocol, 222
Nonzero GFC fields, 267
North American arrangements for band 17.7 to 19.7 GHz, 202
NPC (network parameter control), ATM, 294, 295
NPI (null pointer indication), 14
NRM (network resource management), ATM, 294
NSAP (network service access point), transport protocol, 223
NSDU (network service data unit), 224
NTN (network terminal number), 255
Null pointer indication (NPI), 14
Number of digits, 251
Numbering plan, characteristics and application, 249–255
Numbering plan interworking, 254–255
Numbering plan for public data networks, 248–261
Nx3600/Nx4800 bps to intermediate rate, 183

OAM cells, 294
OAM F5, F4, 298
2-octet framing, ATM PLCP, 302
Octet of option-kind, TCP header, 218
Off-axis angles, 119
Offset fed antenna, XPD, 145, 146
Offset number of AU-*n* pointer, 12
150-ohm STP-A backbone cabling, 58
150-ohm STP cable, 61–64
100-ohm unshielded twisted pair (UTP) cabling systems, 58–61
100-ohm UTP multipair backbone cable, 64–67
On-axis spurious, 119
One's complement of the one's complement sum, 210
One-way propagation time, 3
ON to OFF state transitions, V.110, 191
Open circuit measurements, EIA-562, 170
Operating sequence, V.110, 192–195
Operating wavelength, mono-mode fiber, 70
105/106 operation, V.110, 189
Optimum scattering regions, 128
Options, TCP header, 217–218
Orbit efficiency, 121
Originating countries, 254
Origins of cell delay variation, 298
Output signal applied waveform test set up, EIA-562, 172
Outside plant, metallic systems, 55–67
Outslot signaling assignments, 27, 30, 31, 32
Overhead communication channel bits, 24
Overload point, PCM, 36
Overspeed/underspeed, V.110, 188

PAD, CLNAP, 287
Padding, TCP header, 218
PAD-length, CLNAP, 285
Pair assembly:
  backbone UTP cable, 64
  150-ohm horizontal STP-A cable, 61
  100-ohm UTP cable, 58
Parabolic antennas, passive reflectors, twist and sway values, 140
  coefficients, 137
Parallel sessions, SNA, 242
Parameter code field, transport protocol, 226
Parameter length indication, transport protocol, 226
Parameter problem message, 210
Parity bits, V.110, 188
PAR (positive acknowledgement with retransmission), 212–213
Passive reflector, wind force coefficients for, 139
Path control element, SNA, 246
Path information unit (PIU), SNA, 245
Pattern, idle cell, 270
Payload type field, ATM, 268
PCI (protocol control information), 279
PCM crosstalk, 48–52
PCM equipment and test ports, 35
PCM quiet signal, 34
PCM sampling rate, 33
PCM transmission performance characteristics, 32–54
PCR (peak cell rate), 298–299
Peak cell rate, 296
Peak cell rate, 301
Peak cell rate (PCR), 298, 299
Peer-oriented network configurations, 239–240
Peer-oriented roles, SNA, 232–236
Peer-session protocols, SNA, 241
Percentages of time with and without diversity, rain attenuation, 84
Percentage of useful trails as function of scattering position, 129
Performance requirements for digital radio-relay systems, 105–106
Peripheral border nodes, SNA, 236
Peripheral node, SNA, 234
Peripheral nodes, SNA, 233
Phase difference relative to R1, R2, 196
Phase quadrant change, 154

Philosophy of voice and data, 263
Physical layer, ATM, 273–274
Physical medium sublayer, ATM, 273
Physical temperature of non-radiating elements, 116
Physical unit (PU), SNA, 241
Picture layer, conference TV, 314
Picture layer structure, conference TV, 315
Picture start code, 314
Plane of propagation, meteor burst, 126
Planning rule, qdu, 8
Planning values for attenuation and group-delay distortion, 9–10
Planning values for propagation time, 4
Planning values for quantizing distortion, 7
PLCP (physical layer convergence protocol), 302
1-point CDV, 293
2-point CDV, 293
Pointer generation, 15–16
Pointer interpretation, 16
Pointers, SDH, 11–17
Pointer value, 15
  SDH, 13
Point of interface:
  EIA-561, 159, 160
  EIA-574, 157
Point rainfall rate, 79
Point-to-point signaling virtual channels, 290
POI (path overhead indicator), ATM, 302
Position indicator byte, 26, 30, 31
Positioning of luminance and chrominance samples, 311
8-position nonsynchronous interface between DTE and DCE, 159–161
Positions of regions of optimum scatter, 128–129
Positive acknowledgement with retransmission (PAR), 212–213
Positive acknowledgement with retransmission (PAR), 215
Positive justification, AU-4 pointer adjustment operation, 15
Positive justification opportunity, 12, 13, 14
Positive/negative compensation, V.110, 197
Power-off impedance, EIA-562, 172
Power spectral density estimates for a geostationary satellite, scintillation, 90
Power sum NEXT loss, backbone UTP cable, 66
Pre- and post-temporal processing, 326
Precedence, IP, 206
Predefined header field values, 275
Predicted attenuation exceeded, 81
Prediction, conference TV, 311
Pre-established VPC between two UNIs, 289
Prefixes for data numbering, 252
PRESYNCH state, ATM, 271
Principal Fresnel zone, 123
Priority control, ATM, 294
Private ATM switch, 277
Private UNI, 291
Processing and ARQ delays, 263
Proofstress level:
  dispersion shifted fiber, 75
  optical fiber, 71
Propagation, electromagnetic wave, 79–96
Propagation time, 3
Propagation time, planning values, 4
Protective ground, EIA-530, 163
Protective materials:
  dispersion shifted fiber, 75
  optical fiber, 71
Protective tapes, wire pair cable, 56
Protocol architecture supporting connectionless service, 285
4-PSK like modulation, 100
PTI coding, 268
PTI field, AAL-5, 283
Public data network, 255, 256
Publicly administered individual addresses, 286
Public UNI, 291
PU (physical unit), SNA, 241
Push service, TCP, 213
PVC (permanent virtual circuit), 276

16-QAM like modulation, 100
QCIF, conference TV, 316, 323, 325
QCIF format, conference TV, 311
qdu, 7, 8
qdu and digital processes, 6, 8
qdu (quantizing distortion unit), 5
qdu (quantizing distortion unit), 6
QoS classes, ATM, 295
Quadrant coding for 4800 bps, V.32 bis, 154
Quality of service, network performance and cell loss priority, 295–296
Quality of service (QoS), ATM, 291–294
Quality of service (QoS) selection, 284
Quality of service requested, transport protocol, 221
Quantization, conference TV, 313
Quantization distortion, 5–6
  sources of, 6
Quantizer information, conference TV, 316
Quantizer (MQUANT), conference TV, 317
Quantizing distortion, 46–47
Quantizing distortion, planning values, 7
Quantizing distortion unit (qdu), 5, 6
QUANT ranges, conference TV, 322
Quarter-CIF (QCIF) format, 311
Quiet code, 48, 49
Quiet receiving locations, 130
Quiet signal, PCM, 34

Radiated interference field strength, VSAT, 119
Radiating temperature, 115
Radiating temperature of absorbing medium, 114
Radio-frequency channel arrangements, 97
Radio-frequency channel arrangements in 18 GHz band, 98–103
Radio-frequency channel arrangements for 23 GHz band, 103
Radio-frequency channel arrangements for 25.25–27.5 GHz and 27.5 to 29.5 GHz, 103–104
Radio-frequency separation, 97, 98
Radiometer, 114
Radio systems, 97–131
Rain fading and ionospheric scintillation, 96
Rapid-transport protocol (RTP), SNA, 232
RARP (reverse address resolution protocol), 209
Rate adaptation of 48- and 56-kbps to 64 kbps, 185–186
Rate negotiation, V.110, 181
Rate repetition information, V.110, 181
Rates greater than 19.2 kbps, V.110, 192
Rates less than or equal to 19.2 kbps, V.110, 192
Ray geometry of meteor burst path, 124
Received power and basic transmission loss, meteor burst, 124–127
Receiver, V.14, 199
Receiver characteristics, EIA-562, 173
Receiver electrical equivalent, EIA-562, 173
Receiver element signal timing, 164
Receiver fail-safe protection, EIA-562, 174
Receiver interface point signal definition, EIA-562, 174
Receiving equipment noise temperature values, 118
Reconstruction level into inverse transform, 323
Reconstruction levels for INTRA mode DC coefficient, 323
Reconstruction levels (REC), conference TV, 322
Rectangular aperture deflection angle, nomogram, 142
Redirect message, 210
Reduction factor, satellite path, 80
Reference configuration for provision of connectionless data service, 284
Reference frequency of homogeneous pattern, 104
Regions of optimum scatter, 128–129
Relating OSI to TCP/IP and associated protocols, 202
Relationship between VC, VP and transmission path, 287
Relative levels and R and T pads, 2
Relative levels at 2-wire/4-wire ports, PCM, 36
Release, transport protocol, 221
Relock time for error corrector framing, conference TV, 325
Request for comment (RFC), 218
Request header, SNA, 244
Requirements for absolute group delay, 411
Requirements for short- and long-term variation of level with time, PCM, 37
Requirements of signaling virtual channels, 290
Requirements of a TA supporting flow control, V.110, 190

Requirements for weighted idle channel noise, 44
Residual error rate, 222
Residual time stamp, ATM, 279
Resistance unbalance for VGM pairs, 57
Response header, SNA, 244
Response unit, SNA, 244
Return loss, PCM, 37, 38
Reverse address resolution protocol (RARP), 209
RFC (request for comment), 219
Ring-around-the-rosy, 206
Rip cord and binder, twisted pair cable, 56
Router, 208
Route setup protocol, SNA, 246
Routing functions for VPs, 289
RPOA (recognized private operating agency), 248, 255, 256
R and T pads, examples, 1–3
R and T pads for various countries, 2
RTP (rapid-transport protocol), SNA, 232

Sampling frequency, conference TV, 309
S and X status, V.110, 193
SAP (service access point) identifier, ATM, 275
SAR-PDU, 282
  AAL, 278, 279
SAR-PDU format, 279, 281
SAR-PDU format for AAL-2, 280
SAR (segmentation and reassembly) payload, 279
Scattering properties, meteor burst, 128
Scenario notation for TCP walk-through, 214
Scintillation, 89
Scintillation, 89–90
Scintillation and rain fading, 96
Scintillation event, 90, 91
Scintillation index, 92
Scintillation occurrence based on observations at L-band, 93
Scrambled data stream, V.32 bis, 150
Scrambler and descrambler, V.32 bis, 156–157
Scrambler specification, ATM, 270
SDH-based physical layer, 273
SDH bit rates, 11
SDH with ATM cells, 304–305
SDH pointers, 11–17
SDU (service data unit), transport protocol, 221
SEAL (simple and efficient AAL layer), 282
Search for frame synchronization, V.110, 185
Seasonal-longitudinal dependence, scintillation, 93
Seasonal variations, worst month, rain attenuation, 82
Seasonal weighting functions for stations in different longitudes, 92
61-sec minute, 336
Segment, unit of exchange between TCPs, 211
Segmentation and reassembly, IP, 206
Segmenting and reassembly, transport protocol, 221
Selection of synchronous or asynchronous modes of operation, 199
Selective broadcast signaling virtual channels, 291
Semipermanent connections, ATM, 289
Sequence number, TCP, 216
Serial binary interchange, 159
Service classes for AAL, 278
Service classification for AAL, 278
Service contract, conformance to, 275
Service data unit (SDU), 273
  transport protocol, 221
Service primitive, 220
Services, connection-oriented and connectionless, 283–285
Services assumed from the network layer, 220
Services provided and assumed, transport protocol, 220
Services provided by the transport layer, 220
SES objective, 106
SES (severely errored seconds), 105
Session-activation requests, SNA, 241
Session partners, SNA, 241
Session traffic, SNA, 237
Setup and release of VCCs, 289
Severely errored cell block outcome, 292
Severely errored cell block ratio, 293–294
Severely errored cell blocks, 293, 294
Severely errored seconds, 104
Shield(s):
  EIA-530, 163
  wire pair, 56
Short-circuit measurement, EIA-562, 172
Shower meteors, 122
Signal characteristics, EIA-530, 161–164
Signal characteristics:
  EIA-530, 164
  EIA-561, 161
  EIA-574, 159
Signal common, EIA-562, 171
Signal common and grounding arrangement, EIA-530, 165
Signal constellations, V.32 bis, 152, 153, 154, 155
Signaled errors, underlying error rate, transport protocol, 223
Signal element coding, 14,400 and 12,000 bps, 150
Signal element coding for 4800 bps, V.32 bis, 152
Signal element coding for 9600 bps and 7200 bps, V.32 bis, 151
Signaling function in normal speaking condition, 33
Signaling interference contributions, measurements, 52, 53
Signaling performance criteria, 107
Signaling phase indicator, 30
Signaling requirements, ATM, 289–291
Signaling virtual channels, ATM, 290
Signal reception inside buildings, 85–86
Signal space diagram and mapping for modulation at 4800 bps, V.32 bis, 155
Signal space diagram and mapping for modulation at 7200 bps, 154
Signal space diagram and mapping for modulation at 12,000 bps, 9600 bps, 153
Signal space diagram and mapping for modulation at 14,400 bps, 152
Signal-to-distortion ratio, PCM, 46
Simple TCP connection opening, diagram, 214
Simplified electrical equivalent circuit, EIA-562, 175
Single-domain subarea network, SNA, 239
Single-entry interference, 121
Single frequency noise, PCM, 43
Single mode optical fiber, 69–73
Single mode optical fiber for buildings, 67, 68
Single-mode transmission at 1550 nm, 75
Single octet of option kind, TCP, 218
Single-segment message (SSM), 281
Site diversity, 82–84
Slant-path length, 80
Slant-path rain attenuation, 79
Small-capacity digital transmission, 102
Small-capacity systems, 101
SNA data formats, 244–248
SNA layered architecture, 229
SNA layered architecture, 230–232
SNA logical units (LUs), 242–244
SNA network architectural components, 232–236
SNA network configurations, 238–240
SNA networking blueprint, 231
SNAP (subnetwork access protocol), 202
SNA (system network architecture), 228–248
Solar-geographical and temporal dependence of ionospheric scintillation, 94
Source address, IP datagram, 206
Source address field, CLNAP, 286
Source coder, conference TV, 310–314
Source coder block diagram, 312
Source coding algorithm, conference TV, 309
Source format, conference TV, 310–311
Source port, TCP, 216
Source routing, IP, 209
Sources of quantization distortion, 6
Source traffic descriptor, 296
  ATM, 301
  quality of service and cell loss priority, 297
Spare information (GSPARE), 316
Spare information (PSPARE), conference TV, 314
Spatial filter, conference TV, 312
Spatial variation in meteor flux, 123
Specification of figure of merit, 115
Spectrum behavior, scintillation, 92–93
Speech performance and digital processes, 5
Speech quality and data performance, 5
Splice cases, 71
Splitting and recombining, transport protocol, 221
Sporadic meteor flux rate, 122
Spurious EIRP, 119
Spurious emissions from VSATs, 119–120

Spurious impulse affecting timing, 106
Spurious in-band signals at the channel output port, PCM, 46
Spurious signals at the channel output port, 45–46
SSCP-independent LU, SNA, 241
SSCP-PU session, SNA, 241
SSCP (system services control point), SNA, 238, 239, 240
SSM (single-segment message), 281
Stability loss, PCM, 54
Standard digital analyzer, 33
Standard digital analyzer, 34
Standard frequencies and time signals, 327–334
Standard frequency and time, 327–340
Standard frequency and time signal emissions in the allocated bands, 328, 329, 330, 331, 332, 333
Standard time, 335–337
Standard time and date conventions, 327, 335–340
Start element nominal length, 188
Start polarity, 188
Start-stop character format, 198–199
Status bits, V.110, 178
STM-4 ATM cell mapping, 305
STP-A 150-ohm wire pair cable, 61
Stray capacitance to earth, 174
Streaming mode service, AAL-3/4, 282
Structural return loss, backbone UTP cable, 65, 66
Structural return loss (SRL), 60, 64
Structure and coding of TPDUs, 224–227
Structure of block layer, conference TV, 320
Structure of CLNAP-PDU, 286
Structure of group of blocks (GOB) layer, 316
Structure of macroblock layer, 317
Structure of picture layer, 315
Subarea network, SNA, 234
Subarea nodes, SNA, 233
Subareas, SNA, 238–239
Sublayering of AAL, 277–278
Subnetwork, SNA, 236
Subnetwork access protocol (SNAP), 202
Successful cell transfer outcome, 292
Sunspot cycle, scintillation dependence on, 95
Supported asynchronous user rates, 187
Sustainable cell rate (SCR), 301
SVC (signaling virtual channel), 290
SVC (switched virtual circuit) bandwidth parameter value, 290
Swept frequency input impedance measurements, 65
Switch buffer overflow, ATM, 293
Switching to data mode, V.110, 193
Symmetry of transmission, conference TV, 309
Synchronization by TA of entry to and exit from data transfer phase, 184
Synchronize control flag, TCP, 216
Synchronous data signaling rates, 149
Synchronous interfacing, V.32 bis, 154–155
Synchronous mapping of 139,264 kbps, 20
Synchronous structured payloads into a VC-m, 31
SYNCH state, ATM, 271
Syntax diagram for video multiplex coder, 315
Systematic convolutional encoder, 150
System bandwidth, optical fiber for buildings, 67, 68
System parameters for bands 30 to 60 GHz, LOS microwave, 111
Systems with capacity of 140 and 280 Mbps, 99
System services control point (SSCP), SNA, 238, 239, 240

TA duplex operation, V.110, 192
TA half-duplex operation, V.110, 195
Tandem digital processes, 8
Tandem operation, V.14, 200
TAs supporting asynchronous terminal equipment, V.110, 178
TA synchronization of entry to and exit from data transfer phase, 184
TAT-L, 301
TAT (theoretical arrival time), ATM, 301
TCC (telephone country code), 253
T-CONNECT service primitives, 222
TCP connection, 212
TCP defined, 211
TCP entity state summary, 218
TCP header format, 216–218
TCP/IP and data link layers, 201–204
TCP/IP and OSI reference model, 201
TCP/IP stack with ULP, 204
TCP/IP (transmission control protocol/internet protocol), 201–219
TCP layer provides for reliability of delivery that IP lacks, 204
TCP mechanisms, 212–213
TCP relationship with other layered protocols, 212
TCP walk-through, 214–216
TDMA/TV, 116, 117
Technical overview, EIA-561, 159
Telephony, transmission factors in, 1–10
Television transmission, 309–326
Temporal changes in meteor flux, 123
Temporal reference, conference TV, 314
Temporal subsampling, conference TV, 313
Temporal variations in meteor flux, 122–123
Terminal balance return loss (TRBL), 53–54
Test driving circuits, 38
Testing facilities, V.14, 200
Test ports, PCM equipment, 35
Test signal, PCM, 36
Test termination measurement, EIA-562, 172
TE2/TA (DTE/DCE) interface, V.110, 192
Theoretical arrival time (TAT), 301
Three-step rate adaptation, 187
Three-way handshake, TCP, 213
Three-way handshake mechanism, 215
Thresholds and response times, V.32 bis, 156
Time offsets, 337
Time signals and standard (broadcast) frequencies, 327–334
Time to live, 210
Time-to-live (TTL), IP datagram, 206
Timing arrangements, V.110, 190
Timing and control interface circuits, EIA-562, 174
Timing information transfer, EIA-530, 164
Timing jitter, 106
Timing slips, DS3 ATM, 302
Total distortion, including quantizing distortion, 46–47
Total system noise power, 120
TPDU code, 224, 225
TPDU coding and structure, 224–227
TPDU size, 228
TPDUs (transport protocol data units), 221
TPDU structure, 225
Tracer marking, twisted pair cable, 55
Traffic conformance and traffic contract, ATM, 300
Traffic contract, ATM, 295
Traffic contract, defining, 297
Traffic contract parameter specification, 301
Traffic control and congestion control, 294–301
Traffic control and congestion control, reference configuration, 295
Traffic descriptors, ATM, 296
Traffic descriptors and parameters, 296–297
Traffic parameters, ATM, 296
Traffic parameters PCR and SC and burst tolerance, 299
Traffic policing, 275
Traffic shaping, 275
Trail diffusion, 125
Trail height, meteor burst, 127–128
Trail radius, 124
Transform coefficients, conference TV, 320
Transformer, conference TV, 313
Transit countries, 254
Transition from binary 1 to data mode, V.110, 193
Transition time, EIA-562, 172
Transmission coder, 323–325
Transmission control layer, SNA, 230
Transmission control protocol/internet protocol (TCP/IP), 201–219
Transmission control protocol (TCP), 211–219
Transmission convergence sublayer functions, 274
Transmission convergence (TC) sublayer, ATM, 273
Transmission delays, V.110, 179
Transmission factors in telephony, 1–10
Transmission group (TG), SNA, 237
Transmission header, SNA, 245, 246
Transmission impairments due to digital processes, 5–9
Transmission levels, PCM, 35
Transmission loss, meteor burst, 124–127
Transmission order for transform coefficients, 320
Transmission performance characteristics of PCM, 32–54

Transmission requirements, 100-ohm backbone multipair cable, 65–67
Transmission requirements:
150-ohm STP-A cable, 62–64
100-ohm UTP cable, 59
Transmission of start-stop characters over synchronous bearer channels, 197–200
Transmission time, digital network, 3
Transmit control information sampled, V.110, 181
Transmitted spectrum, V.32 bis, 150
Transmitter, V.14, 199
Transport connection endpoints, 223
Transport entity, 223
Transport header (THDR), SNA, 246
Transporting ATM cells, 301–307
Transport network and network accessible units (NAUs), SNA, 240–242
Transport of ATM cells, 289
Transport protocol data unit (TPDU) code, 224
Transport protocol data units (TPDUs), 221
Transport service access point identifier (TSAP-ID), 227
Transport service primitives, 219
Transport of SMDS, ATM, 302
TRBL (terminal balance return loss), 53–54
Trellis-coded, 149
Trellis encoder, diagram, 151
Tributaries, mapping into VC-4, 18–21
True height distribution of meteors, 129
Trunk junctions, attenuation and group delay distortion, 10
TSAP-ID (transport service access point identifier), 227, 228
TU-1 and TU-2 sizes, 19
TU-12 payload, 30
TU pointer bytes, 19
TU-1/TU-2 level, SDH, 17
Twisted pair data communications cable, 55–57
Twisted pairs of VGM media, 55
Twisting moment, 133
Twist of pairs, 55
Twist and sway, 141
Twist and sway for cross-polarization limited systems, 146
Twist and sway values for parabolic antennas, passive reflectors—, 140
Two-step adaptation, V.110, 178
Two-way TCP data transfer, diagram, 214
Two-wire local and primary exchanges, attenuation and group-delay distortion, 9
Type information (MTYPE), conference TV, 317
Type information (PTYPE), conference TV, 314
Type of service, IP header, 206

UHF signal attenuation through timber-framed buildings, 86
ULP synchronization, TCP, 213
ULP (upper layer protocol), 211
Unassigned cells, adding, 275
Unbalanced digital interface, electrical characteristics, 169–176
Unbalanced interchange circuit, EIA-530, 165
Uncontrolled access, mode, ATM, 267
Underdense echo ceiling and average trail height, 127–128
Underdense echoes, 127
Underdense and overdense trails, 123
Underdense and overdense trails, 130
Underdense trails, 124, 127
Unintegrated digital processes, 9
Units of transmission impairment, 8
UNI (user-network interface), 264, 266
Universal time, 336
Unshielded twisted pair, 100-ohm cabling system, 58–61
UPC enforcement, ATM, 297
UPC function, ATM, 299
UPC/NPC, ATM, 296, 297
UPC/NPC mechanism, 298
UPC/NPC (usage/network parameter control), ATM, 294
U-plane, 264
Upper-layer protocols (ULPs), 211
Urgent pointer, TCP header, 217
Usable bandwidth, meteor burst, 131
Usage/network parameter control (UPC/NPC), 294
Use of channel capacity, V.110, 190
Use of explicit flow control with class 2, transport protocol, 222
Useful burst rate, estimating, 128–129
User information, CLNAP, 287
User-network interface configuration, ATM, 265
User-network interface (UNI), 289
User-network interface (UNI) configuration and architecture, 264–266
User-to-user signaling procedures, ATM, 289
Use of SNA data formats, 247
UTC (coordinated universal time), 336
UTC seconds marker sequence, 337
UTC time scale, 337

Variability in space and time statistics, rain attenuation, 82
Variable bit rate service (VBR), 26, 43
Variable bit rate video, 278, 279
Variable length codes, conference TV, 320
Variable part, TPDU format, 225–226, 227, 228
Variation of gain with input level, PCM, 47, 48
Variation of loss with time, PCM, 37
VBR (variable bit rate service), 263, 264
VC-3, SDH, 21
VC and VP switching, 288
VC, VP and the transmission path, 287
VCC endpoints, 289
VC-4 container, ATM cell mapping, 304
VCC service, 277
VCI value, 276
VCI value translated, 287
VCI (virtual channel identifier), 266, 287
VC-n alignment, 16
VC-1 POH, 24, 26
VC switch/cross-connect, 288
VC-11 to VC-12 conversion for transport by a TU-12, 30
V.110 data rate adaptation to 64-kbps digital channel, 177–197
Version field, IP header, 206
VGM (voice grade media), 55
Video CODEC block diagram, 310
Video coding delay, 324
Video data buffering, 323
Video input/output, 309
Video multiplex arrangement, 314
Video multiplex coder, 314–323
Video multiplex coder, syntax diagram, 315
Video source coding, 311–313
Virtual channel connection, 290
Virtual channel identifier (VCI), 266
Virtual channel (VC) level, 287
Virtual path identifier (VPI), 266
Virtual path level, 289
Virtual path (VP) and virtual circuit (VC) identifier, 275
Virtual path (VP) visibility, 277
Virtual scheduling algorithm, 299, 300
VLC table for macroblock addressing, 318
VLC table for MTYPE, conference TV, 318
VLC table for MVD, conference TV, 321
Voice grade media (VGM), 55
Volume density of meteor trails, 129
VPC (virtual path connection), 276
VPI and VCI fields, 268
VPI/VCI unallocated, 269
VPI (virtual path identifier), 266, 287
VP switching, 288
VP at the UNI, 290
VSAT spurious emissions, 119–120
VTAM (virtual telecommunication access method), 228, 229

Wavelength margin, 70
Week number, 338
Weighted noise, 43, 44
Wideband channels, 102
Wind force coefficients for paraboloid with cylindrical shroud, 137
Wind force coefficients for paraboloid without radome, 136
Wind force coefficients for typical grid antenna without ice, 138
Wind force coefficients for typical horn reflector antenna, 138
Wind force coefficients for typical paraboloid with radome, 137
Wind force coefficients for typical passive reflector, 139
Wind forces on conical horn, 134
Wind forces on flat-plate passive reflectors, 135
Wind forces on paraboloids, 133, 134
Wind load on typical microwave

antenna/reflectors, 133–141
Window, TCP header, 217
Window for flow control, 213
Worst-month, 86–88
WWV/WWVH, 336

X.121 format, data numbering, 253
XON/XOFF operation, V.110 flow control, 189
XPD (cross polarization discrimination), 97–98
XPD for center-fed antenna, 145
XPD for offset fed antenna, 145

y-factor method, 112, 114

Zenith angle dependence, scintillation, 93
Zenith sky temperature, 115
Zero-dispersion slope, 72
Zero-dispersion wavelength, 76, 77
Z overheads removed, ATM, 302